Nanoscale Networking and Communications Handbook

Nanoscale Networking and Communications Handbook

Edited by
John R. Vacca

CRC Press
Taylor & Francis Group
Boca Raton London New York

CRC Press is an imprint of the
Taylor & Francis Group, an **informa** business

CRC Press
Taylor & Francis Group
6000 Broken Sound Parkway NW, Suite 300
Boca Raton, FL 33487-2742

International Standard Book Number-13: 978-1-4987-2731-0 (Hardback)

Visit the Taylor & Francis Web site at
http://www.taylorandfrancis.com

and the CRC Press Web site at
http://www.crcpress.com

Printed and bound in Great Britain by
TJ International Ltd, Padstow, Cornwall

This book is dedicated to my wife, Bee.

Contents

Foreword

AFTER A QUARTER CENTURY of nanoscience research and development, nanotechnologies are starting to improve products and services in many industry sectors, including information technology, medical technology and treatment methods, transportation efficiency and safety, energy production and distribution, environmental management, and even nanodrops in eye-drop solutions.

Nanotechnology has contributed to advances in computing and electronics, which will result in faster, smaller, and more portable systems that can manage and store larger and larger amounts of data. Lawrence Berkeley National Laboratory demonstrated a 1-nm transistor in 2016, and with the deployment of magnetic random access memory (MRAM), computers will boot almost instantly. In addition, ultra-high definition displays that use quantum dots to produce incredibly vibrant colors are more energy efficient.

For nanotechnologies to reach their full potential, they must be networkable and capable of communicating. This book addresses the methods and technology to create and manage nanoscale networks and communications. This effort has all the challenges of networking and communications technology that we rely on every day in our relatively giant-sized world. The design of the nanomachines necessary to enable nanoscale networking is addressed in both the nanomaterial- and biological-based nanomachine realms of networking and communications.

Nanomachines will have their place in the Internet of Things, and that will require signal propagation and processing and synchronization, as well as switching and routing, which are covered in this book. Communication across different scales of technology such as nano-to-micro and micro-to-nano is also covered, along with communications between living and nonliving systems.

The potential future of nanotechnology is as fascinating as it is amazing; that is, if the world's regimes decide not to use their nuclear arsenals to demonstrate their machismo, realms like quantum entanglement photon swapping, nanoscale photon detectors, quantum dot networks, and, especially, quantum nanorobotics will take us where no geek or geekess has ever gone before.

Michael Erbschloe
Information Security Consultant

Preface

SCOPE OF COVERAGE

This comprehensive handbook serves as a professional reference, as well as a practitioner's guide to today's most complete and concise view of nanoscale networking and communications. It offers in-depth coverage of nanoscale networking and communications theory, technology, and practice as they relate to established technologies, as well as recent advancements. It explores practical solutions to a wide range of nanoscale networking and communications issues. Individual chapters are authored by leading experts in the field and address the immediate and long-term challenges in the authors' respective areas of expertise.

The primary audience for this handbook consists of: engineers/scientists interested in monitoring and analyzing specific measurable nanoscale networking and communications environments, which may include anything from intrabody wireless nanosensor networks for advanced health monitoring systems to terabit wireless network-on-chip for ultra-high-performance computer architectures; professionals in nanoscale networking and communications and related areas interested in a cutting-edge research field, at the intersection of nanotechnologies and information and communication technologies; other individuals with an interest in using nanoscale networking and communications to understand specific environments; undergraduates, graduates, academia, government, and industry; anyone seeking to exploit the benefits of nanoscale networking and communications technologies, including assessing the architectures, components, operation, and tools of nanoscale and quantum computing; and anyone involved in the technology aspects of nanoscale networking and communications who has knowledge at the level of the introduction to nanoscale computing or equivalent experience. This comprehensive reference and practitioner's guide will also be of value to students in upper-division undergraduate and graduate-level courses in nanoscale networking and communications.

ORGANIZATION OF THIS BOOK

The book is organized into four sections composed of 20 contributed chapters by leading experts in their fields, as well as four appendices, including an extensive glossary of nanoscale networking and communications terms and acronyms, as follows:

Section I: Introduction

Section I discusses graphene-enabled wireless nanoscale networking communications in the terahertz band, as well as graphene-based antenna design for communications in the terahertz band. It also covers terahertz programmable metasurfaces: networks inside networks. For instance:

Chapter 1, "Graphene-Enabled Wireless Nanoscale Networking Communications in the Terahertz Band," sets the stage for the rest of the book, by revisiting the background and current design trends. Then, it describes a theoretical framework that aims to enable the systematic evaluation of the unique features of graphene antennas in a THz link, including the peculiarities of the THz wireless channel. The methodology is exemplified in a simple use case of indoor communications, and its relevance is discussed in the context of advanced protocols at the physical and link layers of design.

Chapter 2, "Graphene-Based Antenna Design for Communications in the Terahertz Band," provides a detailed description of the state of the art of graphene-based antennas for terahertz band communications, by depicting current design trends, outlining the theoretical foundations behind them, and evaluating a representative set of implementations. The qualitative comparison made in this chapter may, in the future, guide the design of antennas for area-constrained applications, depending on their specific requirements in terms of miniaturization, functional reconfigurability, and efficiency.

Chapter 3, "Terahertz Programmable Metasurfaces: Networks Inside Networks," focuses on the communication requirements and implementation constraints of the controller network within the metasurface. Guidelines for the design of appropriate wired and wireless solutions are extracted from the analysis, paving the way to the realization of programmable terahertz metasurfaces. In other words, this chapter focuses on the challenge of designing nanonetworks for internal software-defined metamaterials (SDM) communication.

Section II: Nanoscale, Molecular Networking Communications

Section II covers channel modeling for nanoscale communications and networking, as well as channel modeling and capacity analysis for nanoscale communications and networking. In addition, the section also covers nanoscale channel modeling in highly integrated computing packages, synchronization for molecular communications and nanonetworking, multiple access control strategies for nanoscale communications and networking, media access control for nanoscale communications and networking, and signal processing for nanoscale communication and networking. For instance:

Chapter 4, "Channel Modeling for Nanoscale Communications and Networking," presents and analyzes the development of an interference model, utilizing time spread on-off keying (TS-OOK) as a communication scheme of the THz communication channel inside the human body, as well as the probability distribution of signal-to-interference-plus-noise ratio (SINR) for THz communication within different human tissues such as the blood, skin, and fat. In addition, this chapter evaluates performance degradation by investigating the mean values of SINR under different node densities in the area and the probability of transmitting pulses. The chapter concludes that the interference restrains the achievable communication distance to approximately 1 mm and a more specific range that depends on the particular transmission circumstance. The results presented in this chapter also show that by controlling the pulse transmission probability and node density, the system performance can be ameliorated.

Chapter 5, "Channel Modeling and Capacity Analysis for Nanoscale Communications and Networking," aims to investigate physical transmission rates and communication ranges reachable in human tissues, starting from the formulation of a sophisticated channel model that takes into account the frequency and spatial dependence of the skin permittivity. Next, the chapter presents the channel model formulated for the stratified media stack describing human tissues. In addition, it describes the investigated transmission techniques. The chapter also discusses the SNR measured in the frequency domain and illustrates physical transmission rates and communication ranges achievable in human tissues as a function of transmission techniques, distance, and position of both transmitter and receiver.

Chapter 6, "Nanoscale Channel Modeling in Highly Integrated Computing Packages," reviews recent efforts that, with the aim of bridging this gap, identify possible propagation paths, define modeling and simulation methodologies, and provide first frequency-domain evaluations of the nanoscale wireless channels within computing packages. In this chapter, the authors also lay down the fundamentals of channel modeling in highly integrated environments. They then review the computer package environments, both in the wireless network-on-chip (WNoC) and SDM, by taking into full consideration realistic chip or system packages and the potential antenna placement. Then, the authors perform a frequency-domain study of the wireless channel within a set of computing packages relevant to both scenarios. They also show that by deriving communication metrics from the channel exploration, architects can accurately estimate the performance and cost of future nanoscale wireless communications in highly integrated computing packages. Finally, the authors provide a summary of results and an outlook of this research field.

Chapter 7, "Synchronization for Molecular Communications and Nanonetworking," delves into the problem of synchronization in molecular communication systems and networks. The authors emphasize that the synchronization purpose is twofold—to perform coordinate actions among a network of nanomachines and to provide the correct sampling time for data detection. The naturally available biological oscillators can be valid tools to achieve the coordinated actions, as the oscillations can be used to approximate the clock signals. They have also seen that extending the conventional synchronization concepts tuned to adapt to the unique challenges in molecular communication allows these concepts to be valid tools for symbol synchronization and data detection.

Chapter 8, "Multiple Access Control Strategies for Nanoscale Communications and Networking," reviews the existing well-known approaches to medium access control (MAC) strategies, outlining their advantages and disadvantages, as well as their expected performance. The next section in the chapter provides a vision of existing approaches in today's networks, in order to better understand the background knowledge on MAC. Other sections address two technological scenarios and the respective MAC solutions existing in the literature: nanoscale communications and molecular communications. Then, the chapter introduces nanoscale communications in the terahertz band and surveys the available MAC strategies for such a scenario. In addition, the chapter introduces the concept of bio-nanocommunications, while presenting proposed techniques for MAC, with their pros and cons. Furthermore, the chapter lists the main simulators for nano- and biocommunications, which are currently available in the research community for experimentation. Finally, the chapter briefly discusses the integration between nano- and molecular communications within the Internet, in the so-called internet of bio-nano things (IoBNT).

Chapter 9, "Media Access Control for Nanoscale Communications and Networking," reviews the problem of MAC for nanoscale communication networks. In essence, the authors provide a brief overview of the main characteristics of different nanonetwork contexts and analyze how they affect the design of appropriate methods at the MAC level. They set the focus on three representative scenarios: First, the authors examine electromagnetic (EM) nanonetworks in the terahertz (THz) band, which is seen as an extremely downscaled version of personal area networks (PANs) or wireless sensor networks (WSNs). Next, the authors analyze EM nanonetworks located within computing packages. Then, they delve into the biocompatible molecular communications scenario. Finally, the authors summarize the main conclusions of the analysis and provide a brief outlook of future perspectives.

Chapter 10, "Signal Processing for Nanoscale Communication and Networking," reviews the theoretical frameworks for the neural communication in manmade nanonetworks. More specifically, the authors give an overview and identify the deficiencies of the existing solutions in the following areas: neural compartments relevant to the signaling aspects of neurons, communication system models for neural communication, and the motivation and justification for novel research efforts in signal processing for neural communication. The remainder of the chapter is organized as follows: reviews of the latest developments in computational neuromodeling and an overview of system models for neural communication.

Section III: Molecular Nanoscale Communication and Networking of Bio-Inspired Information and Communications Technologies

Section III discusses the communication between living and nonliving systems, as well as the molecular communication and cellular signaling from an information-theory perspective. Next, the section explains the design and applications of optical near-field antenna networks for nanoscale biomolecular information, the basis of pharmaceutical formulation, and the droplet-based microfluidics: communications and networking. For instance:

Chapter 11, "Communication Between Living and Nonliving Systems," focuses on the fundamentals of establishing communication between living and manmade (nonliving) systems. First, the authors explore the basics and characteristics of communication among living systems. Second, they discuss the communication basics and potentials of communication between living and nonliving systems. Third, the authors present some exemplary application concepts for the living-to-nonliving systems communication. Finally, they highlight issues such as toxicity and biocompatibility, which arise from the introduction of a manmade nonliving system into the microenvironment of the cells.

Chapter 12, "Molecular Communication and Cellular Signaling from an Information-Theory Perspective," provides some level of initiation to those interested in using information in the study of biological systems. While we cannot hope that a single chapter will treat the entire breadth of applications that might interest biologists (e.g., this chapter does not even touch upon the use of information theory in understanding the collective behavior of animals or the self-assembly of supramolecular structures), it uses several exactly solvable examples to help the reader build a functional intuition about how to think about biological systems from an information theoretic perspective and how to avoid making some common errors in the interpretation of information-theory results. Despite the coarseness of some of the assumptions made in the name of computational facility, the examples presented in this chapter are nonetheless reasonable baseline models for the study of certain classes of molecular communication processes. Finally, this chapter demonstrates how even much more complicated signaling scenarios can often be reduced to one of these simpler problems.

Chapter 13, "Design and Applications of Optical Near-Field Antenna Networks for Nanoscale Biomolecular Information," discusses theoretical backgrounds of a nanoantenna. Performance of various nanoantenna designs and the effects of design parameters on the channel characteristics are also explored in this chapter. Next, the chapter recommends the microwave and radiofrequency (RF) antennas and the optical nanoantennas that may be considered for coupled operations with fluorescent molecules and quantum dots for a broad range of utilities and improved performances. The chapter also explains how the receiver and/or transmitter in a nanoantenna network can be any object, even on a quantum scale, that may absorb or excite photons. Finally, the chapter explains why nanoantennas have emerged as a platform in biomedical imaging and sensing applications.

Chapter 14, "Basis of Pharmaceutical Formulation," covers why microcapsules and nanocoating are the two drug-delivery technologies for delivering the drug interleukin-12. Next, the chapter discusses how nanotechnology is currently taking a big place in the pharmaceutical industry, but, along with advantages, it also has disadvantages. This chapter also discusses the research that is being done through the observation of the nanoparticle–biological interface, as well as how to recognize the possible hazards of wangled nanomaterials, along with the discovery of a new experimental design, and the methods that would guide researchers in the development of harmless and more effectual nanoparticles that could be used in a variety of treatments and products. The chapter also explains how nanoparticles can react with proteins, membranes, cells, DNA, and organelles, because of the formation of protein coronas, particle wrapping, intracellular uptake, and biocatalytic processes that result in positive or negative outcomes, owing to the nanoparticles–biological interfaces, which depend on colloidal forces and biophysico-chemical interactions. Furthermore, this chapter also analyzes how these interfaces show that the size, shape, surface chemistry, roughness, and surface coatings of nanomaterials play an important role in the development of prognostic relationships between structure and activity. Finally, the chapter discusses why knowledge about the interface is important for getting a better understanding about the interrelationship between the intracellular activity and the function of designed and built nanomaterials, which is essential for the development of nanoparticle drug-delivery systems.

Chapter 15, "Droplet-Based Microfluidics: Communications and Networking," introduces the emerging field of microfluidic communications and networking, where tiny volumes of fluids, so-called droplets, are used for communications and/or addressing purposes in microfluidic chips. This chapter starts with the basics of droplet-based microfluidics.

In particular, the authors discuss the analogy between microfluidic and electric circuits and the resistance increase brought by droplets in microfluidic channels. Moreover, they describe different droplet-generation methods, including concepts for the accurate droplet creation at prescribed times, and with a certain size, at a certain distance. Then, the authors discuss different information-encoding schemes for droplet-based microfluidic systems and compare them in terms of encoding/decoding complexity, error performance, and the resulting channel capacity. Furthermore, they characterize the microfluidic communication channel, showing that the noise can be modeled as a Gaussian random variable. This chapter also serves as a basis for switching and addressing in microfluidic networks. Next, the authors introduce the concept of microfluidic networking. After discussing some constraints and the main switching principle, they present single- and multiple-droplet switches that control the path of single or multiple droplets within the network, respectively. Finally, the authors discuss various addressing schemes for a microfluidic bus network, which is a promising network topology, owing to its simplicity, flexibility, and scalability.

Section IV: Advances in Nanoscale Networking-Communications Research and Development

Section IV discusses the nanostructure-enabled high-performance silicon-based photodiodes for future data-communication networks, as well as nanoscale materials and devices for future communication networks. Next, this section covers the microwave-absorbing properties of single- and multilayer materials: microwave-heating mechanism and theory of material–microwave interaction, dynamic mechanical and fibrillation behaviors of nanofibers of LCP/PET blended droplets by repeated extrusion, and nanoscale wireless communications as enablers of massive manycore architectures.

Chapter 16, "Nanostructure-Enabled High-Performance Silicon-Based Photodiodes for Future Data-Communication Networks," provides a brief description of approaches to realize high-speed and high-efficiency Si-based photodiodes. In this chapter, Si-based photodiodes with innovative light-trapping nanostructures will be introduced for low-cost high-bandwidth links for future communication networks. Next, the chapter presents CMOS-compatible materials (Si, SiGe, and Ge-on-Si), their implementation in high-speed photodiodes, trade-off between speed and efficiency of devices based on Si, and how innovative nanostructures break this trade-off by light trapping. A study that uses rigorous coupled-wave theory is included in this chapter to analyze light trapping by periodic nanostructures. Also, in this chapter, recently demonstrated Si photodiodes for short-reach optical fiber links and Ge-on-Si photodiodes for long-reach optical fiber links are reviewed, and light-trapping and bandwidth-enhancement properties of integrated nanostructures are discussed. Then, Si photodiodes in the avalanche mode, as well as for long-haul optical communication and potential applications in single-photon detectors for quantum communications, are presented in this chapter. Finally, a brief summary is provided to highlight key outcomes, limitations, and integration of high-speed Si photodiodes with nanostructures.

Chapter 17, "Nanoscale Materials and Devices for Future Communication Networks," reviews the recent research progress in modulator technologies for the future communication technologies in academia and industry. Next, the authors introduce a design of plasmonic modulators, using vanadium dioxide (VO_2) as modulating material, realized on a silicon on insulator (SOI) wafer with only a 200 nm × 140 nm modulating section within a 1 µm × 3 µm device footprint. Then, they demonstrate a plasmonic bandpass filter integrated with materials exhibiting phase transition, which can be used as a thermally reconfigurable optical switch. In this chapter, the authors first present a plasmonic modulator with only a 200 nm × 140 nm modulating section within a 1 µm × 3 µm device footprint, using vanadium oxide as a modulating material by taking advantage of the large refractive index contrast between the semiconductor and metallic phases. Then, they move on to demonstrate a photonic switch using vanadium dioxide as a switching material, with a hexagonal nanohole array structure, with a maximum 37% transmission at the optical communication band. In conclusion, the authors' work demonstrated in this chapter is one of the many solutions in the world tackling this challenge, which will lay down the foundation for next-generation nanoscale-photonics circuits on a CMOS chip integrated with electronics, in order to handle the internet of things (IoT), big data, machine-learning applications, and cloud computing for 5G and beyond.

Chapter 18, "Microwave-Absorbing Properties of Single- and Multilayer Materials: Microwave-Heating Mechanism and Theory of Material–Microwave Interaction," investigates the microwave-absorption properties of single- and multilayer composites, which are based on composite thickness, number of layers, and size of fillers. Next, the other important phenomena, such as polarization and loss mechanism, are also evaluated for microwave-absorption properties. Finally, this chapter also offers experimental data for single- and multilayer composites.

Chapter 19, "Dynamic Mechanical and Fibrillation Behaviors of Nanofibers of LCP/ PET Blended Droplets by Repeated Extrusion," analyzes the distribution and components of droplets, as well as the miscibility of liquid crystal polymer:polyethylene terephthalate (LCP:PET) blending. Next, the chapter discusses how the droplet-behavior change is supposed to relate with the flow property, miscibility, and surface property of LCP and PET. In this chapter, droplet-behavior change by repeating extrusion a number of times is observed; in that, the process of analysis on blending condition and weight ratio is also observed. In the case of blending, high LCP content is also shown in this chapter; in comparison with PET content and LCP, it becomes a matrix, with PET representing a droplet shape. By repeated extrusion, as shown in this chapter, the LCP:PET blended droplet shows an enlarged droplet size and the PET droplet wraps LCP. Next, the chapter covers the observation of the viscosity difference of two immiscible materials, which is the cause of this phenomenon. Finally, the chapter addresses the blend of two immiscible materials, how the agglomerative phenomenon leads to a reduction of material properties, and how the droplet size correlates with the mechanical properties.

Chapter 20, "Nanoscale Wireless Communications as Enablers of Massive Manycore Architectures," reviews the state of the art of the field, by focusing on nanoscale technologies for wireless on-chip communication in the millimeter-wave and terahertz bands. The scaling of trends for these interconnects are analyzed in the chapter from the physical, link, and network levels, in order to assess the practicality of the idea in the extremely demanding manycore era.

This chapter reviews the nanonetwork challenges associated with manycore processors, as well as several alternative technologies, and makes the case for wireless on-chip interconnects, which are enabled by emerging nanoscale technologies in the millimeter-wave (mmWave) and terahertz (THz) bands. Through different scalability analyses, the authors of this chapter demonstrate the wireless nanoscale interconnects, which are capable of delivering the flexibility and the fast and low-power broadcast capabilities that are required to enable the next generation of manycore processors. Finally, the authors discuss the future prospects in the pathway of actually implementing the vision.

John R. Vacca
Managing and Consulting Editor
TechWrite
Pomeroy, Ohio

Acknowledgments

There are many people whose efforts have contributed to the successful completion of this book. I owe each a debt of gratitude and want to take this opportunity to offer my sincere thanks.

A very special thanks to my executive editor, Rick Adams, without whose continued interest and support I could not make this book possible, and editorial assistant, Jessica Vega, who provided staunch support and encouragement when it was most needed. Thanks to my production editor, Paul Boyd; project manager, John Shannon; and copyeditor, Swati Sharma, whose fine editorial work has been invaluable. Thanks also to my project coordinator, Sherry Thomas, whose efforts on this book have been greatly appreciated. Finally, thanks to all the other people at CRC Press (Taylor & Francis Group), whose many talents and skills are essential to a finished book.

Thanks to my wife, Bee Vacca, for her love, help, and understanding during my long work hours. Also, a very special thanks to Michael Erbschloe for writing the foreword. Finally, I wish to thank all the following authors who contributed chapters that were necessary for the completion of this book: Sergi Abadal, Josep Solé-Pareta, Eduard Alarcón, Albert Cabellos-Aparicio, Seyed Ehsan Hosseininejad, Max Lemme, Peter Haring Bolivar, Christos Liaskos, Andreas Pitsillides, Vasos Vassiliou, Ke Yang, Rui Zhang, Qammer H. Abbasi, Akram Alomainy, V. Musa, G. Piro, P. Bia, L. A. Grieco, D. Caratelli, L. Mescia, G. Boggia, Xavier Timoneda, Anna Tasolamprou, Odysseas Tsilipakos, Maria Kafesaki, Eleftherios N. Economou, Costas Soukoulis, Alexandros Pitilakis, Nikolaos V. Kantartzis, Mohammad Sajjad Mirmoosa, Fu Liu, Sergei Tretyakov, Ethungshan Shitiri, Ho-Shin Cho, Fabrizio Granelli, Cristina Costa, Riccardo Bassoli, Hossein Moosavi, Francis M. Bui, Uche K. Chude-Okonkwo, A.V. Vasilakos, B. T. Maharaj, Reza Malekian, Preetam Ghosh, Kevin R. Pilkiewicz, Pratip Rana. Michael L. Mayo, Hongki Lee, Hyunwoong Lee, Gwiyeong Moon, Donghyun Kim, Shahnaz Usman, Karyman Ahmed Fawzy, Rayisa Beevi, Anab Usman, Werner Haselmayr, Andrea Zanella, Giacomo Morabito, Hilal Cansizoglu, Cesar Bartolo Perez, Jun Gou, M. Saif Islam, Miao Sun, Mohammad Taha, Sumeet Walia, Madhu Bhaskaran, Sharath Sriram, William Shieh, Ranjith Rajasekharan Unnithan, Yuksel Akinay, and Han-Yong Jeon.

Editor

John R. Vacca is an information technology consultant, professional writer, editor, reviewer, researcher, and internationally known, best-selling author based in Pomeroy, Ohio. Since 1982, John has authored/edited 80 books; some of his most recent books include the following:

- *Computer and Information Security Handbook, 3E* (*Publisher:* Morgan Kaufmann [an imprint of Elsevier Inc.] [June 10, 2017])

- *Security in the Private Cloud* (*Publisher:* CRC Press [an imprint of Taylor & Francis Group, LLC] [September 1, 2016])

- *Cloud Computing Security: Foundations and Challenges* (*Publisher:* CRC Press [an imprint of Taylor & Francis Group, LLC] [August 19, 2016])

- *Handbook of Sensor Networking: Advanced Technologies and Applications* (*Publisher:* CRC Press [an imprint of Taylor & Francis Group, LLC] [January 14, 2015])

- *Network and System Security, 2E* (*Publisher:* Syngress [an imprint of Elsevier Inc.] [September 23, 2013])

- *Cyber Security and IT Infrastructure Protection* (*Publisher:* Syngress [an imprint of Elsevier Inc.] [September 23, 2013])

- *Managing Information Security, 2E* (*Publisher:* Syngress [an imprint of Elsevier Inc.] [September 23, 2013])

- *Computer and Information Security Handbook, 2E* (*Publisher:* Morgan Kaufmann [an imprint of Elsevier Inc.] [May 31, 2013])

- *Identity Theft (Cybersafety)* (*Publisher:* Chelsea House Pub [April 1, 2012])

- *System Forensics, Investigation, and Response* (*Publisher:* Jones & Bartlett Learning [September 24, 2010])

- *Managing Information Security* (*Publisher:* Syngress [an imprint of Elsevier Inc.] [March 29, 2010])

- *Network and Systems Security* (*Publisher:* Syngress [an imprint of Elsevier Inc.] [March 29, 2010])

- *Computer and Information Security Handbook, 1E* (*Publisher:* Morgan Kaufmann [an imprint of Elsevier Inc.] [June 2, 2009])

- *Biometric Technologies and Verification Systems* (*Publisher:* Elsevier Science & Technology Books [March 16, 2007])

He has also authored more than 600 articles in the areas of advanced storage, computer security, and aerospace technology (copies of articles and books are available upon request).

John was also a configuration management specialist, computer specialist, and the computer security official (CSO) for NASA's space station program (Freedom) and the International Space Station Program, from 1988 until his retirement from NASA in 1995.

In addition, John is also an independent online book reviewer. Finally, John was one of the security consultants for the MGM movie titled *AntiTrust*, which was released on January 12, 2001. A detailed copy of his author bio can be viewed at http://www.johnvacca. com. John can be reached at john2164@windstream.net.

Contributors

Sergi Abadal
NaNoNetworking Center in Catalunya
  (N3Cat)
Universitat Politecnica de Catalunya
Barcelona, Spain

Qammer H. Abbasi
School of Engineering
University of Glasgow
Glasgow, Scotland

Karyman Ahmed Fawzy
RAK Medical and Health Sciences
  University
Jinnah Sindh Medical University
Karachi, Pakistan

Yuksel Akinay
Karabük University
Karabük, Turkey

Eduard Alarcón
NaNoNetworking Center in Catalunya
  (N3Cat)
Universitat Politecnica de Catalunya
Barcelona, Spain

Akram Alomainy
School of Electronic Engineering and
  Computer Science
Queen Mary University of London
London, United Kingdom

Cesar Bartolo Perez
University of California
Davis, California

Riccardo Bassoli
University of Trento
Trento, Italy

Rayisa Beevi
RAK Medical and Health Sciences
  University
Jinnah Sindh Medical University
Karachi, Pakistan

Madhu Bhaskaran
Functional Materials and Microsystems
  Research Group and the Micro Nano
  Research Facility
RMIT University
Melbourne, Australia

P. Bia
Elettronica S.p.A
Design Solution Department
Rome, Italy

G. Boggia
DEI, Politecnico di Bari
Bari, Italy

and

Consorzio Nazionale Interuniversitario per
le Telecomunicazioni (CNIT)
Parma, Italy

Peter Haring Bolivar
Department of Electrical Engineering
and Computer Science
University of Siegen
Siegen, Germany

Francis M. Bui
University of Saskatchewan
Saskatoon, Canada

Albert Cabellos-Aparicio
NaNoNetworking Center in Catalunya
(N3Cat)
Universitat Politecnica de Catalunya
Barcelona, Spain

Hilal Cansizoglu
University of California
Davis, California

D. Caratelli
The Antenna Company Nederland BV
Eindhoven, the Netherlands

Ho-Shin Cho
Kyungpook National University
Daegu, South Korea

Uche K. Chude-Okonkwo
University of Pretoria
Pretoria, South Africa

Cristina Costa
FBK-CREATE NET
Trento, Italy

Eleftherios N. Economou
Foundation for Research and
Technology – Hellas (FORTH)
Heraklion, Greece

Preetam Ghosh
Virginia Commonwealth University
Richmond, Virginia

Jun Gou
University of California
Davis, California

Fabrizio Granelli
University of Trento
Trento, Italy

L. A. Grieco
DEI, Politecnico di Bari
Bari, Italy

and

Consorzio Nazionale Interuniversitario per
le Telecomunicazioni (CNIT)
Parma, Italy

Werner Haselmayr
Institute for Communications Engineering
and RF Systems
Johannes Kepler University Linz
Linz, Austria

Seyed Ehsan Hosseininejad
Department of Electrical Engineering
Yazd University
Yazd, Iran

Han-Yong Jeon
Department of Chemical Engineering
Inha University
Incheon, South Korea

Maria Kafesaki
Foundation for Research and
    Technology – Hellas (FORTH)
Heraklion, Greece

Nikolaos V. Kantartzis
Aristotle University of Thessaloniki
    (AUTH)
Thessaloniki, Greece

Donghyun Kim
School of Electrical and Electronic
    Engineering
Yonsei University
Seoul, South Korea

Hongki Lee
School of Electrical and Electronic
    Engineering
Yonsei University
Seoul, South Korea

Hyunwoong Lee
School of Electrical and Electronic
    Engineering
Yonsei University
Seoul, South Korea

Max Lemme
RWTH Aachen University
Aachen, Germany

Christos Liaskos
Institute of Computer Science
Foundation for Research and
    Technology – Hellas (FORTH)
Heraklion, Greece

Fu Liu
Aalto University
Aalto, Finland

B. T. Maharaj
University of Pretoria
Pretoria, South Africa

Reza Malekian
University of Pretoria
Pretoria, South Africa

Michael L. Mayo
U.S. Army Engineer Research and
    Development Center
Vicksburg, Mississippi

L. Mescia
DEI, Politecnico di Bari
Bari, Italy

and

Consorzio Nazionale Interuniversitario per
    le Telecomunicazioni (CNIT)
Parma, Italy

Mohammad Sajjad Mirmoosa
Aalto University
Aalto, Finland

Gwiyeong Moon
School of Electrical and Electronic
    Engineering
Yonsei University
Seoul, South Korea

Hossein Moosavi
Senior Software Engineer
DevX Tooling Team
Cisco Systems
Montréal, Quebec, Canada

Giacomo Morabito
Department of Electrical, Electronics and
    Computer Engineering
University of Catania
Catania, Italy

V. Musa
DEI, Politecnico di Bari
Bari, Italy

and

Consorzio Nazionale Interuniversitario per
    le Telecomunicazioni (CNIT)
Parma, Italy

Kevin R. Pilkiewicz
U.S. Army Engineer Research and
    Development Center
Vicksburg, Mississippi

G. Piro
DEI, Politecnico di Bari
Bari, Italy

and

Consorzio Nazionale Interuniversitario per
    le Telecomunicazioni (CNIT)
Parma, Italy

Alexandros Pitilakis
Aristotle University of Thessaloniki
    (AUTH)
Thessaloniki, Greece

Andreas Pitsillides
Networks Research Laboratory (NetRL)
University of Cyprus
Nicosia, Cyprus

Pratip Rana
Virginia Commonwealth University
Richmond, Virginia

M. Saif Islam
University of California
Davis, California

William Shieh
Electrical and Electronic Engineering
    Department
University of Melbourne
Melbourne, Australia

Ethungshan Shitiri
Kyungpook National University
Daegu, South Korea

Josep Solé-Pareta
NaNoNetworking Center in Catalunya
    (N3Cat)
Universitat Politecnica de Catalunya
Barcelona, Spain

Costas Soukoulis
Foundation for Research and
    Technology – Hellas (FORTH)
Heraklion, Greece

Sharath Sriram
Functional Materials and Microsystems
    Research Group and the Micro Nano
    Research Facility
RMIT University
Melbourne, Australia

Miao Sun
Electrical and Electronic Engineering
    Department
University of Melbourne
Melbourne, Australia

Mohammad Taha
Functional Materials and Microsystems
Research Group and the Micro Nano
Research Facility
RMIT University
Melbourne, Australia

Anna Tasolamprou
Foundation for Research and
Technology – Hellas (FORTH)
Heraklion, Greece

Xavier Timoneda
NaNoNetworking Center in Catalunya
(N3Cat)
Universitat Politecnica de Catalunya
Barcelona, Spain

Sergei Tretyakov
Aalto University
Aalto, Finland

Odysseas Tsilipakos
Foundation for Research and
Technology – Hellas (FORTH)
Heraklion, Greece

Ranjith Rajasekharan Unnithan
Electrical and Electronic Engineering
Department
University of Melbourne
Melbourne, Australia

Anab Usman
RAK Medical and Health Sciences
University
Jinnah Sindh Medical University
Karachi, Pakistan

Shahnaz Usman
RAK Medical and Health Sciences
University
Jinnah Sindh Medical University
Karachi, Pakistan

A. V. Vasilakos
Luleå University of Technology
Luleå, Sweden

Vasos Vassiliou
Networks Research Laboratory (NetRL)
University of Cyprus
Nicosia, Cyprus

Sumeet Walia
Functional Materials and Microsystems
Research Group and the Micro Nano
Research Facility
RMIT University
Melbourne, Australia

Ke Yang
School of Marine Science and Technology
Northwest Polytechnical University
Fremont, California

Andrea Zanella
Department of Information Engineering
University of Padova
Padova, Italy

Rui Zhang
School of Electronic Engineering and
Computer Science
Queen Mary University of London
London, United Kingdom

I

Introduction

1

Introduction

Graphene-Enabled Wireless Nanoscale Networking Communications in the Terahertz Band

Sergi Abadal, Josep Solé-Pareta, Eduard Alarcón, and Albert Cabellos-Aparicio

CONTENTS

THE UNIQUE PROPERTIES OF GRAPHENE plasmonic antennas open new perspectives on wireless communications in the terahertz (THz) band, with great promise in a world eager for high bandwidths, small form factors, and operational flexibility. This chapter first revisits the background and current design trends in the area. Then, it describes a theoretical framework that aims to enable the systematic evaluation of the unique features of graphene antennas in a THz link, including the peculiarities of the THz wireless channel. The methodology is exemplified in a simple use case of indoor communications and its relevance discussed in the context of advanced protocols at the physical and link layers of design.

1.1 INTRODUCTION

Terahertz (THz) technology has made remarkable advances in the fields of spectroscopy, imaging, and, more recently, wireless communications. In the latter case, the use of this frequency band between 0.1 and 10 THz becomes extremely attractive, owing to the abundance of bandwidth and the potential for low area and power footprints. These features render THzs wireless communications highly desirable for a multitude of applications by virtue of the ultra-broad bandwidths (for cellular networks beyond 5G, terabit local, or personal networks and secure military communications [1]) or the miniaturization potential (for communications within computing packages or nanosensor networks [2]).

The realization of THz wireless communications and networks currently presents several challenges related to, among others, the large propagation losses in this frequency band, the existence of frequency-dependent effects in the THz channel, or the lack of mature devices and circuits for THz operation. These issues are nowadays being tackled in a variety of works aiming to cope with the particularities of the scenario and driving innovation in the development of antennas and transceivers necessary for the THz communication [1]. This chapter discusses these points from the perspective of graphene, which appears as an excellent candidate for the implementation of new devices in the pursuit of fast and efficient means to transmit THz signals.

Graphene is expected to have a definitive impact on the design of future THz transceivers. Its plasmonic properties can be leveraged to implement compact THz modulators [3,4]. The reported high carrier mobility, saturation velocity, and nonlinear properties of graphene have also been leveraged in recent millimeter-wave (mmWave) and THz radiofrequency (RF) devices and circuits [5,6]. Graphene transistors with cut-off frequencies over 350 GHz [7], high-performance diodes [8], and varactors for flexible electronics [9] have been also implemented and tested. This has led to the conception of circuits such as frequency multipliers [10], mixers [11], and power detectors [12]; some of them have even been integrated into fully operational transceiver prototypes in the microwave [13] and mmWave bands [14].

Graphene has also shown a great promise for the implementation of THz antennas, owing to its ability to support the propagation of surface plasmon polariton (SPP) waves in this particular frequency range [15]. Since SPPs are *slow waves*, graphene antennas show a considerable miniaturization potential [16]. SPPs are also tunable and provide graphene plasmonic devices with unique reconfiguration capabilities, thereby enabling both wide resonance frequency tuning and ultra-fast adaptive matching or beam steering [17]. We refer the interested reader to Chapter 2 in this book for more details.

The unique properties of graphene-based antennas and circuits can represent a game changer in future THz networks. On the one hand, the miniaturization potential becomes a key enabler of wireless applications in area-constrained scenarios [1,18,19]. On the other hand, the functional reconfigurability of graphene plasmonic devices could be used to address challenges faced in the design of physical layer (PHY) and medium access control (MAC) protocols for THz networks, where directionality leads to the deafness problem [20] or molecular absorption leads to highly distance-dependent transmission windows [21]. For instance, the works by Lin et al. [20] and Han et al. [22] leverage the reconfigurability of graphene antennas to develop beam-switching and beam-multiplexing schemes.

Although the potential of graphene-enabled THz wireless communications is huge in a world eager for high bandwidths, small form factors, and operational flexibility, the literature is missing a design methodology, allowing to leverage such potential in a systematic way. To cite an example, the works by Lin et al. [20] and Han et al. [22] describe protocols that build upon the reconfigurability of graphene, but they do not connect their findings to critical aspects such as the necessary graphene quality or the feasibility of the underlying biasing scheme. This chapter aims to provide tools to bridge this methodological gap.

The contents of this chapter are set forth as follows. We first review the theoretical background related to graphene antennas and THz propagation in Section 1.2. These considerations help to explain the main contribution, a cross-cutting methodology to model and evaluate graphene-enabled wireless communications, which is presented in Section 1.3. The proposed methodology is intended to relate the communications performance of graphene-enabled wireless links with the nanoscale principles of the underlying design. By connecting the two ends, explorations can be performed during early development stages, well before experimental testbeds become available. Other works have tried to achieve this objective, but only partially by either inspecting the impact of graphene's technological parameters upon the antenna performance only [23,24] or by introducing graphene antenna performance into a link-level communication analysis, without taking the technological parameters of graphene into consideration [25,26].

The proposed methodology is employed in Section 1.4 to assess the suitability of graphene-enabled links for indoor communications through a vertical design space exploration. We also argue that the usefulness of the proposed methodology goes beyond the examples laid out here and could, for instance, allow engineers to determine the graphene quality requirements based on application-dependent communication metrics. Other higher-level options are discussed in Section 1.5, where we also provide a summary of the future prospects in the field of graphene-enabled wireless communications.

1.2 BACKGROUND

1.2.1 Graphene-Based Antennas

Antennas based on carbon materials were first discussed in works proposing dipoles made of carbon nanotubes [27]. In light of the issues of this option in terms of tuning and compatibility with planar fabrication processes, Jornet et al. instead examined the potential use of micrometric graphene patch antennas [28,29]. It was demonstrated that a micrometric graphene sheet would resonate in the THz band, around two orders of magnitude below the resonance frequency of same-size metallic antennas.

The reason behind the subwavelength behavior of graphene antennas is their support of SPP waves in the THz band. SPPs not only exhibit *slow wave* properties, eventually leading to miniaturization, but are also tunable, as their properties can be changed electrically. Driven by these unique features, recent years have witnessed a continued growth of this research field. Multiple designs can be found in the literature [17]: some examples are illustrated in Figure 1.1, yet we refer the interested reader to Chapter 2 for a much broader review of the fundamentals of graphene antennas as well as recent designs.

For the purposes of this chapter, which are to highlight nanoscale particularities of graphene antennas and their frequency dependency, here, we only briefly mention the two technological parameters of graphene with the greatest impact on the antenna properties: the chemical potential and the carrier mobility. On the one hand, the chemical potential refers to the level in the distribution of electron energies at which a quantum state is equally likely to be occupied or empty. In graphene antennas, higher chemical potentials translate to higher resonance frequencies with higher radiation efficiency, and since the value can be changed electrically, the antenna becomes tunable [24]. On the other hand, the carrier mobility defines the average speed at which electrons can move within the material. Its value is static and mostly depends on the quality of the graphene sheets. Samples with high carrier mobility eventually lead to higher antenna efficiency, without a shift in the frequency [16]. Again, we refer the reader to Chapter 2 for more details.

1.2.2 Terahertz Propagation

Shifting frequencies up to the THz band has deep repercussions on the propagation of electromagnetic waves. Aspects such as the molecular composition of the medium or the roughness of the reflection surfaces, which are generally overlooked at mmWave

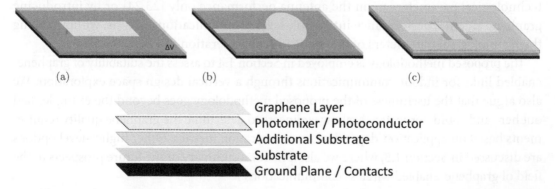

FIGURE 1.1 Schematic representation of a representative set of graphene-based antennas proposed in the literature: (a) Patch. (From Jornet, J.M., and Akyildiz, I.F., *IEEE J. Sel. Areas Commun.*, 31, 685–694, 2013.) (b) circular monopole. (From Zhang, X. et al., Graphene reconfigurable coplanar waveguide (CPW)-fed circular slot antenna, in *Proceedings of the APS/URSI'15*, pp. 2293–2294, 2015.) and (c) dipole. (From Khalid, N., and Akan, O.B., Wideband THz communication channel measurements for 5G indoor wireless networks, in *Proceedings of the ICC'16*, Kuala Lumpur, Malaysia, pp. 16–21, 2016.)

frequencies, gain significance as the wavelengths become commensurate to the size of the molecules or the tiny irregularities at the surfaces. Their effects thus need to be taken into consideration and, as such, were first measured in indoor scenarios with the objective of their standardization within the IEEE 802.15 THz Interest Group [30] and later modeled both in the frequency and time domains [31,32]. We next outline the existing models.

1.2.2.1 Molecular Absorption and Particle Scattering

Molecular absorption is the process by which part of the wave energy is converted into heat as it excites molecules present in the medium [30]. Standard media contain molecules with several resonances at THz frequencies, which implies that the associated attenuation α_M to radiated waves will be significant. At frequency f and for a receiver at distance d, we have that [31]

$$\alpha_M(f,d) = e^{k_A(f)d}, \tag{1.1}$$

where $k_A(f)$ is the medium absorption coefficient, which is highly dependent-selective (see Figure 1.2) and depends on the molecular composition of the medium. It also becomes clear from Equation (1.1) that molecular absorption increases with the distance, that is, with the total number of excited molecules.

Another effect similar to molecular absorption is particle scattering. As wavelengths become commensurate to certain large particles, waves may be reflected and scattered. The associated attenuation α_S is again frequency-selective,

$$\alpha_S(f,d) = e^{k_S(f)d}, \tag{1.2}$$

where the frequency dependence is modeled through the particle scattering coefficient $k_S(f) = \sum_j N_s^j \sigma_s^j$ [32]. This coefficient requires knowledge of the density of particles N and of the scattering cross-section σ of each type of particle j. In the THz band, the scattering cross-section becomes proportional to $\sigma_s^j \sim x_d^6 f^4$, where x_d is the diameter of the particle. In summary, scattering increases with the frequency and the particle size.

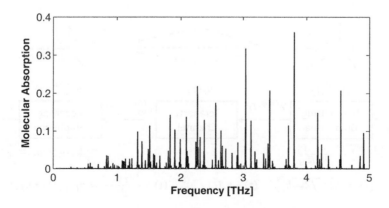

FIGURE 1.2   Molecular absorption of the terahertz channel at a distance of 1 cm.

1.2.2.2 Reflections and Rough-Surface Scattering

The behavior of reflections depends not only on the material but also on the roughness of the surface. In this regard, the main issue is that few materials have been characterized in the THz band so far [33]. Since the reflection coefficient depends on the refractive indices of the incident and reflection media, $n_1(f)$ and $n_2(f)$, respectively, this hinders the modeling of reflections. Luckily, more data are expected to appear as THz wireless communications emerge [34]. Besides, and as wavelengths reach the micrometer scale, the formerly negligible roughness of certain materials may produce scattering [35]. To evaluate such effect, it is necessary to obtain the surface height distribution of the material. Generally, statistic models are used to this end, thereby defining a probabilistic reflection pattern.

1.3 VERTICAL DESIGN METHODOLOGY FOR GRAPHENE-ENABLED WIRELESS COMMUNICATIONS

Progressively scaling the size of the communication units (and the wavelengths) down to the microscale suggests that nanoscale phenomena cannot be neglected even if communication occurs at much larger scales. Here, we describe a cross-cutting methodology that aims to explicitly bridge the conceptual gap between nanoscale phenomena and macroscale communication performance.

The methodology consists of the calculation of the impulse response of the graphene antennas and of the THz channel toward a complete channel model. We follow the considerations employed in [36], which stem from the work in [37]. The model, illustrated in Figure 1.3, relates the voltage $u_{RX}(t)$ at the output of the receiving antenna with the input voltage $u_{TX}(t)$ at transmitting antenna terminals. The methodology and consequent analysis are done in the time domain, given the very high radiation frequency and potentially wideband nature of graphene antennas; however, they can also be performed in the frequency domain. Without loss of generality, we introduce the notation for a single polarization.

1.3.1 Impulse-Response Formulation

Consider a time-dependent voltage $u_{TX}(t)$ at the antenna terminals through a feed of characteristic impedance Z_{TX}. The radiated signal $e_{TX}^{(\theta,\phi)}(t)$ at the $\{\theta,\phi\}$ direction is determined by

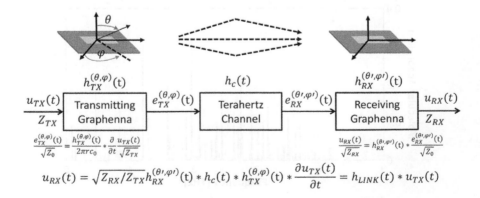

FIGURE 1.3 End-to-end channel model of a graphene-enabled wireless link in the time domain.

$$\frac{e_{TX}^{(\theta,\phi)}(t)}{\sqrt{Z_0}} = \frac{\delta(t-r/c_0)}{2\pi rc_0} * h_{TX}^{(\theta,\phi)}(t) * \frac{\partial}{\partial t} \frac{u_{TX}(t)}{\sqrt{Z_{TX}}}, \tag{1.3}$$

where $*$ represents convolution, $h_{TX}^{(\theta,\phi)}(t)$ is the impulse response of the antenna at the propagation direction, $\partial u_{TX}(t)/\partial t$ is the derivative of the input, r is the distance to the antenna, c_0 is the speed of light, and Z_0 is the free space impedance [36]. From this equation, we infer that *the impulse response of an antenna is the relation between the excitation voltage of the antenna (input) and the radiated field strength (output)*. The impulse response can thus be obtained by applying a voltage and measuring the radiated fields in the time domain. The same strategy can be adopted in the frequency domain by using an equivalent expression [16].

To calculate the voltage $u_{RX}(t)$ at a receiving antenna when impinged by a field $e_{RX}(t,\theta',\phi')$ arriving from the propagation direction, we have

$$\frac{u_{RX}(t)}{\sqrt{Z_{RX}}} = h_{RX}^{(\theta',\phi')}(t) * \frac{e_{RX}^{(\theta',\phi')}(t)}{\sqrt{Z_0}}, \tag{1.4}$$

where $h_{RX}^{(\theta',\phi')}(t)$ is the impulse response of the antenna at the incident direction and Z_{RX} is the characteristic impedance at the receiver. With the definition of impulse response given in Equations (1.3) and (1.4), the reciprocity theorem yields $h_{TX} = h_{RX} = h_A$ for the same antenna [36], which also applies to graphene antennas, as the same plasmonic principles explain their operation in both ends [29]. An equivalent expression can be obtained in the frequency domain [16].

Now, let us consider the radiated signal $e_{TX}(t)$ to be incident at the receiving antenna. By combining Equations (1.3) and (1.4), we can obtain a single expression that describes the end-to-end voltage dependence

$$\frac{u_{RX}(t)}{\sqrt{Z_{RX}}} = h_{RX}^{(\theta',\phi')}(t) * \frac{\delta(t-r_{TR}/c_0)}{2\pi r_{TR}c_0} * h_{TX}^{(\theta,\phi)}(t) * \frac{\partial}{\partial t} \frac{u_{TX}(t)}{\sqrt{Z_{TX}}}, \tag{1.5}$$

where r_{TR} is the distance between the two antennas. The expression clearly separates the contributions of the antennas and the channel. In fact, the term representing the channel can be replaced by an impulse response $h_C(t)$, including all the effects described in Section 1.2. Equation (1.5) would then become:

$$\frac{u_{RX}(t)}{\sqrt{Z_{RX}}} = h_{RX}^{(\theta',\phi')}(t) * h_C(t) * h_{TX}^{(\theta,\phi)}(t) * \frac{\partial}{\partial t} \frac{u_{TX}(t)}{\sqrt{Z_{TX}}}. \tag{1.6}$$

To model the effects of a nontrivial channel, the following generic expression can be used:

$$h_C(t) = \sum_{i=1}^{L} \Gamma_i \alpha_i^{-1}(t,d) e^{j\varphi_i(t,d)} \delta(t-\tau_i), \tag{1.7}$$

where d denotes the path length and L is the number of components that reaches the receiver. Along each path, signals will suffer a given attenuation α, phase shift φ, delay τ, and reflection effects Γ (if any).

Finally, the impulse response of the whole link $h_L(t)$, which includes the channel and the two antennas, can be calculated by assuming that $u_{TX}(t) = \delta(t)$ and applying the properties of the convolution operation:

$$h_L(t) = \sqrt{\frac{Z_{RX}}{Z_{TX}}} \frac{\partial}{\partial t}\left[h_{RX}^{(\theta',\phi')}(t) * h_C(t) * h_{TX}^{(\theta,\phi)}(t) \right]. \qquad (1.8)$$

1.3.2 Outline of the Methodology

The proposed methodology (Figure 1.4) provides a means to characterize the antennas and the channel separately or to perform a joint evaluation of a complete graphene-enabled wireless link in the time domain, as follows:

1. **Model the antennas:** The conductivity of a graphene sheet is obtained as function of the chemical potential E_F and carrier mobility μ, to the integrated the graphene sheet with the rest of the antenna elements within a field solver. This allows to obtain $h_A(t)$ by using Equations (1.3) and (1.4). More details are given in Section 1.3.3.

2. **Model the channel:** The propagation medium is modeled by evaluating, among others, the molecular absorption and particle scattering coefficients, as summarized in Section 1.2. If reflections are significant, the coefficient Γ and roughness scattering

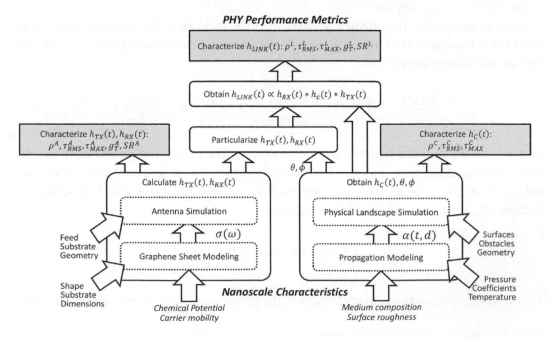

FIGURE 1.4 Vertical methodology for the time-domain characterization of graphene-enabled wireless links as functions of nanoscale phenomena.

shall be assessed as well. All these effects are thrown into Equation (1.7) that, together with a geometrical model of the environment, will allow us to obtain $h_C(t)$. More details are given in Section 1.3.4.

3. **Model the full link:** After obtaining the antenna and channel responses with steps (1) and (2) above, we particularize the transmitting and receiving antenna responses with the suitable angles of radiation and arrival, respectively. The particularized responses are used to derive $h_L(t)$ via Equation (1.8).

4. **Characterize the elements:** The antenna, channel, and link responses can then be characterized using a predefined set of time-domain metrics summarized in Section 1.3.5. It is important to emphasize that, by following these steps, we can explicitly evaluate the different metrics as functions of the parameters representing nanoscale characteristics of the antennas, the channel, or the whole link.

1.3.3 Modeling the Antennas

The first step to model a graphene antenna consists in calculating the conductivity $\sigma(\omega)$ of the graphene sheets. The complexity of the models used to this end depend on the frequency band of interest and the characteristics of the graphene sheet [38]. For the conditions considered here, we use the expression

$$\sigma(\omega) = \frac{2e^2}{\pi\hbar} \frac{k_B T}{\hbar} \ln\left[2\cosh\left[\frac{\mu_c}{2k_B T} \right] \right] \frac{i}{\omega + i\tau^{-1}}, \qquad (1.9)$$

which takes the chemical potential μ_c and carrier mobility $\mu \approx \frac{\tau \cdot v_F^2}{\mu_c}$ as the main parameters. The terms T, e, $\hbar$, k_B, and τ refer to the temperature, elementary charge, reduced Planck, Boltzmann constants, and the relaxation time of graphene. Once the conductivity is obtained, graphene is modeled as an infinitesimally thin surface, with an equivalent impedance of $Z(\omega) = \frac{1}{\sigma(\omega)}$. The graphene layer needs to be shaped according to antenna geometry and then integrated with the whole antenna structure.

Once the antenna is modeled, we can obtain the impulse response via full-wave solving in the time domain. If the feeder is properly defined, the simulator can calculate the radiated fields as a function of the input voltage. Reciprocally, one can also consider a wave incident to the antenna to then calculate the received voltage. In both cases, the response of the antenna can be derived by relating the voltage and the field by using Equations (1.3) and (1.4), respectively. Most simulators allow to solve the fields in the frequency domain, in which case the inverse Fourier transform may be applied to derive the impulse response.

1.3.4 Modeling the Channel

As summarized in Figure 1.4 and according to Equation (1.7), calculating the impulse response of the channel implies modeling the propagation medium to assess the attenuation and phase shift per unit of distance and then simulating the physical landscape in order to obtain the number, distribution, and characteristics of the multipath components.

When considering free space propagation through air, we normally have that $\alpha(t,d) = (2\pi dc_0)^{-1}$, $\phi(t,d) = 0$, and $\tau(t,d) = d/c_0$, with c_0 representing the speed of light [36]. However, THz waves are susceptible to molecular absorption and particle scattering, as summarized in Section 1.2.2, Equations (1.1) and (1.2). These effects are incorporated in the path loss as

$$\alpha(f,d) = \frac{e^{k(f)d}}{2\pi dc_0}, \quad \alpha(t,d) = \mathcal{F}^{-1}(\alpha(f,d)), \tag{1.10}$$

where $k(f) = \sum_j k_A^j + k_S^j$ is the extinction loss coefficient that accounts for both absorption k_A and scattering k_S. We refer the reader to [32] for more details and refined models.

The scenario must then be modeled and simulated to obtain the multipath behavior of the channel. If the transmission range is large in terms of wavelengths, full electromagnetic (EM) simulation becomes too costly, and the use of ray tracing is considered instead. This method obtains the most relevant rays arriving at the receiver, determining their angle of arrival (to fix the antenna response) and reflections along the way (to properly calculate the reflection coefficient Γ and rough surface scattering, if any [34,35]).

1.3.5 Time-Domain Performance Metrics

The final step of the methodology is to employ the impulse responses to characterize the antenna, the channel, or the whole link. This process requires a set of time-domain metrics different from the ones used in narrowband systems. We inspire in ultra-wideband (UWB) works [36,39], where some metrics are based on the envelope of the analytic response of the antenna $|h^+(t)|$, as it faithfully represents the antenna dispersion. The analytic response is given by

$$h^+(t) = h(t) + j\mathcal{H}, \tag{1.11}$$

where $\mathcal{H}$ is the Hilbert transform of the impulse response. To model the channel, the Power-Delay Profile (PDP) is widely used for similar reasons. The PDP $p(t)$ is obtained as

$$p(t) = |h(t)|^2. \tag{1.12}$$

Regardless of whether the analytic response or the PDP is used, the metrics presented here evaluate attenuation and dispersion by quantifying the amplitude and temporal length of the response. For the sake of brevity, next paragraphs present the metrics for the antenna case—we refer the reader to Table 1.1 for the channel and link versions. In any case, the novelty of our proposal resides in that it allows to express all these metrics as functions of nanoscale phenomena such as the chemical potential of graphene and the medium composition.

The **response peak** ρ of a response is defined as the maximum value of its envelope, which is

$$\rho = \max_t |h^+(t)|[m/ns] \tag{1.13}$$

TABLE 1.1 Summary of Performance Metric Formulations

Metric	Antenna	Channel	Link
Response Peak	$\max_t \lvert h^+(t) \rvert$	$\max_t p(t)$	
Response Energy	$\lVert h(t) \rVert^2$	$\int p(t)$	
Transient Gain	$\dfrac{\lVert h(t) * u'_{TX}(t) \rVert^2}{\lVert \sqrt{\pi}\, c_0 u_{TX}(t) \rVert^2}$	$\dfrac{\lVert h(t) * e_{TX}(t) \rVert^2}{\lVert e_{TX} \rVert^2}$	$\dfrac{Z_{TX}}{Z_{RX}} \dfrac{\lVert h(t) * u_{TX}(t) \rVert^2}{\lVert u_{TX} \rVert^2}$
Response Width	t_{FWHM}		
Response RMS	$\sqrt{\int (t - \tau_m)^2 \cdot \lvert h^+(t) \rvert^2}$	$\sqrt{\int (t - \tau_m)^2 \cdot p(t)}$	
Response Duration	$t_\alpha - t_\rho$		
Stretch Ratio	$\dfrac{W(h * u'_{TX})}{W(u_{TX})}$	$\dfrac{W(h * e_{TX})}{W(e_{TX})}$	$\dfrac{W(h * u_{TX})}{W(u_{TX})}$

in the case of the antenna and $\rho = \max_t p(t)$ in dB in the channel and link cases. A high peak value could mean that the energy is highly concentrated around a given time instant. Receiving a strong peak allows for a precise detection of the pulse position, which would be desirable if coherent detection is used. In noncoherent communications, the peak is less important, as receivers are generally based on energy detection of the whole pulse.

The **response energy** E is the integral of the instantaneous power of the response over its duration:

$$E = \lVert h(t) \rVert^2, \tag{1.14}$$

where the norm is $\lVert f(x) \rVert^k = \int_{-\infty}^{\infty} \lvert f(x) \rvert^k dx$ or the integral of the PDP. The value of E indicates how the antenna and channel effects impact the radiated and propagated energy, and, therefore, a higher value is preferred.

The **transient gain** g_T is the time-domain version of the antenna gain or the channel path loss and is an indicator of how efficiently an antenna or a link is able to transmit a given input signal u_{TX}. The original definition of transient gain is taken from [37] as the ratio of radiation intensity of the antenna to that of an isotropic radiator, which leads to

$$g_T(u_{TX}) = \frac{\lVert h(t) * \dfrac{\partial u_{TX}(t)}{\partial t} \rVert^2}{\lVert \sqrt{\pi} c_0 u_{TX}(t) \rVert^2} \tag{1.15}$$

in the case of the antenna [16]. As shoswn in Table 1.1, the link transient gain does not use the derivative of the input waveform as it is included in our impulse-response formulation.

The **response width** τ_W is defined as the full width at half maximum (FWHM), or

$$\tau_W = t_{h2} - t_{h1}, \tag{1.16}$$

where $t_{h1} = t'$, so that $|h^+(t')| = \rho/2$ or $p(t') = \rho/2$, and $t_{h2}^A = t''$, so that $t'' > t_{h1} \wedge |h^+(t'')| = \rho/2$ or $p(t'')| = \rho/2$. The envelope width is a clear indicator of the broadening that transmission signals will suffer and, therefore, a lower value is preferred. Another way to model the dispersion is through the root-mean-square (RMS) delay spread, which is the square root of the second central moment of the response

$$\tau_{RMS} = \sqrt{\int (t - \tau_m)^2 \cdot |h^+(t)|^2}, \tag{1.17}$$

where $\tau_m = \int t \cdot |h^+(t)|^2$ is the mean excess delay of the antenna. As shown in Table 1.1, the channel and link versions use the PDP instead. A low value is preferred.

The **response duration** τ_R is generally defined with respect to a parameter α that represents a portion of energy that can be considered negligible, leading to

$$\tau_R(\alpha) = t_\alpha - t_\rho, \tag{1.18}$$

where $t_\rho = t'$, so that $|h^+(t')| = \rho$ or $p(t') = \rho$, and $t_\alpha = t''$, so that $t'' > t_\rho \wedge |h^+(t'')| = \alpha \cdot \rho$ or $p(t'') = \alpha \cdot \rho$. In an antenna, the response duration is often referred to as *ringing duration,* owing to the ringing tail of the response caused by the resonant behavior of a given structure [36]. In a channel, the response is referred to as *maximum excess delay,* and it can be lengthened by long multipath components. A short response duration is, in all cases, desirable to minimize intersymbol interference at high speeds.

The **pulse width stretch ratio** SR quantifies the broadening of a given waveform u_{TX} caused during the transmission. Let the normalized cumulative energy function of a given signal $s(t)$ be defined as $E_s(t) = \frac{\int_{-\infty}^t |s(t)|^2}{\|s(t)\|^2}$. Assuming that a certain fraction α of ringing energy can be neglected, the width of the signal $W(s)$ can then be obtained as $W(s) = E_s^{-1}(1 - \alpha/2) - E_s^{-1}(\alpha/2)$. Then, the stretch ratio is obtained by dividing the width of the output signal by the width of the input signal [39], which becomes

$$SR(u_{TX}) = \frac{W\left(h * \dfrac{\partial u_{TX}}{\partial t}\right)}{W(u_{TX})} \tag{1.19}$$

for the antenna case. Channel and link versions can be found in Table 1.1. A value close to 1 is desired, which means that the antenna has a nearly flat response in the frequency band of the input signal. Otherwise, the pulse width would increase, leading to reduced transmission data rates.

1.4 WIRELESS INDOOR COMMUNICATIONS AS A USE CASE

This section performs a design space exploration, with the methodology presented in the previous section. The exploration aims to investigate graphene-based wireless communications in the context of indoor communications, which to date remains one of the most promising scenarios for THz wireless networks [20,25,34,35,40]. Unless noted, the link to be explored has the characteristics summarized in Table 1.2.

TABLE 1.2   Parameters for the Design Space Exploration

Parameter	Value
Antenna Dimension ($W \times L$)	$5 \times 1\,\mu m^2$
Chemical Potential (μ_c)	0.1–2 eV
Carrier Mobility (μ)	0.5–6 m^{2}V^{-1}s^{-1}
Feed	10 kΩ photomixer
Substrate (ε_r)	Air (freestanding, $\varepsilon_r = 1$)
Frequency Range	0.1–10 THz
Input Pulse Types	Gaussian
Input Pulse Bandwidth	1 THz
Propagation	Line-of-sight
Vapor Concentration (VC)	0.5%–50%
Distance (d)	1 mm–10 m

We use FEldberechnung für Körper mit beliebiger Oberfläche (FEKO) [41] to evaluate the family of graphene antennas and in-house scripts to implement the models and extract the performance metrics. We focus on the impact of the carrier mobility and chemical potential at the antennas and of the frequency-dependent response of the molecular absorption at the channel. We assume that particle and rough surface scattering are negligible. The simplicity of the target scenario is deliberately simple, so that we can focus on the methodological details rather than on the specific antenna design or channel. However, it is clearly stated in previous sections that the methodology allows to plug in geometrical or behavioral models to describe more complex scenarios.

We evaluate the antennas and the channel separately in Sections 1.4.1 and 1.4.2, to then extract the impulse response of the whole link and provide a joint assessment in Section 1.4.3.

1.4.1 Antenna

Figure 1.5 compares the envelope of the analytic response of two graphene antennas of the same size: the first one with high carrier mobility and low chemical potential, and the second one with exactly the opposite. Table 1.3 summarizes the performance of the antennas, confirming that the resonance frequency and the amplitude of the response improve with higher chemical potential, as mentioned in Section 1.2. The stronger resonance behavior, however, lengthens the response, as indicated by the τ_W and τ_R values.

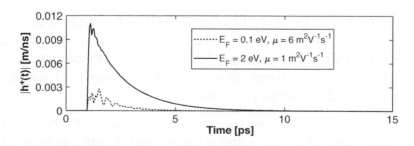

FIGURE 1.5   Envelope of the analytic response $|h^+(t)|$ of two graphene antennas.

TABLE 1.3 Performance Comparison of Metallic and Graphene Antennas

Performance Metric	Graphene 1 $E_F = 0.1, \mu = 6$	Graphene 2 $E_F = 2, \mu = 1$	Gold Antenna 1	Gold Antenna 2
Area	$5\ \mu m^2$	$5\ \mu m^2$	$1125\ \mu m^2$	$5\ \mu m^2$
f_{res}	1.74 THz	7.48 THz	1.52 THz	22.5 THz
p	0.003 m/ns	0.01 m/ns	0.12 m/ns	0.04 m/ns
E (arb. units)	1.76	50.23	264.94	15.04
τ_W	1 ps	1.2 ps	0.45 ps	0.06 ps
$\tau_R(10\%)$	1.95 ps	3.6 ps	1 ps	0.1 ps
g_T	−31.56 dBi	−4.55 dBi	−8.17 dBi	−9.45 dBi
SR	1.013	1.026	0.9957	0.9843

Table 1.3 also compares the performance of the graphene antennas with that of two gold antennas: a patch dimensioned to have a similar resonant frequency as that of the graphene antenna and a patch of the same size (see [16] for details on their modeling). On the one hand, in order to resonate around 1 THz, the gold antenna must be about two orders of magnitude larger. This size difference allows this antenna to radiate more energy, resulting in a higher envelope peak. However, this fact does not ensure the best performance in terms of relative radiation efficiency (transient gain). On the other hand, the small gold patch has the shortest response, as it resonates at a much higher frequency. We presume that this behavior is due to the larger absolute bandwidth of the structure. However, and although remarkable performance is observed in terms of envelope peak, the reduced transient gain implies that graphene antennas of the same size will be able to radiate with higher efficiency. For all this, we conclude that the moderately worse performance of graphene antennas will be compensated by the unique size and tunability of graphene antennas.

To further investigate the impact of the carrier mobility and chemical potential, we obtained the impulse response for different values within the parametric design space: from 0.5 to 6 $m^2V^{-1}s^{-1}$ and from 0.1 to 2 eV. These have been achieved in the literature and are below graphene's electrical breakdown [42]. We next analyze the characteristics of the different responses by using the metrics explained in Section 1.3.5 in contour plots that illustrate the full parametric exploration.

Response peak: Figure 1.6a shows the results regarding the response peak, confirming that the peak value is proportional to the chemical potential, since contour lines are parallel to the Y-axis. The weak dependence with respect to the carrier mobility confirms that the raise in antenna efficiency impacts the impulse-response width and ringing length rather than the peak value.

Response energy: Figure 1.6b shows the energy of the response, illustrating that the two technological parameters have a similar impact, as they both lead to an enhancement of the resonant behavior. However, the change in chemical potential has a larger influence, especially at low values.

Transient gain: Figure 1.6c shows the transient gain when a gaussian pulse is radiated. The behavior is similar to that for the response peak and response energy metrics. Apparently, whether this energy surge revolves around the response peak or not, does not make a difference when evaluating the transient gain–at least for this test signal.

Peak width: Figure 1.6d plots the peak width. Narrow responses are obtained for low chemical potentials and carrier mobilities. The response then widens with the carrier mobility, owing to the increase of the ringing effects. The response also widens as the chemical potential increases up to ~1 eV, because, in those cases, the peak value is rather low, and the ringing effects dominate. Beyond 1 eV, the peak value is high, and the impact of increasing ringing diminishes, leading to a reduction of the relative peak width.

Ringing duration: Figure 1.6e shows the ringing duration, assuming $\alpha = 10\%$. Since this metric depends on the peak value, a similar behavior is shown: the ringing duration increases with the carrier mobility and with the chemical potential until ~1 eV, to plateau beyond that value.

Pulse width stretch ratio: Figure 1.6(f) shows the pulse width stretch ratio assuming a gaussian pulse as input. The trend is similar to that of the ringing: a clear and undesired rise of the stretch ratio is observed when the chemical potential is increased below 0.9 eV, with a particularly strong transition around 0.5 eV and high carrier mobilities. In other words, a clear and undesired rise of the stretch ratio is observed within the range between 0.5 eV and 1 eV, especially for high carrier mobilities.

1.4.2 Channel

Let us now evaluate the impact of molecular absorption on the channel response for a standard atmosphere. We first consider a vapor concentration (*VC*) of 10% [43]. Figure 1.7a shows the spectrum of losses associated with molecular absorption for distances of 1 and 10 cm, illustrating a notable increase in the number of absorption peaks and their amplitude in the latter case. This behavior is due to the exponential scaling of the molecular absorption losses with distance.

We can also observe in Figure 1.7a that the rising peaks limit the bandwidth free of distortion. In order to quantify this effect, we consider the frequency span at which the molecular absorption is at all times below 10 dB (the actual threshold will depend on the application). For the distances considered here, the available bandwidth would be of approximately 27 and 9 THz for 1 and 10 cm, respectively, as shown in blue and red. A generalization of this analysis, shown in Figure 1.7b, reveals that the whole spectrum is free of molecular absorption at the millimeter scale, whereas transmissions at several meters would have less than 6 THz available for a molecular absorption-free transmission.

The impact of molecular absorption can also be evaluated in the time domain. Successive attenuation peaks are expected to cause the channel impulse response to lower the peak

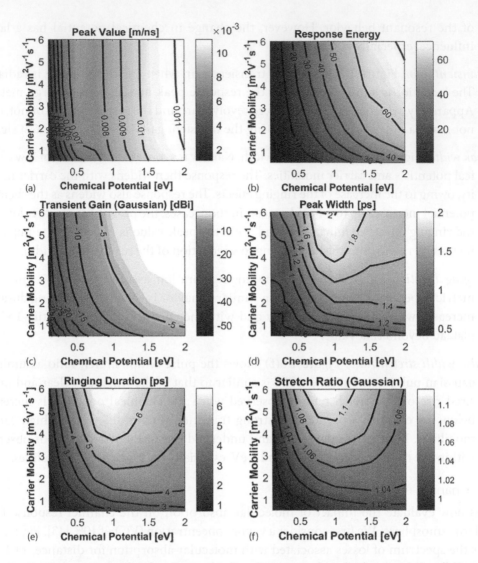

FIGURE 1.6 Response amplitude metrics (top) and response length characteristics (bottom) as functions of the carrier mobility and chemical potential: (a) Response peak, (b) response energy, (c) transient gain, (d) peak width, (e) ringing (10%), and (f) stretch ratio (Gaussian).

amplitude and lead to irregular shapes. Figure 1.8a exemplifies this by showing how the width of the impulse response τ_w^{abs} scales with the distance d as $O(\sqrt[5]{d})$.

As a specific example, we observe that, for a transmission distance of 3 m, the distortion introduced by molecular absorption is limited to 0.05 ps. If we assume that received pulses need to be less than 1-ps long to achieve a bandwidth of around 1 THz, then this channel will only reduce the maximum achievable throughput by 5%.

Another important metric to consider is the energy E of the molecular absorption response. Figure 1.8b shows that the scaling of the response energy with respect to the distance follows a $O(1/\sqrt[3]{d})$ trend. This further confirms that, at short scales, the effect of molecular absorption is negligible.

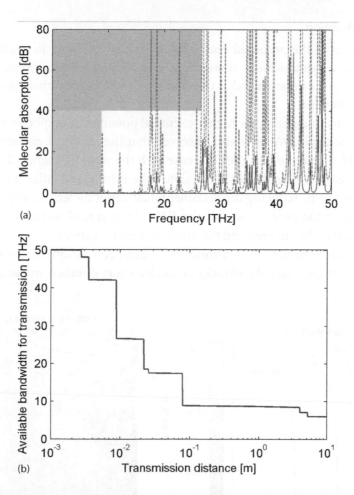

(a)

(b)

FIGURE 1.7   (a) Molecular absorption for 1 cm (gray lines) and 10 cm (dotted gray lines). The light and dark gray backgrounds indicate the available bandwidth for distances of 1 and 10 cm, respectively. (b) Available bandwidth in the frequency band up to 50 THz. In both cases, $VC = 10\%$.

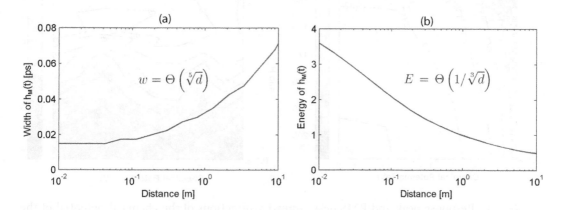

FIGURE 1.8   Semi-log plots of the different characteristics of the channel impulse response due to molecular absorption: (a) Response width and (b) response energy.

1.4.3 Joint Exploration

One of the virtues of the proposed methodology is that it allows to evaluate graphene-enabled wireless communication links from end to end. Thus, interactions between nanoscale effects at the antenna and at the channel can be put together toward a complete design space exploration.

We have seen that increasing the higher chemical potential of graphene implies a higher efficiency of the antenna. We have also observed that the effects of molecular absorption appear at around 1 THz and that they increase with the wave frequency. Thus, it becomes apparent to inspect whether the path loss should be minimized either by increasing the chemical potential or by reducing the detrimental effects of molecular absorption.

Figure 1.9 shows the peak value and the RMS delay spread as functions of both the chemical potential of the antenna and the amount of water vapor in the channel. Plots 3.9a and b represent these metrics for a transmitter–receiver separation of 3 cm to model proximity communications (e.g., a data kiosk). At such a short distance, molecular absorption

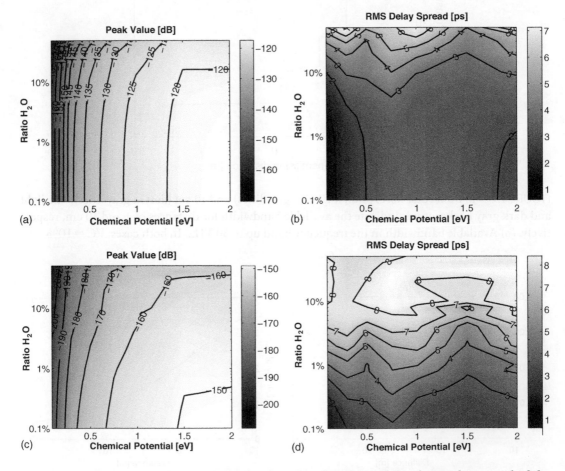

FIGURE 1.9 Response peak and RMS delay spread as functions of the chemical potential of the antenna and the amount of water vapor in the channel for a distance of 3 cm (top) and 1 m (bottom). $\mu = 4\,\text{m}^2\text{V}^{-1}\text{s}^{-1}$: (a) Peak, 3 cm; (b) RMS, 3 cm; (c) Peak, 1 m; and (d) RMS, 1 m.

is far from being an impairment. However, a dependence between the chemical potential and the molecular absorption can still be noticed: the absorption gains significance at high chemical potential values, whereas the RMS shows a more stable value when the molecular absorption is low.

To extend this joint evaluation to the indoor scenario, we plotted the same metrics for a distance of 1 m in Figure 1.9c and d. Here, the difference in terms of molecular absorption is clear: increasing the water vapor concentration does not yield significant changes in terms of peak value for $\mu_c = 0.5$ eV, whereas for $\mu_c = 2$ eV, there is a gap of almost 10 dB when considering low and high water vapor concentrations. The RMS delay spread shows a rather irregular pattern due to the highly frequency-dependent distribution of the molecular absorption.

1.5 SUMMARY

Graphene-based antennas present unique miniaturization and tunability properties in the THz band. In this chapter, we have observed these features under the prism of a methodology that makes explicit the connection between nanoscale phenomena and the performance of a graphene-enabled wireless communication link. We have explored antenna and channel characteristics under a wide range of parameter values. However, the usefulness of the proposed approach becomes apparent when putting both the antenna and the channel together in the analysis: the methodology allows to detect the chemical potential value for which a given antenna resonance point will fall within a molecular absorption peak.

It is worth noting that the compromise between chemical potential and molecular absorption is just one of the trade-offs that can be examined by means of the proposed cross-cutting methodology. For instance, since multipath is a frequency-dependent phenomenon, one could explore the multipath effects with multiple chemical potentials and, thus, multiple frequencies of resonance. In simple terms, reflections that might be harmless for an antenna resonating at a certain frequency may destroy signals when the antenna is tuned at another frequency. Another example would involve scenarios such as wireless on-chip networks [19], where maximum performance is sought and multipath components are fixed. In this context, the vertical methodology can be extremely useful to avoid certain performance-degrading situations when antennas are tuned to different frequencies.

Joining the tunability of graphene with the particularities of the THz wireless channel could also be useful to develop advanced PHY and link-level schemes. As mentioned previously, the distance-aware multicarrier (DAMC) modulation selects the transmission windows, depending on the distance between nodes to avoid emerging absorption peaks [21]. Once the transmission window is fixed, the use of highly directional antennas is still required to combat the large propagation loss. In this context, the ultra-fast reconfigurability of graphene represents an opportunity to develop advanced PHY based on beam multiplexing, as proposed in [20,22]. High directionality also represents a challenge due to the existence of the deafness problem at the MAC layer. To address this, the graphene properties can again be leveraged to implement fast beam scanning methods for neighbor discovery and synchronization [17].

Another work worth highlighting at this point is that of [44], where the authors propose a design strategy analogous to the vertical methodology developed here. In that case, their approach exposes the reconfigurability potential of graphene antennas to the PHY and MAC layers of design. The main idea is to connect the antenna to a digital controller that translates directives coming from the upper layers to a set of voltages that determine the antenna state (frequency and beam). Then, an array of actuators applies the precise voltage level to the antenna elements to achieve the targeted state. This design approach is very relevant to the contents summarized here, as it allows to develop advanced graphene-enabled PHY designs in a systematic way, not only easing the final implementation of current proposals [20,22] but also paving the way to new massive programmable multichannel multibeam protocols for graphene-enabled wireless communications.

ACKNOWLEDGMENTS

This work has been partially funded by the Spanish Ministry of Economia y Competitividad under grants PCIN-2015-012 and TEC2017-90034-C2-1-R (ALLIANCE project) and receives funding from Fondo Europeo de Desarrollo Regional (FEDER), and the Catalan Institution for Research and Advanced Studies (ICREA).

REFERENCES

1. I. F. Akyildiz, J. M. Jornet, and C. Han. Terahertz band: Next frontier for wireless communications. *Physical Communication*, 12:16–32, 2014.
2. S. Abadal, S. E. Hosseininejad, A. Cabellos-Aparicio, and E. Alarcon. Graphene-based terahertz antennas for area-constrained applications. In *Proceedings of the TSP'17*, Barcelona, Spain, 2017.
3. B. Sensale-Rodriguez, R. Yan, M. M. Kelly, T. Fang, K. Tahy, W. S. Hwang, D. Jena, L. Liu, and H. G. Xing. Broadband graphene terahertz modulators enabled by intraband transitions. *Nature Communications*, 3:780, 2012.
4. P. K. Singh, G. Aizin, N. Thawdar, M. Medley, and J. M. Jornet. Graphene-based plasmonic phase modulator for terahertz-band communication. In *Proceedings of the EuCAP'16*, Davos, Switzerland, 2016.
5. T. Palacios, A. Hsu, and H. Wang. Applications of graphene devices in RF communications. *IEEE Communications Magazine*, 48(6):122–128, 2010.
6. Y. Wu, D. B. Farmer, F. Xia, and P. Avouris. Graphene electronics: Materials, devices, and circuits. In *Proceedings of the IEEE*, 101(7):1620–1637, 2013.
7. F. Schwierz. Graphene transistors: Status, prospects, and problems. In *Proceedings of the IEEE*, 101(7):1567–1584, 2013.
8. M. Shaygan, Z. Wang, M. S. Elsayed, M. Otto, G. Iannaccone, A. H. Ghareeb, G. Fiori, R. Negra, and D. Neumaier. High performance metalinsulatorgraphene diodes for radio frequency power detection application. *Nanoscale*, 9:11944–11950, 2017.
9. M. Saeed, A. Hamed, C.-Y. Fan, E. Heidebrecht, R. Negra, M. Shaygan, and Z. Wang. Millimeter-wave graphene-based varactor for flexible electronics. In *Proceedings of the EuMIC'17*, pages 117–120, 2017.
10. H. Wang, D. Nezich, J. Kong, and T. Palacios. Graphene frequency multipliers. *IEEE Electron Device Letters*, 30(5):547–549, 2012.

11. C. V. Antuna, A. I. Hadarig, S. V. Hoeye, M. F. Garcja, R. C. Djaz, G. R. Hotopan, A. F. Las Heras, and H. Andrés. High-order subharmonic millimeter-wave mixer based on few-layer graphene. *IEEE Transactions on Microwave Theory and Techniques*, 63(4):1361–1369, 2015.

12. A. A. Generalov, M. A. Andersson, X. Yang, A. Vorobiev, and J. Stake. A 400-GHz graphene FET detector. *IEEE Transactions on Terahertz Science and Technology*, 7(5):614–616, 2017.

13. S.-J. Han, A. V. Garcia, S. Oida, K. A. Jenkins, and W. Haensch. Graphene radio frequency receiver integrated circuit. *Nature Communications*, 5, 2014.

14. O. Habibpour, Z. S. He, W. Strupinski, and N. Rorsman. Wafer scale millimeter-wave integrated circuits based on epitaxial graphene in high data rate communication. *Scientific Reports*, 7:41828, 2017.

15. F. H. L. Koppens, D. E. Chang, and F. J. G. D. Abajo. Graphene plasmonics: A platform for strong light-matter interactions. *Nano Letters*, 3370–3377, 2011.

16. S. Abadal, I. Llatser, A. Mestres, H. Lee, E. Alarcon, and A. Cabellos-Aparicio. Time-domain analysis of graphene-based miniaturized antennas for ultra-short-range impulse radio communications. *IEEE Transactions on Communications*, 63(4):1470–1482, 2015.

17. D. Correas-Serrano and J. S. Gomez-Diaz. Graphene-based antennas for terahertz systems: A review. *FERMAT*, 2017.

18. S. Abadal, C. Liaskos, A. Tsioliaridou, S. Ioannidis, A. Pitsillides, J. Sole-Pareta, E. Alarcon, and A. Cabellos-Aparicio. Computing and communications for the software-defined metamaterial paradigm: A context analysis. *IEEE Access*, 5:6225–6235, 2017.

19. S. Abadal, J. Torrellas, E. Alarcon, and A. Cabellos-Aparicio. OrthoNoC: A broadcast-oriented dual-plane wireless network-on-chip architecture. *IEEE Transactions on Parallel and Distributed Systems*, 29(3):628–641, 2018.

20. J. Lin and M. A. Weitnauer. Pulse-level beam-switching for terahertz networks. *Wireless Networks*, 7:1–16, 2018.

21. C. Han and I. F. Akyildiz. Distance-aware multi-carrier (DAMC) modulation in Terahertz Band communication. In *Proceedings of the ICC'14*, pages 5461–5467, 2014.

22. C. Han, W. Tong, and X.-W. Yao. MA-ADM: A memory-assisted angular-division-multiplexing MAC protocol in Terahertz communication networks. *Nano Communication Networks*, 13:51–59, 2017.

23. I. Llatser, C. Kremers, A. Cabellos-Aparicio, J. M. Jornet, E. Alarcón, and D. N. Chigrin. Graphene-based nano-patch antenna for terahertz radiation. *Photonics and Nanostructures—Fundamentals and Applications*, 10(4):353–358, 2012.

24. M. Tamagnone, J. S. Gomez-Diaz, J. R. Mosig, and J. Perruisseau-Carrier. Reconfigurable terahertz plasmonic antenna concept using a graphene stack. *Applied Physics Letters*, 101:214102, 2012.

25. C. Han and I. F. Akyildiz. Three-dimensional end-to-end modeling and analysis for graphene-enabled terahertz band communications. *IEEE Transactions on Vehicular Technology*, 66(7):5626–5634, 2017.

26. C. Zhang, C. Han, and I. F. Akyildiz. Three dimensional end-to-end modeling and directivity analysis for graphene-based antennas in the terahertz band. In *Proceedings of the GLOBECOM'15*, pages 1–6, 2015.

27. P. Russer, N. Fichtner, P. Lugli, W. Porod, J. A. Russer, and H. Yordanov. Nanoelectronics-based integrated antennas. *IEEE Microwave Magazine*, 11(7):58–71, 2010.

28. J. M. Jornet and I. F. Akyildiz. Graphene-based nano-antennas for electromagnetic nanocommunications in the terahertz band. In *Proceedings of the EuCAP'10*, pages 1–5, 2010.

29. J. M. Jornet and I. F. Akyildiz. Graphene-based plasmonic nano-antenna for terahertz band communication in nanonetworks. *IEEE Journal on Selected Areas in Communications*, 31(12):685–694, 2013.

30. R. Piesiewicz, T. Kleine-Ostmann, N. Krumbholz, D. Mittleman, M. Koch, J. Schoebel, and T. Kurner. Short-range ultra-broadband terahertz communications: Concepts and perspectives. *IEEE Antennas and Propagation Magazine*, 49(6):24–39, 2007.

31. J. M. Jornet and I. F. Akyildiz. Channel modeling and capacity analysis for electromagnetic wireless nanonetworks in the terahertz band. *IEEE Transactions on Wireless Communications*, 10(10):3211–3221, 2011.

32. J. Kokkoniemi, J. Lehtomaki, K. Umebayashi, and M. Juntti. Frequency and time domain channel models for nanonetworks in terahertz band. *IEEE Transactions on Antennas and Propagation*, 63(2):678–691, 2015.

33. C. Ronne, L. Thrane, P.-O. Astrand, A. Wallqvist, K. V. Mikkelsen, and S. R. Keiding. Investigation of the temperature dependence of dielectric relaxation in liquid water by THz reflection spectroscopy and molecular dynamics simulation. *The Journal of Chemical Physics*, 107(14):5319, 1997.

34. J. Kokkoniemi, V. Petrov, D. Moltchanov, J. Lehtomaki, Y. Koucheryavy, and M. Juntti. Wideband terahertz band reflection and diffuse scattering measurements for beyond 5G indoor wireless networks. In *Proceedings of the EW'16*, 2016.

35. C. Jansen, S. Priebe, and C. Moller. Diffuse scattering from rough surfaces in THz communication channels. *IEEE Terahertz Science and Technology*, 1(2):462–472, 2011.

36. W. Wiesbeck, G. Adamiuk, and C. Sturm. Basic properties and design principles of UWB antennas. In *Proceedings of the IEEE*, 97(2):372–385, 2009.

37. E. G. Farr and C. E. Baum. Time Domain Characterization of Antennas with TEM Feeds. In *Sensor and Simulation Notes*, page Note 426, 1998.

38. A. H. Castro Neto, F. Guinea, N. M. R. Peres, K. S. Novoselov, and A. K. Geim. The electronic properties of graphene. *Reviews of Modern Physics*, 81(1):109–162, 2009.

39. D.-H. Kwon. Effect of antenna gain and group delay variations on pulse-preserving capabilities of ultrawideband antennas. *IEEE Transactions on Antennas and Propagation*, 54(8):2208–2215, 2006.

40. N. Khalid and O. B. Akan. Wideband THz communication channel measurements for 5G indoor wireless networks. In *Proceedings of the ICC'16*, Kuala Lumpur, Malaysia, pages 16–21, 2016.

41. Altair, FEKO—Electromagnetic Simulation Software, 2019. Available: http://www.feko.info.

42. L. Banszerus, M. Schmitz, S. Engels, J. Dauber, M. Oellers, and P. Gr. Ultra-high mobility graphene devices from chemical vapor deposition on reusable copper. *Science Advances*, 1(6):e1500222, 2015.

43. I. Llatser, A. Mestres, S. Abadal, E. Alarcon, H. Lee, and A. Cabellos-Aparicio. Time-and frequency-domain analysis of molecular absorption in short-range terahertz communications. *IEEE Antennas and Wireless Propagation Letters*, 14:350–353, 2015.

44. S. E. Hosseininejad, S. Abadal, M. Neshat, R. Faraji-Dana, M. C. Lemme, C. Suessmeier, P. Haring Bolívar, E. Alarcon, and A. Cabellos-Aparicio. MAC-oriented programmable terahertz PHY via graphene-based yagi-uda antennas. In *Proceedings of the WCNC'18*, 2018.

45. X. Zhang, X. Huang, T. Leng, G. Auton, and E. Hill. Graphene reconfigurable coplanar waveguide (CPW)-fed circular slot antenna. In *Proceedings of the APS/URSI'15*, pages 2293–2294, 2015.

CHAPTER 2

Graphene-Based Antenna Design for Communications in the Terahertz Band

Sergi Abadal, Seyed Ehsan Hosseininejad,

Max Lemme, Peter Haring Bolivar, Josep Solé-Pareta,

Eduard Alarcón, and Albert Cabellos-Aparicio

CONTENTS

G RAPHENE IS REGARDED as an excellent candidate for the implementation of antennas in the terahertz band, where other technologies have failed to provide valid alternatives. In particular, graphene exhibits plasmonic behavior in this frequency range, which can be leveraged to develop miniaturized and tunable antenna designs. This chapter reviews the state of the art of graphene-based antennas for terahertz band communications by depicting current design trends, outlining the theoretical foundations behind them and evaluating a representative set of implementations.

2.1 INTRODUCTION

Graphene has garnered the attention of the scientific community owing to its extraordinary mechanical, electronic, and optical properties [1]. The remarkable properties of this two-dimensional material, unveiled during the last two decades, continues to open the door to a wide variety of applications such as nanoscale integrated circuits, spectroscopy, imaging, transformation optical devices, absorbers, and metamaterials, among many others [2–6].

Graphene has also been examined closely owing to its suitability for terahertz (THz) communications (0.1–10 THz). The THz band, sitting in between the millimeter-wave and optical ranges, promises to be a key enabler of a plethora of applications in both classical networking scenarios and novel nano-communication paradigms [7], and, in this band, graphene acts as a tunable plasmonic material in the THz band, thereby offering excellent conditions for the implementation of miniaturized and reconfigurable antennas [8].

These properties are crucial in novel applications that may require smaller forms of wireless communication or certain functional flexibility. Three clear cases that fall within this category are wireless nanosensor networks (WNSNs) [9], Wireless Networks-on-Chip (WNoCs) [10], and software-defined metamaterials (SDMs) [11]. Their main communication requirements are estimated in Table 2.1 based on recent data (Figure 2.1).

The goal of WNSNs is to exploit the novel capabilities of nanosensors in terms of sensitivity or ubiquity by means of wireless communication among nanosensors, forming a distributed and collaborative network. The WNoC paradigm, on the other hand, consists of the communication of different components of a single-chip multiprocessor via on-chip antennas, which allows to solve the performance bottlenecks of current interconnects (see more details in Chapter 20 of this handbook). Finally, SDMs are metamaterials whose electromagnetic behavior can be programmed, to then adapt to changes in the environment or without the intervention of the user. To achieve this, SDMs incorporate a network of

TABLE 2.1 Characteristics of Area-Constrained Applications

	WNSN [9]	**WNoC [10]**	**SDM [11]**
Node Size	1–100 μm^2	0.01–1 mm^2	0.01–100 mm^2
Transmission Range	0.1–100 cm	0.1–10 cm	0.1–100 cm
Data Rate	0.001–0.1 Gbps	10–100 Gbps	0.01–1 Gbps
Latency	1–10 ms	1–10 ns	1–10 μs
Bit Error Rate (BER)	10^{-3}	10^{-15}	10^{-6}
Energy	0.1–1 pJ/bit	1–10 pJ/bit	1–10 pJ/bit

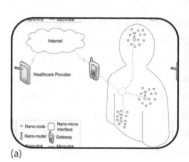  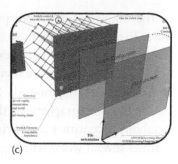

(a)   (b)   (c)

FIGURE 2.1 Artistic representation of applications suitable to the properties of miniaturized antennas: (a) Wireless nanosensor networks (WNSNs). (From Akyildiz, I.F. and Jornet, J.M., *IEEE Wireless Commun.*, 17, 58–63, 2010), (b) Wireless Networks-on-Chip (WNoCs). (From Abadal, S. et al., *IEEE MICRO*, 35, 52–61, 2015), and (c) Software-defined metamaterials (SDMs). (From Abadal, S. et al., *IEEE Access*, 5, 6225–6235, 2017.)

controllers within the metamaterial, where controllers may wirelessly communicate to each other to attain the programmed behavior (see more details in Chapter 3 of this handbook).

Both the miniaturization and tunability properties of graphene antennas stem from the ability of graphene to support surface plasmon polaritons (SPPs) in the THz band. The fact that metals may act as lossy nonplasmonic conductors in this frequency range [12], coupled to the high potential of the emerging area-constrained applications, has been driving a growing body of research around graphene-based antennas in the recent years [13]. This chapter aims to provide a comprehensive view of the field by explaining the fundamental theory behind graphene antenna design and surveying current practices.

The remainder of the chapter is organized as follows: Section 2.2 reviews the main research lines of graphene-based antenna design for the last decade. Then, Section 2.3 lays down the fundamental theory behind the modeling of graphene, the plasmonic effects, and their guided-wave modes—key to understand the feasible radiation modes of the antennas. Then, a representative set of antennas is outlined and evaluated in Section 2.4. The choice of design points aims to exemplify the evolution of graphene-based antenna development over the years. Finally, we give a few practical considerations regarding the feeding mechanisms and electrostatic biasing in Section 2.5.

2.2 REVIEW OF THE STATE OF THE ART

The first antenna theory involving novel carbon nanomaterials proposed to employ carbon nanotubes as dipole antennas. Theoretical efforts analyzed their transmission line properties and radiation pattern [14,15], whereas experimental tests led to the implementation of a nanotube-based receiver [16]. Nanotube antennas, however, face significant challenges in terms of manufacturing, tuning, and placement on planar implementation processes. Interest on planar antennas based on graphene, another carbon of nanomaterial, grew in the following years, thereby avoiding the tuning and placement issues of nanotubes. This is not the only reason, as we will see next.

The following subsections present a short review of works that consider graphene in several ways. To taxonomize them, we will distinguish between three main groups, depending on the property of graphene that is exploited and/or the role of graphene within the antenna structure. Since other extensive reviews are available in the literature [13,17], here, we will focus on a subset of relevant proposals: Section 2.2.1 summarizes efforts that mainly exploit miniaturization by using graphene as the main radiator; Section 2.2.2 focuses on designs that exploit tunability, oftentimes employing graphene as an auxiliary passive element; and Section 2.2.3 outlines several works where graphene is the basis of mechanically flexible antennas.

2.2.1 Miniaturization

Jornet et al. pioneered the field by deriving the quantum properties of antennas based on graphene nanoribbons [18,19], estimating that the first resonant frequency for a micrometric patch would be around the THz band. This suggested that plasmonic effects could not only lead to effective miniaturization of graphene-based antennas but also to a scaling of the resonance frequency of $L^{-1/2}$ instead of L^{-1}, where L is the antenna length [20]. This is further illustrated in Section 2.4.1.

A first surge of graphene antenna proposals for THz band communication revolved around the concept of resonant plasmonic sheets. In essence, the antennas would consist of a number of finite-size graphene layers (the radiating elements) mounted over a metallic flat surface (the ground plane), with a dielectric material in between and a source. This is schematically represented in Figure 2.2, including a variety of shapes, which could be lithographically defined with the existing techniques [21], easing the compatibility with planar fabrication processes. Patch antenna configurations [22–24] and dipole-like designs [25,26], where the feed is placed in the middle of two identical graphene patches, have been the main currents of design. Most proposals suggested the use of high-impedance sources to better match with the high-input impedance of graphene. More details are given in Section 2.5.1.

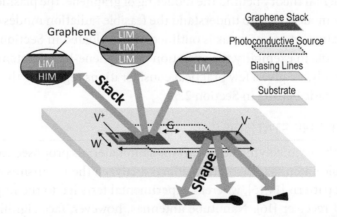

FIGURE 2.2 Schematic representation of resonant antennas fed with a photoconductive source exploiting miniaturization. The graphene-based plasmonic radiating elements can not only take different shapes but also be formed by different stacks combining graphene with High Index Material (HIM) and Low Index Material (LIM) materials.

The main issue with antennas based on pure plasmonic graphene structures is the low radiation efficiency caused by the losses of the supported SPP modes [25,27,28]. Several proposals have tried to combat this problem by means of different techniques. The work in [29] proposes to use few-layer graphene, which would theoretically have higher conductivity than the single-layer counterpart, thereby increasing the performance but limiting the miniaturization capability. In [30], hybrid plasmonic guided-wave structures are shown to provide a better balance between miniaturization and efficiency than the pure plasmonic modes. This is schematically represented in Figure 2.2 as different *stacks*. Following such proposals, other hybrid structures such as dielectric resonant antennas coupled to graphene dipoles [31] and Yagi–Uda arrays [32] have been proposed to increase the antenna gain. Finally, Akyildiz et al. proposed to increase the gain by taking advantage of the miniaturization of graphene antennas to create ultra-massive arrays [33], taking a 1024 × 1024 array as an extreme case.

2.2.2 Tunability and Switchability

The tunability of graphene plasmons confers graphene antennas with unique reconfigurability properties. With proper means of biasing (see Section 2.5.2), the conductivity of graphene changes significantly. This feature has been applied to graphene antennas in numerous works, to change either the frequency of resonance or the impedance at a fixed frequency, depending on the use to the graphene sheet. This is numerically illustrated in Section 2.4.

Take, for instance, the Yagi–Uda antenna structure from Figure 2.3 [32], where both the radiators and the parasitic elements are graphene sheets. On the one hand, the radiating elements can exploit tunability by changing the resonance frequency if connected to a wideband source. On the other hand, parasitic elements will or will not couple with the radiators, depending on the applied voltage: effectively, the parasitics are switched ON or OFF to change the radiation pattern [34]. In antenna arrays, this same principle can be applied to generate ultra-massive multiple-input-multiple-output (MIMO) with multiple beams or carriers that can be tuned according to the communication requirements [35,36].

Other research lines have continued to focus on the tunability for elements surrounding the antenna rather than the radiator. For instance, Tamagnone et al. place graphene sheets between the source and a metallic radiating element to retain the tunability while having a fair efficiency [37]. Aldrigo et al. demonstrated an antenna structure that switches

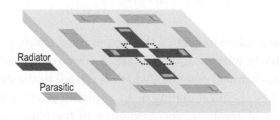

FIGURE 2.3 Schematic representation of a graphene-based Yagi–Uda antenna exploiting tunability and switchability.

between omnidirectional and broadside radiation, depending on the bias applied to a graphene sheet that acts as the ground plane [38]. A generalization of this idea leads to low-complexity and reconfigurable graphene-based reflectarrays, which are able to control the beam direction [39,40] or to generate a backscattered modulated wave [41]. Beyond that, a variety of papers have considered graphene as a tunable building block for the creation of high-performance metasurfaces in the THz band. The applications are manifold and include tunable absorption [42], ultrafast beam steering [43], beam modulation [44], controllable focusing [45], and dynamic vorticity control [46].

2.2.3 Mechanical Flexibility

Last but not least, we note that graphene's excellent mechanical and electrical properties have been leveraged to build mechanically flexible antennas for radio frequency (RF) applications. These do not operate in the THz band and are strictly out of the scope of this chapter, but a brief summary has been provided. Huang et al. [47]. have published a methodology to create graphene nanoflake ink with high conductivity, which is used to print antennas for wearable applications [47]. In parallel, researchers from other institutions have also developed their own formula to achieve low-cost, flexible, printed antennas based on graphene compounds [48,49].

2.3 FUNDAMENTALS OF GRAPHENE-BASED ANTENNA MODELING

This section outlines the fundamentals of graphene modeling and graphene antenna design and analysis. Section 2.3.1 reviews the conductivity models of graphene, paying attention to the technological parameters affecting the conductivity. Section 2.3.2 discusses the plasmonic modes supported by graphene, whereas Section 2.3.3 presents the analysis approaches to study graphene-based guided-wave structures that can be employed as basic blocks of advanced graphene antennas.

2.3.1 Conductivity Model of Graphene

The conductivity of graphene has been a subject of numerous studies in the last 15 years, enabling a precise modeling of the plasmonic phenomena occurring in graphene antennas [19,26]. In general form, a graphene sheet can be modeled as a nonlocal two-sided surface characterized by a complex conductivity tensor [50] as

$$\sigma(\omega, \mu_c(E_0), \tau, T, B_0) = \begin{bmatrix} \sigma_d & \sigma_o \\ \sigma_o & \sigma_d \end{bmatrix} \qquad (2.1)$$

where ω, T, σ_d, and σ_o are radian frequency, the temperature, the diagonal, and off-diagonal conductivities, respectively, whereas E_0 and B_0 are the electrostatic and magnetostatic bias fields. Variable μ_c is the chemical potential, which is related to the density of charged carriers and can be tuned by chemical doping or by applying electrostatic bias field. τ is the electron relaxation time of graphene, corresponding to the phenomenological scattering rate Γ, as $\tau^{-1} = 2\Gamma$. The elements of this tensor can be extracted from microscopic, semiclassical, and quantum mechanical considerations. Here, we just review the well-known model in which the conductivity is calculated by means of the Kubo formula [51]. More details about μ_c and τ are also given below in Equation (2.2).

Assuming no magnetic bias field ($B_0 = 0$), the off-diagonal terms vanish, and graphene can be considered as an isotropic material. Following the Kubo formula, the conductivity can be evaluated with the assumption of harmonic time dependence $exp(+j\omega t)$ as

$$\sigma(\omega, \mu_c, \tau, T) = \sigma_{intra} + \sigma_{inter}$$

$$= \frac{-j}{\omega - j\tau^{-1}} \frac{e^2 k_B T}{\pi \hbar^2} \left(\frac{\mu_c}{K_B T} + 2\ln\left(e^{-\frac{\mu_c}{K_B T}} + 1 \right) \right)$$

$$+ \frac{-j(\omega - j\tau^{-1})e^2}{\pi \hbar^2} \int_0^\infty \frac{f(-\varepsilon) - f(+\varepsilon)}{(\omega - j\tau^{-1})^2 - 4(\varepsilon / \hbar)^2} d\varepsilon$$

(2.2)

where e, $\hbar$, and k_B are constants corresponding to the charge of an electron, the reduced Planck constant, and the Boltzmann constant, respectively [51,27]. $f(\varepsilon) = 1/(1 + exp[(\varepsilon - \mu_c)/(k_B T)])$ is the Fermi-Dirac distribution function. The first term in Equation (2.2) refers to the contribution of the intraband transition of an electron to the conductivity of graphene, while the second term accounts for the interband transition. For this, the intraband contribution dominates approximately in the THz region. In the near-infrared and visible regions (>50 THz), phenomena such as interband damping and electron-phonon interactions start appearing and the interband process dominates instead.

As is clearly observed in Equation (2.2) and discussed next, the graphene conductivity strongly depend on the chemical potential and the relaxation time, which in turn depend on the carrier mobility. The chemical potential μ_c, also referred to as Fermi energy, refers to the level in the distribution of electron energies at which a quantum state is equally likely to be occupied or empty. Since its value can be changed at will by means of electrostatic biasing or chemical doping, the chemical potential can be considered as a design parameter for graphene antennas. As implied in Figure 2.4a, higher chemical potential leads to an increase of the conductivity.

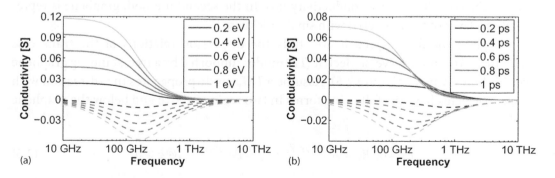

FIGURE 2.4 Graphene conductivity for different chemical potential and relaxation time values. Solid lines represent the real part of the conductivity $\Re[\sigma]$, whereas dotted lines represent the imaginary part of the conductivity $\Im[\sigma]$. $\tau = 1$ ps and (b) $\mu_c = 0.6$ eV.

On the other hand, the relaxation time τ is the interval required for a material to restore a uniform charge density after a charge distortion is introduced. At the band of interest (<10 THz), where the intraband part of the conductivity dominates, the relaxation time becomes

$$\tau \approx \tau_{DC} = \mu \hbar \sqrt{n\pi} \, / \, (ev_F), \tag{2.3}$$

where μ is the carrier mobility, n is the carrier density, and v_F is the Fermi velocity [52]. The carrier density depends on the chemical potential, whereas the Fermi velocity is independent of the Fermi energy [53], but their expressions can be simplified to

$$\tau \approx \mu \frac{\mu_c}{v_p^2}, \tag{2.4}$$

as long as $\mu_c \gg K_B T = 26 \text{ meV}$, which typically holds in THz antennas. Here, the carrier mobility μ has a key role to determine the relaxation time, as it somehow depends on the quality of the graphene sheet. The carrier mobility is defined as the average speed at which electrons move within graphene; large values imply a high-quality material with few defects and therefore large relaxation time, whereas low values entail a defective material. As illustrated in Figure 2.4b, sheets of high-quality graphene will exhibit a high conductivity, and we will see in next sections that this leads to high radiation efficiency. Eventually, the value of the carrier mobility will depend on the manufacturing and encapsulation processes [54,55], and therefore, it can be considered as a design parameter of graphene antennas.

2.3.2 Graphene Plasmonics

Graphene has become a novel platform to implement highly integrated plasmonic devices thanks to the unique properties of its conductivity [56]. Let us investigate the basis of plasmonics based on graphene by supposing the simplest guided-wave plasmonic structure. Two modeling approaches can be considered to solve Maxwell's equations in the presence of graphene. In the first method, a graphene layer is modeled as a boundary condition that includes the complex surface conductivity [51]. In the second method, graphene is represented as a layer of bulk material with small thickness [5].

To explain the first approach, we study the dispersion relations of the transverse magnetic (TM) and transverse electric (TE) modes supported by a free-standing graphene layer in the $x-y$ plane (Figure 2.5). Assume a TM_y mode propagates in y-direction with the electric fields have the following forms in two regions above and below the graphene sheet:

$$E_x = 0; \; E_y = Ae^{-j\tilde{\gamma}_y y - \tilde{\gamma}_z z}; \; E_z = Be^{-j\tilde{\gamma}_y y - \tilde{\gamma}_z z}, @ \, z > 0 \tag{2.5}$$

$$E_x = 0; \; E_y = Ce^{-j\tilde{\gamma}_y y + \tilde{\gamma}_z z}; \; E_z = De^{-j\tilde{\gamma}_y y + \tilde{\gamma}_z z}, @ \, z < 0 \tag{2.6}$$

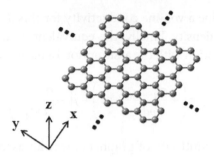

FIGURE 2.5   A free-standing graphene layer in x–y plane.

where $\tilde{\gamma}_y$ is the complex wave number in y-direction and $\tilde{\gamma}_z = \sqrt{\tilde{\gamma}_y^2 - k_0^2}$ is the attenuation constant in z-direction. Furthermore, the boundary condition related to graphene sheet with the complex conductivity $\tilde{\sigma}_g$ is written as

$$Z \times (\bar{H}^+ - \bar{H}^-) = \bar{J}_s = \tilde{\sigma}_g \bar{E} \tag{2.7}$$

Applying Maxwell's equations and the boundary condition, the dispersion relation is derived as

$$\frac{2}{\tilde{\gamma}_z} = \frac{j\tilde{\sigma}_g}{\omega\varepsilon_0} \tag{2.8}$$

Consequently, the complex effective index ($\tilde{n}_{eff} = \tilde{\gamma}_y / k_0$) for TM_y is calculated as

$$\tilde{n}_{eff} = \sqrt{1 - \left(\frac{2}{\eta_0 \tilde{\sigma}_g}\right)^2}, \text{ for } TM_y \tag{2.9}$$

where k_0 and η_0 are the wave number and the intrinsic impedance of free space, respectively. Similarly, the complex effective index for TE_y is extracted as

$$\tilde{n}_{eff} = \sqrt{1 - \left(\frac{\eta_0 \tilde{\sigma}_g}{2}\right)^2}, \text{ for } TE_y \tag{2.10}$$

For TM mode, if $\Im[\tilde{\gamma}_z] > 0$, we have a slow surface wave, an SPP. According to Equation (2.8), we must have $\Im[\tilde{\sigma}_g] < 0$. It means that the graphene supports SPP wave when the intraband contribution dominates. But, for TE mode, if $\Im[\tilde{\sigma}_g] > 0$ (when interband contribution dominates), the mode is a weakly surface wave whose wavenumber is near free-space wavenumber.

To explain the second approach, we try to find the same dispersion relation for the mentioned configuration. Graphene is represented as a layer of bulk material with small thickness (d_G). We can define a volume conductivity for this d_G-thick monolayer and then consider a volume current density. Finally, the equivalent permittivity $\tilde{\varepsilon}_G$ is calculated by recasting the Maxwell's equation with the assumption of harmonic time dependence $e^{+j\omega t}$ as

$$\tilde{\varepsilon}_g = \left(+\frac{\sigma_{g-imag}}{\omega d_g} + \varepsilon_0 \right) + j \left(-\frac{\sigma_{g-real}}{\omega d_g} \right). \tag{2.11}$$

Knowing the equivalent permittivity of graphene, we can assume the free-standing graphene layer as insulator-metal-insulator (IMI) waveguide. Here, graphene layer plays the role of plasmonic material instead of noble metals (e.g., silver and gold). Therefore, the slab can support an odd TM mode, with dispersion relation reported by [57] as

$$\cot\left(\sqrt{\omega^2 \mu_0 \tilde{\varepsilon}_g - \tilde{\gamma}_y^2} \, \frac{d_g}{2} \right) = -\frac{\tilde{\varepsilon}_g}{\varepsilon_0} \frac{\sqrt{\tilde{\gamma}_y^2 - \omega^2 \mu_0 \varepsilon_0}}{\sqrt{\tilde{\gamma}_y^2 - \omega^2 \mu_0 \tilde{\varepsilon}_g}} \tag{2.12}$$

By substituting $\tilde{\varepsilon}_g$ with the equivalent permittivity $\tilde{\varepsilon}_g = (\tilde{\sigma}_g / j\omega d_g + \varepsilon_0)$ and taking limit when $d_g \to 0$, the dispersion relation is simplified as

$$1 = \frac{j\tilde{\sigma}_g}{2\omega\varepsilon_0} \sqrt{\tilde{\gamma}_y^2 - k_0^2} \tag{2.13}$$

which is exactly same as Equation (2.8). Moreover, we can obtain the same dispersion relation with Equation (2.10) for *TE* mode.

To wrap up the discussion, graphene supports TM SPP waves with high confinement when $\Im[\tilde{\sigma}_g] < 0$ ($\Re[\tilde{\varepsilon}_g] < 0$). Therefore, it is a promising plasmonic material to implement compact electromagnetic devices.

2.3.3 Graphene Guided-Wave Structures

Surveying the characteristics of guided-wave structures, including one-dimensional and two-dimensional waveguides, facilitates the construction of the antennas as the three-dimensional radiated-wave structures. First, we present several analytical and numerical tools to extract the characteristics of guided-wave structure. Then, a quantitative study of a couple of graphene antennas is performed based on the performance of their fundamental structures.

The transfer matrix method is one of the most efficient methods for the analysis of multilayer waveguides (Figure 2.6). In order to calculate the complex effective index of guided modes in the graphene-integrated structures, the formulations of transfer matrix theory provided in [58] are applied. The dispersion relation of TM mode propagated in a general multilayer one-dimensional structure is defined as follows:

$$+j\left(\frac{\tilde{\gamma}_{xS}}{\varepsilon_{rS}} m_{11} + \frac{\tilde{\gamma}_{xC}}{\varepsilon_{rC}} m_{22} \right) = \frac{\tilde{\gamma}_{xS}\tilde{\gamma}_{xC}}{\varepsilon_{rS}\varepsilon_{rC}} m_{12}, \tag{2.14}$$

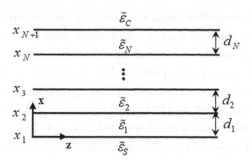

FIGURE 2.6   Multilayer waveguide structure for guided-wave analysis.

where ε_{rC} and ε_{rS} are the dielectric constants of cover and substrate layers, respectively, m_{ij} are the elements of the total transfer matrix M, as defined in [58].

$$\tilde{\gamma}_{xC} = \sqrt{\tilde{\gamma}_{eff}^2 - k_0 \varepsilon_{rC}},$$
$$\tilde{\gamma}_{xS} = \sqrt{\tilde{\gamma}_{eff}^2 - k_0 \varepsilon_{rS}}. \tag{2.15}$$

The zeroes of Equation (2.14), which are the guided mode complex propagation constants

$$\tilde{\gamma}_{eff} = k_0 \tilde{n}_{eff} = k_0 (n_{eff} - j k_{eff}) = \beta_{eff} - j\alpha_{eff}, \tag{2.16}$$

are obtained analytically.

For the two-dimensional structures, the effective index method (EIM) is applied to obtain propagation constants of the guided modes [12]. Considering a refractive index profile, which depends on two coordinates $n = n(x,y)$ and z direction as the propagation direction, the wave equation can be written as

$$\nabla_{xy}^2 \psi(x,y) + (k_0^2 n^2(x,y) - \tilde{\gamma}_{eff}^2)\psi(x,y) = 0, \tag{2.17}$$

where $\Psi(x,y,z) = \psi(x,y)e^{-j\tilde{\gamma}_{eff}z}$ can be any of the fields components, $k_0 = 2\pi/\lambda_0$ is the wave number in free space, $\tilde{\gamma}_{eff} = k_0(\tilde{n}_{eff})$ is the complex propagation constant, and $\tilde{n}_{eff}$ is the complex effective index of the guided modes. In EIM, an approximate solution can be written as

$$\psi(x,y) = \psi_1(x,y)\psi_2(y), \tag{2.18}$$

wherein $\psi_1(x,y)$ is a slowly varying function of y ($\partial\psi_1/\partial y = 0$). Therefore, a system of two coupled differential equations is obtained as

$$\begin{cases} \dfrac{1}{\psi_1}\dfrac{\partial^2 \psi_1}{\partial x^2} + k_0^2 n^2(x,y) = k_0^2 n_{eff}^2(y) \\[2ex] \dfrac{1}{\psi_2}\dfrac{d^2 \psi_2}{dy^2} - \tilde{\gamma}_{eff}^2 = -k_0^2 n_{eff}^2(y) \end{cases} \tag{2.19}$$

The first step in the EIM procedure consists of solving the first equation of Equation (2.19) that gives the eigenvalue solution $n_{eff}(y)$. In order to find the solution of this equation, the two-dimensional waveguide is divided into one-dimensional waveguides for which we can assume the refractive index profile, that is, $n(x, y)$, as being independent of y, that is, $n(x)$. In the second step, using the function $n_{eff}(y)$, we can solve the second equation of Equation (2.19) that results in the propagation constant. This equation is for a one-dimensional waveguide with refractive index profile $n_{eff}(y)$.

Furthermore, the full-wave solver such as COMSOL [59] could be used to verify the results of analytical methods. As shown in previous works [12,58], both methods yield results with little to no difference.

2.4 PERFORMANCE EVALUATION

Next sections depict a set of representative antenna designs in the THz band and evaluate their performance. Unless noted, all antennas are modeled and simulated in CST Microwave Studio [60].

2.4.1 Designs Based on Resonant Sheets

We first evaluate a resonant plasmonic dipole, as shown in Figure 2.2. Graphene patches are 8-μm wide and deposited over an LIM (quartz with $\varepsilon_r = 3.8$, in this case), leaving a gap of 3 μm in between the two patches. We analyze how the antenna resonance and performance scale with respect to the dipole length L, the chemical potential μ_c and the relaxation time τ. By default, these values are $L = 20\,\mu m$, $\mu_c = 0.2\,eV$, and $\tau = 1\,ps$. The resonance point is defined as the frequency, where the impedance Z shows $\Im[Z] = 0$ and high $\Re[Z]$ for a better matching with high-impedance sources.

Figure 2.7a plots the resonance frequency as a function of the dipole length. It is shown how longer dipoles lead to lower resonance points, as one would expect, with much lower value than same-size metallic dipoles. This is consistent with previous works [23,27]. The maximum gain (IEEE) and the simulated radiation efficiency are relatively low in the considered cases, around −10 dB and 4.5%, respectively. As shown in Figure 2.7b, increasing the chemical potential leads to better efficiency (52% and −0.46 dB for $\mu_c = 0.8\,eV$) but at the expense of having to apply an increasing bias. More remarkably, the change in chemical potential leads to a significant shift in the resonant frequency, thereby illustrating the outstanding tunability of graphene dipoles. Finally, we consider a dipole with $\mu_c = 0.6\,eV$ and variable τ. The results in Figure 2.7c confirm that the radiation efficiency improves with the relaxation time (32% and −2.34 dB for $\tau = 1\,ps$), owing to the better carrier mobility of the graphene sheets. The resonance frequency remains unchanged.

Although not shown, the use of graphene does not change the radiation pattern of the dipole at resonance, as proved in other works [8,26]. This suggests that graphene does not modify the *electrical shape* of the antenna, opening the door to applying shape optimizations [61], possibly enabled by lithographic techniques [21]. For instance, graphene bowtie antennas can offer increased bandwidth [25].

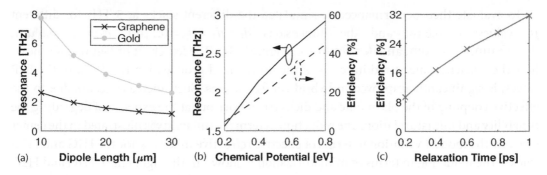

(a) Dipole Length [μm]   (b) Chemical Potential [eV]   (c) Relaxation Time [ps]

FIGURE 2.7   Scaling of resonant plasmonic dipoles. (a) Dipole length, (b) Chemical potential, and (c) Relaxation time.

2.4.2 Graphene Stacks

This section exemplifies how the use of different vertical stacks in the radiating elements can help to balance the properties of graphene. The main concept is the hybridization of the surface plasmon and dielectric wave modes: the former leads to compact and tunable devices, whereas the latter delivers higher efficiency. We start by comparing the stack configurations depicted in Figure 2.8a: the pure plasmonic structure studied in the previous section (1G), a pure plasmonic structure composed of two monolayers separated by thin dielectric (2G), a hybrid structure with a single graphene monolayer (H1G), and a hybrid structure with two graphene monolayers (H2G).

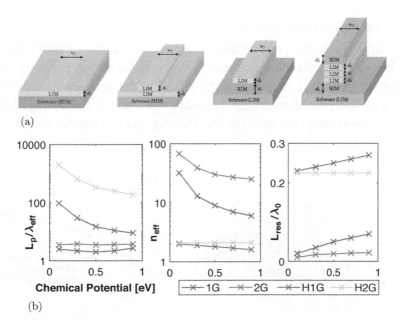

FIGURE 2.8   Different stack configurations evaluated from different perspectives as functions of the chemical potential. (a) Evaluated stacks (1G, 2G, H1G, H2G), (b) Efficiency (left), tunability (center), and miniaturization (right).

To evaluate their performance, we simulated the different stacks at 3 THz for different μ_c values; $\tau = 0.6$ ps; and the dimensions $d_1 = d_2 = d_3 = 0.1 \, \mu m$, $d_4 = d_6 = 0.5 \, \mu m$, $d_5 = 15 \, \mu m$, $d_7 = 2 \, \mu m$, and $d_8 = 9 \, \mu m$ (see Figure 2.8a). The left chart of Figure 2.8b shows how the efficiency, measured in terms of the normalized propagation length L_p of the SPP waves, is significantly improved in hybrid configurations. Such increase occurs due to the effective coupling of the plasmonic and dielectric modes but at the expense of degrading the tunability and miniaturization. The reduction in terms of tunability is illustrated in the center chart by the modest variation in terms of effective refractive index n_{eff} for the H1G and H2G stacks. The reduction in terms of miniaturization is shown in the right chart: H1G and H2G have a much larger resonant length L_{res}. The main takeaway of these results is that, depending on the requirements of the final application, we can navigate the different miniaturization-tunability-efficiency trade-offs by means of these graphene-based structures.

To further exemplify these results in the context of a particular antenna design, we next depict a design of a dielectric resonator antenna (DRA) coupling the dielectric wave mode to the plasmonic graphene mode [31]. An instantiation of such design is outlined in Figure 2.9: the graphene dipole is deposited on top of a thin layer of LIM and is placed below a $a \times b \times d_H$ rectangular resonator. The resonator is composed by a low-loss HIM over an LIM spacer. The height d_H of the resonator is chosen so that the dipole and dielectric wave modes are coupled. Generally, thicker resonators lead to higher-order modes with higher gain. As shown in Figure 2.9, the structure with $l = 20 \, \mu m$, $w = 5 \, \mu m$, $g = 2 \, \mu m$, $\mu_c = 0.8 \, eV$, $\tau = 0.6$ ps, $d_{L1} = d_{L2} = 100$ nm, and $a = b = 20 \, \mu m$ achieves a gain improvement of 3.5 dB for the TE_y^{111} mode ($d_H = 60 \, \mu m$) and of 6.5 dB for the TE_y^{111} mode ($d_H = 120 \, \mu m$) at 2.5 THz. The radiation efficiency is maintained around 65%.

2.4.3 Arrays and Metasurface Structures

Another way to improve the performance of graphene antennas besides improving the radiator stack is to form antenna arrays. Given the miniaturization property of plasmonic antennas, a higher density of elements can be placed in the antenna area. Before jumping into array designs, however, let us exemplify the approach with a unique Yagi–Uda antenna, where all the elements are plasmonic dipoles. This concept has been explored in [32,34] to

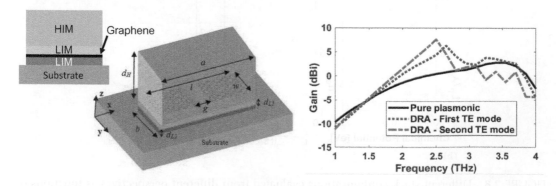

FIGURE 2.9 Schematic of a dielectric resonator antenna coupled to a graphene dipole (left) and gain of its first two TE modes compared with that of a graphene dipole.

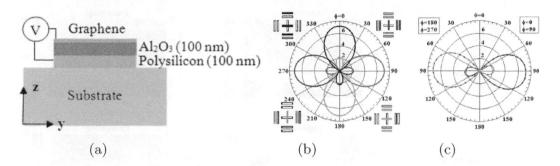

FIGURE 2.10   Directivity patterns of the directional modes of a graphene-based Yagi–Uda antenna: (a) Cross section, (b) XY plane, and (c) YZ plane.

control the radiation pattern through graphene's tunability. This small (passive) array leads to a modest increase in gain. For instance, the scheme depicted in Figure 2.3 can be implemented with the materials shown in Figure 2.10a. With the appropriate voltages applied to the different elements, the gain can be increased by around 3.6 dB (from isolated dipole to Yagi–Uda) with a beam width of 69°.

The Yagi–Uda approach serves to exemplify the benefits of graphene plasmonic arrays: miniaturization allows to reduce the area footprint of the array, whereas tunability opens the door to reconfigurable designs. Such attributes were studied in [33], where the authors proposed ultra-massive arrays, with each graphene patch connected to its own source. This configuration provides very high gain density for beamforming and multiple space-frequency channels in multiband designs. The data from [33], shown in Figure 2.11, are based on theoretical approximations with no mutual coupling and a miniaturization of around one order of magnitude. This can be leveraged to fit more antennas in the same space and thus to increase the achievable gain by 15–40 dB for a fixed antenna area.

Another possibility described in the previous sections is to use the plasmonic properties of graphene as a metamaterial, which, in the end, can be regarded as a very dense and passive antenna array. To illustrate the potential of graphene-based metamaterials, here, we

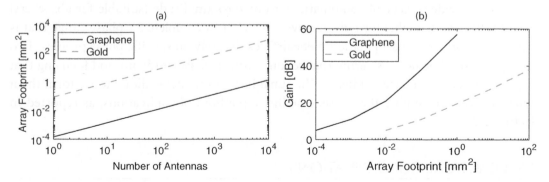

FIGURE 2.11   Number of antennas and achievable gain per unit of area for metallic and graphene antennas at 1 THz: (a) Number of antennas and (b) Array footprint (mm^2). (From Akyildiz, I.F. and Jornet, J.M., *Nano Commun. Netw.*, 8, 46–54, 2016.)

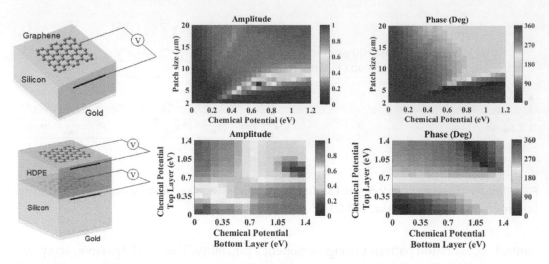

FIGURE 2.12   Amplitude and phase response of graphene-based metamaterial unit cells. These can be used as building blocks of reconfigurable metamaterials in the THz band, with numerous applications. (From Wang, X.-C. and Tretyakov, S.A., Tunable perfect absorption in continuous graphene sheets on metasurface substrates *arXiv preprint*, 2017; Orazbayev, B. et al., Ultrafast beam steering based on graphene metamaterial. *Proceedings of the EuCAP 2017*, 3896–3899, 2017; Momeni, A. et al., *Sci. Rep.*, 8, 6200, 2018; Hosseininejad, S.E. et al., *Sci. Rep.*, 9, 2868, 2018; Shi, Y. and Zhang, Y., *IEEE Access*, 6, 5341–5347, 2017.)

report the performance of two unit cell designs for their use in reflectarray metasurfaces. As shown in the leftmost charts of Figure 2.12, we compare two resonant cavities: graphene-substrate-metal and graphene-insulator-graphene-substrate-metal. By default, graphene patches are square with lateral size of $d_G = 16\,\mu m$ in 20-μm unit cells, with $\mu_c = 0.2$ eV, $\tau = 0.6$ ps, and $f = 2$ THz. The substrate is silicon and the insulator is high-density polyethylene (HDPE), with low losses in the THz band.

The results in Figure 2.12 explore the design space of $d_G \sim \mu_c$ in the first case and $\mu_{c,1} \sim \mu_{c,2}$ in the second case. The reflection amplitude and phase are reported. Both unit cells show a wide variety of combinations from zero amplitude (suitable for absorbers) to full reflection (suitable for reflectarrays), as well as a complete 2π phase range that is crucial for waveform manipulation metasurfaces that rely on specific phase profiles. Two examples would be beam steering that requires linear phase gradients and focusing that requires concentric phase gradients. The tunability of graphene allows to perform them dynamically by changing the phase, enabling a plethora of applications, as reported in Section 2.2.

2.5 PRACTICAL CONSIDERATIONS

This section provides a summary of practical considerations regarding the elements that are required to deliver the appropriate direct current (DC) bias and THz signals to the antenna. As we will see next, the use of graphene has an effect on their design and properties.

2.5.1 Feeding Mechanisms

Several works have proposed to approach the antenna with a photomixer, since its impedance, typically of several $k\Omega$, leads to a reasonable matching with the relatively high impedance of the graphene dipole in open circuit resonance [26,62]. The THz signal is a continuous wave that results from the coherent mixing of two laser sources with a relatively small frequency shift and has a moderate bandwidth, so it needs to be carefully tuned to the resonant frequency of the antenna.

Cabellos et al. have proposed to use a photoconductor working in pulsed mode to excite the antenna [25]. In this configuration, a photoconductive material is placed below the antenna and illuminated with a pulsed laser. With an appropriate electrostatic bias, the photocarriers generated at the dipole gap can be accelerated, and the resulting photocurrent enters the antenna. This process has been traditionally very inefficient, but recent years have seen dramatic improvements thanks to the development of plasmonic contact electrodes [63]. In this respect, a remarkable difference with respect to the photomixer option is the input impedance: in this case, the impedance of the gap reaches very low values during illumination [64], and therefore, the antenna matches better at the frequency of short-circuit resonance. Moreover, pulsed photoconductive sources have a huge bandwidth, turning them into excellent candidates to test tunable graphene antenna designs.

Although the use of photoconductive sources may be useful for the experimental validation of graphene antennas, it is impractical for area-constrained applications. Instead, compact electronic THz sources are required. In this direction, Jornet et al. propose the use of a high-electron-mobility transistor (HEMT) with graphene-based gate. If the transistor is sized properly, the application of voltage generates a THz plasma wave at the gate [65]. This wave can be fed into the graphene antenna without contact issues, reducing the impedance mismatch problems. However, a direct comparison of the efficiency of HEMT-based and photoconductive sources becomes challenging owing to their radically different working principles.

2.5.2 Biasing Schemes

Another important practical aspect is the biasing scheme required to achieve a given Fermi energy and, if possible, a wide tuning range. Several possibilities exist in the literature. Huang et al. proposes to place a thin layer of Al_3O_2 on top of the graphene sheet and then sandwich the stack between two metallic electrodes [40]. In [66], two graphene layers are separated by an insulator layer and contacted by two different electrodes. The authors then test different configurations by changing the bias applied to each of the electrodes.

In essence, both approaches apply some sort of capacitive coupling to achieve the biasing. With this process, the main goal is to bring graphene as close as possible to the Dirac point with low conductivity, so that the tuning range can be maximized. Under those conditions, the voltage V_g required to shift the working point of graphene depends on the characteristics of the *capacitor* that is formed. Assuming single-layer graphene, the chemical potential variation $\Delta\mu_c$ relates to the change of voltage ΔV_g as

$$\Delta V_g^{SLG} = \frac{e\mu_c^2 t}{\pi\hbar^2 v_F^2 \varepsilon_0 \varepsilon_r},$$

(2.20)

where e is the elementary charge, $\hbar$ is the reduced Planck constant, $v_F \approx 10^6$ is the Fermi velocity, and ε_0 is the vacuum permittivity, whereas ε_d and t are the permittivity and thickness of the material below graphene, respectively [67]. The relation $V_g \sim \mu_c$ becomes linear for bilayer graphene

$$\Delta V_g^{BLG} = \frac{2e\mu_c m^* t}{\pi \hbar^2 \varepsilon_0 \varepsilon_r}, \tag{2.21}$$

where m^* is the mass of a carrier in bilayer graphene [67]. The $V_g \sim \mu_c$ relation is currently unclear for trilayer graphene and above. In any case, the equations above suggest that the biasing scheme should employ thin dielectrics with high permittivity, conditions that are not always available owing to technological or cost constraints.

2.6 SUMMARY

Graphene-based antennas present unique miniaturization and tunability features in the THz band. These, however, will come at the cost of reduced radiation efficiency, unless the quality of the graphene monolayers is improved in the near future. We have observed that stacks containing multiple graphene monolayers and thin dielectric layers exhibit similar trade-offs between the different performance characteristics. Similarly, the adoption of graphene as the building block of ultra-compact antenna arrays or metasurfaces opens the door to new unique designs with unprecedented frequency-beam reconfigurability. The qualitative comparison made in this chapter may, in the future, guide the design of antennas for area-constrained applications, depending on their specific requirements in terms of miniaturization, functional reconfigurability, and efficiency.

ACKNOWLEDGMENTS

This work has been partially funded by the Iran National Science Foundation (INSF), the German Research Foundation (DFG) under grants HA 3022/9-1 and LE 2440/3-1, the Spanish Ministry of Economia y Competitividad under grants PCIN-2015-012 and TEC2017-90034-C2-1-R (ALLIANCE project) that receives funding from FEDER, and the Catalan Institution for Research and Advanced Studies (ICREA).

REFERENCES

1. A. H. Castro Neto, F. Guinea, N. M. R. Peres, K. S. Novoselov, and A. K. Geim. The electronic properties of graphene. *Reviews of Modern Physics*, 81(1):109–162, 2009.
2. Q. Bao and K. P. Loh. Graphene photonics, plasmonics, and broadband optoelectronic devices. *ACS Nano*, 6(5):3677–3694, 2012.
3. F. H. L. Koppens, D. E. Chang, and F. J. G. D. Abajo. Graphene plasmonics: A platform for strong light-matter interactions. *Nano Letters*, pages 3370–3377, 2011.
4. F. Schwierz. Graphene transistors: Status, prospects, and problems. *Proceedings of the IEEE*, 101(7):1567–1584, 2013.
5. A. Vakil and N. Engheta. Transformation optics using graphene. *Science*, 332(6035):1291–1294, 2011.
6. Y. Wu, D. B. Farmer, F. Xia, and P. Avouris. Graphene electronics: Materials, devices, and circuits. *Proceedings of the IEEE*, 101(7):1620–1637, 2013.

7. I. F. Akyildiz, J. M. Jornet, and C. Han. Terahertz band: Next frontier for wireless communications. *Physical Communication*, 12:16–32, 2014.

8. S. Abadal, I. Llatser, A. Mestres, H. Lee, E. Alarcon, and A. Cabellos-Aparicio. Time-domain analysis of graphene-based miniaturized antennas for ultra-short-range impulse radio communications. *IEEE Transactions on Communications*, 63(4):1470–1482, 2015.

9. I. F. Akyildiz and J. M. Jornet. The internet of nano-things. *IEEE Wireless Communications*, 17(6):58–63, 2010.

10. S. Abadal, B. Sheinman, O. Katz, O. Markish, D. Elad, Y. Fournier, D. Roca et al. Broadcast-enabled massive multicore architectures: A wireless RF approach. *IEEE MICRO*, 35(5):52–61, 2015.

11. S. Abadal, C. Liaskos, A. Tsioliaridou, S. Ioannidis, A. Pitsillides, J. Solé-Pareta, E. Alarcon, and A. Cabellos-Aparicio. Computing and communications for the software-defined metamaterial paradigm: A context analysis. *IEEE Access*, 5:6225–6235, 2017.

12. S. Hosseininejad and N. Komjani. Comparative analysis of graphene-integrated slab waveguides for terahertz plasmonics. *Photonics and Nanostructures—Fundamentals and Applications*, 20:59–67, 2016.

13. D. Correas-Serrano and J. S. Gomez-Diaz. Graphene-based Antennas for Terahertz Systems: A Review. *FERMAT*, 2017.

14. P. J. Burke, S. Li, and Z. Yu. Quantitative theory of nanowire and nanotube antenna performance. *IEEE Transactions on Nanotechnology*, 5(4):314–334, 2006.

15. G. Hanson. Fundamental transmitting properties of carbon nanotube antennas. *IEEE Transactions on Antennas and Propagation*, 53(11):3426–35, 2005.

16. K. Jensen, J. Weldon, H. Garcia, and A. Zettl. Nanotube radio. *Nano Letters*, 7(11):3508–3511, 2007.

17. R. Wang, X. Ren, Z. Yan, L. Jiang, W. E. I. Sha, and G. Shan. Graphene based functional devices: A short review. *Frontiers of Physics*, 14(1):9–18, 2018.

18. J. M. Jornet and I. F. Akyildiz. Graphene-based nano-antennas for electromagnetic nanocommunications in the terahertz band. In *Proceedings of the EuCAP'10*, pages 1–5, 2010.

19. J. M. Jornet and I. F. Akyildiz. Graphene-based plasmonic nano-antenna for terahertz band communication in nanonetworks. *IEEE Journal on Selected Areas in Communications*, 31(12):685–694, 2013.

20. I. Llatser, C. Kremers, A. Cabellos-Aparicio, E. Alarcon, and D. N. Chigrin. Comparison of the resonant frequency in graphene and metallic nano-antennas. *AIP Conference Proceedings*, 1475:143–145, 2012.

21. L. Zakrajsek, E. Einarsson, N. Thawdar, M. Medley, S. Member, and J. M. Jornet. Lithographically defined plasmonic graphene antennas for terahertz-band communication. *IEEE Antennas and Wireless Propagation Letters*, 15:1553–1556, 2016. J. S. Gomez-Djaz, C.

22. S. A. Amanatiadis, T. D. Karamanos, and N. V. Kantartzis. Radiation efficiency enhancement of graphene THz antennas utilizing metamaterial substrates. *IEEE Antennas and Wireless Propagation Letters*, 16:2054–2057, 2017.

23. I. Llatser, C. Kremers, D. Chigrin, J. M. Jornet, M. C. Lemme, A. Cabellos-Aparicio, and E. Alarcon. Radiation characteristics of tunable graphennas in the terahertz band. *Radioengineering Journal*, 21(4):946–953, 2012.

24. X. Zhang, G. Auton, E. Hill, and Z. Hu. Graphene THz ultra wideband CPW-fed monopole antenna. In *1st IET Colloquium on Antennas, Wireless and Electromagnetics*, 2013.

25. A. Cabellos-Aparicio, I. Llatser, E. Alarcón, A. Hsu, and T. Palacios. Use of THz photoconductive sources to characterize tunable graphene RF plasmonic antennas. *IEEE Transactions on Nanotechnology*, 14(2):390–396, 2015.

26. M. Tamagnone, J. S. Gomez-Diaz, J. R. Mosig, and J. Perruisseau-Carrier. Analysis and design of terahertz antennas based on plasmonic resonant graphene sheets. *Journal of Applied Physics*, 112:114915, 2012.

27. I. Llatser, C. Kremers, A. Cabellos-Aparicio, J. M. Jornet, E. Alarcon, and D. N. Chigrin. Graphene-based nano-patch antenna for terahertz radiation. *Photonics and Nanostructures—Fundamentals and Applications*, 10(4):353–358, 2012.

28. M. Tamagnone, J. S. Gomez-Diaz, J. R. Mosig, and J. Perruisseau-Carrier. Reconfigurable terahertz plasmonic antenna concept using a graphene stack. *Applied Physics Letters*, 101:214102, 2012.

29. S. E. Hosseininejad, M. Neshat, R. Faraji-Dana, M. C. Lemme, P. Haring Boljvar, A. Cabellos-Aparicio, E. Alarcón, and S. Abadal. Reconfigurable THz plasmonic antenna based on few-layer graphene with high radiation efficiency. *MDPI Nanomaterials*, 8(8):577, 2018.

30. S. E. Hosseininejad, E. Alarcon, N. Komjani, S. Abadal, M. C. Lemme, P. Haring Boljvar, and A. Cabellos-Aparicio. Surveying of pure and hybrid plasmonic structures based on graphene for terahertz antenna. In *Proceedings of the NANOCOM'16*, page Art. 1, 2016.

31. S. E. Hosseininejad, M. Neshat, R. Faraji-Dana, S. Abadal, M. C. Lemme, P. Haring Bolívar, E. Alarcon, and A. Cabellos-Aparicio. Terahertz dielectric resonator antenna coupled to graphene plasmonic dipole. In *Proceedings of the EuCAP'18*, 2018.

32. S. E. Hosseininejad, S. Abadal, M. Neshat, R. Faraji-Dana, M. C. Lemme, C. Suessmeier, P. Haring Boljvar, E. Alarcon, and A. Cabellos-Aparicio. MAC-oriented programmable terahertz PHY via graphene-based yagi-uda antennas. In *Proceedings of the WCNC'18*, 2018.

33. I. F. Akyildiz and J. M. Jornet. Realizing ultra-massive MIMO (10241024) communication in the (0.06–10) terahertz band. *Nano Communication Networks*, 8:46–54, 2016.

34. S. E. Hosseininejad, K. Rouhi, M. Neshat, R. Faraji-Dana, A. Cabellos-Aparicio, S. Abadal, and E. Alarcon. Reprogrammable graphene-based metasurface mirror with adaptive focal point for THz imaging. *Scientific Reports*, 9:2868, 2019.

35. Z. Xu, X. Dong, and J. Bornemann. Design of a reconfigurable MIMO system for THz communications based on graphene antennas. *IEEE Transactions on Terahertz Science and Technology*, 4(5):609–617, 2014.

36. L. M. Zakrajsek, D. A. Pados, and J. M. Jornet. Design and performance analysis of ultra-massive multi-carrier multiple input multiple output communications in the terahertz band. *Proceedings of the SPIE'17*, 2017.

37. M. Tamagnone, J. S. Gomez-Djaz, J. Mosig, and J. Perruisseau-Carrier. Hybrid graphene-metal reconfigurable terahertz antenna. In *Proceedings of the IMS'13*, pages 9–11, 2013.

38. M. Aldrigo, M. Dragoman, and D. Dragoman. Smart antennas based on graphene. *Journal of Applied Physics*, 116(11):114302, 2014.

39. E. Carrasco and J. Perruisseau-Carrier. Reflectarray antenna at terahertz using graphene. *IEEE Antennas and Wireless Propagation Letters*, 12:253–256, 2013.

40. Y. Huang, L. Wu, M. Tang, and J. Mao. Design of a beam reconfigurable THz antenna with graphene-based switchable high-impedance surface. *IEEE Transactions on Nanotechnology*, 11(4):836–842, 2012.

41. M. Donelli and F. Viani. Graphene-based antenna for the design of modulated scattering technique (MST) wireless sensors. *IEEE Antennas and Wireless Propagation Letters*, 15:1561–1564, 2016.

42. X.-C. Wang and S. A. Tretyakov. Tunable perfect absorption in continuous graphene sheets on metasurface substrates. *arXiv preprint*, 2017.

43. B. Orazbayev, M. Beruete, and I. Khromova. Ultrafast beam steering based on graphene metamaterial. In *Proceedings of the EuCAP 2017*, pages 3896–3899, 2017.

44. A. Momeni, K. Rouhi, H. Rajabalipanah, and A. Abdolali. An information theory-inspired strategy for design of re-programmable encrypted graphene-based coding metasurfaces at terahertz frequencies. *Scientific Reports*, 8(1):6200, 2018.

45. S. E. Hosseininejad, K. Rouhi, M. Neshat, A. Cabellos-Aparicio, S. Abadal, and E. Alarcon. Reprogrammable metasurface based on graphene for THz imaging: Flat metamirror with adaptive focal point. *Scientific Reports*, 9(1):2868, 2018.

46. Y. Shi and Y. Zhang. Generation of wideband tunable orbital angular momentum vortex waves using graphene metamaterial reflectarray. *IEEE Access*, 6:5341–5347, 2017.

47. X. Huang, T. Leng, X. Zhang, J. C. Chen, K. H. Chang, A. K. Geim, K. S. Novoselov, and Z. Hu. Binder-free highly conductive graphene laminate for low cost printed radio frequency applications. *Applied Physics Letters*, 106(20):203105, 2015.

48. A. Lamminen, K. Arapov, G. De With, S. Haque, H. G. Sandberg, H. Friedrich, and V. Ermolov. Graphene-flakes printed wideband elliptical dipole antenna for low-cost wireless communications applications. *IEEE Antennas and Wireless Propagation Letters*, 16(c):1883–1886, 2017.

49. T. Leng, X. Huang, K. Chang, J. Chen, M. A. Abdalla, and Z. Hu. Graphene nanoflakes printed flexible meandered-line dipole antenna on paper substrate for low-cost RFID and sensing applications. *IEEE Antennas and Wireless Propagation Letters*, 15:1565–1568, 2016.

50. V. Gusynin, S. Sharapov, and J. Carbotte. Magneto-optical conductivity in graphene. *Journal of Physics: Condensed Matter*, 19(2):026222, 2006.

51. G. W. Hanson. Dyadic green's functions for an anisotropic, non-local model of biased graphene. *IEEE Transactions on Antennas and Propagation*, 56(3):747–757, 2008.

52. M. Jablan, H. Buljan, and M. Soljačić. Plasmonics in graphene at infrared frequencies. *Physical Review B*, 80(24):245435, 2009.

53. C. H. Gan, H. S. Chu, and E. P. Li. Synthesis of highly confined surface plasmon modes with doped graphene sheets in the midinfrared and terahertz frequencies. *Physical Review B*, 85(12):125431, 2012.

54. L. Banszerus, M. Schmitz, S. Engels, J. Dauber, M. Oellers, and P. Gr. Ultra-high mobility graphene devices from chemical vapor deposition on reusable copper. *Science Advances*, 1(6):e1500222, 2015.

55. H. Hirai, H. Tsuchiya, Y. Kamakura, N. Mori, and M. Ogawa. Electron mobility calculation for graphene on substrates. *Journal of Applied Physics*, 116(8):083703, 2014.

56. A. Grigorenko, M. Polini, and K. Novoselov. Graphene plasmonics. *Nature photonics*, 6(11):749, 2012.

57. S. A. Maier. *Plasmonics: Fundamentals and Applications*. Springer Science & Business Media, Berlin, Germany, 2007.

58. S. E. Hosseininejad, N. Komjani, and M. Talafi Noghani. A comparison of graphene and noble metals as conductors for plasmonic one-dimensional waveguides. *IEEE Transactions on Nanotechnology*, 14(5):829–836, 2015.

59. COMSOL Multiphysics Modeling Software, ver. 5.0, 2014. Available: https://www.comsol.com/.

60. CST Microwave Studio. 2019. Available: http://www.cst.com.

61. K. Q. Costa, V. Dmitriev, C. M. Nascimento, and G. L. Silvano. Graphene nanoantennas with different shapes. *2013 SBMO/IEEE MTT-S International Microwave & Optoelectronics Conference (IMOC)*, pages 1–5, 2013.

62. M. Tamagnone and J. Perruisseau-carrier. Predicting input impedance and efficiency of graphene reconfigurable dipoles using a simple circuit model. *IEEE Antennas and Wireless Propagation Letters*, 13:313–316, 2014.

63. N. T. Yardimci, S. H. Yang, C. W. Berry, and M. Jarrahi. High-power terahertz generation using large-area plasmonic photoconductive emitters. *IEEE Transactions on Terahertz Science and Technology*, 5(2):223–229, 2015.

64. N. Khiabani, Y. Huang, Y.-C. Shen, and S. J. Boyes. Theoretical modeling of a photoconductive antenna in a terahertz pulsed system. *IEEE Transactions on Antennas and Propagation*, 61(4):1538–1546, 2013.

65. J. M. Jornet and I. F. Akyildiz. Graphene-based plasmonic nano-transceiver for terahertz band communication. In *Proceedings of the EuCAP'14*, pages 492–496, 2014.

66. Moldovan, S. Capdevila, J. Romeu, L. Bernard, A. Magrez, A. Ionescu, and J. Perruisseau-Carrier. Self-biased reconfigurable graphene stacks for terahertz plasmonics. *Nature Communications*, 6(6334):1–8, 2015.

67. Y.-J. Yu, Y. Zhao, S. Ryu, L. E. Brus, K. S. Kim, and P. Kim. Tuning the graphene work function by electric field effect. *Nano Letters*, 9(10):3430–3434, 2009.

Terahertz Programmable Metasurfaces

Networks Inside Networks

Sergi Abadal, Christos Liaskos, Andreas Pitsillides,
Vasos Vassiliou, Josep Solé-Pareta, Albert
Cabellos-Aparicio, and Eduard Alarcón

CONTENTS

T UNABLE METASURFACES ARE ULTRA-THIN, artificial electromagnetic (EM) components that provide engineered and adjustable functionalities up to the terahertz (THz) band and beyond. The concept of software-defined metamaterials (SDMs) has recently emerged as tunability has been applied at the unit cell level and exposed to the user via programmatic commands. To realize this approach, SDMs require the integration of a network of controllers within the structure of the metasurface, where each controller interacts locally and communicates globally to obtain the programmed behavior. At the same time, the approach allows to interconnect several metasurfaces to address complex EM problems in a distributed way. Within this *networks inside networks* context, the present chapter focuses on the communication requirements and implementation constraints of the controller network within the metasurface. Guidelines for the design of appropriate wired and wireless solutions are extracted from the analysis, paving the way to the realization of programmable THz metasurfaces.

3.1 INTRODUCTION

Metamaterials have drawn a great deal of attention since their conception, as they enable unprecedented levels of EM control [1]. They have led to significant breakthroughs in various fields such as imaging, radar, and wireless communications, to name a few [2,3]. Metasurfaces, the thin-film planar analog of metamaterials, are composed of an array of subwavelength resonators and inherit the unique properties of their three-dimensional counterparts while addressing their issues related to bulkiness, losses, and cost. For this, metasurfaces operating at the microwave [4], THz [5], or optical [6] bands have been widely proposed to achieve attractive features such as negative refraction, cloaking, superlensing, and holographic behavior.

The main downturn of the existing metasurfaces is that they are specifically designed for a single application working under preset conditions (e.g., steering to a fixed angle) and cannot be reused. In response to these drawbacks, metasurface designs with tunable or switchable elements in the unit cell have emerged [7]. The resulting reconfigurable metasurfaces can be globally or locally tunable, depending on the specific scheme, and with appropriate control means, they become programmable [8,9].

A step further toward the compelling vision, fully adaptive metasurfaces with multiple concurrent functionalities form the concept of SDM [10,11]. The SDM paradigm builds upon the description of EM functions in reusable software modules. Such *software-defined* approach allows authorized (possibly nonexpert) users to easily change the behavior of the metasurface by sending preset commands. With appropriate means of communication, SDMs could be even interconnected, enabling the control of EM waves with unprecedented degrees of freedom and in a distributed and coordinated way. Novel applications such as the so-called programmable wireless environments at millimeter-wave (mmWave) and THz bands [12,13] will emerge from this new paradigm.

Figure 3.1 represents the functional and physical architecture of an SDM. The target EM function, described in software, is an abstraction of the tuning state of the metasurface cells. Therefore, when the user sends a set of commands to change the EM function, the SDM modifies the internal state of its unit cells. The gap between these two ends is

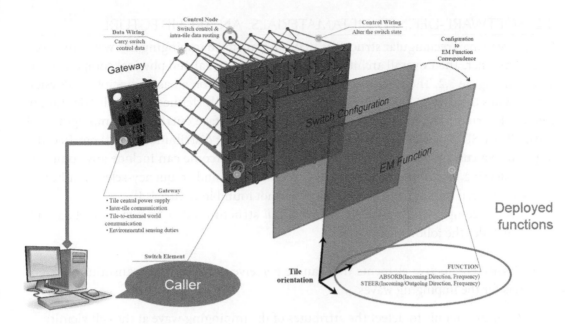

FIGURE 3.1　The functional and physical architecture of an SDM tile.

bridged by the key enabler of the approach: an integrated network of miniaturized controllers within the metasurface structure. This integrated network receives programmatic commands from the user via a gateway, to then disseminate, interpret, and apply each of these commands to achieve the desired EM behavior.

The controller network lies at the heart of an SDM and poses significant implementation and cointegration challenges [10]. This is especially so at mmWave and THz frequencies, because the unit cells scale proportionally to the EM wavelength. In the particular case of the THz range, controllers have to be tens of micrometers in size, and the interconnection network effectively becomes a nanonetwork [11]. At this scale, the controller nanonetworks are expected to be simple and ultra-efficient, yet powerful enough to enable real-time adaptivity and support multiple ways of interacting locally, globally, and with external entities. However, this combination of constraints and requirements poses important challenges, thus requiring a careful definition of the computation and communication mechanisms that will drive the operation of SDMs.

This chapter focuses on the challenge of designing nanonetworks for internal SDM communication. The main aim is to provide an extensive context analysis that clarifies the characteristics of the scenario. This is provided in Sections 3.2 and 3.3. There, we depict the canonical architecture of an SDM and its different functional layers, to then delve into the communication layer and shed light on its requirements and limitations. This analysis allows to extract design guidelines and other insights, which are later leveraged in Section 3.4 to discuss the design of intra-SDM networks. Finally, we conclude the chapter in Section 3.5.

3.2 SOFTWARE-DEFINED METAMATERIALS: AN ARCHITECTURE

SDMs are planar, rectangular structures that can transform impinging EM waves in a controlled manner. Their overall architecture, comprising virtual and physical components, is shown in Figure 3.2. The core functionality of SDMs relies on a basic principle in Physics, which states that the EM emissions from a surface are fully defined by the distribution of electrical current over it. The sources producing the surface currents are impinging EM waves. Thus, SDMs seek to control and modify the current distribution over them, in order to produce a custom EM emission as a response. This outcome can include any combination of steering, absorbing, polarity and phase altering, and frequency-selective filtering over the original impinging wave, even in ways not found in nature [14].

An SDM comprises a massively repeated *cell* structure (also known as meta-atom), which includes the following:

1. Passive conductive elements that can be perceived as receiving/transmitting antennas for the impinging waves

2. A sensor module to detect the attributes of the impinging wave at the cell vicinity

3. An actuation module, which can regulate the local current flow within the cell vicinity

4. A computation and communication module responsible for controlling the sensory and actuation tasks within the cell, as well as exchanging data with other cells (inter-cell networking), to perform synergistic tasks with other cells and communicate with SDM-external entities

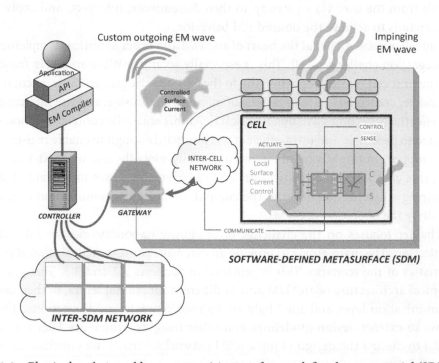

FIGURE 3.2   Physical and virtual layers comprising a software-defined metamaterial (SDM).

Each SDM unit has a gateway that handles its connectivity to the external world. The gateway participates in the intercell network as a peer to the external world via any common protocol (e.g., WiFi and Ethernet). Its overall role is to (i) aggregate and transfer sensory data from the SDM to an external controller and (ii) receive cell actuation commands and diffuse them for propagation within the intercell network.

Finally, a regular computer can act as the external entity that gathers the sensory information from all SDMs within an environment and subsequently calculates their configurations that fit a given application scenario. For instance, programmable wireless environments use multiple SDMs to customize the wireless propagation for multiple mobile devices, thus achieving state-of-the-art communication quality [13].

SDMs come with software libraries that facilitate the creation of applications. This software suite comprises the SDM application programming interface (API) [15] and the SDM EM Compiler [16]. The SDM API contains software descriptions of the metasurface EM functions and allows the programmer to customize, deploy, or retract them on demand via a programming interface with appropriate callbacks. The API serves as a strong layer of abstraction. It hides the internal complexity of the HyperSurface and offers general-purpose access to metasurface functions, without requiring the underlying hardware and physics. The EM Compiler handles the translation of the API callbacks into SDM actuation directives, in an automatic manner, transparently to the user.

Having provided an overview of the SDM architecture and control flow, we proceed to detail the operation of SDM following an operation layer approach.

3.2.1 The Metasurface Layer

The metasurface layer describes the behavior of the SDM from the aspect of Physics. As such, it comprises the elements that can be used to control the SDM surface currents, abstracting all other concerns. Therefore, in the most general view, the metasurface layer comprises the passive SDM elements (e.g., the passive rectangular patches in Figure 3.1), as well as the set of local impedance values that can be enforced at each cell, in order to affect the local surface current. The specific manner in which these impedance values are attained is not a concern of this layer, and these values are derived from the chosen manufacturing technology. For instance, the work of [17] considers complementary metal–oxide–semiconductor (CMOS) switches at each cell, which can be at the ON (most conductive) or OFF (most insulating) state. Each state then corresponds to a local surface impedance value. We note that dynamic meta-atom designs constitute an extensively studied subject in the literature, offering a wide variety of choices in terms of active/passive element combinations [14].

The cell size and the thickness of the SDM are important design factors that define the maximum frequency for EM wave interaction. As a rule of thumb, meta-atoms are bounded within a square region of $\lambda/10 \leftrightarrow \lambda/5$, λ being the EM interaction wavelength. The minimal SDM thickness is also in the region of $\lambda/10 \leftrightarrow \lambda/5$. Thus, for an interaction frequency of 5 GHz, the meta-atom would have a size of 7.5 mm, and the SDM would have a similar thickness. Moreover, a minimum of 30×30 meta-atoms are usually required to exert a consistent EM functionality over the impinging wave.

3.2.2 The Control and Sensing Layer

The control and sensing layer is composed of the specific electronic hardware components that enable the operation of the metasurface layer, as well as the sensory duties described earlier. It comprises the SDM cells, that is, their specific, low-level implementation, inclusive of all actuation, sensing, and computation hardware.

The actuation modules can include any kind of switching element that can affect the local surface current at a cell, provided that this effect can be electronically controlled. In this sense, the aforementioned CMOS switches can be complimented by microelectromechanical systems (MEMS) or microfluidic structures [18]. Each specific actuation technology may offer a different set of attainable impedance values, as well as different response times and power drain. The impedance values can be a discrete or continuous set or vary in their range. An actuator technology may offer faster transitions between impedance values than any other (e.g., CMOS vs. microfluidics). Finally, an actuator may require constant power supply to retain a state (CMOS) or can be state-preserving and require power only during an impedance transition (e.g., microfluidic).

The sensing module comprises hardware sensors for detecting an attribute of the impinging wave that needs to be modified. These attributes can include the wave's direction of arrival, phase, polarity, power, and frequency. These attributes can be derived by a specialized hardware sensor or be inferred by local current measurements at the cell vicinity. In any case, however, sensing is hindered by the strong mutual couplings among cells. In other words, the sensed attribute values at given cell are the product of not only the wave impinging locally but also the inductive phenomena between cells.

Finally, the computation module handles the control over actuation and sensing, as well the computations required for performing synergistic tasks. The latter include the data exchange between coupled cells, to deduce the actual attribute values of the impinging wave at each cell. Another synergistic task is data routing. The compute module of each cell takes decisions to facilitate the propagation of data for cell-to-cell and cell-to-external entity communication.

3.2.3 The Communications Layer

The communications layer specifies the hardware, medium, topology, and protocols that enable the bidirectional communication for the intercell networking, as well as between the cell network and the external world.

Early SDM designs considered a grid topology of cells with rectangular wired connectivity [10,12,19]. In this case, the data communication and routing protocols from the field of network-on-chip (NoC) have been considered for adaption to the SDM case [20]. The XY-routing and its variations for latency and error resilience constitute extremely lightweight and effective candidate solutions that can be easily implemented in the cell hardware. Moreover, studies have been initiated for wirelessly communicating cell designs. In such cases, SDMs include a dedicated substrate layer that acts as the propagation medium for wireless cell communications [21]. The wireless approach can add versatility in assembling SDMs, since cells are treated as more autonomous units that can be put together and connect automatically. The wired approach, on the other hand, constitutes the easiest and most inexpensive approach to implement SDMs, resulting, however, in a fixed intercell network. More details on both approaches are given in Section 3.4. In either case, the SDM gateway can be simply considered as an IoT device that participates as a peer to the intercell network.

We note that future advancements in electronics manufacturing may allow for the mass production of miniaturized, powerful, and inexpensive cell hardware. Such SDM cells may be able to directly support advanced communication protocols. Therefore, they may be able to communicate with the external world directly, without requiring a gateway to act as a mediator. Depending on their capabilities, such cells may also act as ambient intelligence, handling the decisions for the SDM configuration in an autonomous fashion, without the need for an external compiling service.

3.3 METASURFACE COMMUNICATIONS CONTEXT

The communications layer is a crucial enabler of the SDM paradigm, as it allows the global coordination of the cells of a metasurface tile toward obtaining the desired EM function. The challenge here resides in the unique application context, which leads to an exotic combination of communication requirements, implementation constraints, and optimization opportunities. Since these characteristics will largely determine the communications approach, it is key to understand the context to better drive the design decisions.

Toward a better understanding of the SDM scenario from a communications perspective, we describe the SDM workflow in Section 3.3.1 and the resulting communication flows in Section 3.3.2, to then analyze the physical constraints that will limit the intratile network implementation in Section 3.3.3.

3.3.1 The SDM Workflow

We proceed to describe the SDM operational workflow, involving its hardware and software components. In a coarse manner, the workflow is as follows. An external application incorporating the SDM API executes a supported callback, to deploy an EM function to the SDM. The callback is translated to data packets that are sent to the SDM gateway, using a given communication protocol. The gateway diffuses the information within the SDM cells, using an intercell communication protocol. The cells receive the information pertaining to them and set the states of the actuating elements accordingly. In a parallel operation track, cells send their sensory data back to the application, which employs the data to continuously update the EM function deployed at the SDM.

Based on this coarse description, we identify the following conceptual entities involved in the SDM API workflow:

- *The Application:* The entity that calls the SDM API, while also receiving and handling. interrupts, for example, sensory messages sent by the SDM.

- *The Callback:* A single function of the SDM API.

- *The Configuration:* A data entry containing the information for mapping a Callback to a specific set of cell actuation module states.

- *The EM Compiler:* The software entity that converts callbacks into configurations.
- *The Actuator:* A switch element of the SDM. It can be set to any state within a set of possible states.
- *The Actuator State:* The state of a switch element. It can comprise several impedance values, that is, discrete combinations of resistance and capacitance.
- *The gateway external communication protocol:* It is the protocol that transfers data between the SDM and the device that hosts the Application.
- *The gateway:* The SDM gateway hardware, as previously described.
- *The SDM intercell communication protocol:* The protocol that transfers data between the gateway and the cells, as well as between the cells themselves.
- *The SDM Tile:* The complete, assembled SDM unit.

We now define the tentative form of the API callbacks. These callbacks have the following general form:

$$outcome \leftarrow callback(action\_type, parameters)$$

The *action_type* is an identifier denoting the intended function, for example, STEER, ABSORB, POLARIZE, and FILTER. Each action type is associated with a set of parameters. For instance, STEER commands require (i) an incident wave direction, *I*, (ii) an intended reflection direction, *O*, and (iii) the applicable wave frequency, *F*. ABSORB commands require no *O* parameter.

We proceed to illustrate the API Callback process for setting the state of an actuating element, as shown in Figure 3.3. The Application executes a Callback function, which in

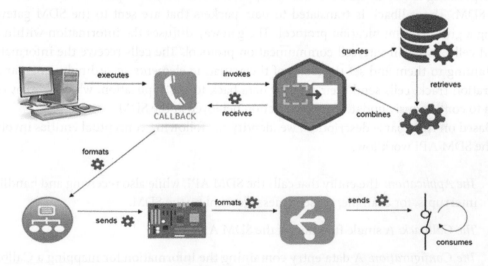

FIGURE 3.3   The workflow for configuring an SDM for a required electromagnetic functionality.

turn invokes the Compiler. Depending on whether the Callback requests a combination of functionalities, the Compiler may return one or more Configurations that are combined to match the request. The final Configuration is conveyed to the gateway by using the corresponding protocol. The gateway reformats the received Configuration and diffuses it within the intercell network. Each cell receives actuation directives addressed to it, thereby "consuming" the corresponding part of the Configuration.

Apart from setting the state of the cell actuator, the SDM workflow (and the API) include processes for getting the state of the cells (i.e., sensory data and other hardware state). We refer to these functions as monitor Callbacks owing to their immediate usage to monitor the state of an SDM. In Figure 3.4, the Application executes a monitor Callback to get the state of a cell. Notice that this Callback type requires no interaction with the Compiler. This monitoring technique is called polling, since the cells are queried for their state.

In addition, the cells themselves may trigger a monitoring event, such as sending a sensory event and informing of a local malfunction. This approach, called reporting, is illustrated in Figure 3.5. Notice that this approach requires a service to be active at the Application side, ready to receive incoming reports from the cells.

In both polling and reporting monitoring approaches, time-outs or other transmission failures may occur for a variety of reasons (malfunction, interference, and power supply issue). Such events can be handled by either the application or the SDM, for example, by employing an additional synergy between cells and the SDM gateway.

3.3.2 SDM-Communication Flows

The methods described previously clearly outline the possible communication flows within an SDM, thereby representing the first step toward knowing which kind of networks would

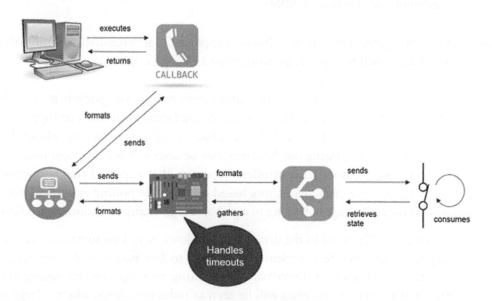

FIGURE 3.4   The workflow for polling the state of an SDM cell.

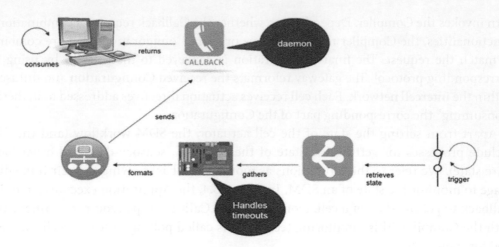

FIGURE 3.5   The self-reporting work of an SDM cell.

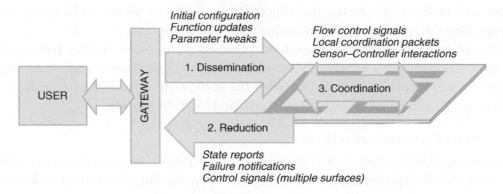

FIGURE 3.6   Communication flows in SDMs.

better suit this new application context. It seems apparent that three distinct kinds of communication patterns will take place, as summarized in Figure 3.6.

- *Dissemination flows:* The SDM configuration steps require the gateway to send directives to all the metasurface cells. Subsequent functionality updates or input parameter tweaks may need to be directed to a subset of cell controllers but always have a global scope. Finally, polling mechanisms may be seen as point-to-point interactions from the gateway to specific controllers; however, it is expected that gateways will implement fault-detection algorithms based on multiple polling operations to large subsets of controllers, again leading to a global few-to-many communication scheme.

- *Reduction flows:* Opposed to the dissemination flows, reactions to gateway directives are expected to generate complementary many-to-few patterns. Acknowledgment packets to receiving those directives or to polling messages or the aggregation of multiple self-reporting processes will be seen as reduction flows, where a large set of nodes communicates with one or very few sinks.

- *Local coordination flows:* Future SDM will integrate sensors within the metasurface to measure external events and, eventually, react to them, without having to communicate with the gateway. In addition, it is expected that the network within the SDM will implement local flow control and fault-tolerance mechanisms in order to gradually approach a fully autonomous SDM paradigm.

  Knowledge on the communication flows provides insight about the possible spatiotemporal characteristics of traffic, which will need to be confirmed as prototypes and appropriate simulator tools appear. A further quantification of the traffic can be estimated if designers take into consideration the desired spatial granularity (cells per unit of area) and temporal granularity (reconfigurations per second) of control within the SDM. This will eventually depend on the final application, as it determines the following:

- *The spatial accuracy:* Some functionalities require a very-fine-grained discretization of space, leading to more dense integration (more cells per unit of area). For a fixed area, this means more cells and thereby both a denser network topology and higher communication density.

- *The cell complexity:* The controllers need to fix the state of a cell, choosing among a finite set of options. Some applications may require a fine control of the cell, thereby leading to more options per cell. This, in the end, scales the amount of information that each packet needs to carry and, consequently, the bandwidth requirements.

- *The temporal scale:* The time discretization is contingent on the incident phenomena that the metasurface needs to control. As a rule of thumb, the metasurface needs to operate at twice the rate of the fastest-changing input. For instance, a beam steering SDM shall operate at a frequency relative to the speed at which a person or a vehicle moves toward or away from the SDM device.

3.3.3 Physical Implementation Constraints

The communication network within an SDM needs to be designed while accounting for a series of physical limitations. The size restriction is evident, as controllers (and the integrated communications devices) need to be significantly smaller than λ. This may not suppose a problem at the gigahertz (GHz) range, where cells are close to the centimeter range, but becomes very stringent at THz frequencies.

Owing to the required integration density and also seeking to maximize the usability of the SDM paradigm, both the controllers and the nanonetwork should have very strict power and energy budgets. Depending on the final application, SDMs may need to be powered with batteries or even via harvesting devices. This *energy first* constraint suggests the employment of ultra-low-power approaches that, as much as possible, consume energy only when used. This points toward asynchronous schemes that, unlike clock-based networks, do not waste energy while the SDM is idle.

In relation to this last point, it is very important to note that the electronic implementation underlying the SDM should not interfere with the EM functionality. The switching activity of the different circuits generates a low-frequency EM noise whose harmonics can reach the SDM range and interfere with its operation. This, again, encourages the use of asynchronous clockless schemes. Since such systems do not limit their switching activity to the instants marked by a common clock, surges of EM noise are completely avoided.

Last but not least, the communication approach taken in SDM needs to carefully consider the restrictions of the available hardware. In this category, we consider pin limitation as the most incapacitating feature of SDM systems. If we assume that a chip hosts a subset of controllers, the bottleneck of the system will then be the interchip communication and will be determined by the amount of input/output chips that can be assigned to communications. This sets a very restrictive bound on the available bandwidth, the number of links that can be implemented, and the acceptable complexity. We will see in Section 3.4.1 that such hard constraints can be overcome even in the case where the number of available pins is less than 25.

3.4 NETWORK DESIGN WITHIN SOFTWARE-DEFINED METASURFACES

Existing context analyses, summarized in the previous part, point out remarkable resemblances between multiprocessors and SDMs in terms of layout, levels of integration, or communication patterns, among others [10]. As such, it has been suggested that on-chip communication techniques may be a valid approach for SDM. The NoC paradigm essentially refers to packet-switched networks of routers and links cointegrated with the clients, which are the processor cores in the multiprocessor case or the controllers in the SDM paradigm.

Most multiprocessors have a much larger power budget than what SDM can offer, and, consequently, existing NoC designs prioritize performance over power [22]. Although this trend is changing in the dark silicon era, NoCs still rely on fairly complex routers and wide links to meet the multiprocessor requirements [20]. As such, they are not directly portable to the SDM scenario, owing to the limitations outlined in the previous sections, especially as we scale the SDM designs to the THz band.

This section describes two complementary approaches that could be employed in the networks within future SDMs: wired or wireless. On the one hand, Section 3.4.1 focuses on the wired option, which essentially consists an extremely simple NoC with a minimal grid topology and single-bit asynchronous links. The discussion revolves around making the most of the very limited resources of the SDM scenario to afford features such as adaptive routing and active acknowledging. On the other hand, Section 3.4.2 examines a more advanced technology based on miniaturized antennas, which leverages the enclosed nature of the SDM to implement efficiently a fast means of broadcast communication.

3.4.1 Integrated Asynchronous Network

Previous sections have discussed the need for an integrated network within an SDM, discussed the stringent limitations of the scenario, and suggested the use of asynchronous logic to implement the communication due to power and EM reasons. These guidelines are leveraged in the following subsections, which detail the design principles of an interconnect prototype for the first generations of SDM. To this end, we assume widespread 25-pin chips hosting a single controller and interchip communication to achieve programmability in a 5-GHz SDM. Although this is far from the THz range, the network design principles are the same as we scale down dimensions to reach THz operation: chips would host more than a single controller, so that part of communication would occur in a NoC, while the rest would occur among chips, with the same bottleneck arising from pin limitation.

Next subsections depict the pin allocation, asynchronous communication protocols, topology and router microarchitecture, and schemes for fault identification and tolerance.

3.4.1.1 Wired Physical Design

A total of 23 pins are needed by each intratile controller for the purposes of intratile communication and operations and for metamaterial configuration. The pins of the controller are graphically illustrated in Figure 3.7a. The pins are allocated as follows:

- *Channel endpoint allocation:* As required by the operation of the asynchronous interconnects (see Section 3.4.1.2), 12 pins will be allocated for communication on the intratile controller.

- *Pin allocation for configuring metamaterial patches:* Each controller is configured for four metamaterial patches via eight analog signals (two analog signals for each patch). Therefore, eight pins on the intratile controller will be allocated for the purpose of configuring the metamaterial patches of each controller.

- *Pin allocation for additional operations:* The intratile controller also allocates one pin for a global reset signal; the gateway controller can use the global reset signal to perform a hard reset on each intratile controller. Hard reset means that all the intratile controller will reset to their initial state (delete any buffered packets and initialization information).

- *Two additional pins* are to be allocated on the intratile controller to be used by the gateway controller to issue a global state, designating the operation mode of every intratile controller in the network.

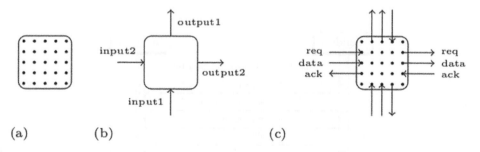

FIGURE 3.7 Controller: (a) Number of controller chip pins, (b) Position of two unidirectional inputs and two unidirectional outputs, and (c) Pin allocation for communication.

3.4.1.2 Asynchronous Interconnects

By default, most NoC designs are synchronous. This requires the distribution of a clock signal throughout the chip, which takes substantial area and power. To avoid it, one can adopt the globally asynchronous locally synchronous (GALS) approach consisting of the use of asynchronous links to communicate synchronous cores [22]. In a synchronous controller design, an interface is required to connect with the clockless network, whereas in an asynchronous or event-based approach, no further adaptation is required.

The chosen hardware protocol for asynchronous bit-level communication is the four-way asynchronous communication protocol. The protocol is used for the communication between intratile controllers and between an intratile controller and a gateway controller. The protocol requires three separate signals for a correct implementation of a unidirectional bit exchange. Therefore, for the implementation of asynchronous communication, each intratile controller needs to allocate three pins per unidirectional communication channel. At the chip level, three signals correspond to the allocation of three chip pins, as shown in Figure 3.7c.

Four-way asynchronous communication is implemented using three signals between the interconnected nodes:

1. The data line or data signal

2. The request or req signal

3. The acknowledgment or ack signal

A description of the four-way protocol is given in terms of sequence diagram in Figure 3.8. The three signals in the four-way asynchronous communication protocol are used for the exchange of a single bit. The initial state requires that the req and ack signals have value zero. The transmission is initiated when the sender sets the value of the data line to the bit to be transmitted. The sender then sets the value of the req signal to 1. The req signal lets the receiver know that the bit on the data line is valid for reception. After the reception, the receiver sets the ack signal to one. It is then safe for the sender to reset the req signal to 0. When the receiver senses a zero req signal, it also resets the ack signal to zero, and thus, the signals return to the initial protocol state.

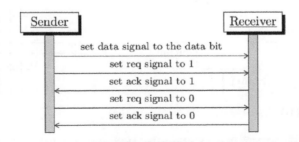

FIGURE 3.8   Sequence diagram of a four-way asynchronous communication.

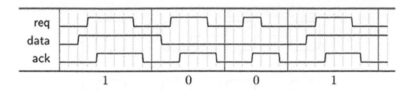

FIGURE 3.9   Timing diagram in the transmission of the bit stream "1001."

An iterative application of the protocol allows for the exchange of a stream of bits, which, in our proposed metasurface, will form packets of data. An example of a transmission of a stream of bits over time is shown in the timing diagram in Figure 3.9. The diagram shows the value of the three signals over time for the transmission of the bit stream 1001. Initially, all three signal values are 0. At some point in time the sender sets the data signal to 1, which corresponds to the first bit of the stream. The next value change happens on the req signal, where the sender sets the signal to 1. Then, the receiver reads the data signal and raises the ack signal. The interaction for the first bit finishes when observing the sequence of req and ack signals to be set to 0. Before sending the next bit 0, the state of the channel has the req and ack signals set to 0 and the data signal set to 1, because the last transmitted bit was 1. So, the data line needs to be set to 0 and then follow the entire protocol once more to transmit bit 0. The next bit to be transmitted is also 0. This means that the data line will not change value, and the entire protocol will take place for a third time. The last bit to be transmitted is 1. As demonstrate in the timing diagram, each bit may require a different amount of time to be transmitted, owing to the asynchronous nature of the four-way asynchronous communication.

3.4.1.3 Topology and Router Microarchitecture

As in chip multiprocessors, a bidimensional mesh seems a natural fit for SDMs, owing to its ease of layout and performance. Yet still, even simpler topologies such as a ring [23] are an intelligent choice, since they allow the use of minimalistic router microarchitectures. Application-specific topologies such as the custom trees built with microswitches [24] could also be an ideal fit for the scatter-gather communication flows of this scenario but are harder to fit in a regular layout, like the grid structure of SDM.

We implement a grid topology with four unidirectional communication channels that interconnect a controller with its four vertical and horizontal neighbor controllers. Owing to the scarcity of pins, the four channels are unidirectional, separated into two input directions and two output directions that are distributed as shown in Figure 3.7b. Starting from the bottom connection in clockwise direction, two input channels (input1 and input2) are allocated for a vertical and a horizontal connection, respectively. Continuing the clockwise direction, two output channels (output1 and output2) are allocated for a vertical and a horizontal connection, respectively.

The network topology depends on the requirements of the project and the design of the controller. On the one hand, the topology needs to allow for packet routing from/to the gateway to/from the controllers. It also needs to allow for the detection of faults,

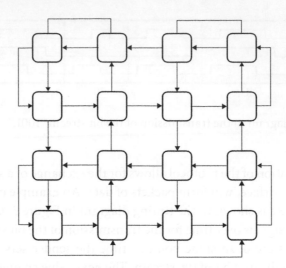

FIGURE 3.10 Manhattan topology with wraparounds.

and it must be robust enough to implement simple fault-tolerant protocols. On the other hand, the network topology depends on the intratile controller design and limitations. For example, as discussed next, it becomes a necessity to allow for interconnection wraparound at the edges of the network topology to increase the connectivity and robustness of the network.

Based on the controller channel allocation, the adopted topology for the intratile controller network is the Manhattan network topology with bidirectional channels at the edges, as presented in Figure 3.10.

Additional considerations are the number and placement of the gateways. The Manhattan topology with bidirectional channels at the edges allows for connecting gateways with no additional cost, just by replacing wraparounds. Replacing wraparounds with gateways comes with no additional allocation of pins or without additional hardware logic on the controllers connected with the gateway. Wraparound replacement implies that a gateway controller is connected with a controller with an output channel endpoint and an input channel endpoint, allowing to send and receive packets to and from the controller network, respectively. The adopted solution utilizes two gateways. The first gateway replaces the southwestern-most wraparound of the grid, and the second gateway replaces the northeastern-most corner of the grid, as in Figure 3.11.

3.4.1.4 Communication Solution and Fault Tolerance

Given the physical layer selection of wired connections, utilizing an asynchronous mode of communication, over a modified Manhattan network topology with bidirectional channels at the edges, we now describe the initialization and addressing of the intratile controllers, the routing within a tile, and the provisions made for reporting.

3.4.1.4.1 Addressing Protocol and Initialization of Operation   The purpose of the addressing protocol is to inform each controller about its coordinates. At the initialization mode,

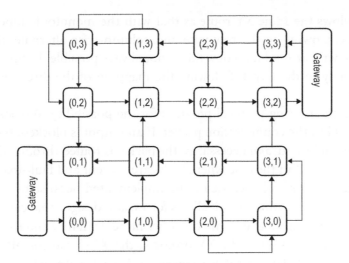

FIGURE 3.11   Connection of the gateways with the intratile network.

each intratile controller has no knowledge about its address and its orientation. Each controller that operates at the initialization mode and has no address knowledge, upon the receipt of a packet, will set its address to the coordinates described by the received packet. After setting its address, each intratile controller has all the information to calculate and set its type and therefore map its local routing directions. A controller that has knowledge of its address operates as in the normal mode of operation, that is, routes packets toward their destination. The initialization process is achieved via a predefined sequence of packets sent by the gateway controller, beginning in an increasing order of their distance from the gateway. The protocol is designed to be robust enough to avoid routing paths that contain faulty controllers and at the same time address all connected intratile controllers. Figure 3.11 shows the proposed addressing scheme. The bottom left intratile controller has (0,0) as its xy-coordinates. The x-axis increases from left to right, and the y-axis increases from bottom to top.

3.4.1.4.2 Gateway Controllers   The design proposes two gateway controllers. Two gateway controllers serve the requirement to send configuration packets from bottom left gateway controller to the intratile controller network and send report-acknowledgment packets from the intratile controller network to the top-right gateway controller.

3.4.1.4.3 Routing Protocol   The routing protocol adopted for implementation is a variant of the XY routing protocol. The XY protocol [25] in a monotonic topology requires the target coordinates as parameters. Routing takes place as follows: The packet is routed on the x-axis until it reaches the x-column. Then, it is routed on the y-axis until it reaches the target. The proposed XY protocol, however, is adapted for the specific requirements of the Manhattan edge wraparound topology.

The XY routing procedure takes the packet to be routed as a parameter and the coordinates of the target controller. In the Manhattan edge wraparound topology, the protocol follows the same XY route as that with the monotonic topology, whenever possible. However, owing to the row-wise and column-wise alternate directions of the Manhattan edge wraparound topology, this is not always possible. In this case, the packet is routed indirectly to the target, following the mapping of the directions "up," "down," "left," and "right."

Furthermore, the routing protocol provides for the possibility to block outputs, using the roadblock field in the configuration packet. If an output is blocked, then the XY routing protocol is not followed, and necessarily, the packet is routed through the other output. Roadblocks can be used in the case where faults are present and fault-tolerant techniques are needed. The XY protocol is chosen to be implemented because of its simplicity and deadlock freedom. At the same time, it allows for variant designs that can be implemented in future implementation iterations and that offer more flexibility and robustness.

For example, an extension of the XY protocol, the XYi protocol, allows for multiple routing path options at the controller gateway level. Each path determined by an extra input that sets the routing column. A gateway can compute, based on several factors, the best XYi routing path for a packet and set it through the routing column parameter. Factors that might be taken into account by the gateway for choosing a routing path areas follows: avoid faulty intratile controllers, if any; reduce congestion, since different routing paths, are expected to balance the network load; etc. [19]. This approach differs from other protocols that, for instance, consider a general routing protocol that treats the network as an arbitrary topology and, through long-enough addresses, are capable of dodging faulty nodes [26].

3.4.1.4.4 *Reporting Protocol* The XY protocol can be used to send packets from a gateway to a controller or from a controller to a controller. To facilitate the reporting functionality, the XY protocol is also used to send acknowledgment packets from a controller to the acknowledgment gateway: [10]

1. Each controller, after receiving a configuration packet, creates an acknowledgment packet.

2. Each controller that creates or receives an acknowledgment packet routes toward the gateway designated to receive acknowledgment packets (the gateway at the top right corner of the tile).

3. The controller at the top-right corner of the tile is reached by setting the x and y routing coordinates of the report packet to maximum value.

3.4.2 Wireless Interconnects

Communication among the controllers of an SDM can be wireless. *A priori*, wired means are preferable, because vast knowledge from similar scenarios like low-power embedded systems can be reused [20,22]. When scaling the SDM paradigm toward THz operation, however, conventional NoC may encounter issues in the form of complex layouts or poor performance in very dense networks [27].

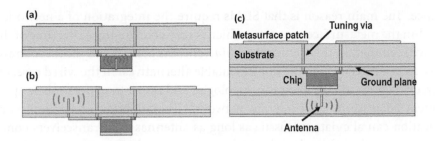

FIGURE 3.12 Potential intra-SDM wireless channels leveraging (a) Chip layer propagation, (b) Metasurface layer propagation, and (c) dedicated layer propagation.

Figure 3.12 represents the cross-section of the unit cell of an SDM, together with three ways to implement wireless communications within the structure. In all cases, the controller chips are connected to the metasurface patches through a tuning via that adjusts the patches' impedance. Each controller chip is also connected to an antenna that takes the form of a monopole and whose placement depends on the specific layer used to guide the wave propagation. We refer the reader to Chapter 6 for more details.

On the one hand, since a path infrastructure is not required among controllers, the wireless option is not affected by pin limitations. The communication bandwidth is therefore determined by the data modulation rather than by the number of wires laid down between controllers. Moreover, the inherently broadcast nature of wireless communications eases the implementation of the one-to-many gateway-to-controllers pattern required for command dissemination. On the other hand, such solution has the disadvantage of a higher complexity, as tiny antennas and transceivers need to be developed and cointegrated within the SDM. Fortunately, recent advances in on-chip antennas in mmWave and THz bands, as well as the constant miniaturization of RF transceivers for short-range applications, are leading to multi-Gbps designs with footprints as small as 0.1 mm^2 and 1 pJ/bit [21,28–30].

Existing methods for wireless chip-scale communications are not optimized for the SDM scenario. Hence, smaller area and energy footprints than the ones shown earlier could be a possibility if custom designs trade off performance for cost. Since communication in SDMs is expected to be occasional and much less latency-sensitive than in NoCs, the requirements cast upon the antenna and transceiver can be highly relaxed. Electrically small antennas or, in the extreme nanonetwork case, miniaturized graphene THz antennas (see Chapter 1) can be employed to reduce the footprint of the wireless interfaces. Other unique features such as resilience to inexact tuning of unit cells (i.e., high bit error rates are tolerated) and the monolithic design of the complete system (i.e., new optimization approaches arise) are additional factors that further suggest that unprecedented levels of performance and efficiency can be obtained in intra-SDM wireless networks.

3.5 SUMMARY

The potential of the SDM concept is vast, given the plethora of potential applications in the microwave, THz, and optical ranges. However, their feasibility is currently limited to the development of proof-of-concept devices maintaining a simple architecture and performance. The main reason is that SDMs require the integration of a network of controllers within the metasurface structure, which is very challenging at the time of this writing. We have seen that, in this *networks within networks* context, physical constraints play a definitive role in determining the implementable alternatives. In the wired case, pin limitations force designers to take low-radix topologies and resort to opportunistic approaches to implement adaptive or fault-tolerant networks. The adoption of wireless intra-SDM communication can alleviate this issue as long as antennas and transceivers commensurate with the unit cells are developed, which is again very challenging, especially as SDMs are scaled toward THz operation.

ACKNOWLEDGMENTS

This work was supported in part by the European Unions' Horizon 2020 research and innovation programme Future Emerging Topics (FETOPEN) under grant agreement no. 736876.

REFERENCES

1. N. Engheta and R. W. Ziolkowski. *Metamaterials: Physics and Engineering Applications.* Piscataway, NJ: IEEE Press, 2006.
2. S. B. Glybovski, S. A. Tretyakov, P. A. Belov, Y. S. Kivshar, and C. R. Simovski. Metasurfaces: From microwaves to visible. *Physics Reports*, 634:1–72, 2016.
3. S. Vellucci, A. Monti, M. Barbuto, A. Toscano, and F. Bilotti. Satellite applications of electromagnetic cloaking. *IEEE Transactions on Antennas and Propagation*, 65(9):4931–4934, 2017.
4. H. Yang, X. Cao, F. Yang, J. Gao, S. Xu, M. Li, X. Chen, Y. Zhao, Y. Zheng, and S. Li. A programmable metasurface with dynamic polarization, scattering and focusing control. *Scientific Reports*, 6(35692), 2016.
5. S. W. Qu, H. Yi, B. J. Chen, K. B. Ng, and C. H. Chan. Terahertz reflecting and transmitting metasurfaces. *Proceedings of the IEEE*, 105(6):1166–1184, 2017.
6. Z. Li, K. Yao, F. Xia, S. Shen, J. Tian, and Y. Liu. Graphene plasmonic metasurfaces to steer infrared light. *Scientific Reports*, 5:1–9, 2015.
7. G. Oliveri, D. Werner, and A. Massa. Reconfigurable electromagnetics through metamaterials—A review. *Proceedings of the IEEE*, 103(7):1034–1056, 2015.
8. T. J. Cui, M. Q. Qi, X. Wan, J. Zhao, and Q. Cheng. Coding metamaterials, digital metamaterials and programming metamaterials. *Light: Science & Applications*, 3:e218, 2014.
9. F. Liu et al. Programmable metasurfaces: State of the art and prospects. In *Proceedings of the ISCAS'18*, 2018.
10. S. Abadal, C. Liaskos, A. Tsioliaridou, S. Ioannidis, A. Pitsillides, J. Sole-Pareta, E. Alarcon, and A. Cabellos-Aparicio. Computing and communications for the software-defined metamaterial paradigm: A context analysis. *IEEE Access*, 5:6225–6235, 2017.
11. C. Liaskos, A. Tsioliaridou, A. Pitsillides, I. F. Akyildiz, N. V. Kantartzis, A. X. Lalas, X. Dimitropoulos, S. Ioannidis, M. Kafesaki, and C. M. Soukoulis. Design and development of software defined metamaterials for nanonetworks. *IEEE Circuits and Systems Magazine*, 15(4):12–25, 2015.
12. C. Liaskos, S. Nie, A. Tsioliaridou, A. Pitsillides, S. Ioannidis, and I. Akyildiz. A new wireless communication paradigm through software-controlled metasurfaces. *IEEE Communications Magazine*, 56:162–169, 2018.
13. C. Liaskos, N. Shuai, A. Tsioliaridou, A. Pitsillides, S. Ioannidis, and I. Akyildiz. A novel communication paradigm for high capacity and security via programmable indoor wireless environments in next generation wireless systems. *Ad Hoc Networks*, 87:1–16, 2019.

14. A. Li, S. Singh, and D. Sievenpiper. Metasurfaces and their applications. *Nanophotonics*, 7(6):989–1011, 2018.
15. C. Liaskos et al. Initial UML definition of the HyperSurface programming interface and virtual functions. *European Commission, H2020-FETOPEN-2016-2017*, Project VISORSURF: Accepted Public Deliverable D2.1, 31-Dec-2017, [Online:] http://www.visorsurf.eu/m/VISORSURF-D2.1.pdf, 2017.
16. C. Liaskos et al. Initial UML definition of the HyperSurface compiler middle-ware. *European Commission, H2020-FETOPEN-2016-2017*, Project VISORSURF: Accepted Public Deliverable D2.2, 31-Dec-2017, [Online:] http://www.visorsurf.eu/m/VISORSURF-D2.2.pdf, 2017.
17. L. Zhang et al. Space-time-coding digital metasurfaces. *Nature Communications*, 9(1), 2018.
18. V. Kaajakari et al. Practical mems: *Design of Microsystems, Accelerometers, Gyroscopes, Rf mems, Optical Mems, and Microfluidic Systems*. Las Vegas, NV: Small Gear Publishing, 2009.
19. T. Saeed et al. Fault adaptive routing in metasurface network controllers. In *Proceedings of the NoCArc'18*, 2018.
20. D. Bertozzi, G. Dimitrakopoulos, J. Flich, and S. Sonntag. The fast evolving landscape of on-chip communication. *Design Automation for Embedded Systems*, 19(1):59–76, 2015.
21. A. C. Tasolamprou et al. Intercell wireless communication in software-defined metasurfaces. In *Proceedings of the ISCAS'18*, 2018.
22. T. Bjerregaard and S. Mahadevan. A survey of research and practices of Network-on-chip. *ACM Computing Surveys*, 38(1):1–51, 2006.
23. J. Kim and H. K. H. Kim. Router microarchitecture and scalability of ring topology in on-chip networks. In *Proceedings of the NoCArc'09*, New York: ACM, pp. 5–10, 2009.
24. H. Kwon and T. Krishna. Rethinking NoCs for Spatial Neural Network Accelerators. In *Proceedings of the NoCS'17*, page Art. 19, 2017.
25. S. D. Chawade, M. A. Gaikwad, and R. M. Patrikar. Review of *xy* routing algorithm for network-on-chip architecture. *International Journal of Computer Applications*, 43(21):975–8887, 2012.
26. P. Stroobandt, S. Abadal, W. Tavernier, E. Alarcon, D. Colle, and M. Pickavet. A general, fault-tolerant, adaptive, deadlock-free routing protocol for network-on-chip. In *Proceedings of the NoCArc'18*, 2018.
27. J. Balfour and W. J. Dally. Design tradeoffs for tiled CMP on-chip networks. In *Proceedings of the ICS'06*, page 187, New York, 2006.
28. H. M. Cheema and A. Shamim. The last barrier: On-chip antennas. *IEEE Microwave Magazine*, 14(1):79–91, 2013.
29. O. Markish, B. Sheinman, O. Katz, D. Corcos, and D. Elad. On-chip mmwave antennas and transceivers. In *Proceedings of the NoCS'15*, page Art. 11, 2015.
30. X. Yu, J. Baylon, P. Wettin, D. Heo, P. Pratim Pande, and S. Mirabbasi. Architecture and design of multi-channel millimeter-wave wireless network-on-chip. *IEEE Design & Test*, 31(6):19–28, 2014.

II

Nanoscale, Molecular Networking Communications

CHAPTER 4

Channel Modeling for Nanoscale Communications and Networking

Ke Yang, Rui Zhang, Qammer H. Abbasi, and Akram Alomainy

CONTENTS

4.1 GENERAL CHANNEL CHARACTERIZATION FOR NANO-COMMUNICATION AND NETWORK

4.1.1 Common Description of the Medium Characterization

Before discussing the channel performance, it is important to investigate the medium of the channel. Usually, the electromagnetic parameters, that is, permittivity ε and permeability μ, are used to describe medium in microwave and radiofrequency (RF); while at optical frequency, all information is commonly delivered in terms of refractive index (or index of refraction):

$$\tilde{n}(f) = n_r(f) - jn_i(f) \tag{4.1}$$

The real part n_r can be usually obtained directly from the measurement, while the imaginary part n_i (which would also be referred to as κ, named extinction coefficient) can be calculated from the absorption coefficient α. The relationship between the two can be given by

$$n_i(f) = \frac{\alpha(f)\lambda_o}{4\pi} \tag{4.2}$$

where $\lambda_o = c/f$ is the wavelength in free space.

The relationship between both can be described as

$$\tilde{n}^2 = \varepsilon\mu \tag{4.3}$$

Usually, the relative permeability of the medium can be regarded as 1, because it is non-magnetic; thus, we can obtain the relative permittivity constant ε as

$$\varepsilon(f) = \tilde{n}(f)^2 = n_r(f)^2 - \kappa(f)^2 - j2n_r(f)\kappa(f) \tag{4.4}$$

Thus, we can obtain

$$\begin{cases} \varepsilon' &= n_r(f)^2 - \kappa(f)^2 \\ \varepsilon'' &= 2n_r(f)\kappa(f) \end{cases} \tag{4.5}$$

where, ε' and ε'' are the real and imaginary parts, respectively, of the permittivity ε.

The absorption coefficient of some human tissues, that is, blood, skin, and fat, is shown in Figure 4.1a ([1,2]), while the EM parameters, that is, permittivity of the corresponding tissues, calculated from Equation 4.5, are shown in Figure 4.1a and c.[1]

---

[1] The refractive index n_r is 1.97, 1.73, and 1.58 for blood, skin, and fat, respectively.

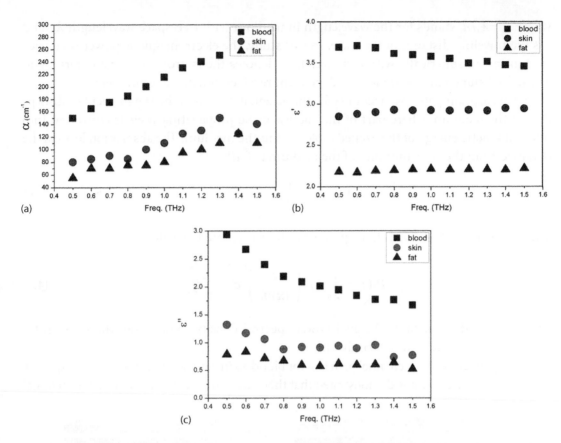

FIGURE 4.1 Optical and electromagnetic parameters of human tissues (blood, skin, and fat). (a) Measured absorption coefficient ©2003 Springer. Reprinted with permission from [1,2], (b) Real part of the relative permittivity (calculated from Equation 4.5), and (c) Imaginary part of the relative permittivity (calculated from Equation 4.5). ©2015 IEEE. (Reprinted with permission from Yang et al., *IEEE Trans. Thz. Sci Techn.*, 5, 419–426, 2015.)

4.1.2 Path Loss

Numerous papers have presented channel models for the terahertz (THz) wave propagating in the atmosphere [3–6], followed by the study of the in vivo nano-communication channel models at the THz [7–9]. The path loss of the THz wave inside human tissues can be divided into two parts: the spread path loss PL_{spr} and the absorption path loss PL_{abs}:

$$PL_{total}[dB] = PL_{spr}(f,r)[dB] + PL_{abs}(f,r)[dB] \qquad (4.6)$$

where f stands for the frequency, while r is the path length.

The spread path loss is introduced by the expansion of the wave in the medium, which is defined as:

$$PL_{spr}(f,r) = \left(\frac{4\pi r}{\lambda_g}\right)^2 = \left(\frac{4\pi n_r f r}{c}\right)^2 \qquad (4.7)$$

where $\lambda_g = \lambda_o/n_r$ stands for the wavelength in medium with free-space wavelength λ_o, and r is the traveling distance of the wave. In this study, the electromagnetic power is considered to spread spherically with distance. $4\pi r^2$ denotes the isotropic expansion term, and $4\pi(n_r f/c)^2$ stands for the frequency-dependent receiver antenna aperture term.

The absorption path loss accounts for the attenuation caused by the molecular absorption of the medium, where part of the energy of the propagating wave is converted into internal kinetic energy of the excited molecules in the medium. The absorption loss can be obtained from the transmittance of the medium $\tau(f, d)$:

$$PL_{abs} = \frac{1}{\tau(f,r)} = e^{\alpha(f)r} \tag{4.8}$$

Thus, the expected received signal power can be represented by [10]

$$P_R(r) = \int_B S(f) \left(\frac{c}{4\pi n f r} \right)^2 e^{-\alpha(f)r} df \tag{4.9}$$

where $S(f)$ is the transmitted signal power spectral density (p.s.d) from the transmitter antenna and B is the channel bandwidth.

The dependency of the channel path loss for blood, skin, and fat on distance and frequency is shown in Figure 4.2. It is demonstrated that there are some fluctuations in each individual

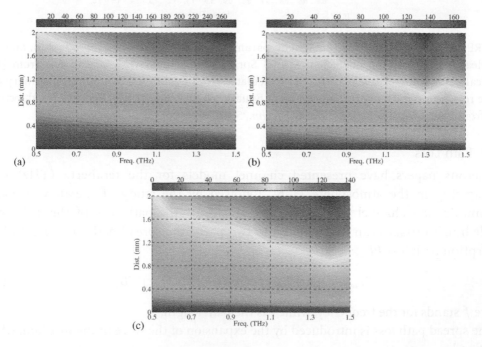

FIGURE 4.2 Total path loss as a function of the distance and frequency for different human tissues. (a) Blood, (b) Skin, and (c) Fat. (From K. Yang, *Characterisation of the In-vivo Terahertz Communication Channel within the Human Body Tissues for Future Nano-Communication Networks*, Queen Mary University of London, 2016. ©2016 IEEE. With permission.)

figure because absorption path loss is related to the extinction coefficient, κ, which is not a monotonous function along the required frequency band, in addition to the expected increase in path loss values with larger distances and higher frequency components. For different tissues, the path loss varies, with blood experiencing the highest losses, followed by the skin, owing to the water concentration, which contributes a significant absorption path loss. At the level of the millimeters, the path loss of the blood is around 120 dB, while that of the skin is around 90 dB and of the fat is around 70 dB. Compared with the channel attenuation of the molecular communication [11], the future of the EM paradigms is promising, because at 1 kHz (here, the frequency is the operation frequency of the RC circuit that depicted the emission and absorption process of the diffusion-based particle communication) and at a distance of 0.05 mm, the molecular channel attenuation is above 140 dB, which is substantially higher than the case for blood at the distance of 1 mm, applying THz EM communication mechanism. In [12], the capacity was also compared between the two paradigms, showing that the EM communication keeps extremely high data rate until the distance is shorter than 10 mm, while the molecular communication scheme provides much lower capabilities.

4.1.3 Noise

In this section, the analysis of the end-to-end (including channel) noise model, including transmission, propagation, and reception of the EM wave, is performed. Considering the complexity of the real human tissues, two assumptions are made here:

- A spherically symmetric propagation environment is assumed, with the receiver at the center of the sphere and the transmitter at the distance r from the receiver.

- The antennas of the transmitter are assumed to be ideal isotropic ones.

4.1.3.1 Molecular Absorption

Molecular absorption is a process in which the EM energy is partially transformed to the kinetic energy internal to vibrating molecules [4], which can be described by the absorption coefficient. Because the vibration frequencies at which a given molecule resonates change with the internal structure of the molecule [4], this quantity depends on the frequency and gives the THz band a unique frequency-selective spectral absorption profile. Given the absorption coefficient, the amount of incident EM radiation that is capable of propagating through the absorptive medium at a given frequency can be calculated. This parameter is defined by transmittance, which is obtained by using the Beer–Lambert law as [13,14]:

$$\tau(r,f) = e^{-\alpha(f)r} \tag{4.10}$$

where f is the frequency of the EM wave, r stands for the total path length, and $\alpha(f)$ is the absorption coefficient.

Molecular absorption causes attenuation to signals, which can be obtained from the transmittance of the medium τ given by (1), when traveling a distance r, as [3,4]:

$$A_{abs}(r,f) = \frac{1}{\tau(r,f)} = e^{\alpha(f)r} \tag{4.11}$$

In this chapter, body-centric nano-networks are focused on human blood, skin, and fat tissues, and their absorption coefficients at the frequency band of interest are shown in Figure 4.1. The details on the calculation of the absorption coefficient for human blood, skin, and fat tissues can be found in [7,9]. Compared with the absorption coefficient in the atmosphere provided in [4], on the one hand, the absorption coefficient in human tissues can be thousands of times that in air at the same frequency. On the other hand, different from the thousand resonant peaks of water vapor over the THz band, the absorption coefficient of human tissues increases with frequency much more steadily. It gives in vivo THz communication some peculiar behaviors, which will be presented later.

4.1.3.2 Noise Model

The noise in the THz band is primarily contributed by the molecular absorption noise. This kind of noise is caused by vibrating molecules, which partially reradiate the energy that has been previously absorbed [4]. Thus, this noise is dependent on the transmitted signal. In [15], the total molecular absorption noise as p.s.d. S_N is written as the summation of the atmospheric noise S_{N0} and the self-induced noise S_{N1}:

$$S_N(r,f) = S_{N0}(r,f) + S_{N1}(r,f) \tag{4.12}$$

$$S_{N0}(f) = \lim_{r \to \infty} k_B T_0 \left(1 - e^{-\alpha(f)r}\right) \left(\frac{c}{\sqrt{4\pi} f_0}\right)^2 \tag{4.13}$$

$$S_{N1}(r,f) = S(f) \left(1 - e^{-\alpha(f)r}\right) \left(\frac{c}{4\pi f r}\right)^2 \tag{4.14}$$

where r refers to the propagation distance, f stands for the frequency of the EM wave, k_B is the Boltzmann constant, T_0 is the reference temperature of the medium, $\alpha(f)$ is the absorption coefficient, c is the speed of light in vacuum, f_0 is the design center frequency, and S is the p.s.d of the transmitted signal.

The atmospheric noise is caused by the temperature of the absorbing atmosphere, making the atmosphere (or any medium) an effective black-body radiator in homogeneously absorbing medium (in the frequency domain) [16]. This atmospheric noise is therefore known as a background noise, which is independent of the transmitted signal. However, the noise model in Equation 4.13 is only a special case for THz communication in air. Without loss of generality, the term $k_B T_0$ in Equation 4.13 should be replaced with the Planck's function, since it is a general radiative function of the surface of the black body [16]. Consequently, it is believed that the molecular absorption noise should be contributed by the background noise $S_{Nb}(r,f)$, the self-induced noise $S_{Ns}(r,f)$, and other noise $S_{No}(r,f)$:

$$S_N(r,f) = S_{Nb}(r,f) + S_{Ns}(r,f) + S_{No}(r,f) \tag{4.15}$$

4.1.3.2.1 Background Noise   The background noise caused by the radiation of the medium can be described by the Planck function [17]:

$$B(T_0, f) = \frac{2h\pi f^3}{c^2} \left(e^{\frac{hf}{k_B T_0}} - 1 \right)^{-1}$$

(4.16)

where k_B is the Boltzmann constant and h is the Planck constant. The Planck function is multiplied with π to transform the unit from $W/Hz/cm^2/sr$ to $W/Hz/cm^2$.

For simplicity, the transmission medium is assumed to be an isothermal and a homogeneous layer with the thickness r. As mentioned previously, this background noise is generated by the radiation of the local sources of the medium, and it is assumed that this radiation is only from the original energy state of the molecules before transmission happens; thus, it is independent of the transmitted signal. The background noise can be described as [17]:

$$S_{Nb}(f) = \int_0^r B(T_0, f)\alpha(f)e^{-\alpha(f)s}ds$$
$$= B(T_0, f)(1 - e^{-\alpha(f)r}) \simeq B(T_0, f)$$

(4.17)

The integral in Equation 4.17 describes the noise intensity at the center of a sphere with a radius r, given that all the points s in the medium contribute to the noise intensity. Since this is obtained by the Planck function, the unit of the background noise p.s.d is $W/Hz/cm^2$. The background noise can be further approximated by taking into account the (ideal) antenna aperture term $c^2/(4\pi f_0)^2$ to get the background noise p.s.d with the unit W/Hz [16]:

$$S_{Nb}(f) = B(T_0, f)\frac{c^2}{4\pi f_0^2}$$

(4.18)

4.1.3.2.2 Self-induced Noise   In terms of the induction mechanism of the self-induced noise, the internal vibration of the molecules turns into the emission of EM radiation at the same frequency of the incident waves that provoked this motion [3,4]. It is obtained with the assumption that all the absorbed energy from the transmitted signal received at the receiver would turn into molecular absorption noise:

$$S_{Ns}(r, f) = S(f)\left(1 - e^{-\alpha(f)r}\right)\left(\frac{c}{4\pi fr}\right)^2$$

(4.19)

where $(4\pi rf/c)^2$ accounts for the spreading loss.

4.1.3.2.3 Other Noises   In addition to the molecular absorption noise, there are other noise sources that can affect the communication performance, such as the device noise.

A number of prototypes of antennas emitting at the THz band available today are built with conventional materials. In this case, the Johnson–Nyquist thermal noise should be taken into account as a source of noise [18]. However, with the development of new materials, it is possible that the thermal noise can be neglected with the use of graphene and its derivatives to create THz antennas. Graphene-based nanostructures allow the ballistic transport of electrons, leading to very low thermal noise in the device; thus, it is reasonably expected that the noise is minor [19,20]. Therefore, the molecular absorption noise in the transmission is the dominant contributor to noise at the receiver.

With regard to the in vivo scenario, the speed of light in the human body should change with the composition of the medium and the frequency of the THz wave. Therefore, c is replaced with c/n in Equations 4.18 and 4.19, which results in that the total molecular absorption noise p.s.d. S_N is contributed by the background noise S_{Nb} and the self-induced noise S_{Ns} and can be represented as:

$$S_N(r, f) = S_{Nb}(r) + S_{Ns}(r, f) \tag{4.20}$$

$$S_{Nb}(f) = B(T_0, f) \left(\frac{c}{\sqrt{4\pi n_0 f_0}} \right)^2 \tag{4.21}$$

$$S_{Ns}(r, f) = S(f) \left(1 - e^{-\alpha(f)r} \right) \left(\frac{c}{4\pi nrf} \right)^2 \tag{4.22}$$

where n is the corresponding refractive index of the THz wave in the medium, when the frequency is f, f_0 is the design center frequency, and n_0 is the corresponding refractive index.

It is clearly shown that the background noise depends on the temperature and composition of the medium in Equation 4.22. It is assumed that the human tissues are isothermal; thus, the background noise changes slightly with the refractive index in different transmission mediums. Moreover, the self-induced noise is dependent on the transmitted signal, and for simplicity in this section, only transmitted signal with flat p.s.d over the entire frequency is considered to comparatively illustrate the difference between these two kinds of noise. To keep the numerical results realistic, and in light of the state of the art in nano-transceivers, the flat power is adopted with the total energy equal to 1 pJ and the pulse duration equal to 100 fs [21,22]. The background noise and self-induced noise p.s.d for human blood, skin, and fat tissues are shown in Figures 4.3 and 4.4, respectively.

From Figure 4.3, it can be clearly seen that the background noise p.s.d is almost the same in different kinds of tissue, because the slight difference of refractive index does not play a significant role in Equation 4.22. Besides, when sharing the same transmitted signal power, the self-induced noise slightly increases from blood to fat, because the absorption coefficient and the refractive index increase with the water concentration in the medium, and comparatively, blood has higher water proportion than skin and fat. More importantly, it can be seen from Figure 4.4 that the noise p.s.d has a steady change with frequency,

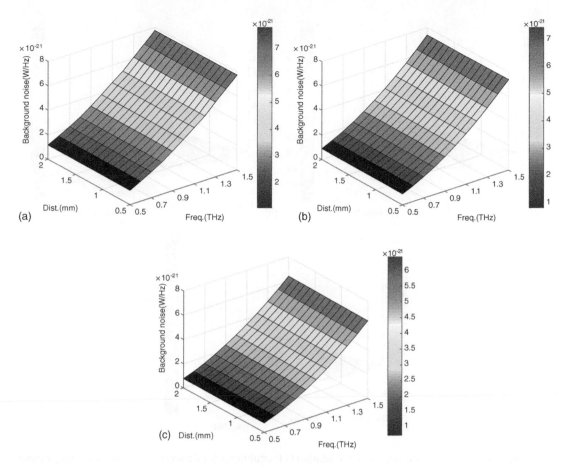

FIGURE 4.3   Background noise p.s.d at terahertz frequencies for different human tissues: (a) Blood, (b) skin, and (c) fat.

which is different from the abrupt fluctuation of THz communication in air, as shown in [4]. The reason is that the molecular absorption coefficient has a steady increase over the frequency of interest. Furthermore, the self-induced noise p.s.d decreases with distance. Because the self-induced noise is directly proportional to the transmitted signal, and the signal is inversely proportional to the transmission distance. Thus, the self-induced noise is inversely proportional to the distance.

Moreover, comparing Figure 4.3 with Figure 4.4, the self-induced noise p.s.d is about seven orders of magnitude higher than the background noise in all these different human tissues. The main reason is that the transmitted pulse energy is chosen to be high enough for better information transmission. In this case, it can be concluded that the self-induced noise is the dominant noise source and the background noise is negligible in the THz band for in vivo nano-networks.

The molecular absorption noise p.s.d for these considered tissues is illustrated in Figure 4.5, which is almost the same with the self-induced noise p.s.d. This highlights the fact that the background noise can be discarded in our analysis.

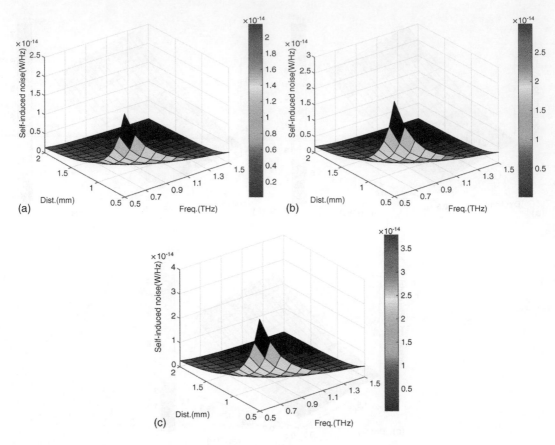

FIGURE 4.4   Self-induced noise p.s.d at terahertz frequencies for different human tissues: (a) Blood, (b) skin, and (c) fat.

4.2 TIME SPREAD ON-OFF KEYING FOR IN VIVO NANO-NETWORKS IN THE THz BAND

Owing to the fact that nano-devices in wireless nano-sensor networks (WNSNs) are highly energy constrained with limited capabilities, it is technologically challenging for a nano-transceiver to generate a high-power carrier frequency in the THz band. Thus, the best modulation option for WNSNs is carrier-less pulse-based modulation [23]. In light of the state of the art in graphene-based nano-electronics, a transmission scheme for nano-devices, based on the transmission of 100-femtosecond-long pulses by following an on-off keying modulation spread in time, is proposed [24], named time spread on-off keying (TS-OOK). These very short pulses can be generated and detected with nano-transceivers based on graphene and high-electron-mobility materials such as gallium nitride and indium phosphide [21].

TS-OOK is a communication scheme assuming that a nano-machine needs to transmit a binary stream. A logical "1" is transmitted by using a femtosecond-long pulse, and a logical "0" is transmitted as silence. The time between symbols T_s is much longer than the symbol duration T_p. So far, TS-OOK is the most promising communication scheme for resource-constrained nano-networks.

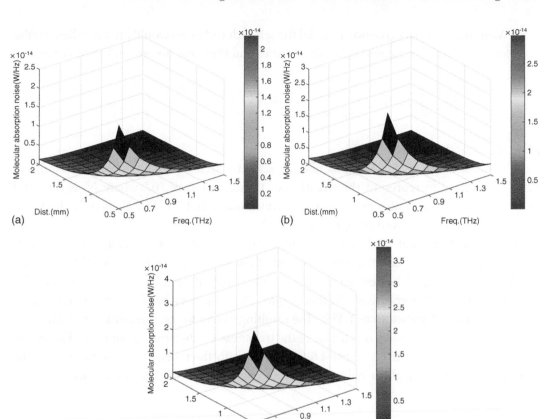

FIGURE 4.5 Molecular absorption noise p.s.d at terahertz frequencies for different human tissues: (a) Blood, (b) skin, and (c) fat.

When adopting TS-OOK as the communication scheme for nano-networks, the nano-transmitters can start transmitting at any time, without being synchronized or controlled by any type of network central entity. Hence, the traffic in the network can drastically increase at specific times, owing to correlated detection in several nano-sensors, and collisions between symbols can occur [24]. In order to proceed with the system-level investigation of in vivo nano-networks, it is significantly important to confirm the feasibility of TS-OOK for nano-communication inside the human body.

To investigate TS-OOK communication scheme inside human tissues, a comparative illustration of the results for both the received signal power and the noise power is presented. The received signal power is denoted as P_1, as shown in Figure 4.9.

Meanwhile, two scenarios of the noise characteristics should be considered here:

- Only the background noise shows up by transmitting silence for a logical "0", and the noise power can be obtained by integration of Equation 4.21 over the bandwidth as:

$$N_0 = \int_B B(T_0, f) \frac{c^2}{4\pi (n_0 f_0)^2} df \qquad (4.23)$$

- While both the background noise and the self-induced noise would present when a pulse for a logical "1" is transmitted in the channel, and the noise power can be written as:

$$N_1 = \int_B \left(B(T_0,f)\frac{c^2}{4\pi(n_0 f_0)^2} + S(f)\left(1-e^{-\alpha(f)r}\right)\left(\frac{c}{4\pi rnf}\right)^2 \right) df \qquad (4.24)$$

where B is the channel bandwidth.

To summarize, N_1 is used to describe the scenario that a pulse has been transmitted, while N_0 is used to represent the situation of transmission of "0," where N_1 is the total molecular absorption noise power derived from Equation 4.20, and N_0 is only the background noise power derived from Equation 4.21.

To keep the numerical results realistic, and in light of the state of the art in molecular electronics, the total energy and the pulse duration of the Gaussian power are equal to 1 *pJ* and 100 *fs*, respectively [4]. The derivative order and the standard deviation of the Gaussian pulse are set to 6 and 0.15, respectively. In this chapter, the frequency band of interest is 0.5 THz to 1.5 THz, with a bandwidth of 1 THz. The resulting power for communication inside three human tissues is represented as function of the transmission distance, as shown in Figure 4.6.

It can be clearly seen from Figure 4.6, compared with the received signal power, that the noise power is big enough to influence the link quality. According to the results presented

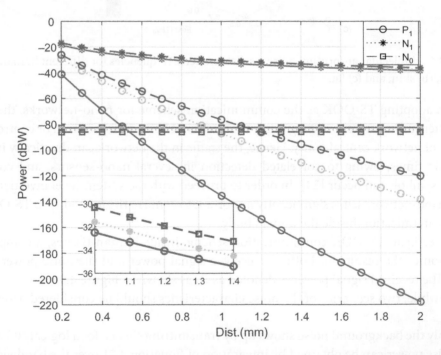

FIGURE 4.6 Power versus path length for three different human tissues, blood (solid line), skin (dotted line) and fat (dashed line). P_1 (noise power when pulse is transmitted), N_1 (noise power associated with a pulse) and N_0 (noise power associated with silence). The graph in the insets shows magnifications of N_1 with the distance from 1 to 1.4 mm.

in [25], the received power when transmitting a pulse tends to be lower than the noise power when the transmission distance increases to 1 cm. For the case within in vivo nano-networks at the THz band, the results are evidently different. The received signal power P_1 when a pulse is transmitted inside human tissues is always much lower than the molecular absorption noise power N_1, regardless of the type of the transmission medium. This difference could be explained by the fact that the molecular absorption coefficient inside human tissues is thousand times of that in air [4,7], which causes a much more significant attenuation to the signal. In addition, it is noticed that the noise power associated with the transmission of silence is constant with distance, because only the background noise is present in this case and the background noise is independent of the signal transmission. More importantly, even if the power of the received signal tends to zero, the noise power caused by the propagating pulse keeps much higher than the background noise power. In other words, when a pulse is transmitted, the received power, including both the targeted signal power P_1 and the molecular absorption, noise power N_1 is always much higher than the power of the silence case, whatever human tissues are considered. Thus, it is clear for the receiver that a pulse is transmitted, provided that the molecular absorption noise power is detected, which means that the logical "1" and logical "0" can be clearly detected, which reduce the error probability for in vivo nano-networks. The obtained results demonstrate the feasibility of TS-OOK as a communication scheme for the THz communication inside the human body. Moreover, since the received signal power is directly proportional to the transmitted signal power, reasonable transmission power needs to be chosen to keep the power at the receiver much higher with transmitting pulses than the power with transmitting silence, to make the signal detection more accurate.

4.3 END-TO-END SINGLE-USER SCENARIO CHANNEL MODELING FOR IN VIVO NANO-COMMUNICATION

In this section, the scenario of the end-to-end single user case would be described in detail based on the previous channel model. Different power allocations schemes are considered to obtain SNR and the information rate of the single used cases for the nano-communication inside human tissues and the properties of TS-OOK communication scheme.

4.3.1 Power Allocation Scheme

For the general communication systems, besides the effect of both path loss and noise, communication capabilities are strictly influenced by the distribution of power transmission P_T in the frequency domain [26]. As discussed in Section 1.3.2, it is clear that the molecular absorption noise is directly related to the transmitted signal, and when the transmitted signal is high enough, the self-induced noise could be the only noise source in the channel; thus, in this section, two communication schemes (namely, flat and pulse-based) are considered to investigate the signal-to-noise ratio (SNR) of the in vivo communication at the THz band. In this section, the scenario of the end-to-end single user case would be described in detail based on the previous channel model. Different power allocations schemes are considered to obtain SNR and the information rate of the single used cases for the nano-communication inside human tissues and the properties of TS-OOK communication scheme.

Flat communication: In the simplest case, the total transmitted signal power P_T is uniformly distributed over the entire operative band (0.5–1.5 THz). Thus, the corresponding transmitted signal p.s.d is:

$$S_{flat}(f) = P_T / B \text{ for } f \in B, 0 \text{ otherwise} \tag{4.25}$$

Pulse-based communication: The transmitted signal can be modeled with an nth derivative of a Gaussian shape: $\phi(f) = (2\pi f)^2 \, ne^{(-2\pi\sigma f)^2}$ [27]. Thus, the signal p.s.d can be expressed as [4],

$$S_P^{(n)}(f) = a_0^2 \phi(f) \tag{4.26}$$

where σ and a_0^2 are the standard deviation of Gaussian pulse and a normalizing constant, respectively. Considering that $\int_{f_m}^{f_M} S_P^{(n)}(f) df = P_T$, the normalizing constant is obtained as [27]:

$$a_0^2 = \frac{P_T}{\int_{f_m}^{f_M} \phi(f)} \tag{4.27}$$

In the subsequent analysis, the total energy and the pulse duration of the flat power are equal to 1 *pJ* and 100 *fs*, respectively. Meanwhile, for the Gaussian pulse-based transmission scheme, the derivative order n and the standard deviation of the Gaussian pulse σ are set to 4 and 0.15, respectively.

4.3.2 Signal-to-Noise Ratio

In order to analyze the communication performance, especially the achievable communication range, SNR of the in vivo communication channel is investigated, which can be written as a function of the transmission distance and frequency:

$$SNR(r, f) = \frac{S(f)}{S_N(f,r) PL(f,r)} \tag{4.28}$$

where S stands for the p.s.d of the transmitted signal, PL denotes the channel path loss, and S_N refers to the molecular absorption noise p.s.d.

In our analysis, because the pulse energy is high enough to enable the background noise to be neglected and the self-induced noise is the dominant noise source for in vivo nano-networks and the self-induced noise is directly proportional to the transmitted signal, with the simplification of Equation 4.28, it can be easily concluded that SNR is independent of the transmitted signal under this circumstance.

For the considered human tissue types, that is, blood, skin, and fat, the two communication schemes (flat and Gaussian-shaped pluses) result in similar SNR values, as shown in Figure 4.7. It allows us to conclude that the SNR can be independent of the transmitted signal power, when high pulse energy is transmitted, which enables the self-induced noise to be the dominant noise source for the in vivo nano-networks at the THz band.

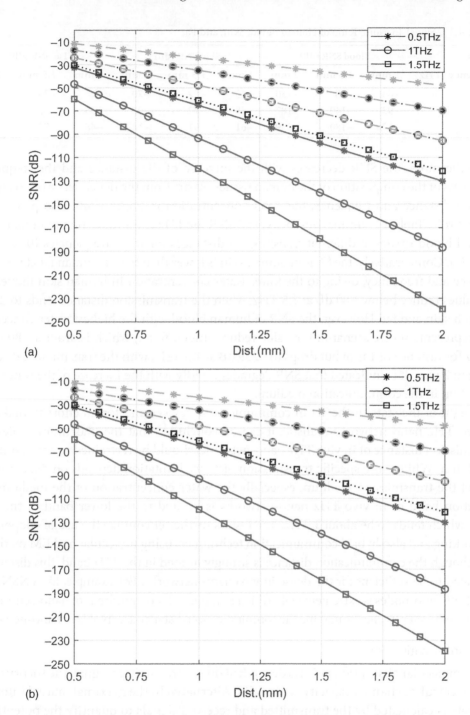

FIGURE 4.7 Channel SNR as a function of transmission distance for different power allocation schemes in human blood (solid line), skin (dotted line) and fat (dashed line) tissues. (a) Flat pulse and (b) Gaussian pulse.

TABLE 4.1　Signal-to-Noise Ratio of Human Blood, Skin, and Fat

Tissues	Blood SNR (dB)			Skin SNR (dB)			Fat SNR (dB)		
Frequency (THz)	1 mm	1.5 mm	2 mm	1 mm	1.5 mm	2 mm	1 mm	1.5 mm	2 mm
0.5	−65	−98	−130	−35	−52	−70	−24	−36	−48
1	−93	−140	−187	−48	−72	−96	−35	−52	−70
0.5	−120	−180	−239	−61	−91	−122	−48	−72	−96

Furthermore, the SNR decreases with the increase of the distance and the frequency, regardless of the composition of the human tissues. Apart from the dependence on the transmission distance and frequency, the water concentration in the medium also plays a significant role. Table 4.1 summarizes the typical SNR for THz communication in human skin, fat, and blood tissues at different transmission distances for three frequencies (0.5, 1, and 1.5 THz). Comparatively, the SNR in human skin is lower than that in human fat at the same distance and frequency, owing to the lower water concentration in human skin tissue, and the values reduce below −80 dB at 1.5 THz, when the transmission distance tends to 2 mm for both skin and fat. However, the SNR of human blood with the highest water concentration represents severe scenario: the value reduces much faster, which is around −80 dB at 1 THz for distance of 1 mm but drops to −239 dB at 1.5 THz, with the transmission distance going to 2 mm. It is indicated that SNR degrades rapidly with the increase of the water concentration in the communication medium.

In light of the state of the art in communication devices and for an effort to make the in vivo THz nano-networks realistic, it can be concluded that the maximum achievable transmission distance of in vivo THz nano-networks should be restrained to approximately 1 to 2 mm. While more specific transmission distance limitation depends on the composition of the transmission medium, especially the water concentration of the medium, the operation band of in vivo THz nano-networks is limited to the lower band of the THz band, which tends to be about 1 THz. The results further encourage the use of cooperative networking and also hybrid communication techniques, using molecular and EM methods.

Although the communication distance is strongly limited in the THz band, this distance is estimated to be sufficient for the dense in vivo nano-networks. For example, in WNSNs, the density of nano-nodes is extremely high, which is in the order of hundreds of nano-sensors per square millimeter [6], hence making the communication distance acceptable in nano-networks.

4.3.3 Information Rate

As the molecular absorption noise is dependent on the transmitted signal, the theory using SNR to calculate channel capacity is not valid. Alternatively, the maximal mutual information rate is calculated by the transmitted and received signals to quantify the potential of THz band for communication inside the human body with the use of TS-OOK communication paradigm, stated in Section 4.2.

The maximum achievable information rate in bit/symbol of a communication system for a specific modulation scheme is given by the well-known Shannon Limit Theorem [28]:

$$IR = \max_X \{H(X) - H(X \mid Y)\} [\text{bit} / \text{symbol}] \tag{4.29}$$

where X refers to the source of information, Y stands for the output of the channel, $H(X)$ refers to the entropy of the source X, and $H(X|Y)$ stands for the conditional entropy of X given Y or the equivocation of the channel. When using TS-OOK, the source X can be modeled as a discrete binary random variable. Therefore, the entropy of the source $H(X)$ is given by [28]:

$$H(X) = -\sum_{m=0}^{1} p_X(x_m) \log_2 p_X(x_m) \tag{4.30}$$

where $p_X(x_m)$ refers to the probability of transmitting the symbol $m = 0,1$, that is, the probability to stay silent or to transmit a pulse, respectively.

Since the molecular absorption noise can be modeled as additive colored Gaussian noise (ACGN) [29], the probability density function (p.d.f) N of the molecular absorption noise at the receiver conditioned to the transmission of symbol x_m is given by [25]:

$$f_N(n|X = x_m) = \frac{1}{\sqrt{2\pi N_m}} e^{-\frac{n^2}{2N_m}} \tag{4.31}$$

where n refers to noise and N_m refers to the molecular absorption noise power when symbol m is transmitted in Equations 4.23 and 4.24.

When considering a 1-bit hard receiver based on power detection, the system becomes a binary asymmetric channel (BAC) and Y is a discrete random variable. This channel can be fully characterized by the four transition probabilities [25]:

$$\begin{aligned}
p_Y(Y = 0 | X = 0) &= \int_{th1}^{th2} f_Y(y | X = 0) dy \\
p_Y(Y = 1 | X = 0) &= 1 - p_Y(Y = 0 | X = 0) \\
p_Y(Y = 0 | X = 1) &= \int_{th1}^{th2} f_Y(y | X = 0) dy \\
p_Y(Y = 1 | X = 1) &= 1 - p_Y(Y = 0 | X = 1)
\end{aligned} \tag{4.32}$$

where th_1 and th_2 are two threshold values and $f_Y(y | X = x)$ is the p.d.f of the channel output Y conditioned to the transmission of the symbol $X = x$, which is given by [25]:

$$f_Y(y | X = x_m) = \delta(y - a_m) * f_N(n = y | X = x_m) = \frac{1}{\sqrt{2\pi N_m}} e^{-\frac{(n - a_m)^2}{2N_m}} \tag{4.33}$$

where δ stands for the Dirac delta function, and a_m refers to the received symbol amplitude, obtained from Equation 4.9.

Contrary to the classical symmetric additive Gaussian noise channel, in the asymmetric channel, there are two points at which $f_Y(y | X = 0)$ and $f_Y(y | X = 1)$ intersect.

These thresholds are considered to be defined for the case without interference [25]. Therefore, th_1 and th_2 can be analytically computed from the intersection between two Gaussian distribution $N(0, N_0)$ and $N(0, N_1)$, respectively, which results in [25]:

$$th_{1,2} = \frac{a_1 N_0}{N_0 - N_1} \pm \frac{\sqrt{2 N_0 N_1^2 \log \dfrac{N_1}{N_0} - 2 N_0^2 N_1 \log \dfrac{N_1}{N_0} + a_1^2 N_0 N_1}}{N_0 - N_1} \tag{4.34}$$

The equivocation of the channel $H(X|Y)$ for the BAC is given by [25]:

$$H_{BAC}(X|Y) = \sum_{y=0}^{1} \sum_{x=0}^{1} p_Y(Y = y | X = x) p_X(X = x)$$

$$log_2 \left(\frac{\displaystyle\sum_{q=0}^{1} p_Y(Y = y | X = q) p_X(X = q)}{p_Y(Y = y | X = x) p_X(X = x)} \right) \tag{4.35}$$

Finally, the maximum achievable information rate in bit/second is obtained by multiplying the rate in bit/symbol Equation 4.29 by the rate at which symbols are transmitted, $R = 1/T_s = 1/(\beta T_p)$, where T_s is the time between symbols, T_p is the pulse length, and β is the ratio between them. If we assume that the $BT_p; 1$, where B stands for the channel bandwidth, the rate in bit/second is given by [25]:

$$IR_u = \frac{B}{\beta} IR_{u_{sym}} \tag{4.36}$$

If $\beta = 1$, that is, all the symbols (pulses or silences) are transmitted in a burst, and the maximum rate per nano-device is achieved, the incoming information rate and the read-out rate to and from the nano-transceiver can match the channel rate. By increasing β, the single-user rate is reduced, but the requirements on the transceiver are greatly relaxed.

In this chapter, the bandwidth of the THz communication channel is considered to be 1 THz, and the information rate for three considered human tissues has been studied, and the results are shown in Figure 4.8. It can be clearly seen that the information rate tends to be 1 terabits per second (Tbps) for communication in three kinds of human tissue, when the communication distance is 0.5 mm. The information rate decreases steadily with the transmission distance, regardless of the medium type, and reaches about 0.96 Tbps when the transmission distance is further increased to 2 mm. Differently, the information rate presents the best case in human fat, because it has the least water concentration, compared with human blood and skin. The main reason for such high information rate is the extremely high bandwidth for the THz communication. The obtained information rate indicates that complex tasks can be completed in the envisioned nano-communication inside the human body by the high ability of successfully transmitting information over the communication channel.

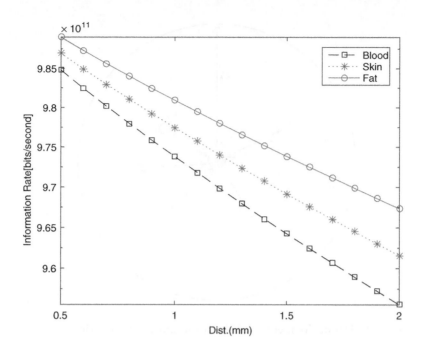

FIGURE 4.8 Information rate as a function of transmission distance for different human tissues.

4.4 END-TO-END MULTIUSER SCENARIO CHANNEL MODELING FOR IN VIVO NANO-COMMUNICATION

In this section, an interference model that is valid to any kind of power allocation scheme is introduced. Besides, the probability distribution and mean values of SINR are derived for THz communication inside the human body with the presence of multiple interferer in dense nano-networks. The investigation captures the unique characteristics of the THz band channel inside human tissues and the properties of TS-OOK communication scheme.

4.4.1 System Model

Generally, a random nodes deployment in R^2, as shown in Figure 4.9, is used to perform the assessment of cellular, ad hoc, and device-to-device networks [30]. The targeted receiver is assumed to be at the center of the disc, while the targeted transmitter locates at a distance r_0 from the receiver. All the other nodes in this field are considered as interfering nodes for the targeted receiver. Following most studies, Poisson point process (PPP) is utilized to provide first-order approximation of nodes positions within a disc of radius R [31]. Thus, the probability of finding M nodes in the area $A(R)$ can be represented as [32]:

$$P[M \mid A(R)] = \frac{(\lambda \pi R^2)^M}{M!} e^{-\lambda \pi R^2} \qquad (4.37)$$

where λ refers to the node density in $nodes/m^2$. The mean and variance of the number of interferer M can be written as [32]:

$$E[M] = \lambda \pi R^2, \sigma^2[M] = \lambda \pi R^2 \qquad (4.38)$$

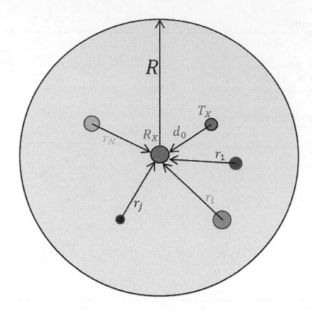

FIGURE 4.9   Nano-devices deployment inside body for multiuser scenario.

4.4.2 Moments and Distribution of SINR

Signal to Interference plus Noise Ratio (SINR) is an important metric to evaluate the system performance of the dense nano-networks. The instantaneous frequency-dependent SINR is defined as:

$$SINR(\vec{r}, S, f, \lambda) = \frac{P_R(r_0, S, f)}{I(\vec{r}, S, f, \lambda) + N(\vec{r}, S, f, \lambda)} \tag{4.39}$$

where $\vec{r}$ is the vector of distances $r_i, i = 1, 2, ..., M$, standing for the separation distances between the interferer and the targeted receiver; S is the p.s.d of the transmitted signal power; and M is the number of interfering nodes. f is the operating frequency, and λ is the intensity of interferer. $P_R(r_0, S, f)$ refers to the targeted receiver signal power, while $I(\vec{r}, S, f, \lambda)$ denotes the aggregate power of the interfering signals at the targeted receiver. $N(\vec{r}, S, f, \lambda)$ is the noise power at the targeted receiver, including all the noise power caused by the targeted transmitter and interferer. In this chapter, it is assumed that there is no power control capabilities, which means that $S_i = S_j = S, i, j = 0, 1, ..., M$ [33,34]. For simplicity, in the following, we drop arguments of notation that are often silently assumed, f, S, and λ. The aggregate interference from M sources is given by [33]:

$$I(\vec{r}) = \sum_{i=1}^{M} \int_B S(f) \left(\frac{c}{4\pi n f} \right)^2 r_i^{-2} e^{-\alpha(f) r_i} df \tag{4.40}$$

where $S(f)$ is the transmitted signal p.s.d from the transmitter antenna. The noise power is written as a summation of the noise caused by both the targeted transmitter and the interfering nodes in the communication area:

$$N(\vec{r}) = N_m + \sum_{i=1}^{M} \int_B S(f) \left(\frac{c}{4\pi nf} \right)^2 r_i^{-2} \left(1 - e^{-\alpha(f) r_i} \right) df \qquad (4.41)$$

where N_m is the molecular absorption noise power obtained from Equation 4.20. Substituting Equations 4.40 and 4.41 into Equation 4.39 gives:

$$SINR(\vec{r}) = \frac{P_R}{N_m + \sum_{i=1}^{M} \int_B S(f) \left(\frac{c}{4\pi nf} \right)^2 r_i^{-2} df} \qquad (4.42)$$

Whereas in Equation 4.42, P_R is a constant value that can be estimated for any given distance r_0. The second term in the denominator of Equation 4.42 is the only random term, which is given as:

$$X(\vec{r}) = \sum_{i=1}^{M} A r_i^{-2}, A = \int_B S(f) \left(\frac{c}{4\pi nf} \right)^2 df \qquad (4.43)$$

where X is caused by the presence of interferer in the communication medium, which includes both the received signal power generated by the interferer nodes and the noise power caused by the interferer signal.

Another important observation is that the distances from any interferer to the targeted receiver are independent and identically distributed (IID). For a sufficiently large number of users, the central limit theorem can be invoked and the Gaussian assumption can be made for X, when estimating the aggregate interference [24]. Therefore, the moments of the interference from a single node are first determined, and then, the central limit theorem is applied to approximate the aggregated interference.

With respect to the dense nano-networks with the adoption of TS-OOK as the communication scheme, a collision between symbols will occur when they reach the receiver at the same time and overlap. The probability of having an arrival during T_s seconds is a uniform random probability distribution, with probability density function (pdf) equal to $1 / T_s$ [24]. Therefore, for a specific transmission, a collision will happen with a probability $2T_p / T_s$ (with an assumption that a correlation-based energy detector is used at the receiver) [24].

It is noted that not all types of symbols harmfully collide; only pulses (logical "1"s) create interference, because the molecular absorption noise is signal power-dependent. It is assumed that all nano-nodes in the transmission area share the same pulse transmitting probability. Therefore, the node density parameter λ in Equation 4.37 can be replaced by [25]:

$$\lambda' = \lambda_T (2T_p / T_s) p_1 \qquad (4.44)$$

where λ_T refers to the density of active nodes in nodes/m^2, T_p is the symbol duration, T_s is the time between symbols, and p_1 denotes the probability of a nano-machine to transmit a pulse. This expression highlights the fact that transmission of logical "0" does not generate interference to other ongoing transmissions. Both the interference caused by the interfering nodes in the transmission area and the interference generated by the utilization of TS-OOK are taken into account in Equation 4.44.

4.4.2.1 Single Node Interference Model

For a PPP, a given number of interferers in a disc of radius R are independently and uniformly distributed. Therefore, the distance to the targeted receiver, denoted as a random variable D, has the same pdf for any interferer node:

$$f_D(r) = 2r / R^2, 0 < r < R \tag{4.45}$$

Now, consider a random variable $G = 1/D^2$. Under Poisson assumption, the moments of G can be written as [35]:

$$E[G^\theta] = \int_0^R \frac{2x}{R^2} \left(\frac{1}{x^2}\right)^\theta dx \tag{4.46}$$

The integral does not converge because it is unbounded, approaching zero from the right. To deal with this issue, it is assumed that the transmitters cannot be located closer than a certain very small distance a from the receiver. This assumption is warranted from the practical point of view, especially taking into account that a can be chosen as small as required [34]. Thus, the distribution of the distance to the targeted receiver could be approximated as the distance from a point arbitrarily distributed in the region bounded by two concentric circles of radius a and R, $R > a$, to their common center. It is known to be [33]:

$$f_D(r) = 2r / (R^2 - a^2), a < r < R \tag{4.47}$$

The first moment of variable G is computed as:

$$E[G] = \int_a^R \frac{2x}{(R^2 - a^2)} \frac{1}{x^2} dx = \frac{2(\ln R - \ln a)}{R^2 - a^2} \tag{4.48}$$

Similarly, the variance of G is calculated to be:

$$\sigma^2[G] = \frac{1}{a^2 R^2} - 4 \left(\frac{\ln R - \ln a}{R^2 - a^2}\right)^2 \tag{4.49}$$

The interference model for a single node has been obtained, and then, it is moved to the aggregate interference. The stochastic sum of random variables G is considered in Equation 4.43.

4.4.2.2 Aggregate Interference Model

It is assumed that the number of interferers is exactly k, which results in conditional moment of X to be [33]:

$$E[X(\vec{r}) \mid M = k] = A \sum_{i=1}^{k} E[G_i] = AkE[G] \tag{4.50}$$

Denoting $P_r(M = k) = p_k$ and unconditioning Equation 4.50 gives:

$$E[X(\vec{r})] = A \sum_{k=0}^{\infty} p_k kE[G] = AE[G]E[M] \tag{4.51}$$

Similarly, the second conditional moment of X is given by:

$$E[X^2(\vec{r}) \mid M = k] = A^2 E\left[\left(\sum_{i=1}^{k} r_i^{-2}\right)^2\right]$$

$$= A^2 \sum_{i=1}^{k} \sum_{j=1}^{k} (E[G_i]E[G_j] + K_{ij}) \tag{4.52}$$

where $K_{ij} = Cov(G_i, G_j)$ is the pairwise covariance. Since G_i and G_j are pairwise independent, $K_{ij} = 0, i = 1, 2, k, ij$ and $K_{ii} = \sigma^2[G_i], i = 1, 2, ..., k$ [25]. Further, since all G_i are identically distributed,

$$E[G_i] = E[G_j], \sigma^2[G_i] = \sigma^2[G_j]. \tag{4.53}$$

Thus, after unconditioning Equation 4.52,

$$E[X^2(\vec{r})] = A^2((E[G])^2 E[M^2] + \sigma^2[G]E[M]). \tag{4.54}$$

Then, the variance of X can be found as:

$$\sigma^2[X(\vec{r})] = A^2((E[G])^2 \sigma^2[M] + \sigma^2[G]E[M]) \tag{4.55}$$

In Equation 4.42, the moments of random variable X are being calculated; thus, pdf of SINR in dB can be obtained based on conventional methods of finding distributions of functions of random variables. Recall that pdf of a random Y, $w(f)$, expressed as monotonous function $y = \phi(x)$ of another random variable X, with pdf $f(x)$, is given by [36]:

$$w(y) = f(\psi(y))|\psi'(y)| \tag{4.56}$$

where $x = \psi(y) = \phi^{-1}(x)$ is the inverse function.

The inverse of $y = \phi(x) = 10 log_{10}(P_R / (N_m + x))$ is unique and monotonous and given by $x = \psi(y) = P_R 10^{-y/10} - N_m$. The modulo of the derivative is $|\psi'(y)| = |P_R 10^{-y/10} ln(10^{-1/10})|$. Substituting these into Equation 4.56, pdf of logarithm of SINR is:

$$w_{logS}(y) = \frac{|P_R 10^{-y/10} ln(10^{-1/10})|}{\sqrt{2\pi}\sigma} e^{-\frac{\left(P_R 10^{-y/10} - N_m - \mu\right)^2}{2\sigma^2}} \qquad (4.57)$$

where $\mu = E[X(\vec{r})]$, $\sigma^2 = \sigma^2[X(\vec{r})]$ can be obtained in Equations 4.23 and 4.55.

4.4.3 Analytical Results

Based on the aforementioned derived models, it is noted that the distribution and average values of SINR are directly dependent of the node density and probability of transmitting pulses. In this section, we analytically investigate the effect of these parameters on the system performance of in vivo nano-networks inside three human tissues (blood, skin, and fat).

The simulation environment of the following analytical study is summarized in Table 4.2.

With respect to TS-OOK, it is envisioned that $T_s / T_p = 100$ in Equation 4.44, to satisfy the requirement with regard to the time between symbols, which should be much longer than the symbol duration. The node density of WNSNs can be hundreds of nano-sensors per millimeter. Therefore, $\lambda_T = 10, 100$, and 1000 nodes/mm^2 are chosen to evaluate the effect of the interfere density on the system performance.

4.4.3.1 SINR Distribution

Figure 4.10 shows the result for SINR distribution at different node densities, with a specific signal transmission probability $p_1 = 0.5$ for THz wave communicating inside human blood, skin, and fat tissues.

As expected, it can be concluded that for a specific probability of transmitting pulses, SINR decreases significantly with the increase of node density. Specifically, SINR created by a Poisson field of nano-devices with parameter $\lambda_T = 10$ nodes/mm^2, which operate under the previous conditions in human blood, has an average power of approximately −115 dB. When the node density is increased to $\lambda_T = 100$ and $\lambda_T = 1000$ nodes/mm^2, the values reduce to −120

TABLE 4.2   Simulation Environment

Parameters	Definition
$R = 3mm$	The radius of the considered disc.
$r_0 = 1mm$	The distance between the targeted transmitter and the targeted receiver.
$T_0 = 310K$	The temperature of human tissues, and tissues are assumed as isothermal.
$P_T = 1pJ$	The total energy of the transmitted Gaussian pulse.
$B = 1THz$	The bandwidth of the transmitted Gaussian pulse.
$T_p = 100fs$	The pulse duration of the transmitted Gaussian pulse.
$n = 6$	The derivative order of the transmitted Gaussian pulse.
$\sigma = 0.15$	The standard deviation of the transmitted Gaussian pulse.

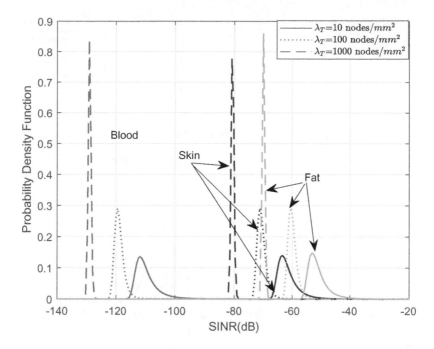

FIGURE 4.10 Probability density function of SINR for different node densities λ_T when $p_1 = 0.5$ in human blood, skin, and fat tissues.

and −130 dB, respectively. The reason is that collisions occur with a higher probability when interferer nodes grow in the communication area. Similarly, THz communication in human skin and fat experiences this trend because more interference could be caused with the presence of larger number of nodes. Specifically, in the human skin scenario, the SINR values decrease from about −63 to −70 and −80 dB with node densities increasing. And for human fat, the values are about −52, −61, and −70 dB when node densities are $\lambda_T = 10$, $\lambda_T = 100$ and $\lambda_T = 1000$ nodes/mm, respectively. Differently, the SINR in human blood is the worst case among these mediums, because blood has the highest water concentration and the molecular absorption is dominantly contributed by the molecules of water vapor [4].

Figure 4.11 illustrates the effect of probability of transmitting pulses on the distribution of SINR for the communication system. For a specific node density $\lambda_T = 100$ nodes/mm^2, by increasing the probability p_1 from 0.1 to 0.9, the average SINR of nano-networks in human blood decreases from −112 to −119 to −122 dB. Similarly, for communication inside human skin, the average SINR value goes down from −65 to −71 dB and then to −73 dB when the probability grows. In fat scenario, a descending trend happens from −56 to −61 dB and then to −63 dB with the increase of the probability. These results emphasize the fact that the molecular absorption noise and multiuser interference are directly dependent on the transmitted signal; thus, the transmitting pulses potentially degrade the communication system performance. These results depict that the degradation effect, caused by both the molecular absorption noise and the interference, can be mitigated by using lower transmission probability and can prevent transmission errors from occurring at the beginning.

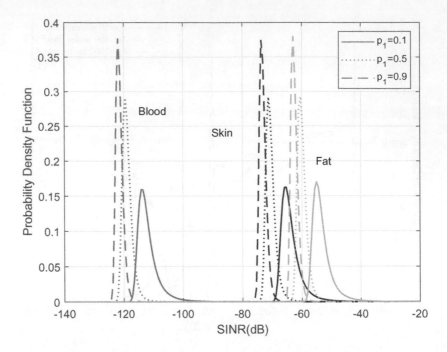

FIGURE 4.11   Probability density function of SINR for different probability of transmitting pulses λ_T when $\lambda_T = 100$ nodes/mm^2 in human blood, skin, and fat.

4.4.3.2 Mean SINR Assessment

In this section, the effect of the node densities and probabilities of transmitting pulses on in vivo nano-communication at the THz band is studied. The communication happens between a deterministic targeted transmitter and receiver, with random interfering nodes in the communication area inside all three human tissues (blood, skin, and fat). The communication performance, especially the achievable communication range, is evaluated, by investigating the dependence of the mean values of SINR on the transmission distance in different human tissues.

4.4.3.2.1   Effect of Node Density   The mean SINR for scenarios with three different node densities in the communication area is illustrated in Figure 4.12. It can be clearly seen that SINR steadily decreases with the transmission distance and the rise of node density for THz communication in all three kinds of communication medium. More specifically, SINR degrades about 10 dB when the node density increases one order.

4.4.3.2.2   Effect of Pulses Probability   The effect of the probability of transmitting pulses on the mean SINR of THz communication inside human tissues is shown in Figure 4.13. It is presented that average SINR degrades with the communication distance and the pulses probability, regardless of the medium type. Specifically, the SINR drops approximate 5 and 2 dB when the pulses transmission probability rises from 0.1 to 0.5 and from 0.5 to 0.9, respectively.

One significant observation from both Figures 4.12 and 4.13 is that the maximum achievable transmission distance to enable simultaneous communications is strictly constrained.

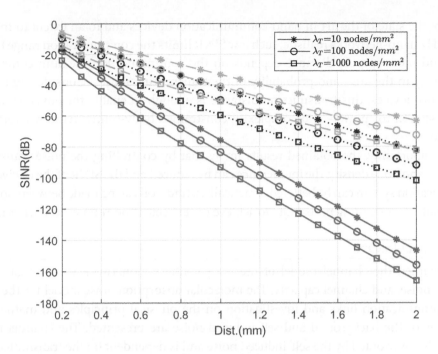

FIGURE 4.12 Average SINR versus communication distance for different node densities in the communication area in human blood (solid line), skin (dotted line) and fat (dashed line) tissues.

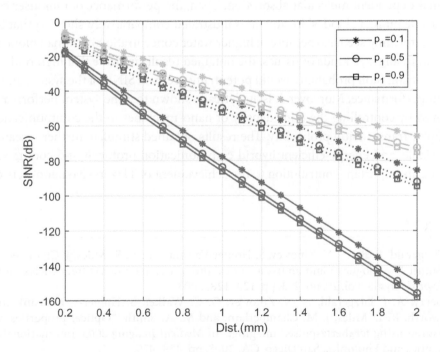

FIGURE 4.13 Average SINR versus communication distance for different probabilities of transmitting pulses of THz communication in human blood (solid line), skin (dotted line) and fat (dashed line) tissues.

In light of the state of the art in nano-communication devices and to an extent to make the in vivo THz nano-networks realistic, such low SINR limits the communication range to about 1 mm, and more specific distance depends on the communication medium composition, node density in the area, and probability of the transmitting pulses. Considering the scale of the nano-devices, this transmission distance is enough to enable the envisioned nano-communication for power- and complexity-constrained body-centric nano-network at the THz band.

More importantly, the obtained results imply that by controlling the pulse transmitting probability and node density, the interference can be reduced and the SINR of the in vivo nano-communication system can be bettered. Practically, a trade-off can be made between node density and pulse transmitting probability to achieve the expected communication performance.

4.5 CONCLUSION

In this chapter, the channel model of the body-centic nanonetwork is investigated from path loss, noise, and channel capacity. The molecular absorption noise model for the in vivo nano-communication links and investigation on the physical principles and mathematical derivations of the background and self-induced noise are presented. The channel noise is dominantly contributed by the self-induced noise and is dependent on the transmitted signal power allocation schemes, which motivates the investigation of SNR on different transmitted power schemes. Based on these studies, end-to-end communication is studied, where the study highlights that the SNR is independent of the power allocation schemes, mainly because of the signal-dependent molecular absorption. Then, the performance of multiuser communication inside human blood, skin, and fat is illustrated comparatively, showing that blood is the worst performing scenario because of higher water concentration in human blood than in skin and fat. In all three kinds of tissues, the obtained results indicate that high node density and pulse transmission probability would potentially decrease SINR of the system and impair the system performance. Moreover, it is analytically shown that the system performance can be improved by controlling the node density of nano-machines in the communication field and the pulse transmission probability. The results obtained stimulate further researches on simple, reliable, and energy-efficient hybrid communication protocols and coding schemes and provide an important contribution to the achievement of THz in vivo nano-networks.

REFERENCES

1. A. Fitzgerald, E. Berry, N. Zinov'ev, S. Homer-Vanniasinkam, R. Miles, J. Chamberlain, and M. Smith, "Catalogue of human tissue optical properties at terahertz frequencies," *Journal of Biological Physics*, vol. 29, no. 2–3, pp. 123–128, 2003.
2. E. Berry, A. J. Fitzgerald, N. N. Zinov'ev, G. C. Walker, S. Homer-Vanniasinkam, C. D. Sudworth, R. E. Miles, J. M. Chamberlain, and M. A. Smith, "Optical properties of tissue measured using terahertz-pulsed imaging," in *Medical Imaging 2003*. International Society for Optics and Photonics, San Diego, CA, 2003, pp. 459–470.
3. J. M. Jornet and I. F. Akyildiz, "Channel capacity of electromagnetic nanonetworks in the terahertz band," in *Communications (ICC), 2010 IEEE International Conference on*. IEEE, 2010, Conference Proceedings, pp. 1–6, the noise temperature and power are calculated in this paper.

4. J. M. Jornet and I. F. Akyildiz, "Channel modeling and capacity analysis for electromagnetic wireless nanonetworks in the terahertz band," *Wireless Communications, IEEE Transactions on*, vol. 10, no. 10, pp. 3211–3221, 2011.

5. P. Boronin, V. Petrov, D. Moltchanov, Y. Koucheryavy, and J. M. Jornet, "Capacity and throughput analysis of nanoscale machine communication through transparency windows in the terahertz band," *Nano Communication Networks*, vol. 5, no. 3, pp. 72–82, 2014.

6. J. M. Jornet and I. F. Akyildiz, "Femtosecond-long pulse-based modulation for terahertz band communication in nanonetworks," *Communications, IEEE Transactions on*, vol. 62, no. 5, pp. 1742–1754, 2014.

7. K. Yang, A. Pellegrini, M. O. Munoz, A. Brizzi, A. Alomainy, and Y. Hao, "Numerical analysis and characterization of thz propagation channel for body-centric nano-communications," *IEEE Transactions on Terahertz Science and Technology*, vol. 5, no. 3, pp. 419–426, 2015.

8. Q. H. Abbasi, H. El Sallabi, N. Chopra, K. Yang, K. A. Qaraqe, and A. Alomainy, "Terahertz channel characterization inside the human skin for nano-scale body-centric networks," *IEEE Transactions on Terahertz Science and Technology*, vol. 6, no. 3, pp. 427–434, 2016.

9. Q. H. Abbasi, A. Alomainy, M. U. Rehman, and K. Qaraqe, *Advances in Body-Centric Wireless Communication: Applications and State-of-the-Art.* The Institution of Engineering and Technolog (IET) Publication, Stevenage, UK, 2016.

10. J. Kokkoniemi, J. Lehtomäki, K. Umebayashi, and M. Juntti, "Frequency and time domain channel models for nanonetworks in terahertz band," *IEEE Transactions on Antennas and Propagation*, vol. 63, no. 2, pp. 678–691, 2015.

11. M. Pierobon and I. F. Akyildiz, "A physical end-to-end model for molecular communication in nanonetworks," *Selected Areas in Communications, IEEE Journal on*, vol. 28, no. 4, pp. 602–611, 2010.

12. S. Bush, J. Paluh, G. Piro, V. Rao, V. Prasad, and A. Eckford, "Defining communication at the bottom," *Molecular, Biological and Multi-Scale Communications, IEEE Transactions on*, vol. PP, no. 99, pp. 1–1, 2015.

13. S. Paine, "The am atmospheric model (sma tech. memo 152; cambridge: Harvard univ.)," 2004.

14. M. Fox, "*Optical Properties of Solids. Oxford Master Series in Condensed Matter Physics*," Oxford University Press, Oxford, UK, 2001.

15. P. Wang, J. M. Jornet, M. A. Malik, N. Akkari, and I. F. Akyildiz, "Energy and spectrum-aware mac protocol for perpetual wireless nanosensor networks in the terahertz band," *Ad Hoc Networks*, vol. 11, no. 8, pp. 2541–2555, 2013.

16. J. Kokkoniemi, J. Lehtomäki, and M. Juntti, "A discussion on molecular absorption noise in the terahertz band," *Nano Communication Networks*, vol. 8, pp. 35–45, 2016.

17. S. Chandrasekhar, *Radiative Transfer.* Courier Corporation, North Chelmsford, MA, 2013.

18. P. Boronin, D. Moltchanov, and Y. Koucheryavy, "A molecular noise model for thz channels," in *Communications (ICC), 2015 IEEE International Conference on*. IEEE, 2015, pp. 1286–1291.

19. A. K. Geim and K. S. Novoselov, "The rise of graphene," *Nature Materials*, vol. 6, no. 3, pp. 183–191, 2007.

20. A. N. Pal and A. Ghosh, "Ultralow noise field-effect transistor from multilayer graphene," *Applied Physics Letters*, vol. 95, no. 8, p. 082105, 2009.

21. J. M. Jornet and I. F. Akyildiz, "Graphene-based plasmonic nano-transceiver for terahertz band communication," in *The 8th European Conference on Antennas and Propagation (EuCAP 2014)*. IEEE, 2014, pp. 492–496.

22. V. Ryzhii, M. Ryzhii, V. Mitin, and T. Otsuji, "Toward the creation of terahertz graphene injection laser," *Journal of Applied Physics*, vol. 110, no. 9, p. 094503, 2011.

23. E. Zarepour, M. Hassan, C. T. Chou, and S. Bayat, "Performance analysis of carrier-less modulation schemes for wireless nanosensor networks," in *Nanotechnology (IEEE-NANO), 2015 IEEE 15th International Conference on*. IEEE, 2015, pp. 45–50.

24. J. M. Jornet and I. F. Akyildiz, "Low-weight channel coding for interference mitigation in electromagnetic nanonetworks in the terahertz band," in *2011 IEEE International Conference on Communications (ICC)*. IEEE, 2011, pp. 1–6.

25. J. M. Jornet, "Low-weight error-prevention codes for electromagnetic nanonetworks in the terahertz band," *Nano Communication Networks*, vol. 5, no. 1, pp. 35–44, 2014.

26. A. Goldsmith, *Wireless Communication*. Cambridge University Press, New York, 2005.

27. G. Piro, K. Yang, G. Boggia, N. Chopra, L. A. Grieco, and A. Alomainy, "Terahertz communications in human tissues at the nanoscale for healthcare applications," *IEEE Transactions on Nanotechnology*, vol. 14, no. 3, pp. 404–406, 2015.

28. C. E. Shannon, "A mathematical theory of communication," *ACM SIGMOBILE Mobile Computing and Communications Review*, vol. 5, no. 1, pp. 3–55, 2001.

29. J. M. Jornet and I. F. Akyildiz, "Information capacity of pulse-based wireless nanosensor networks," in *Sensor, Mesh and Ad Hoc Communications and Networks (SECON), 2011 8th Annual IEEE Communications Society Conference on*. IEEE, 2011, pp. 80–88.

30. H. ElSawy, E. Hossain, and M. Haenggi, "Stochastic geometry for modeling, analysis, and design of multi-tier and cognitive cellular wireless networks: A survey," *IEEE Communications Surveys & Tutorials*, vol. 15, no. 3, pp. 996–1019, 2013.

31. J. G. Andrews, R. K. Ganti, M. Haenggi, N. Jindal, and S. Weber, "A primer on spatial modeling and analysis in wireless networks," *IEEE Communications Magazine*, vol. 48, no. 11, pp. 156–163, 2010.

32. D. R. Cox and V. Isham, *Point processes*. CRC Press, Boca Raton, FL, 1980, vol. 12.

33. V. Petrov, D. Moltchanov, and Y. Koucheryavy, "Interference and sinr in dense terahertz networks," in *Vehicular Technology Conference (VTC Fall), 2015 IEEE 82nd*. IEEE, 2015, pp. 1–5.

34. D. R. Cox and V. Isham, "On the efficiency of spatial channel reuse in ultra-dense thz networks," in *2015 IEEE Global Communications Conference (GLOBECOM)*. IEEE, 2015, pp. 1–7.

35. A. Papoulis and S. U. Pillai, *Probability, Random Variables, and Stochastic Processes*. Tata McGraw-Hill Education, New Delhi, India, 2002.

36. W. Feller, *An Introduction to Probability Theory and its Applications*. Volume I. John Wiley & Sons, London, UK, 1968, vol. 3.

37. K. Yang, *Characterisation of the In-vivo Terahertz Communication Channel within the Human Body Tissues for Future Nano-Communication Networks*. Queen Mary University of London, 2016.

CHAPTER 5

Channel Modeling and Capacity Analysis for Nanoscale Communications and Networking

V. Musa, G. Piro, P. Bia, L. A. Grieco,
D. Caratelli, L. Mescia, and G. Boggia

CONTENTS

5.1 INTRODUCTION

The innovation process triggered by nanotechnology is rapidly concretizing the idea to deploy network architectures at the nanoscale, made up by integrated devices, with size ranging from one to a few hundreds of nanometers. These devices are able to interact with each other by using novel communication mechanisms, thus enabling new pioneering applications in information communications technology (ICT), biomedical, industrial, and military domains [1]. Accordingly, the time is ready to conceive innovative networking methodologies, protocols, and algorithms, which properly embrace the main facets of nanoscale communication systems, while fulfilling the requirements of enabled applications. However, at this embryonic stage of the research, any activity focusing on nanoscale networking should ground its roots to solid studies that carefully describe how the information is really exchanged between transmitter and receiver at the nanoscale. In this context, channel modeling and capacity analysis become key aspects to investigate, before deeply proceeding in this direction of research.

With reference to the healthcare domain, for instance, nanoscale communications and networking could enable advanced immune systems, biohybrid implant solutions, drug-delivering systems, pervasive health monitoring, and genetic engineering [2,3]. It is foreseen, in fact, that biomedical nanodevices can be implanted, ingested, or worn by humans for collecting diagnostic information (e.g., the presence of sodium, glucose, and/or other ions in blood and cholesterol, as well as cancer biomarkers and other infectious agents) and for tuning medical treatments (e.g., insulin and other drugs' injection through underskin actuators). In this context, while graphene-based nanoantennas generating electromagnetic waves in the terahertz band (i.e., from 0.1 to 10 THz) make the communication feasible at the nanoscale and in human tissues [4], the actual physical transmission rates and communication ranges are significantly influenced by many aspects characterizing the communication process [5]. The most important ones include propagation losses, the dispersive nature of the communication channel, the molecular noise, the adopted transmission techniques, and the positions of both transmitter and receiver.

Starting from these premises, this book chapter aims to investigate physical transmission rates and communication ranges reachable in human tissues, starting from the formulation of a sophisticated channel model that takes into account the frequency and spatial dependence of the skin permittivity. First, the communication channel is modeled as a stratified medium, composed of stratum corneum, epidermis, dermis, and hypodermis. Here, the electromagnetic field and the Poynting vector are calculated by using the finite-difference time-domain (FDTD) technique, able to directly solve the Maxwell equations in time domain. Second, starting from the aforementioned channel model, the total path loss (expressed as the sum of spreading and absorption path loss), the molecular noise temperature, and the noise power spectral density are evaluated as a function of the communication frequency and the distance between transmitter and receiver. To make the study more general as possible, two configurations are considered. The first one, namely bottom-up, assumes that the transmitter is implanted in the human body and the receiver is directly positioned on the skin surface. On the contrary, the second configuration, namely top-down, investigates the communication process when the position of both transmitter and receiver is inverted with respect to the previous case. Third, by considering three different transmission mechanisms based on the Time Spread On-Off Keying (TS-OOK) modulation scheme [4,6] (namely flat, pulse-based, and optimal), the signal-to-noise ratio (SNR) is evaluated as a function of communication frequency and distance between transmitter and receiver. All obtained results are finally processed for studying the upper bound of physical transmission rates and communication ranges achievable in human tissues, when the reference communication bandwidth is delimited to the set of frequencies spanning from 0.5 to 1.5 THz.

The proposed study demonstrates that a physical data rate in the order of Tbps can be only reached for transmission ranges less than 2 mm. When the distance between transmitter and receiver exceeds 9 mm, communication capabilities are extremely impaired (i.e., the physical data rate tends to be lower than 1 bps). Moreover, higher performance is measured for the bottom-up configuration, where inner layers of the communication medium produce lower levels of attenuation of the propagating signal.

The rest of this book chapter is organized as follows: Section 5.2 presents the channel model formulated for the stratified media stack describing human tissues. Section 5.3 describes the investigated transmission techniques. Section 5.4 discusses the SNR measured in the frequency domain and illustrates physical transmission rates and communication ranges achievable in human tissues as a function of transmission techniques, distance, and position of both transmitter and receiver. Finally, Section 5.5 draws the conclusions.

5.2 CHANNEL MODELS FOR NANOSCALE COMMUNICATIONS IN HUMAN TISSUES

The modeling of the pulsed electric field (PEF) propagation in biological tissues is a subject of increasing research activities, since they are used in a number of applications in bioelectrics, a new interdisciplinary field combining knowledge of electromagnetic principles and theory, modeling and simulations, physics, material science, cell biology, and medicine [7–9]. Several studies are focused on the use of PEFs for reversible or irreversible electroporation to achieve selective killing of cancer cells, tissue ablation, gene therapy, and DNA-based vaccination [10,11]. Moreover, recent applications employing PEF technology include medical implant communication service, wireless medical telemetry service, body area networks, nanonetworks in a living biological environment, and in-body electromagnetic communications [12–15].

All these technologies involve the interaction of electromagnetic fields with complex dielectric materials. For instance, in the contest of the in-body communications among the nanodevices, the right evaluation of the network performance in term of data rate and transmission range needs an accurate modeling of the electromagnetic field propagation inside human tissues. As a result, the development of theoretical models and computational techniques to determine the propagation properties of electromagnetic pulses is fundamental to gain insight into the several phenomena occurring within complex dielectric materials subject to an imposed PEF. In fact, the complexity of the structure and composition of such matter produces a time-domain response generally nonsymmetric and markedly different from that of dielectric media modeled by the simple dielectric response relationships. As a consequence, the dielectric response in the frequency domain usually requires empirical models exhibiting fractional powers of the angular frequency $j\omega$. Owing to this, the solution of the Maxwell equations in the time domain is not trivial, since it involves the concept of fractional derivatives [16].

5.2.1 Dielectric Dispersion Model

The interaction between PEF and biological tissues takes place in different relaxation processes such as reorientation of dipolar molecules, interfacial polarization, ionic diffusion due to ions of different signs of charges, conductivity of surface cell structures, motion of the molecules, and the nonspherical shape, as well as in different resonant phenomena due to molecular, atomic, or electronic vibrations. Their resulting behavior causes a frequency dispersion pattern of permittivity and conductivity [17].

The analytical theory modeling of the frequency-dependent permittivity of dielectric media is based on the response function through the following relation:

$$\phi(\omega) = \frac{\varepsilon(\omega) - \varepsilon_\infty}{\Delta\varepsilon} \tag{5.1}$$

where $\varepsilon(\omega)$ is the frequency-dependent complex relative permittivity, ε_∞ is asymptotic relative permittivity, and $\Delta\varepsilon$ is the dielectric strength. A useful technique allowing the reproduction of the experimental spectra $\varepsilon(\omega)$ by adjusting some free parameters of a mathematical expression is based on the use of an empirical dielectric function model in conjunction with nonlinear least-square optimization. Using this method, the empirical response functions exhibiting a broad distribution of relaxation times have been proposed [18]:

$$\phi(\omega) = \frac{1}{\left[(j\omega\tau)^\gamma + (j\omega\tau)^\alpha \right]^\beta} \tag{5.2}$$

where the adjusting parameters $0 \le \alpha, \beta, \gamma \le 1$ account for shape and behavioral features of the permittivity function. Equation (5.2) strongly deviates from the conventional Debye law ($\gamma = 0$, $\alpha = \beta = 1$), and it can be used to reproduce other empirical response functions as Cole–Cole ($\gamma = 0$, $\beta = 1$), Cole–Davidson ($\gamma = 0$, $\alpha = 1$), and Havriliak–Negami ($\gamma = 0$). However, its effectiveness may fall when the dielectric response of more complex materials having heterogeneous, inhomogeneous, and disordered structure at both microscopic and mesoscopic scales has to be modeled. To overcome this limitation and to provide an extended model parametrization, as well as a better and flexible fitting of the experimental data over broad frequency ranges, a general fractional polynomial series approximation has been proposed by the authors [7,8,19].

$$\phi(\omega) = \sum_{i=1}^{N} \frac{\Delta\varepsilon_i}{Q_i} \tag{5.3}$$

and

$$Q_i = \sum_{k=0}^{K} b_{k,i} \left(j\omega\tau_{k,i} \right)^{\beta_{k,i}} \tag{5.4}$$

where $b_{k,i}, \beta_{k,i}$ denote suitable real-valued parameters chosen (i) to avoid model singularities, (ii) to fulfill the consistency of the representation, and (iii) to ensure the passivity condition. Applying a dedicated optimization algorithm by employing a suitable relative error function, the free parameters $b_{k,i}, \beta_{k,i}, \tau_{k,i}, K, N$ can be evaluated [20]. This method is versatile because it is capable of dealing with every data, it can reproduce fine details, and it proved to feature superior effectiveness in terms of convergence rate and accuracy.

5.2.2 FDTD Modeling

The FDTD technique is a well-known numerically robust and appropriate method for the computer technology of today. It is well known that the solution of the Maxwell equations in dispersive media is a stiff problem, and the development of FDTD methods to study the transient wave propagation in such media is an area of active interest. In detail, the FDTD implementation of the dispersion characteristics described by Equations (5.3) and (5.4) is difficult, and it requires special treatments. However, the nature of the fractional-order operators modeling the dielectric response described by Equation (5.4) enables its incorporation into time-domain Maxwell equations, using nonlocal pseudodifferential operators of noninteger order.

For a nonmagnetic and isotropic dispersive dielectric material with response function described by Equation (5.4), the frequency-domain Maxwell equations can be written as:

$$\nabla \times \mathbf{H} = j\omega\varepsilon_0 \left(\varepsilon_\infty - j\frac{\sigma}{\omega\varepsilon_0} \right) \mathbf{E} + \sum_{i=1}^{N} \mathbf{J}_i \tag{5.5}$$

$$\nabla \times \mathbf{E} = -j\omega\mu_0 \mathbf{H} \tag{5.6}$$

where σ is the static conductivity, $\mathbf{H}$ is the magnetic field, and the pth term of the auxiliary displacement current density $\mathbf{J} = \sum_{i=1}^{N} \mathbf{J}_i$ is given by:

$$\mathbf{J}_p = j\omega\varepsilon_0 \frac{\Delta\varepsilon_p}{Q_p} \mathbf{E} \tag{5.7}$$

where

$$Q_p = \sum_{k=0}^{K} b_{k,p} \left(j\omega \tau_{k,p} \right)^{\beta_{k,p}} \tag{5.8}$$

Taking the inverse Fourier transform of Equation (5.7) and following the procedure well illustrated in [7,21–23], it is possible to find the following updating equations, in time domain, for the magnetic and electric fields, as well as for the displacement current density:

$$\mathcal{H}\big|^{m+1} = \mathcal{H}\big|^{m} - \frac{\Delta t}{\mu_0} \nabla \times \mathcal{E}\big|^{m+1/2} \tag{5.9}$$

$$\mathcal{E}\big|^{m+1/2} = \frac{2\varepsilon_0 \varepsilon_\infty - \sigma \Delta t}{2\varepsilon_0 \varepsilon_\infty + \sigma \Delta t} \mathcal{E}\big|^{m-1/2} + \frac{2\Delta t}{2\varepsilon_0 \varepsilon_\infty + \sigma \Delta t} \left[\nabla \times \mathcal{H}\big|^{m} - \frac{1}{2} \sum_{i=1}^{N} \left(\mathcal{J}_i\big|^{m-1/2} + \mathcal{J}_i\big|^{m+1/2} \right) \right] \tag{5.10}$$

$$\left(\varepsilon_\infty + \frac{\sigma \Delta t}{2\varepsilon_0} \right) \frac{C^{(\beta_{k,p})}}{\Delta \varepsilon_p} \mathcal{J}_p\big|^{m+1/2} + \frac{1}{2} \sum_{i=1}^{N} \mathcal{J}_i\big|^{m+1/2}$$

$$= \nabla \times \mathcal{H}\big|^{m} - \sigma \nabla \times \mathcal{E}\big|^{m-1/2} +$$

$$- \frac{1}{2} \sum_{i=1}^{N} \mathcal{J}_i\big|^{m-1/2} - \frac{1}{\Delta \varepsilon_p} \left(\varepsilon_\infty + \frac{\sigma \Delta t}{2\varepsilon_0} \right) \tag{5.11}$$

$$\left[\sum_{k=0}^{K_i} \sum_{s=1}^{v} \xi_{k,s}^{(\beta_{k,p})} \mathcal{J}_p\big|^{m-s+1/2} + \right.$$

$$\left. + \sum_{k=0}^{K_i} \sum_{s=1}^{v} \sum_{q=1}^{Q_{k,\alpha_p}} \eta_{k,s,q}^{(\beta_{k,p})} \psi_q^{(\beta_{k,p})}\big|^{m-s} \right]$$

With the aim to suitably bound the computational domain, a dedicated uniaxial perfectly matched layer (UPML) boundary condition was derived in combination with the basic time-marching scheme, accounting for the electrical conductivity and the multirelaxation characteristics of the dielectric material under analysis. In particular, the developed approach combines the stretched auxiliary electric field and density current vectors with the fractional derivative equation describing the dispersion properties of the medium. In particular, by following the mathematical procedure well detailed in [7,21], the update equations for both electric and magnetic fields within the UPML termination can be written as:

$$\mathcal{E}\big|^{m+1/2} = \frac{2\varepsilon_0 \kappa_x - \sigma_x \Delta t}{2\varepsilon_0 \kappa_x + \sigma_x \Delta t} \mathcal{E}\big|^{m-1/2} + \frac{2\varepsilon_0}{2\varepsilon_0 \kappa_x + \sigma_x \Delta t} \left(\mathbf{e}\big|^{m+1/2} - \mathbf{e}\big|^{m-1/2} \right) \tag{5.12}$$

and

$$\mathcal{H}\big|^{m+1} = \frac{2\varepsilon_0\kappa_x - \sigma_x\Delta t}{2\varepsilon_0\kappa_x + \sigma_x\Delta t}\mathcal{H}\big|^m +$$

$$-\frac{2Y_0\Delta t}{2\varepsilon_0\kappa_x + \sigma_x\Delta t}\nabla\times\mathcal{E}\big|^{m+1/2}$$

(5.13)

where Y_0 is the wave admittance in free space, σ_x, κ_x are the UPML material parameters, and **e** is the auxiliary electric field vector.

5.2.3 Electromagnetic Simulations

Most recent studies of electromagnetic channel at the terahertz band for the body-centric nanonetworks treat the skin tissue as a homogeneous semi-infinite medium resulting by the binary mixture of water and biological background material (bound water, keratin, lipids, and collagen) [1,24]. In general, the skin is a mosaic in which layers of laminated, inhomogeneous cell structure pile up on top of one another. As a result, it can be modeled as a stratified media stack consisting of stratum corneum (SC), epidermis (E), dermis (D), and hypodermis (HYP) (see Figure 5.1). The SC contains corneocytes embedded in a lipid matrix. The corneocyte does not contain a nucleus, as well as the extracellular matrix mainly consists of lipids and proteins and very little bound water. So, the total water volume fraction in the SC is 0.15–0.25 (The volume fraction is an adimensional value (ratio between two volume values)), while 90% of the water is contained within the corneocyte. The SC thickness depends on the body site, and it is typically tens of micrometers. In our modeling, the SC is 20 micrometers thick [25–28]. However, owing to the high lipid and protein and low water content, the

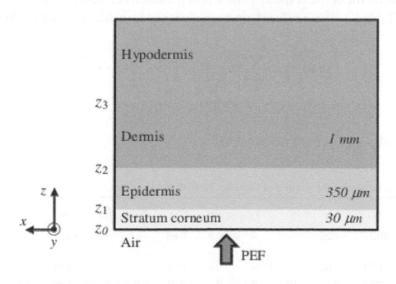

FIGURE 5.1 The considered layered structure describing human tissues.

SC differs significantly from the epidermis layer. In fact, the epidermis mainly consists of keratinocytes and is characterized by an overall water volume fraction of about 0.7, equally distributed among intra- and extracellular spaces. The dermis is the more complex and heterogeneous layer. In fact, the upper 10% consists of a dense collagen network and blood, while the major part is mainly composed of irregular connective tissue, lymphatic vessels, nerves, blood vessels, stromal cells (fibroblasts), and other cellular components (macrophages or plasma) [29,30]. Collagen is a major component embedded in the dermal matrix. In addition to collagen and elastin, the extracellular space is mainly composed of glycosaminoglycans, gelatin, and sugars embedded in water. Finally, the HYP mainly consists of white fat cells and adipocytes, building the subcutaneous fat. The intracellular fat forms a spherical droplet having a volume fraction of about 0.9. Thus, the aqueous phase volume fraction is only 0.1.

Considering the complexity of the skin and with the aim to provide a more detailed modeling of the electromagnetic field propagation inside the skin at the terahertz band, a nonhomogeneous and dispersive model based on stratified media stack illustrated in Figure 5.1 has been taken into account. The dielectric properties of stratum corneum, epidermis, dermis, and fat in a desired frequency range, as well as the thickness of each layer, have been calculated by considering experimental results reported in the literature [25–28]. In particular, the frequency-domain permittivity function has been designed by minimizing the following error function:

$$\text{err} = \frac{\sum_{\omega_{min}}^{\omega_{max}} \varepsilon_{exp}(\omega) - \varepsilon(\omega)^2 \omega}{\sum_{\omega_{min}}^{\omega_{max}} \varepsilon_{exp}(\omega)^2 \omega} \le \delta, \tag{5.14}$$

where δ is the maximum tolerable error, ε_{exp} is the measured permittivity, and ε represents the general dielectric response. The sets of parameters related to each layer have been reported in Table 5.1. These values have been obtained by using the relationship:

$$\varepsilon(\omega) = \varepsilon_\infty + \sum_{i=1}^{2} \frac{\Delta\varepsilon_i}{1 + \sum_{k=1}^{2} b_{i,k}\left(j\omega\tau_{i,k}\right)^{\beta_{i,k}}} - j\frac{\sigma}{\omega\varepsilon_0} \tag{5.15}$$

in a bandwidth ranging from 0.5 to 1.5 THz.

TABLE 5.1 Parameters of the Recovered Complex Permittivity Function

Tissue	$b_{1,1}$	$b_{1,2}$	$b_{2,1}$	$b_{2,2}$	$\beta_{1,1}$	$\beta_{1,2}$	$\beta_{2,1}$	$\beta_{2,2}$	$\Delta\varepsilon_1$	$\Delta\varepsilon_1$	$\sigma(S/m)$	ε_∞
Stratum Corneum	10.11	−9.25	—	—	0.9	0.88	—	—	12.22	—	0.035	2.4
Epidermis	1.04	−0.02	—	—	0.9	0.03	—	—	89.61	—	0.01	3
Dermis	0.88	−0.17	10	−9.1	0.77	0.01	0.9	0.88	5.96	380.4	0.1	4
Hypodermis	0.89	−0.19	0.96	−0.05	0.81	0.01	0.8	0.04	1.14	9.8	0.035	2.5

The electromagnetic source is a plane wave propagating along the positive z-direction, with electric field linearly polarized along the x-axis. In particular, the time-domain signal source is an electric current density $\mathcal{J}_0$ placed at a given position $z = \bar{z}$ inside the computational domain:

$$\mathcal{J}_0(z,t) = \exp - a^2 \left(t - \frac{2}{a} \right)^2 \sin\left[2\pi f_0 \left(t - \frac{4}{a} \right) \right] \delta\left(z - \bar{z} \right) x, \qquad (5.16)$$

Where the parameters $f_0 = 1\,\text{THz}$ and $1/a = 100\,\text{fs}$ have been selected to achieve a bandwidth from 0.5 to 1.5 THz. The considered time and spatial steps are $\Delta t = 10$ fs and $\Delta z = 6$ μm, respectively. The validation of the developed numerical procedure has been illustrated in detail in our previous papers [7,8,19,21,31].

To provide a further insight, the Poynting vector, $S(\omega, z)$, is also reported in Figure 5.4. In particular, it has been calculated as:

$$S(\omega, z) = EH^*,$$

where E and H are the Fourier transforms of $\mathcal{E}$ and $\mathcal{H}$, respectively.

In Figures 5.2 through 5.4, the multiple reflected waves generated by the stratified media stack, as well as the main reflection phenomenon occurring at the air–skin interface, can be observed. Moreover, the wave pulse spreading due to the propagation inside the dispersive biological media is evident.

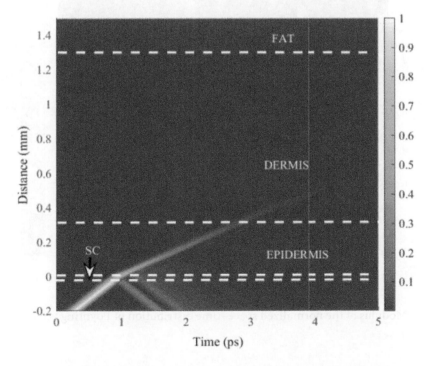

FIGURE 5.2 Modulus of the normalized space–time distribution of electric field.

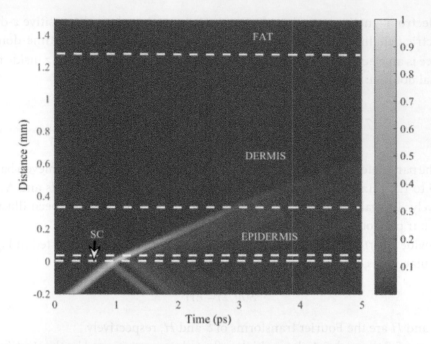

FIGURE 5.3   Modulus of the normalized space–time distribution of magnetic field.

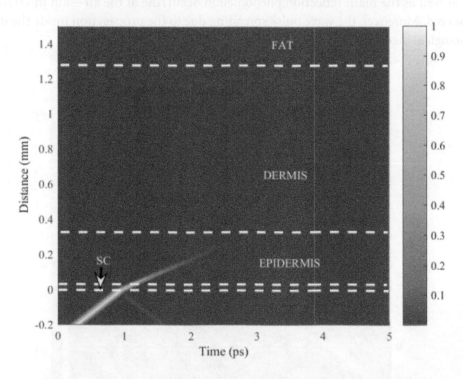

FIGURE 5.4   Modulus of the normalized space–time distribution of Poynting vector.

5.2.4 Path Loss and Noise Models

To deal with nanoscale communications and networking, it is important to know how the signal propagates across the medium. Starting from the FDTD model and related results reported previously, this section develops an accurate channel model, taking into account two possible configurations. The first one refers to the bottom-up communication, where it is assumed that the transmitter is implanted, and the receiver is positioned outside the human body (but attached to the skin). The second one refers to the top-down configuration, where the position of transmitter and receiver is inverted with respect to the previous case. The resulting channel model embraces absorption path loss, spreading path loss, molecular noise temperature, and noise power spectral density. Such models will be used in the next section for evaluating the communication capacity as a function of the distance between transmitter and receiver.

5.2.4.1 Absorption Path Loss

When an electromagnetic wave propagates through the medium, several molecules excite and start to vibrate. In this case, part of the energy carried by the electromagnetic wave is lost or converted to the kinetic energy [1]. The absorption path loss or molecular loss, that is, $A_{abs}(\omega, z)$, describes the attenuation produced by the vibration of molecules as a function of both the distance between transmitter and receiver and the communication frequency, that is:

$$A_{abs}(\omega, z) = 10 \log \frac{S(\omega, z)}{S(\omega, z_0)} = 10 k(\omega) d \log e, \qquad (5.17)$$

where S is the Poynting vectors, z_0 is the z-coordinate of the reference section, $d = z - z_0$ is the considered path length, and $k(\omega)$ is the medium absorption coefficient.

Given the stratified channel model reported in Figure 5.1, the absorption path loss offered by human tissues in the bandwidth from 0.5 to 1.5 THz is shown in Figure 5.5. Without loss of generality, the reported results consider the top-down configuration. The absorption path loss clearly appears as a frequency-selective attenuation, which grows up when both transmission range and communication frequency increase.

5.2.4.2 Spreading Path Loss

The spreading path loss, that is, $A_{spread}(\omega, z)$, refers to the attenuation due to the expansion of an electromagnetic wave propagating through a given medium. It is defined as:

$$A_{spread}(\omega, z) = 20 \log \left(4\pi \int_{z_0}^{z} \frac{dz}{\lambda_g(\omega, z)} \right), \qquad (5.18)$$

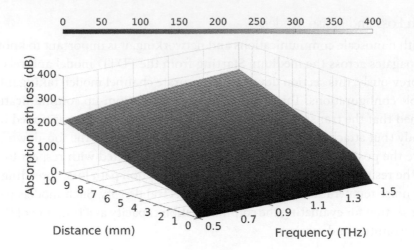

FIGURE 5.5   Absorption path loss as a function of frequency and transmission range, evaluated for the top-down configuration.

where λ_g takes into account the stratified medium and depends on the distance from the air–skin interface. Specifically, λ_g can be expressed as:

$$\lambda_g(\omega,z) = \begin{cases} \lambda_{g,1}(\omega,z) & 0 \leq z \leq z_1 \\ \lambda_{g,2}(\omega,z) & z_1 \leq z \leq z_2 \\ \lambda_{g,3}(\omega,z) & z_2 \leq z \leq z_3 \\ \lambda_{g,4}(\omega,z) & z \geq z_3, \end{cases} \tag{5.19}$$

where

$$\lambda_{g,k} = \frac{\lambda_0}{\sqrt{\dfrac{\varepsilon_k'}{2}\left[\sqrt{1+\left(\dfrac{\varepsilon_k''}{\varepsilon_k'}+\dfrac{\sigma_k}{f\varepsilon_0\varepsilon_k'}\right)^2}+1\right]}} \qquad k=1,2,3,4 \tag{5.20}$$

is the wavelength of the propagating wave within the k-th lossy medium, while λ_0 is the free-space wavelength.

Figure 5.6 depicts the spreading path loss in human tissues as a function of the transmission range and the communication frequency, evaluated for the top-down configuration. In general, the spreading path loss registers considerably high values at the nanoscale. But, in this specific use case, it appears negligible with respect to the level of attenuation provided by the absorption path loss.

5.2.4.3 Total Path Loss

The total signal path loss, that is, $A(\omega,z)$, defines the total level of attenuation offered to the signal that propagates across the medium. According to [1], it jointly takes care of the absorption path loss due to the absorption of human tissues, that is, $A_{abs}(\omega,z)$, and the

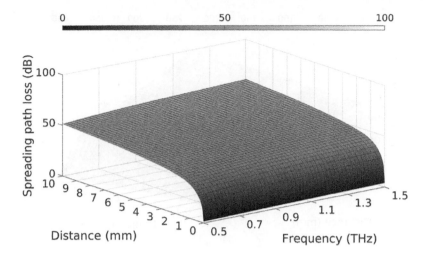

FIGURE 5.6   Spreading path loss as a function of frequency and transmission range, evaluated for the top-down configuration.

spreading path loss generated by the expansion of waves in human body, that is, $A_{spread}(\omega, z)$. Thus, $A(\omega, z)$ can be defined as:

$$A(\omega, z) = A_{spread}(\omega, z) + A_{abs}(\omega, z). \tag{5.21}$$

Figure 5.7 shows the total path loss as a function of the distance between transmitter and receiver and the communication frequency, by taking into account both top-down and bottom-up configurations. Generally, the total path loss grows up when the communication frequency and the transmission range increase. However, the comparison between Figure 5.7a and b shows how the outer skin layers introduce higher attenuation levels than the inner ones. Moreover, by comparing Figure 5.7a with both Figures 5.5 and 5.6, it is evident how the total path loss is mainly influenced by the molecular absorption, which generates a loss up to six times higher than the one introduced by the expansion phenomenon. Just to provide an example, with reference to the considered communication bandwidth, the maximum value of the spreading path loss is 60.3 dB, while the absorption path loss registers a maximum value of 287.2 dB. Moreover, the reported results fully confirm that the terahertz band is strongly frequency-selective; in fact, the propagation loss significantly increases with both communication frequency and transmission range.

5.2.4.4 Molecular Noise Temperature
Molecular absorption also generates the molecular noise. Specifically, the equivalent molecular noise temperature is computed as:

$$T_{eq}(\omega, z) = T_0 \varepsilon(\omega, z) = T_0 \left[1 - \frac{\mathbf{S}(\omega, z)}{\mathbf{S}(\omega, z_0)} \right], \tag{5.22}$$

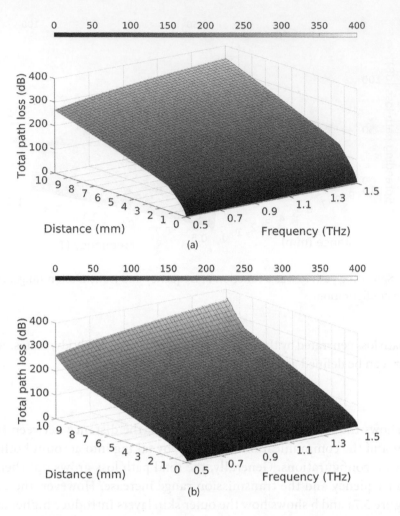

FIGURE 5.7 Total path loss as a function of frequency and transmission range: (a) Top-down configuration and (b) bottom-up configuration.

where T_0 is the reference temperature, equal to normal body temperature (i.e., $T_0 = 310K$), $\mathbf{S}(\omega, z)$ is the Poynting vector, and $\varepsilon(\omega, z)$ is the channel emissivity.

As depicted in Figure 5.8, the molecular noise temperature changes with communication frequency and transmission distance. Indeed, the internal vibrations of the medium molecules absorb the propagating electromagnetic field and convert the carried energy first to the kinetic energy and then to heat. In this context, the emissivity of the channel could be expressed as a function of the absorption path loss and could be set equal to $1 - \mathbf{S}(\omega, z) / \mathbf{S}(\omega, z_0)$.

To provide a further insight, both top-down and bottom-up configurations are taken into account. As already observed, the inner tissue layers register lower attenuation levels. Consequently, the molecular noise temperature increases slower when the bottom-up configuration is considered. On the other hand, in both configurations, it is important to note that, at the level of millimeters, the equivalent molecular noise temperature is not very high, reaching the maximum value approximately equal to 310 K (see Figure 5.8).

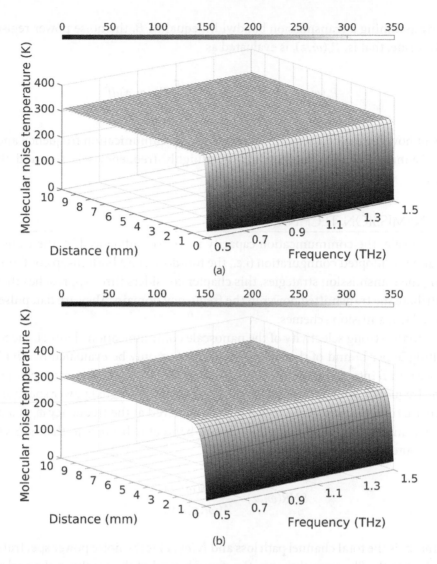

FIGURE 5.8   Molecular absorption noise temperature as a function of frequency and transmission range: (a) Top-down configuration and (b) Bottom-up configuration.

The bounded level of the equivalent noise temperature suggests that a communication link, with tolerable SNR, can be established among transmitter and receiver located inside and outside the human tissues at the terahertz band.

5.2.4.5 Noise Power Spectral Density

Starting from the equivalent noise temperature $T_{eq}(\omega, z)$, it is possible to evaluate the noise power spectral density, that is, $N(\omega, z)$, as reported below:

$$N(\omega, z) = k_B T_{eq}(\omega, z), \qquad (5.23)$$

where k_B is the Boltzmann constant (i.e., $k_B = 1.380658 \times 10^{-23} \, J/K$).

Now, by assuming a transmission bandwidth equal to B, the noise power registered at the receiver side, that is, $P_n(\omega, z)$, is evaluated as:

$$P_n(\omega, z) = \int_B N(\omega, z) df = k_B \int_B T_{eq}(\omega, z) df \tag{5.24}$$

It is evident how the molecular noise increases with communication frequency and transmission distance. This also suggests that it is highly frequency-selective and therefore nonwhite.

5.3 TRANSMISSION TECHNIQUES

At the nanoscale, the communication capacity is strictly influenced by the transmission strategy and the adopted configuration (i.e., the top-down and bottom-up configurations). Regarding the transmission strategies, this chapter considers three approaches that differently distribute the transmitted power in the frequency domain. They are flat, pulse-based, and optimal transmission schemes.

Thanks to the strong selectivity of the nanoscale communication channel, the SNR and the resulting upper bound of the channel capacity can only be evaluated by dividing the total bandwidth B into many narrow sub-bands lasting Δf, where the channel is nonselective in the frequency domain. Let $S(\omega_i, z)$ be the transmitted signal power spectral density of the generic transmission in the i-th sub-band centered at the frequency ω_i at a distance z from the transmitter. The SNR for the i-th sub-band at a distance d, that is, $SNR(\omega_i, z)$, can be computed as:

$$SNR(\omega_i, z) = \frac{S(\omega_i, z)}{A(\omega_i, z)N(\omega_i, z)}, \tag{5.25}$$

where $A(\omega_i, z)$ is the total channel path loss and $N(\omega_i, z)$ is the noise power spectral density.

According to the Shannon theorem, the upper bound of the resulting channel capacity is equal to:

$$C(z) = \sum_i \Delta f \log_2 \left[1 + SNR(\omega_i, z) \right]$$

$$= \sum_i \Delta f \log_2 \left[1 + \frac{S(\omega_i, z)}{A(\omega_i, z)N(\omega_i, z)} \right]. \tag{5.26}$$

As demonstrated earlier, the total path loss $A(\omega_i, z)$ and the noise power spectral density $N(\omega_i, z)$ are influenced by the communication frequency and the distance between transmitter and receiver. On the contrary, the signal power spectral density $S(\omega_i, z)$ is influenced by the adopted transmission strategy. As a result, the channel capacity is strictly influenced by the adopted transmission strategy as well.

5.3.1 Transmission Strategies

In nanoscale communications, the information is generally encoded by using short pulses spread over a large bandwidth. The scientific literature considers the TS-OOK as a promising modulation technique for the nanoscale, able to ensure both high energy and communication efficiency [4,6]. With TS-OOK, a logical 1 is encoded as a short pulse and a logical 0 is encoded as a silence. Moreover, the time interval between two consecutive pulses is much longer than the pulse duration. Consequently, two important advantages are ensured: the nanodevices have not to be synchronized, and the medium can be shared among multiple users, without the risk of collisions.

In what follows, let $B = f_M - f_m$ be the total available bandwidth, where f_m and f_M are the lower and the higher operative frequencies, respectively. When a pulse is transmitted, an amount of power P_{tx} is distributed in the frequency domain according to the three transmission strategies described in the following subsections.

5.3.1.1 Flat Communication

Flat communication represents the simplest technique, where the transmitted power P_{tx} is uniformly distributed in the frequency domain. In this case, the transmitted power spectral density is:

$$S_{flat}(\omega, z) = \begin{cases} S_0 = \dfrac{P_{tx}}{B} & f_m \leq f \leq f_M \\ 0 & \text{otherwise} \end{cases} \tag{5.27}$$

5.3.1.2 Optimal Communication

Differently from the previous case, the optimal communication scheme optimally distributes the transmission power within the operative bandwidth, in order to maximize the channel capacity. Specifically, the signal power spectral density is computed by solving the optimization problem:

$$\text{maximize} \left\{ \sum_i \Delta f \log_2 \left[1 + \frac{S_{opt}(\omega_i, z)}{A(\omega_i, z)N(\omega_i, z)} \right] \right\} \tag{5.28}$$

by jointly taking into account the following three constraints:

1. The total transmission power, expressed as the sum of the signal power spectral density for each sub-bands, multiplied by the subchannel width Δf, cannot exceed the maximum available power over the entire bandwidth P_{tx}:

$$\sum_{i \in \Omega} S_{opt}(\omega_i, z)\Delta f \leq P_{tx}; \tag{5.29}$$

2. The power in a single sub-band should be a fraction, γ, of the total transmission power:

$$S_{opt}(\omega_i, z)\Delta f \leq \gamma P_{tx} \quad \forall i \in \Omega, \gamma \in \left]0,1\right]; \tag{5.30}$$

3. The chosen transmission subchannels must be adjacent:

$$\Omega = \{i \mid i_{min_\Omega} \leq i \leq i_{max_\Omega}\}. \tag{5.31}$$

The optimization problem formulated in Equation (5.28) can be solved by using the method of Lagrange multipliers. First, by considering the first constraint, the optimization problem can be expressed as:

$$\max\left\{\sum_i \Delta f\left(\log_2\left[1 + \frac{S_{opt}(\omega_i, z)}{A(\omega_i, z)N(\omega_i, z)}\right] + \lambda S_{opt}(\omega_i, z)\right) - P_{tx}\right\} \tag{5.32}$$

where λ is the Lagrange multiplier. To find the maximum value, the derivative of the argument of Equation (5.32) with respect to $S_{opt}(\omega_i, z)$ is put as zero, thus obtaining:

$$\ln(2)[S_{opt}(\omega_i, z) + A(\omega_i, z)N(\omega_i, z)] = \lambda^{-1} \quad \forall i \tag{5.33}$$

As a consequence, the overall channel capacity is maximized when:

$$S_{opt}(\omega_i, z) = \beta - A(\omega_i, z)N(\omega_i, z), \tag{5.34}$$

where β is a constant value that can be computed by means of an iterative procedure, following the water-filling principle. In particular, at the n-th step, β is equal to:

$$\beta(n) = \frac{1}{L(n)}\left[\frac{P_{tx}}{\Delta f} + \sum_i A(\omega_i, z)N(\omega_i, z)\right], \tag{5.35}$$

where $L(n)$ is the number of sub-bands at the n-th step. In details, to evaluate $L(n)$ value, the signal power spectral density $S_{opt}(\omega_i, z)$ is computed by considering Equation (5.34). If $S_{opt}(\omega_i, z) \leq 0$, the corresponding value is set to 0 and the considered sub-band is excluded for the following iterative cycles. The procedure can be stopped when there are no other sub-bands with a negative $S_{opt}(\omega_i, z)$ value. At the end, the total signal power appears optimally distributed in the sub-bands, which experience better channel conditions.

It is important to remark that, owing to the monotonic behavior of the total path loss, as demonstrated in the previous section, the selected sub-bands are already adjacent, and consequently, the third constraint is intrinsically respected.

5.3.1.3 Pulse-Based Communication

Flat and optimal transmission schemes are ideal approaches. Nanotechnology (and in particular graphene-based nanoantennas), instead, allows the transmission of pulses having Gaussian shapes in the time domain. Let $p(t) = \frac{1}{s\sqrt{2\pi}}e^{-(t-\mu)^2/(2s^2)}$ be a Gaussian-based pulse, where μ is its mean value and s is the related standard deviation. The pulse-based communication approach models the transmitted pulse through the n-th derivative of $p(t)$, whose representation in the frequency domain is:

$$\phi^{(n)}(\omega,z) = (2\pi f)^{2n} e^{(-2\pi sf)^2} \qquad (5.36)$$

Thus, the signal power spectral density of the n-th time derivative is still Gaussian-shaped and could be described as:

$$S_{pulse}^{(n)}(\omega,z) = a_0^2 \phi^{(n)}(\omega,z) = a_0^2 (2\pi f)^{2n} e^{(-2\pi sf)^2}, \qquad (5.37)$$

where a_0 is a normalizing constant to adjust the pulse total energy, which is obtained as $a_0^2 = P_{tx} / \int_{f_m}^{f_M} \phi(\omega,z) df$.

Figure 5.9 shows examples of Gaussian-based pulses, obtained by setting n in the range from 1 to 3. Note that, while the pulse duration remains constant (i.e., equal to $100\,fs$), the number of oscillations grows up with the derivative order n.

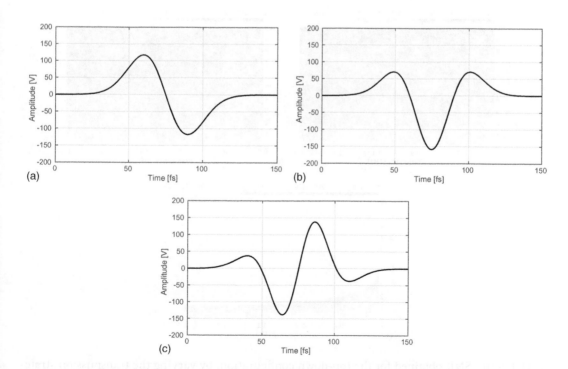

FIGURE 5.9 Examples of pulses generated with the pulse-based communication scheme: (a) $n = 1$, (b) $n = 2$, and (c) $n = 3$.

5.4 ANALYSIS OF PHYSICAL TRANSMISSION RATES AND COMMUNICATION RANGES

In line with [1,3,4,32–34], the performance of a nanoscale communication system is evaluated by assuming that the pulse energy and the pulse duration are equal to 500 pJ and 100 fs, respectively. Thus, the resulting amount of power for each transmitted pulse is set to $P_{tx} = 500/100 = 5$ kW. When the pulse-based communication strategy is evaluated, the derivative order n of the Gaussian pulse is chosen in the range from 1 to 3 (as already shown previously). Whereas, its standard deviation is set to 0.15.

5.4.1 SNR

Figures 5.10 and 5.11 show the SNR estimated for top-down and bottom-up configurations, respectively. First, it is possible to observe that these configurations show a common behavior.

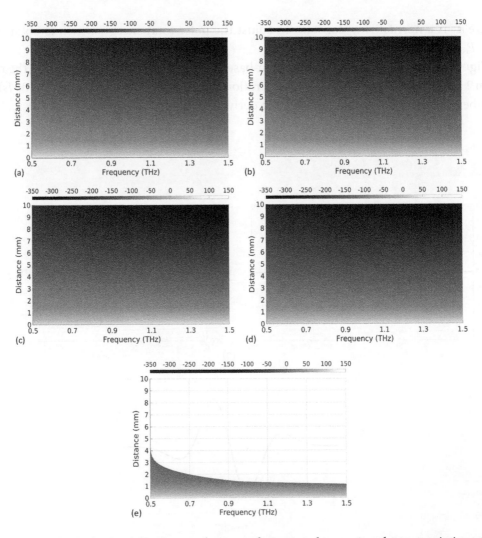

FIGURE 5.10   SNR obtained for the top-down configuration, by varying the transmission strategies. (a) Flat transmission; (b) Pulse-based transmission, $n = 1$; (c) Pulse-based transmission, $n = 2$; (d) Pulse-based transmission, $n = 3$; and (e) Optimal transmission.

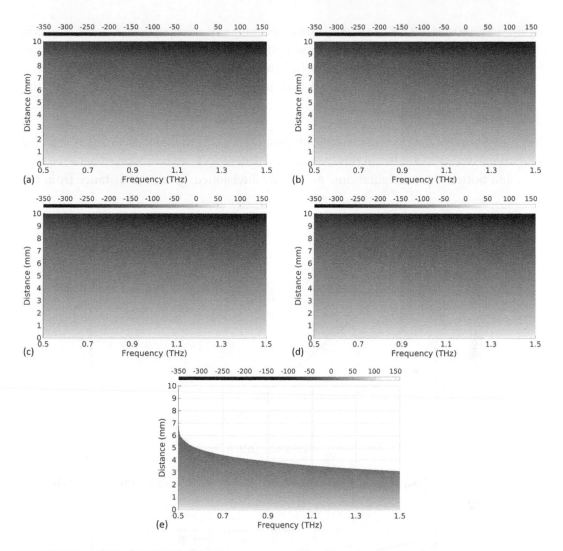

FIGURE 5.11 SNR obtained for the bottom-up configuration, by varying the transmission strategies. (a) Flat transmission; (b) Pulse-based transmission, $n = 1$; (c) Pulse-based transmission, $n = 2$; (d) Pulse-based transmission, $n = 3$; and (e) Optimal transmission.

In both cases, according to the total path loss trend, the SNR grows up when the communication frequency and the distance between source and destination node decrease. On the other hand, given the distance from the source node, the bottom-up configuration registers higher values of SNR than the top-down approach. This is again justified through the total path loss behavior: the inner layers of the stratified medium introduce lower path loss than the outer ones. This anticipates that the direction of the communication in human tissues influences the resulting link capacity.

Furthermore, by taking into account different communication schemes, the behavior of flat and pulse-based transmission strategies is not completely the same, and this will bring to different values of the communication capacity (as illustrated later). Moreover, the study of the optimal transmission scheme leads to two important considerations. First, it is possible to note that the SNR can be evaluated only for a limited portion of the bandwidth, based on the power profile solution of the optimization problem. Second, the adoption of the optimal power profile brings to SNR values that slightly reduce with the distance between communicating nodes, while maintaining similar values in the frequency domain.

5.4.2 Channel Capacity

The upper bound of the channel capacity, obtained according to Equation (5.26), is depicted in Figures 5.12 and 5.13. Also, in this case, several differences can be observed between top-down and bottom-up configurations. As already envisioned, given the distance from the

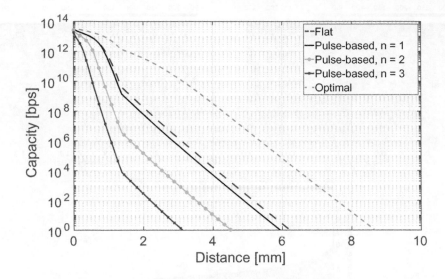

FIGURE 5.12   Channel capacity as a function of the transmission range, evaluated for the top-down configuration.

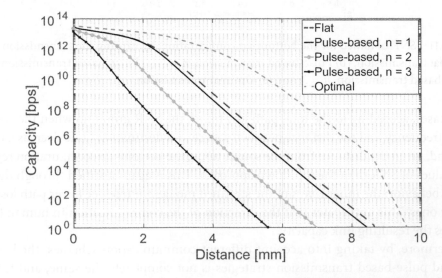

FIGURE 5.13   Channel capacity as a function of the transmission range, evaluated for the bottom-up configuration.

source node, the bottom-up direction ensures higher channel capacity. On the other hand, in any case, the channel capacity decreases with the transmission range. This clearly highlights the destroying effect of the path loss, which is even more evident when the distance between transmitter and receiver increases.

As expected, the optimal transmission scheme reaches optimal performance. At the same time, the more realistic approach, based on the transmission of Gaussian-based pulses, generally achieves the lowest performance, which in turn gets worse when the derivative order n of the Gaussian-based pulse increases. In summary, it is possible to conclude that a channel capacity in the order of Tbps can only be reached for transmission ranges less than 2 and 4 mm when the top-down configuration and the bottom-up configuration are considered, respectively. Furthermore, when the distance between source node and destination device exceeds 9 mm (top-down) and 9.5 mm (bottom-up), communication capabilities are extremely injured (i.e., the channel capacity becomes lower than 1 bps).

5.5 SUMMARY

This chapter investigated physical transmission rates and communication ranges reachable in human skin. To this end, a sophisticated channel model taking into account the spatial dependence of the skin permittivity has been formulated. Indeed, electromagnetic field and the Poynting vector were calculated by using the FDTD technique, while path loss, noise power spectral density, and SNR were evaluated as a function of the communication frequency and the distance between transmitter and receiver. Starting from these models, physical transmission rates and communication ranges have been calculated by varying the transmission techniques and the positions of both transmitter and receiver. The obtained results demonstrate that a physical data rate in the order of Tbps can only be reached for transmission ranges less than 2 mm. When the distance between transmitter and receiver exceeds 9 mm, communication capabilities are extremely impaired (i.e., the physical data rate tends to be lower than 1 bps). These important findings are extremely useful to drive future research activities devoted to the design of innovative networking methodologies, protocols, and algorithms for the nanoscale.

REFERENCES

1. J. M. Jornet and I. F. Akyildiz, "Channel modeling and capacity analysis for electromagnetic wireless nanonetworks in the terahertz band," *IEEE Transactions on Wireless Communications*, vol. 10, no. 10, pp. 3211–3221, 2011.
2. B. Atakan, O. Akan, and S. Balasubramaniam, "Body area nanonetworks with molecular communications in nanomedicine," *IEEE Communications Magazine*, vol. 50, no. 1, pp. 28–34, 2012.
3. G. Piro, G. Boggia, and L. A. Grieco, "On the design of an energy-harvesting protocol stack for Body Area Nano-NETworks," *Nano Communication Networks Journal*, vol. 6, no. 2, pp. 74–88, 2015.
4. J. M. Jornet and I. F. Akyildiz, "Graphene-based plasmonic nano-antenna for terahertz band communication in nanonetworks," *IEEE Journal on Selected Areas in Communications*, vol. 31, no. 12, pp. 685–694, 2013.
5. I. F. Akyildiz, C. Han, and S. Nie, "Combating the distance problem in the millimeter wave and terahertz frequency bands," *IEEE Communications Magazine*, vol. 56, no. 6, pp. 102–108, 2018.

6. J. M. Jornet and I. F. Akyildiz, "Information capacity of pulse-based wireless nanosensor networks," in *Proceedings of IEEE Conference on Sensor, Mesh and Ad Hoc Communications and Networks, SECON*, June, 2011, pp. 80–88.

7. D. Caratelli, L. Mescia, P. Bia, and O. V. Stukach, "Fractional-calculus-based FDTD algorithm for ultrawideband electromagnetic characterization of arbitrary dispersive dielectric materials," *IEEE Transactions on Antennas and Propagation*, vol. 64, pp. 3533–3544, 2016.

8. P. Bia, D. Caratelli, L. Mescia, R. Cicchetti, G. Maione, and F. Prudenzano, "A novel FDTD formulation based on fractional derivatives for dispersive Havriliak-Negami media," *Signal Processing*, vol. 107, pp. 312–318, 2015.

9. H. Akiyama and R. Heller, *Bioelectrics*. 1 em plus 0.5 em minus 0.4 em Springer, Berlin, Germany, 2016.

10. M. L. Yarmush, A. Golberg, G. Sersa, T. Kotnik, and D. Miklavcic, "A novel FDTD formulation based on fractional derivatives for dispersive havriliak-negami media," *The Annual Revision Biomedical Engineering*, vol. 16, pp. 295–320, 2014.

11. M. Sustarsic, A. Plochowietz, L. Aigrain, Y. Yuzenkova, N. Zenkin, and A. Kapanidis, "Optimized delivery of fluorescently labeled proteins in live bacteria using electroporation," *Histochemistry and Cell Biology*, vol. 142, pp. 113–124, 2014.

12. S. Movassaghi, M. Abolhasan, J. Lipman, D. Smith, and A. Jamalipour, "Wireless body area networks: A survey," *IEEE Communications Surveys Tutorials*, vol. 16, no. 3, pp. 1658–1686, 2014.

13. T. Nakano, M. Moore, F. Wei, A. Vasilakos, and J. Shuai, "Molecular communication and networking: Opportunities and challenges," *IEEE Transactions on NanoBioscience*, vol. 11, no. 2, pp. 135–148, 2012.

14. G. Piro, P. Bia, G. Boggia, D. Caratelli, L. A. Grieco, and L. Mescia, "Terahertz electromagnetic field propagation in human tissues: A study on communication capabilities," *Nano Communication Networks*, vol. 10, pp. 51–59, 2016.

15. S. Canovas-Carrasco, A. J. Garcia-Sanchez, and J. Garcia-Haro, "A nanoscale communication network scheme and energy model for a human hand scenario," *Nano Communication Networks*, vol. 15, pp. 17–27, 2018.

16. R. Magin, *Fractional Calculus in Bioengineering*. 1em plus 0.5em minus 0.4em Begell House, Danbury, CT, 2006.

17. C. Polk and E. Postow, *Biological Effects of Electromagnetic Fields*. 1em plus 0.5em minus 0.4em CRC Press, Boca Raton, FL, 1996.

18. V. Raicu, "Dielectric dispersion of biological matter: Model combining debye-type and "universal" responses," *Physical Review. E, Statistical Physics, Plasmas, Fluids, and Related Interdisciplinary Topics*, vol. 60, pp. 4677–4680, 1999.

19. L. Mescia, P. Bia, M. A. Chiapperino, and D. Caratelli, "Fractional calculus based FDTD modeling of layered biological media exposure to wideband electromagnetic pulses," *Electronics*, vol. 6, p. 106, 2017.

20. P. Bia, D. Caratelli, L. Mescia, and J. Gielis, "Analysis and synthesis of supershaped dielectric lens antennas," *IET Microwaves, Antennas & Propagation*, vol. 9, pp. 1497–1504, 2015.

21. L. Mescia, P. Bia, and D. Caratelli, "Fractional derivative based FDTD modeling of transient wave propagation in Havriliak-Negami media," *IEEE Transactions on Microwave Theory and Techniques*, vol. 62, pp. 1920–1929, 2014.

22. P. Bia, L. Mescia, and D. Caratelli, "Fractional calculus-based modeling of electromagnetic field propagation in arbitrary biological tissue," *Mathematical Problems in Engineering*, vol. 2016, pp. 1–11, 2016.

23. L. Mescia, P. Bia, and D. Caratelli, "Fractional-calculus-based electromagnetic tool to study pulse propagation in arbitrary dispersive dielectrics," *Physica Status Solidi A*, 2018.

24. K. Yang, A. Pellegrini, M. Munoz, A. Brizzi, A. Alomainy, and Y. Hao, "Numerical analysis and characterization of THz propagation channel for body-centric nano-communications," *IEEE Transactions on Terahertz Science and Technology*, vol. 5, no. 3, pp. 419–426, 2015.

25. K. Sasaki, M. Mizuno, K. Wake, and S. Watanabe, "Measurement of the dielectric properties of the skin at frequencies from 0.5 GHz to 1 THz using several measurement systems," in *Proceedings of International Conference on Infrared, Millimeter, and Terahertz Waves (IRMMW-THz)*, Hong Kong, China, August 2015.

26. M. Ney and I. Abdulhalim, "Does human skin truly behave as an array of helical antennae in the millimeter and terahertz wave ranges?" *Optics Letters*, vol. 35, pp. 3180–3182, 2010.

27. S. Naito, M. Hoshi, and S. Yagihara, "Microwave dielectric analysis of human stratum corneum in vivo," *Biochimica et Biophysica Acta*, vol. 1381, pp. 293–304, 1998.

28. P. Hasgall, F. D. Gennaro, C. Baumgartner, E. Neufeld, M. Gosselin, D. Payne, A. Klingenbock, and N. Kuster. 2015. IT-IS database for thermal and electromagnetic parameters of biological tissues. [Online]. Available: http://www.itis.ethz.ch/database (accessed April 18, 2019).

29. S. Huclova, D. Erni, and J. Frohlich, "Modeling and validation of dielectric properties of human skin in the MHz region focusing on skin layer morphology and material composition," *Journal of Physics D Applied Physics*, vol. 45, p. 025301, 2012.

30. P. Zakharov, F. Dewarrat, A. Caduff, and M. Talary, "The effect of blood content on the optical and dielectric skin properties," *Physiological Measurement*, vol. 32, pp. 131–149, 2011.

31. L. Mescia, P. Bia, and D. Caratelli, "Fractional-calculus-based FDTD method for solving pulse propagation problems," in *Proceedings of IEEE International Conference on Electromagnetics in Advanced Applications (ICEAA)*, September, 2015, pp. 460–463.

32. G. Piro, K. Yang, G. Boggia, N. Chopra, L. A. Grieco, and A. Alomainy, "Terahertz communications in human tissues at the nano-scale for healthcare applications," *IEEE Transactions on Nanotechnology*, vol. 14, no. 3, pp. 404–406, 2015.

33. S. F. Bush, J. Paluh, G. Piro, V. Rao, V. Prasad, and A. Eckford, "Defining communication at the bottom," *IEEE Transactions on Molecular, Biological, and Multi-scale Communications (T-MBMC)*, vol. 1, no. 1, pp. 90–96, 2015.

34. J. M. Jornet and I. F. Akyildiz, "Joint energy harvesting and communication analysis for perpetual wireless nanosensor networks in the terahertz band," *IEEE Transactions on Nanotechnology*, vol. 11, no. 3, pp. 570–580, 2012.

CHAPTER 6

Nanoscale Channel Modeling in Highly Integrated Computing Packages

Sergi Abadal, Xavier Timoneda, Josep Solé-Pareta,
Eduard Alarcón, Albert Cabellos-Aparicio, Anna
Tasolamprou, Odysseas Tsilipakos, Christos Liaskos, Maria
Kafesaki, Eleftherios N. Economou, Costas Soukoulis,
Alexandros Pitilakis, Nikolaos V. Kantartzis, Mohammad
Sajjad Mirmoosa, Fu Liu, and Sergei Tretyakov

CONTENTS

CONTINUED DOWNSCALING TRENDS ARE enabling the introduction of wireless communications within computing packages. This opens the door to significant improvements in multiprocessor interconnects and exciting new applications. However, the realization of these possibilities requires the understanding of the propagation mechanisms in such dense and integrated scenarios, to later design appropriate wireless communications protocols that adapt to them. This chapter reviews recent efforts that, with the aim of bridging this gap, identify possible propagation paths, define modeling and simulation methodologies, and provide first frequency-domain evaluations of the nanoscale wireless channels within computing packages.

6.1 INTRODUCTION

Constant downscaling of radio frequency (RF) circuits have recently opened the door to the design of antennas and transceivers that can be integrated within chips [24]. Although higher integration was initially driven by a need to lower fabrication costs, recent times have seen the emergence of new wireless applications, where the size of the RF front end plays a critical role. These applications, summarized next, are enabled by advances in nanotechnology that continue to push the limits of miniaturization and frequency of operation—leading to wireless systems in the millimeter-wave (mmWave) (30–300 GHz) and terahertz (THz) bands (0.3–3 THz). Figure 6.1 illustrates this trend by plotting the transceiver area and power of a wide variety of integrated implementations targeting short-range, high-speed wireless communications.

RF technology has indeed reached a point where tens or even hundreds of antennas and transceivers could be integrated within a computing system. This allows to establish

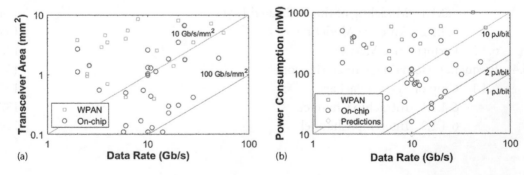

FIGURE 6.1   (a) Transceiver area and (b) power consumption as functions of the achieved data rate in mmWave and THz designs for short-range, high-speed wireless communications. (From Tasolamprou, A.C. et al., Intercell wireless communication in software-defined metasurfaces, in *Proceedings of the ISCAS'18*, 2018.)

wireless links between the modules within a data center rack [2], the different components of a printer [3], the different chips of a multichip module [4], or even the different processor cores of a single-chip multiprocessor [5]. In the extreme downscaling cases, probably most relevant to the nanocommunications community, two novel applications stand out: wireless network-on-chip (WNoC) [6] and software-defined metamaterial (SDM) [7]. In the former, low-latency broadcast-capable links are established to distribute data shared among the processor cores of a multiprocessor. In the latter, those wireless links can be employed to interconnect the different unit cells of a programmable metamaterial to aid in its internal reconfiguration. The reader will find more details on these applications in Chapters 3 and 20.

The adoption of wireless communications in such highly integrated environments poses significant challenges in aspects such as optimal antenna placement, cointegration, interference management, and data modulation. A common problem to several of these challenges is the lack of a proper characterization of the wireless channel within a computing package, even though the theory is well formulated [8], and several works have analyzed the on-chip [9–12], off-chip [13–15], and printed circuit board (PCB) board cases [16–19]. However, very few studies include the chip package in their simulations or measurements, and those that do it are limited to low frequencies or lack proper justifications on the antenna type and placement [20–22]. Thus far, no works have addressed the novel SDM environment yet.

In this chapter, we lay down the fundamentals of channel modeling in highly integrated environments. We review the computer package environments both in the WNoC and SDM, taking into full consideration realistic chip or system packages (Section 6.2) and the potential antenna placement (Section 6.3). Then, we perform a frequency-domain study of the wireless channel within a set of computing packages relevant to both scenarios (Section 6.4). By deriving communication metrics from the channel exploration, architects can accurately estimate the performance and cost of future nanoscale wireless communications in highly integrated computing packages. Finally, a summary of results and an outlook of this research field are given in Section 6.5.

6.2 ENVIRONMENT DESCRIPTION

This chapter focuses on two emerging applications for nanoscale wireless communications within computing packages: WNoC and SDM. The former places antennas within a multiprocessor environment, whereas the latter is located within the structure of a metasurface. However, the two scenarios share several features, and a similar set of representative cases can be applied to both.

On the one hand, Figure 6.2a illustrates the multiprocessor scenario in a generic multichip module package, where different components are placed over package substrate, which is in turn connected to the PCB board of the computer [23]. On top of the components, there is a package material that is generally covered by a metallic layer that acts as a heat sink. In this case, antennas are generally placed within the different components, and communication can be either intrachip or interchip. On the other hand, Figure 6.2b shows a schematic example of an SDM: a programmable metasurface with internal controllers and intercontroller network. The metasurface is basically composed by an array of metallic patches placed over a substrate

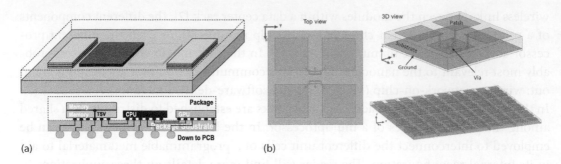

FIGURE 6.2   Schematic diagrams of typical (a) multiprocessor and (b) reconfigurable metasurface environments.

and a common ground plane. A set of vias emerges from below the ground plane, where the controller chips lie, and is used to tune the response of the metamaterial. Each metasurface may account for several chips that, in turn, can host multiple controllers. Therefore, this case also admits intrachip and interchip wireless communication.

Since both scenarios are structurally unalike, the paths through which electromagnetic waves propagate differ depending on the case. However, we find that they share three potential approaches to create a wireless channel within the device, which are summarized in Figure 6.3 and detailed in next sections:

- *Standard:* Through the chip or system package

- *Opportunistic:* Through existing structures that can be exploited as some sort of waveguides

- *Custom:* Through a dedicated layer created specifically for the purpose of wireless intradevice communications

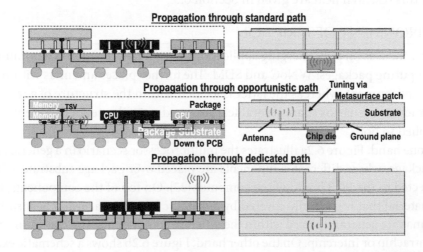

FIGURE 6.3   Schematic diagrams of the standard, custom, and opportunistic paths in the WNoC and SDM environments (left and right, respectively).

6.2.1 Chip or System Package

Top schematics of Figure 6.3 depict the standard environment, where antennas are integrated within the chip or at its close vicinities, and propagation occurs in two regions: (i) the intrachip region, in which the waves radiated by the monopole inside the silicon substrate travel through several layers of the chip (mainly the silicon layer), and (ii) the interchip region, in which the waves that have left the chip travel through the interchip space, until they reach the boundaries of another chip.

The main advantage of this configuration is that the communication channel does not require significant modifications of the original structure. This leads to a very cost-effective implementation that leaves the antenna and transceiver circuits as the only elements that may incur some area overhead. Depending on the actual implementation of the system package, this scenario could lead to a totally enclosed volume, thereby excluding the possibility of any electromagnetic coupling between external electromagnetic phenomena (e.g., metasurface operation and external incoming signals) and the communication. Still, losses may still arise due to reflections, the chip materials, or spreading in undesired areas within the package.

In a multiprocessor, chip packages are typically of the flip-chip type due to thier lower inductance and higher power/bandwidth density than wire bonding. In this configuration, chips are turned over and carefully connected to the package carrier (or interposer) through a set of solder bumps. The packaged chip then takes the canonical form presented in Figure 6.4. On top, the heat sink draws the limits of the package and is interfaced to the chip via the heat spreader material. Below the heat spreader, bulk silicon with low resistivity (10 Ω·cm) serves as the foundation of the digital circuits. The interconnect layers, which occupy the bottom of the silicon die, as shown in the inset of Figure 6.4, are generally made of copper and surrounded by an insulator such as silicon dioxide (SiO_2) [24].

Table 6.1 summarizes the typical dimensions and characteristics of the materials found in the cross-section of a chip package. Note that the lateral dimensions are not indicated, since they depend much on the final application. Also, the actual structure below the bumps may vary to adapt to the system's requirements.

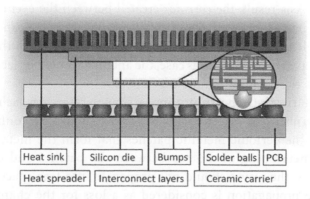

FIGURE 6.4   Schematic of the layers of a flip-chip package.

TABLE 6.1   Characteristics of the Layers in a Computing Package

	Thickness (May Vary)	Material	ε_r	tan(δ)
Heat sink	0.5 mm	Aluminum	—	—
Heat spreader	0.2 mm	Thermal conditions	8.6	$3 \cdot 10^{-4}$
Silicon die	0.7 mm	Bulk Silicon	11.9	0.2517
Interconnections	13 µm	Cu and SiO_2	3.9	0.03
Bumps	87.5 µm	Cu and Sn	—	—
Ceramic carrier	0.5 mm	Alumina	9.4	$4 \cdot 10^{-4}$
Solder balls	0.32 mm	Lead	—	—
PCB	0.5 mm	Epoxy resin	4	—

Although dipoles and other antennas can be printed on-chip, recent works have proposed to build vertical monopoles that radiate in the horizontal directions [25–27]. In this case, the bump array at the bottom of the chip acts as ground plane, and a quarter-wave monopole can be employed. The antennas can be placed within the chip or in close vicinities, in which case the radiation pattern would slightly differ.

6.2.2 Slotted Structures

The schematics in the middle of Figure 6.3 depict the opportunistic implementation of a wireless channel within a computing package. The opportunistic nature resides in the use of already-existing structures to create a sort of waveguide for the electromagnetic waves. The main advantage of this approach is that, if properly engineered, it can offer a much better propagation than in the standard chip package case, without adding any extra overhead. However, these advantages come with a complexity cost arising from the need to avoid undesired interferences with the normal operation of the leveraged structures. The actual cost of the solution will very much depend on the employed propagation path, which, in turn, is conditioned by the application.

In the WNoC paradigm, the opportunity arises in the space left between the microbump array below the insulator layer (typically SiO_2). Below the bumps, the package carrier generally holds a metallic redistribution layer that can be seen as a thin metallic plane within the carrier. As a result, the *corridor* created between the evenly spaced bumps is sandwiched between two layers with dense metallization layers that can act as a waveguide as long as the wavelength of the wireless signals is much lower than the bump pitch and insulator–carrier separation [25]. In this case, the antenna would be placed in the interconnection layer that is closer to the bumps.

In the SDM scenario, the opportunity appears much clearer in the metasurface layer. Based on such channel, the electromagnetic waves should be confined between the ground plane and the periodic metallic patches that form the metasurface structure. With a proper dimensioning of the gap between the patches and depending on the operation frequency, the waves should not propagate into free space. From this point of view, free-space propagation is considered as a loss for the channel. To excite the waves, a monopole can be assumed to be fed from the chips below the ground plane.

This is possible as long as a cylindrical hole with a radius larger than the radius of the monopole is made on the ground plane around the antenna. This hole prevents the connection of the antenna to the ground.

6.2.3 Dedicated Layer

Bottom schematics of Figure 6.3 depict the dedicated layer implementation, whereby it is considered that the communication is enabled by an additional sublayer serving only as a channel for transferring the signals between the communication nodes. The channel is created by introducing two additional metallic plates below the chip, which form a parallel-plate metallic waveguide channel whose thickness, as explained later on, is specified by the desired frequency of operation. The space between the two metallic plates is empty or filled with a uniform dielectric material. In each chip, a probe, that is, an antenna placed through a small hole between the two plates, in the vertical direction, z, is connected, as seen in Figure 6.3. This probe acts as an omnidirectional wire antenna, that is, a device that transmits or receives electromagnetic power isotropically in the horizontal xy plane.

The main advantage of the present configuration is that the communication channel is electromagnetically isolated from the rest of the system by the computing chips in the WNoC paradigm or the parts that contribute to the metasurface operation in the SDM case. In that sense, the possibility of any electromagnetic coupling between the processor/metasurface operation and the communication is excluded by design. Moreover, the parallel plates create an enclosed volume, where no power leakage is allowed (the holes under the chip that host the probe antenna are electromagnetically very small and are not taken into consideration). In addition, there are no obstacles in the propagation space, apart from the probe antennas themselves. For all these reasons, the dedicated layer provides increased robustness and design flexibility. However, it requires additional fabrication effort and increases the overall volume of the unit cell.

6.3 CHANNEL-CHARACTERIZATION METHODOLOGY

6.3.1 Modeling Approach

The multiprocessor and metasurface structures, as well as the monopole antennas, can be readily modeled in full-wave solvers to obtain the fields and other performance metrics characterizing propagation. In this work, we use computer simulation technology (CST) [28] to this end. We consider, unless noted, the dimensions and materials listed in Table 6.1. Most of these materials are well characterized at the bands analyzed in this work.

It is worth noting that the proposed simulation approach can be changed, depending on the conditions of the analysis. For instance, one could consider frequencies around the THz band, without downscaling the scenario. In that case, the environment becomes very large in terms of wavelengths λ and, as such, computationally intractable. This situation suggests using a ray-tracing approach, instead, to reduce the computational cost without compromising accuracy [9]. Electromagnetic field modeling within package can help in this regard [29].

6.3.1.1 Modeling of the Dedicated Layer

While the propagation environment is mostly given in the standard and opportunistic scenarios, the dedicated layer is custom-built to optimize wireless propagation. In this case, we follow a parallel-plate waveguide approach, as presented in Figure 6.5. The metallic plates perfectly reflect the electromagnetic waves, and they can guide a discrete spectrum of transverse electric (TE) and transverse magnetic (TM) modes, depending on the distance of the plates d, the dielectric permittivity of the filling materials ε_r, and the frequency of operation f.

One characteristic property of the parallel-plate waveguide is that the lowest-order supported mode corresponds to transverse electric-magnetic (TEM) waves, in which both the electric and magnetic fields are perpendicular to the propagation direction. This TEM mode can be excited form zero frequency direct current (DC), and it is the only mode that propagates in the waveguide from the DC up to the cut-off frequency of the first higher-order mode. It is desired that the waveguide remains single mode, and this is the case as long as the frequency of operation is smaller than the next mode cut-off frequency [30],

$$f < \frac{c_0}{2d\sqrt{\varepsilon_r}}, \qquad (6.1)$$

where c_0 is the speed of light. The propagation in the waveguide remains single mode, as long as Equation 6.1 is satisfied; this effectively specifies a maximum distance between the plates d, when the frequency of operation is fixed. This would also affect the size of the antennas placed within the parallel-plate waveguide. However, this is not a problem, as we see next.

6.3.1.2 Antenna Modeling

As described in Section 6.2, quarter-wave monopole antennas are a valid option for all the environments, owing to the presence of perpendicular ground planes in several locations. The corresponding frequency of resonance f_0 therefore depends on the antenna length L as:

$$f_0 = \frac{c_0}{4L\sqrt{\varepsilon_r}}, \qquad (6.2)$$

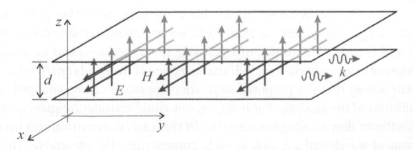

FIGURE 6.5 Parallel-plate waveguide and TEM wave propagation. The waveguide in TEM operation is naturally matched to the free space, assuming there is no dielectric loading between the plates.

where ε_r denotes the relative permittivity of the host material in which the dipole is embedded. At this frequency, we may expect the zero reactance as the imaginary part of the input impedance of the antenna.

There are two ways to drive the antenna in the full-wave simulator. On the one hand, in cases where a hole needs to be made in the ground plane to let the vertical via pass and thus prevent the connection of the antenna to the ground, the cylindrical hole may operate as a coaxial cable. This situation allows to excite the antennas by using the waveguide port in the full-wave simulator. In such case, the diameter of the hole is chosen such that the characteristic impedance of the equivalent coaxial cable becomes 36 Ω, yielding close-to-perfect matching, because this value corresponds to the radiation resistance of a quarter-wavelength antenna. On the other hand, the lumped element port can be also employed to excite the antenna in cases where a coaxial-like feed cannot be built. This port can be positioned between the ground and the dipoles, which should be shifted a bit above the ground. It is worth mentioning, though, that both methods will yield the same results.

To illustrate the validity of the approach, Figure 6.6 shows the schematic of a monopole antenna within a multiprocessor package, its simulated radiation pattern, and the return loss. The return loss is shown for different instances of the monopole, whose length is modified to tune the antenna to frequencies from 50 to 140 GHz in 10-GHz steps. The radiation pattern is omnidirectional in the coplanar direction and tends to radiate away from the ground plane.

Note that the results from Figure 6.6 assume that there is no metallic plane near the monopole's end. In the parallel-plate waveguide case (dedicated layer), for instance, the distance between the plates defines the frequency of TEM operation and at the same time the length and therefore the properties of the probe antenna. To exemplify this, let us consider that the probe antenna is a single-wire antenna (monopole) placed between the two copper plates (separated by air) along the z direction, as shown in Figure 6.7a. It is implemented by a cylindrical rod of radius $R=0.12$ mm, while the length of the wire L varies according to the distance of the plates $L=d-2*d_{gap}$, where d_{gap} is the gap between the antenna and the upper and lower plates, fixed to approximately 0.1 mm. The monopole is assumed to be excited through one plate by means of a discrete wire port of impedance equal to $Z_0 = 50$ Ω that feeds current to the rod and stimulates the radiation.

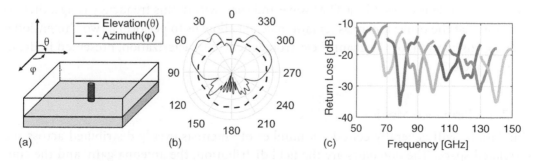

FIGURE 6.6 (a) Schematic representation of an On-chip monopole, (b) Expected radiation pattern, and (c) Return loss for instantiations optimized at 50–140 GHz.

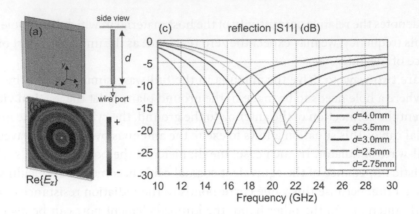

FIGURE 6.7 (a) Schematic of the single monopole antenna standing between the parallel plates, and schematic side view. (b) Radiated field distribution and (c) Reflection coefficient, S_{11}, calculated in the emitter for various parallel-plate distances, d, in the frequency range $f = [10, 30$ GHz$]$ (single-mode operation).

Assuming that the distance of the plates varies in the range $d = [2.5, 4$ mm$]$, according to the cut-off condition of Equation 6.1, the parallel-plate waveguide remains single mode up to 37.5 GHz; the investigation is restricted within the single-mode operation regime where, additionally, there is a natural boundary impedance match between the waveguide and the free space. The distribution of the radiated field is presented in Figure 6.7b. Figure 6.7c presents the amplitude of the probe antenna reflection coefficient S_{11}, as the distance of the parallel plates varies in the range $d = [2.5, 4$ mm$]$ and the length of the antenna is $L = 3.8$ mm. In this case, for large d, it is found that the resonance occurs at $f = 14.2$ GHz ($\lambda \sim L / 4$). At the resonance, the antenna is matched, since $S_{11} < -20$ dB, and remains matched ($S_{11} < -15$ dB) in the range $f = [13.5, 15.5$ GHz$]$, that is, in a frequency span of approximately 14% about the central frequency of operation.

The frequency span can be extended by improving the impedance matching of the antenna, that is, by varying the gap form the plates d_{gap}. As presented in Figure 6.7, decreasing the distance of the plates, and consequently the length of the antenna, nicely tunes the antenna resonance. The facts that the electromagnetic power in the parallel plate single-mode waveguide is carried by a TEM wave and that, within this frequency range, there is a regime that the probe antennas remain matched allow us to perform a two-dimensional (2D) approximation for the channel electromagnetic characterization, presented in detail in Section 6.4.3.

6.3.2 Performance Metrics

Simulations will generally consider a number of antennas evenly distributed across the simulated space. The outcomes are the field distribution, the antenna gain, and the coupling between antennas. To see the electromagnetic coupling (communication) between any antennas, it is enough to observe the magnitude of the scattering matrix component

S_{21} (that is a complex number). Generally speaking, for a two-port network, the scattering matrix is described as

$$S = \begin{bmatrix} S_{11} & S_{12} \\ S_{21} & S_{22} \end{bmatrix}, \tag{6.3}$$

where S_{21} and S_{12} determine the transmitted power from one port to the other port, and S_{11} and S_{22} represent the reflection of the power from each port due to its mismatch to the network. In the structure under study, however, in addition to S_{21}, we also observe the S_{11} component in order to understand how much power is reflected toward the antennas operating as the transmitter/receiver. The best case is to achieve a very low value for the magnitude of the S_{11}, meaning negligible loss due to the reflection, and a high value for the magnitude of S_{21}. However, in the case of noticeable loss (reflection), we can resolve this issue by employing an external matching circuit. In fact, what is indeed important is the transmission coefficient. Here, it is worthwhile to note that due to the reciprocity and symmetry of the problem, it holds that $S_{12} = S_{21}$ and $S_{22} = S_{11}$, respectively.

Besides the S-parameters, a metric closer to channel characterization is the transfer function $|H(f)|$ of the channel, which can be obtained via the following expression:

$$|H(f)|^2 = \frac{|S_{21}|^2}{(1-|S_{11}|^2)G_r G_t}, \tag{6.4}$$

in which G_r and G_t are the gain of the receiving and transmitting antennas, respectively [31]. Once the transfer function has been obtained, one can evaluate the channel attenuation L in the frequency of interest, and, using data from various antennas, a path loss analysis can be performed by fitting the attenuation L over transmission distance d_{TRX} as:

$$L_{dB} = 10n \cdot \log_{10}(d_{TRX}) + C, \tag{6.5}$$

where n is the path loss exponent [12]. The path loss exponent is around 2 in free space, below 2 in guided or enclosed structures, and above 2 in lossy environments.

6.4 FREQUENCY-DOMAIN CHARACTERIZATION

6.4.1 Chip or System Package

6.4.1.1 Intrachip Channel

The first analysis in the standard environment considers a homogeneous array of 4×4 antennas in a single chip with both length and width of 20 mm. The materials and vertical dimensions by default are given in Table 6.1. In this first analysis, we consider that the antennas are initially tuned to 60 GHz and perform two explorations: coupling as a function of the radiation frequency and dimensions of the chip constituents. To this end, let $\bar{S}_{ij}$ be the average of the coupling between transmitter j and receiver i over the targeted frequency band. The minimum of $\bar{S}$ is used as a benchmark to evaluate the worst case between any two antennas of the 4×4 array. It is expressed as:

$$S_{min} = \min_{i,j \neq i} \overline{S_{ij}}. \tag{6.6}$$

Leaving the dimensions fixed, an increase of the frequency is sought due to the potential for smaller transceivers and higher speeds. To understand how this affects the intrachip channel, Figure 6.8 shows how S_{min} scales over frequency. It is observed that the loss between links increases with frequency. This is most likely due to the decreasing aperture of the antennas and the larger propagation losses at the dielectrics. These effects, however, are partially compensated by the enclosed nature of the on-chip scenario, mitigating the impacts of frequency scaling.

We next analyze the impact of the bulk silicon and heat spreader dimensions to the coupling between antennas. The reasons behind this analysis are that silicon is a lossy material from the electromagnetic propagation perspective and that the heat spreader has generally better electrical properties. Also, the thicknesses of these components can be changed at the manufacturing time (e.g., the silicon is thinned down for three-dimensional [3D] stacking architectures [32]), thereby becoming a new design factor to take into account.

To better understand the impact of the two materials, we explore the worst-case coupling as a function of the silicon and heat spreader thicknesses. We reduce the silicon layer down to 100 μm, whereas the heat spreader is aluminum nitride (AIN) in our explorations and is increased up to 0.9 mm. Figure 6.9 confirms the intuitions: thinning down the silicon and increasing the size of the heat spreader lead to a considerable reduction of the losses. The difference between 0.1- and 0.7-mm silicon dies is over 40 dB (Figure 6.9b) and around 33 dB when comparing a thick heat spreader with not having it at all (Figure 6.9b), respectively.

These results suggest that package optimization can be performed to minimize the channel losses. We repeated the simulations multiple times with shallow silicon and thick heat spreader, to then analyze the path loss over distance. Figure 6.10 shows a comparison of three cases: a case close to the standard dimensions specified in Table 6.1, optimal dimensions as obtained in the optimization process, and a quite suboptimal design point.

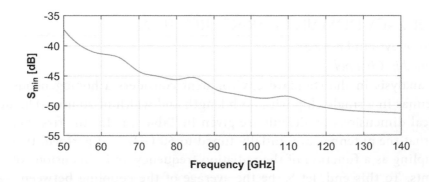

FIGURE 6.8   Worst-case coupling S_{min} as a function of frequency in the single-chip case.

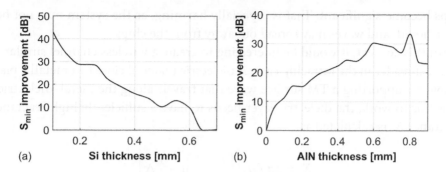

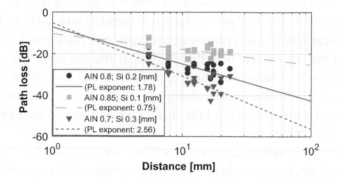

FIGURE 6.9   Average coupling enhancement for different AIN and Si thicknesses: (a) Improvement over 0.7-mm Si and (b) Improvement over no AIN.

FIGURE 6.10   Path loss as a function of distance, including linear regression fitting, for the single-chip case at 60 GHz.

Remember that the path loss decouples the antenna effects and leaves just losses due to propagation. The results show how optimization *reduces not only the path loss overall but also the path loss exponent*. For the default case, the path loss exponent is 1.78, slightly lower than the free-space path loss, thanks to having a confined environment. In the optimal case, we are able to cut the exponent down to 0.75, thereby showing a strong waveguiding effect in propagation. The suboptimal case, with an exponent higher than 2, demonstrates that the losses introduced by silicon cannot be neglected.

6.4.1.2 Interchip Channel

From the environment description, it becomes apparent that systems may rely on wireless communications among different chips. In the WNoC case, it is likely that chips will be close to each other. In this case, the work confirms the existence of two opposed effects [26]. On the one hand, propagation occurring in the interchip space instead of that in lossy silicon leads to better coupling. On the other hand, reflections due to media changes (silicon-package space silicon) leads to a reduction of coupling.

In the SDM case, however, chips may not be close to each other. Each chip may be controlling the tuning of a small subset of metasurface patches and may need to communicate with rather distant controllers. As a result, the losses of propagation in the space between the chips become significant. That is especially harming, as the system may not be completely enclosed, and waves may spread out away from the chips.

Within this context, it would be interesting to create a wireless channel similar to that of the heat spreader in the intrachip case. A dielectric material close to a metallic plane can be capable of supporting a TM surface wave that travels along the metal–dielectric interface. For this to work, the dielectric layer needs to have a sufficiently high reactance [33]. The reactance X_s is calculated as:

$$X_s = 2\pi f \mu_0 \left[\left(\frac{\varepsilon_r - 1}{\varepsilon_r} \right) t + \frac{1}{2} \Delta \right] \tag{6.7}$$

where ε_r is the permittivity of the material, t is the layer thickness, and Δ is the skin depth of the conductor at the frequency of operation, which is given by $\Delta = (\pi f \mu_0 \sigma)^{-1/2}$. σ refers to the conductivity of the metallic plane. Our study targets $f = 60$ GHz and assumes a layer of AIN (like the heat spreader in a processor package) with thickness $t = 0.3253$ mm.

Figure 6.11 shows the electric field distribution at the interchip space. The top small chart illustrates how, as expected, most of the power radiates away from the ground plane without the dielectric layer. The bottom chart, on the other hand, demonstrates that a well-designed dielectric layer is able to bind the surface waves and reinforce propagation along the dielectric, in the path between chips.

To evaluate the improvement in terms of path loss, we obtained the channel attenuation with and without dielectric in a scheme with dies located at distances of 4λ to each other. In that case, the improvement has been of 6.5 dB. A better understanding of the value of surface-wave communications, however, is obtained studying how the path loss scales over distance. The results, shown in Figure 6.12, demonstrate that the surface waves propagate with lesser losses than in free space and than just considering the dielectric loss. Therefore, this technique can be very useful for links between distant chips.

6.4.2 Slotted Structures

The evaluation of the opportunistic approach is performed in the SDM paradigm, based on the structure shown in Figure 6.2b with the following dimensions. Each unit cell comprises four square patches possessing the same thickness (0.0175 mm). Initially, we assume that the width of each patch is equal to 4.2 mm, while the width of one unit cell is 12 mm. This means that the width of the gap between two patches is 1.8 mm. The thickness of the

FIGURE 6.11 Electric field distribution at the surroundings of the radiating chip without and with dielectric layer.

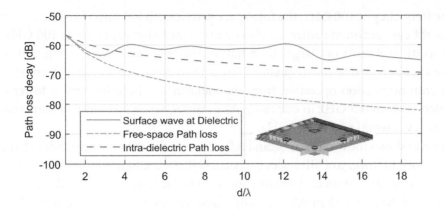

FIGURE 6.12   Path loss using the dielectric layer technique as a function of the relative distance.

substrate (between the ground and the patches) is 1.575 mm, and it is made of the known-material Rogers RT/Duroid 5880 (tm) having a relative permittivity of 2.2 and loss tangent of 0.0009 (indicating that the dielectric loss is negligible in terms of the given communication distance and frequency, and it does not cause a dramatic attenuation of the energy). This corresponds to a metasurface targeting operation at the 5 GHz band.

Each unit cell possesses one monopole antenna that is located under the center of one of the patches in the cell. Notice that the distance between the transmitting antenna in one cell and the receiving antenna in the neighboring cell is exactly 12 mm. The placement of the dipole antenna (under the center) was judiciously chosen so as to minimize unwanted radiation leakage into free space through the gap between patches. The height of the antenna is 1.4 mm.

In order to prevent the electromagnetic fields from leaking outside the waveguide and radiating into the free space, the gap width should be small compared with the wavelength. To exemplify this effect, we simulate the illumination of a thick metallic plane with periodic slots and measure the electric field distribution along the axis of the plane. At low frequencies, the slots are electrically very small compared with the radiation wavelength, and the electric field does not leak. At high frequencies, the slots become comparable in size with the radiation wavelength, and electric field can be measured within the slots (Figure 6.13).

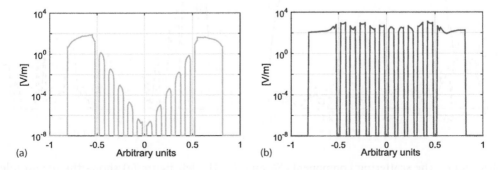

FIGURE 6.13   Field distribution along the axis of a metallic plane with slots. (a) Low frequency and (b) High frequency.

At the frequency of 100 GHz, the free-space wavelength is 3 mm. Therefore, for our particular SDM case outlined earlier, we choose our simulation range up to 100 GHz to keep the gap width (1.8 mm) not larger than half of the free-space wavelength. On the other hand, we are interested in the frequencies larger than 30 GHz because of decoupling of the wireless communication operation from the metasurface functionality at lower frequencies (which is around 5 GHz in our case). As a consequence, we simulate our structure in the frequency interval of [30,100] GHz.

Figure 6.14 shows the transmission and reflection coefficients, corresponding to the initial structure, both in linear and dB scales (gray curves). As the figure indicates, the magnitude of S_{21} is always smaller than 0.1 (−20 dB) above 45 GHz. However, around the frequency of 40 GHz, it can be larger than 0.1 (−20 dB). These values are acceptable from the point of view of communication. Up to −30 dB is satisfactory for wireless communication in this range of distance between the two dipole antennas (i.e., 12 mm). Looking at the magnitude of S_{11}, it has a local minimum at about 40 GHz. The reflection coefficient also has other noticeable local minima around 75 and 95 GHz, but the transmission coefficient at these frequencies is smaller than the transmission coefficient at 40 GHz. Therefore, 40 GHz can be a favorable choice for the frequency of the wireless communication. Here, we emphasize that lower reflection coefficient does not necessarily correspond to a higher transmission coefficient. We should not forget the effect of the complex metasurface environment. Owing to the gap between the patches, the radiated field from the transmitting antenna could penetrate into the free space, for example; this is not reflected in the magnitude of the S_{11}, but S_{21} is significantly affected. This can be seen in Figure 6.14 by comparing

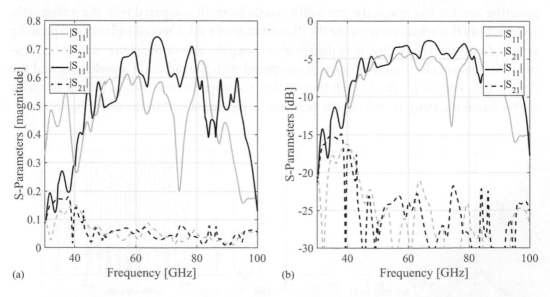

(a)   Frequency [GHz]   (b)   Frequency [GHz]

FIGURE 6.14 The scattering components S_{11} and S_{21}. The left figure (a) shows the magnitude of these components and the right one (b) describes them in dB. Notice that gray curves correspond to the initial structure, and black curves correspond to the second structure under study.

the S-parameters corresponding to 40 and 75/95 GHz. At 75/95 GHz, it seems that the wave radiates into the free space rather than coupling to the receiving antenna. Due to this fact, our structure is not a real two-port network. That is why the known equation, $|S_{11}|^2 + |S_{21}|^2 = 1$, does not hold for the results obtained in Figure 6.14.

In order to investigate whether it is possible to improve the electromagnetic coupling between the transmitting and receiving dipole antennas, we change the initial values of the parameters of the structure and optimize it. We keep in mind that our changes should not be dramatic, because changing these values may correspond to a shift in the functionality of the metasurface from one specific frequency to lower or higher frequencies. However, if the changes are smooth, we can bring back the operational frequency of the metasurface functionality to the desired one by tuning the resistance and capacitance values of the chips connected to the unit cells. One way to reach this goal (optimization) is certainly to decrease the gap width a bit. In this way, the gap width becomes electrically small, and the field leakage outside the waveguide is reduced. Furthermore, we can also vary the thickness of the substrate. We retain the height of the antennas at 1.4 mm. We simulated different structures; here, we only explain the results that are acceptable from the coupling improvement point of view. We increase the substrate thickness to 2.6 mm and reduce the gap width to 1 mm. The S-parameters are shown in Figure 6.14 by gray and black solid curves. The solid curves show the S_{11} parameters, and the dotted curve indicates the S_{21} parameter. As shown, from 30 to 40 GHz, the $|S_{21}|$ is improved (it has a relative maximum of −15 dB, which can be already a very good result). Notice that at the same frequency interval ($f = [30, 40]$ GHz), the reflection coefficient is smaller. After 40 GHz, the coupling reduces similarly to that for the previous structure.

Besides S-parameters, it would be useful to achieve the transfer function of the channel, using Equation 6.4. Notice that both dipole antennas are equal, and therefore, they have the same gain and directivity. We assume that $G_r = G_t = 1.5$. This is an approximation used for the simplicity of the problem. In reality, the gain is a frequency-dependent parameter of the antenna, and it varies as frequency changes. More accurate results can be acquired if we take into account this frequency dependency. Figure 6.15 illustrates the transfer function in linear and dB scales for both first and second structures. The figure completely confirms our previous results, regarding S-parameters, since the magnitude of the transfer function is indeed high (especially for the second structure) between 30 and 40 GHz.

6.4.3 Dedicated Layer

The dedicated layer consists of the parallel-plate waveguide formed by the additional metallic plate behind the chip backplane. The waveguide supports the propagation of TEM waves in single-mode operation according to the cut-off frequency limit defined in Equation 6.1. In addition, as seen in Figure 6.7, the distance between the plates also defines the operation frequency of the monopole antenna, which remains impedance matched ($S_{11} < -20$ dB) in a frequency range of 14% about the central frequency, which also lies within the single-mode operation frequency range.

The fact that the electromagnetic power in the parallel-plate single-mode waveguide is carried by a TEM mode and that the monopole probe is impedance matched in a range

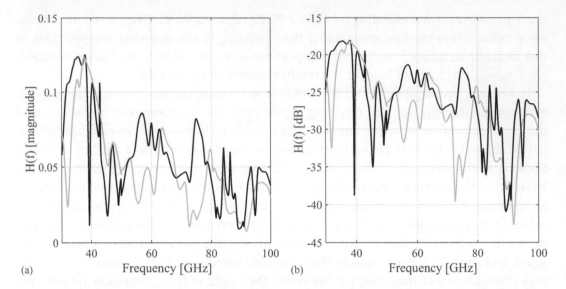

FIGURE 6.15 Transfer function of the channel. The left figure (a) shows the magnitude of the function and the right one (b) describes it in dB scale. The gray curve corresponds to the initial structure, and black curve corresponds to the second (or optimized) structure.

within the single-mode band allows us to perform the following approximation: the wave propagation in the 3D waveguide can be approximated by a 2D analogue, where the monopoles are replaced by finite-size conducting scatterers placed vertically along the antenna probes. Each of the scatterers radiates 2D cylindrical waves in the surrounding space and diffracts the power coming from the environment. The field radiated from the emitter and the diffracted field from the scatterers interfere, creating destructive or constructive patterns in the waveguide. By performing a full-wave numerical analysis, the total field in each position and frequency is calculated. This 2D approximation allows us to solve for large areas in a relatively short computation time and provides a fair qualitative evaluation of the propagation properties in a multiscattering environment. Moreover, as it is a priori assumed that each frequency addressed lies within the resonance of the antenna where it is impedance matched, it is ensured that only the TEM mode is excited.

Since the 2D approximation allows for large-scale calculation, a system of 25×25 nodes or elements is investigated. That is, in each unit cell of a 25×25 metasurface, there is a monopole antenna connected behind the backplate and within the dedicated layer. This translates into a 2D 25×25 rectangular grid of scatters, with nodal distance equal to the size of the metasurface unit cell, here $D = 12$ mm. The schematic of the grid is shown in Figure 6.16. Each antenna/scatterer is a finite-size copper cylinder of radius $R = 0.12$ mm. In this approximation, the impedance characteristics of the probe antennas are not taken into account. The emitter, element no. 17 in Figure 6.16, is simulated as a field source that radiates omnidirectional electromagnetic waves. All the surrounding scatterers reflect the incoming wave. In this way, it is possible to estimate the power profile of the propagating waves in the presence of the reflecting obstacles.

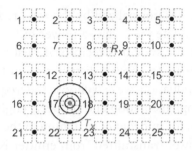

FIGURE 6.16 Schematic of the scatterer grid used in the 2D approximation. In this specific example, antenna no. 17 acts as the emitter and radiates 2D cylindrical waves in the surrounding space. All the scatterers diffract the power coming from the environment, and antenna no. 8 acts as the receiver. The small dotted squares correspond to metallic patches in the other side of the metasurface; each grid of 2×2 patches defines one unit cell.

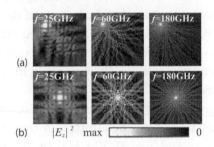

FIGURE 6.17 (a) Power distribution of the field when the emitter is element no. 1, for frequencies $f = 25$, $f = 60$, and $f = 180$ GHz. (b) Power distribution of the field when the emitter is element no. 13 for frequencies $f = 25$, $f = 60$, and $f = 180$ GHz.

Figure 6.17a and b presents the profile of the total power in the plane for frequency $f = 25$, $f = 60$, and $f = 180$ GHz when the emitter is no. 1 and no. 13, respectively. Evidently, the electromagnetic waves interfere either destructively or constructively, producing area patterns of low or high power. Bright spots correspond to high power, and dark spots correspond to low power; consequently, placing a receiver in the dark spots would result in impaired communication, while bright spots would correspond to improved reception position. For example, observing the first of the panels in Figure 6.17a for frequency $f = 25$ GHz, it is obvious that node no. 1 and node no. 21 will not be able to communicate, whereas node no. 1 and node no. 13 can communicate well. Moreover, in Figure 6.17, it is observed that, as expected, the higher frequencies suffer more from multiscattering than the lower frequencies.

As the communication quality depends both on the position of the nodes and the frequency, it is useful to evaluate the connection between separate nodes. This is achieved by estimating, in the position of the receiver, the power that can be captured by the multipath propagation coming from all directions. For the received power calculation alone, we assume that the receiver probe does not reflect. The total power that accumulates in the position of the receiver, $P_{received}$, is normalized by the total radiated power from the emitter, P_0. In this way, we can estimate the received power of the different pairs of emitter N to receiver M, P_{MN}. Note that the system is reciprocal; that is, $P_{MN} = P_{NM}$. As mentioned, this method allows us to evaluate the power distribution in the 2D grid over a large frequency range. Based on this evaluation, we can choose the frequency of operation; then, we can return to the actual 3D implementation and properly engineer the probe antennas, for example, calculate their length for impedance matching in the desired band.

Figure 6.18 presents the power transmitted from emitter N to receiver M over the frequency range of $f = [25, 200 \text{ GHz}]$. In particular, the received power in the communication pairs no. 1–no. 21, no. 1–no. 16, no. 13–no. 1, no. 1–no. 25, no. 13–no. 23, and no. 7–no. 17 is presented. As observed in all cases, the received power remains on average the same for each pair in the entire frequency span. However, for nearly each case, there are some frequency points where the received power drops. For example, in the case of the pair no. 7–no. 17 (panel [f]), three dips in the received power appear at around $f = 45$, $f = 80$, and $f = 115$ GHz. These points correspond to dark spots of destructive wave interference pattern, as seen in Figure 6.18. Moreover, the anticipated trend of decreased received power as the node pair distance increases is observed. For example, in the pair no. 1–no. 21 (panel [a]), the average received power is –15 dB, whereas in the pair no. 1–no. 6 (panel [b]), the received power is on average –8 dB.

Figures 6.19 and 6.20 present the power distribution of the total field with respect to the frequency and the distance from the emitter, in two different frequency ranges: $f = [15, 35 \text{ GHz}]$ in Figure 6.19 and $f = [50, 70 \text{ GHz}]$ in Figure 6.20. In particular, it is

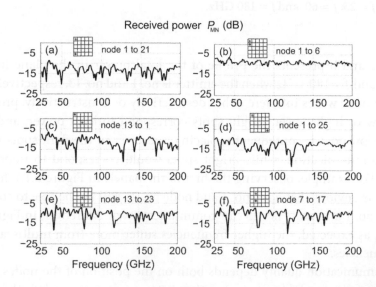

FIGURE 6.18 Received power from the communication between the nodes M and N in frequency range 25–200 GHz. The distance between the neighbouring nodes is 12 mm and the size of the sample is 25 × 25. Six cases of MN node pairs are presented, schematically depicted in the insets (a) to (f).

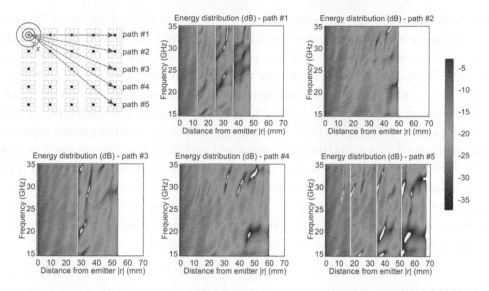

FIGURE 6.19   Power distribution of the fields, frequency $f = [15, 35 \text{ GHz}]$ versus distance from the emitter. The emitter antenna is no. 1, and the power of the total (emitter and scattered) field is calculated along the five paths shown in the schematic.

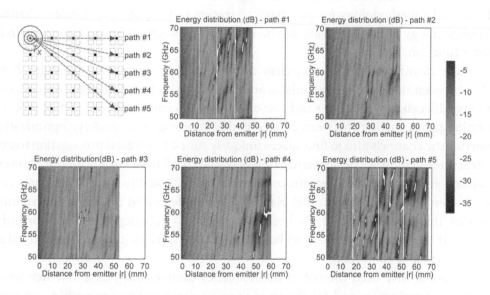

FIGURE 6.20   Power distribution of the fields, frequency $f = [50, 70 \text{ GHz}]$ versus distance from the emitter. The emitter antenna is no. 1, and the power of the total (emitter and scattered) field is calculated along the five paths shown in the schematic.

assumed that element no. 1 radiates, and the power of the total fields is calculated along the marked paths #1, #2, #3, #4, and #5 that connect the position of element no. 1 with elements no. 5, no. 10, no. 15, no. 20, and no. 25, respectively. These paths are shown in the schematics of Figures 6.19 and 6.20. For this calculation, it is assumed that all the elements scatter the incoming field and a specific receiver is not defined. The general trend is that as the distance from the emitter increases the power of the field drops. This stands in accordance with the 2π angular spread of the emitted power and the power decay law. In addition, we observe interference patterns that indicate that some positions are completely "blind." In these positions, the power level drops below −35 dB. In the power profiles, this is indicated by the blank (white) areas. An example may be the distribution of the power in path #1 of Figure 6.19. At the positions of the scatterers, $r = 12$ mm, $r = 24$ mm, and $r = 36$ mm, the field is zero, given the presence of the perfect conductors. At the frequency $f = 24$ GHz, there is a zero-field spot between element no. 3 and no. 4, around $r = 30$ mm. Similarly, there is a zero field at the position $r = 32$ mm at the frequency range $f = [31, -35$ GHz]. This kind of power mapping can be calculated for every emitter, path, and frequency. The study provides us with an initial insight on the power patterns in the multiscattering environment and allows us to select a frequency regime, optimum paths, probe positions, etc., for the full 3D design.

6.5 SUMMARY

This chapter delves into the channel characterization of highly integrated computing packages for applications such as WNoCs and SDMs. After a comprehensive description of these dense and area-constrained environments, we discussed antenna placement and outlined a methodology for the characterization of the potential wireless channels. The methodology is used to evaluate the coupling between antennas and the wireless transfer function.

A rough classification of the wireless channels within computing packages distinguishes between standard, opportunistic, and custom paths. Standard channels seek compatibility with existing structures and ease of antenna placement, but this comes at the cost of a rather poor coupling (between 20 and 40 dB, improvable via package optimization). Opportunistic channels aim to find spaces uniquely suited to wireless propagation to existing structures to improve performance (losses below 20 dB have been achieved); however, this option generally involves certain complexity or functionality trade-offs. Finally, custom solutions allow to create parallel-plate waveguides dedicated to wireless communications only. This approach obtains the best performance, even better than the opportunistic solutions, at the expense of having to build a dedicated layer (with its associated volume overhead and fabrication cost).

These results have been limited to the frequency domain only, as a first step toward channel characterization and feasibility of wireless communications within computing packages. However, it is evident that future works should also inspect the behavior of the wireless channel in the time domain. This would shed light on the amount of multipath components to expect and their impact on the signal delay, to assess the coherence bandwidth of the channel and the potential intersymbol interference effects. In addition, further efforts need to be directed to understanding the propagation mechanisms and further optimizing the performance of the standard and opportunistic cases toward disruptive low-power wireless communications in the multiprocessor and intrametamaterial nanoscale paradigms.

ACKNOWLEDGMENTS

This work was supported in part by the European Unions' Horizon 2020 research and innovation programme Future Emerging Topics (FETOPEN) under grant agreement No 736876.

REFERENCES

1. A. C. Tasolamprou, M. S. Mirmoosa, O. Tsilipakos, A. Pitilakis, F. Liu, S. Abadal, A. Cabellos-Aparicio et al. Intercell wireless communication in software-defined metasurfaces. In *Proceedings of the ISCAS'18*, 2018.
2. J.-Y. Shin, E. G. Sirer, H. Weatherspoon, and D. Kirovski. On the feasibility of completely wireless datacenters. *IEEE/ACM Transactions on Networking*, 21(5):1666–1679, 2012.
3. M. Ohira, T. Umaba, S. Kitazawa, H. Ban, and M. Ueba. Experimental characterization of microwave radio propagation in ICT equipment for wireless harness communications. *IEEE Transactions on Antennas and Propagation*, 59(12):4757–4765, 2011.
4. H. H. Yeh, N. Hiramatsu, and K. L. Melde. The design of broadband 60 GHz AMC antenna in multi-chip RF data transmission. *IEEE Transactions on Antennas and Propagation*, 61(4):1623–1630, 2013.
5. S. Abadal, J. Torrellas, E. Alarcón, and A. Cabellos-Aparicio. OrthoNoC: A broadcast-oriented dual-plane wireless network-on-chip architecture. *IEEE Transactions on Parallel and Distributed Systems*, 29(3):628–641, 2018.
6. S. Deb, A. Ganguly, P. P. Pande, B. Belzer, and D. Heo. Wireless NoC as interconnection backbone for multicore chips: Promises and challenges. *IEEE Journal on Emerging and Selected Topics in Circuits and Systems*, 2(2):228–239, 2012.
7. S. Abadal, C. Liaskos, A. Tsioliaridou, S. Ioannidis, A. Pitsillides, J. Solé-Pareta, E. Alarcón, and A. Cabellos-Aparicio. Computing and communications for the software-defined metamaterial paradigm: A context analysis. *IEEE Access*, 5:6225–6235, 2017.
8. D. Matolak, S. Kaya, and A. Kodi. Channel modeling for wireless networks-on-chips. *IEEE Communications Magazine*, 51(6):180–186, 2013.
9. Y. Chen and C. Han. Channel modeling and analysis for wireless networks-on-chip communications in the millimeter wave and terahertz bands. In *Proceedings of the INFOCOM WKSHPS'18*, IEEE, New York, 2018.
10. V. Petrov, D. Moltchanov, M. Komar, A. Antonov, P. Kustarev, S. Rakheja, and Y. Koucheryavy. Terahertz band intra-chip communications: Can wireless links scale modern x86 CPUs? *IEEE Access*, 5(c):6095–6109, 2017.
11. W. Rayess, D. W. Matolak, S. Kaya, and A. K. Kodi. Antennas and channel characteristics for wireless networks on chips. *Wireless Personal Communications*, 95(4):5039–5056, 2017.
12. Y. P. Zhang, Z. M. Chen, and M. Sun. Propagation mechanisms of radio waves over intra-chip channels with integrated antennas: Frequency-domain measurements and time-domain analysis. *IEEE Transactions on Antennas and Propagation*, 55(10):2900–2906, 2007.
13. P. Baniya, A. Bisognin, K. L. Melde, and C. Luxey. Chip-to-chip switched beam 60 GHz circular patch planar antenna array and pattern considerations. *IEEE Transactions on Antennas and Propagation*, 66(4):1776–1787, 2018.
14. Z. Chen and Y. Zhang. Inter-chip wireless communication channel: Measurement, characterization, and modeling. *IEEE Transactions on Antennas and Propagation*, 55(3):978–986, 2007.
15. R. S. Narde and J. Venkataraman. Feasibility study of transmission between wireless interconnects in multichip multicore systems. In *Proceedings of the APS/URSI'17*, pp. 1821–1822, IEEE, New York, 2017.
16. P. Y. Chiang, S. Woracheewan, C. Hu, L. Guo, H. Liu, R. Khanna, and J. Nejedlo. Short-range, wireless interconnect within a computing chassis: Design challenges. *IEEE Design & Test of Computers*, 27(4):32–43, 2010.

17. K. Guan, B. Ai, A. Fricke, D. He, Z. Zhong, D. W. Matolak, and T. Kürner. Excess propagation loss of semi-closed obstacles for inter/intra-device communications in the millimeter-wave range. *Journal of Infrared, Millimeter, and Terahertz Waves*, 1–15, 2016.
18. S. Kim and A. Zajic. Characterization of 300 GHz wireless channel on a computer motherboard. *IEEE Transactions on Antennas and Propagation*, 64(12):5411–5423, 2016.
19. H.-T. Wu, J.-J. Lin, and K. K. O. Inter-chip wireless communication. In *Proceedings of the EuCAP'13*, pp. 3647–3649, IEEE, San Diego, CA, 2013.
20. J. Branch, X. Guo, L. Gao, A. Sugavanam, J. J. Lin, and K. K. O. Wireless communication in a flip-chip package using integrated antennas on silicon substrates. *IEEE Electron Device Letters*, 26(2):115–117, 2005.
21. K. Kim, W. Bornstad, and K. K. O. A plane wave model approach to understanding propagation in an intra-chip communication system. In *Proceedings of the APS'01*, pp. 166–169, IEEE, New York, 2001.
22. R. S. Narde, N. Mansoor, A. Ganguly, and J. Venkataraman. On-chip antennas for inter-chip wireless interconnections: Challenges and opportunities. In *Proceedings of the EuCAP'18*, IEEE, New York, 2018.
23. X. Zhang, J. K. Lin, S. Wickramanayaka, S. Zhang, R. Weerasekera, R. Dutta, K. F. Chang et al. Heterogeneous 2.5D integration on through silicon interposer. *Applied Physics Reviews*, 2(2), 2015.
24. O. Markish, B. Sheinman, O. Katz, D. Corcos, and D. Elad. On-chip mm wave antennas and transceivers. In *Proceedings of the NoCS'15*, page Art. 11, IEEE/ACM, New York, 2015.
25. X. Timoneda, S. Abadal, A. Cabellos-Aparicio, D. Manessis, J. Zhou, A. Franques, J. Torrellas, and E. Alarcón. Millimeter-wave propagation within a computer chip package. In *Proceedings of the ISCAS'18*, IEEE, Florence, Italy, 2018.
26. X. Timoneda, A. Cabellos-Aparicio, D. Manessis, E. Alarcón, and S. Abadal. Channel characterization for chip-scale wireless communications within computing packages. In *Proceedings of the NOCS'18*, IEEE, Barcelona, Spain, 2018.
27. J. Wu, A. Kodi, S. Kaya, A. Louri, and H. Xin. Monopoles loaded with 3-D-printed dielectrics for future wireless intra-chip communications. *IEEE Transactions on Antennas and Propagation*, 65(12):6838–6846, 2017.
28. X. Timoneda, S. Abadal, A. Cabellos-Aparicio, D. Manessis, and E. Alarcón. Channel characterization for chip-scale wireless communications within computing packages. *Proceedings of the NOCS '18*, pp. 20:1–20:8, Torino, Italy, 2018. http://dl.acm.org/citation.cfm?id=3306619.3306639, IEEE Press, Piscataway, NJ.
29. X. Timoneda, S. Abadal, A. Cabellos-Aparicio, and E. Alarcón. Modeling the EM field distribution within a computer chip package. In *Proceedings of the WCNC'18*, IEEE, Gothenburg, Sweden, 2018.
30. C. A. Balanis. *Antenna Theory: Analysis and Design*. 3rd edition, Hoboken, NJ: Wiley, 2005.
31. J. Lin, H. Wu, Y. Su, L. Gao, A. Sugavanam, and J. Brewer. Communication using antennas fabricated in silicon integrated circuits. *IEEE Journal of Solid-State Circuits*, 42(8):1678–1687, 2007.
32. F. Bieck, S. Spiller, F. Molina, M. Töpper, C. Lopper, I. Kuna, T. C. Seng, and T. Tabuchi. Carrierless design for handling and processing of ultrathin wafers. In *Proceedings of the ECTC'10*, pp. 316–322, AIP Publishing, Melville, NY, 2010.
33. M. Opoku Agyeman, Q.-T. Vien, A. Ahmadnia, A. Yakovlev, K.-F. Tong, and T. Mak. A resilient 2-D waveguide communication fabric for hybrid wired-wireless NoC design. *IEEE Transactions on Parallel and Distributed Systems*, 28(2):359–373, 2016.

Synchronization for Molecular Communications and Nanonetworking

Ethungshan Shitiri and Ho-Shin Cho

CONTENTS

7.1 INTRODUCTION

One of the most critical elements to establish a reliable communication nanonetwork is synchronization, regardless of the communication paradigm, namely *terahertz communication* (TC) and *molecular communication* (MC). In this chapter, we will primarily focus on synchronization in MC-based nanonetworks (MCNs). We first review the purpose of synchronization in MCNs.

7.1.1 Synchronization in MCN

For MCNs, synchronization purpose is two-fold. First, as we know, synchronization provides *timing* and *carrier phase* recovery for detecting incoming symbols successfully. Since the propagation delay, clock skews, and the clock offsets are generally unknown to a receiver, *timing synchronization* determines the time instants at which the received signal has to be sampled by the receiver. *Carrier phase synchronization* determines the frequency and phase of the received signal to whose values the local oscillator adapts. Both these schemes guarantee synchronization, and depending on the accuracy of the synchronization, the performance of a system will be determined. We should note that these techniques have been well studied for traditional communication networks [1,2]. However, given the unique characteristics of the MC channel and the nanomachines, it is pertinent to develop suitable timing and carrier synchronization specific to MCNs.

Second, the limitations posed by the size and power of the nanomachines call for synchronization in the form of *coordination* among the nodes, thereby enhancing their operations. Coordination may be achieved in one of the several ways. One way is to utilize the biological mechanisms that involve *inhibiting* molecules as the drivers for synchronization. Another way is to utilize the biological mechanisms that involve *inducing* molecules as the drivers for synchronization. Both the inhibiting and inducing molecules can regulate the oscillations within the transmitter and the receiver. Coordinated actions are crucial where an MCN has to cooperate, such as to release drug particles at the same time. Therefore, depending on the accuracy of coordination, the success and effectiveness of an MCN will be determined.

We consider the approaches inspired by nature to achieve coordinated actions as *bio-inspired synchronization approaches* and the approaches extending the conventional techniques to derive the timing and carrier synchronization as *traditional synchronization approaches*.

7.2 BIO-INSPIRED SYNCHRONIZATION APPROACHES

Bio-inspired synchronization offers the advantage of biocompatibility and low energy requirements. Under the assumption that nanomachines are embedded with biological oscillators, the oscillations can be used to approximate the clock signals to facilitate synchronization. Here, we briefly look at the working principle of biological oscillators.

7.2.1 Biological Oscillators

Any biological system, wherein a source of excitation, a restorative process, and a delay element exist, with appropriate system parameters that lead to a cyclic behavior can be regarded as a biological oscillator [3]. Oscillations are generated by the periodic fluctuations

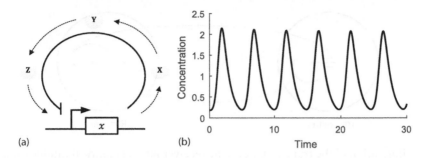

FIGURE 7.1 (a) The Goodwin oscillator model highlighting the transcription and translation pathways. (b) Oscillations in the concentration of X.

in the concentration of molecules through negative and positive feedback loops. Figure 7.1a illustrates the simple yet pioneering model that captures the oscillatory behavior in genetic regulatory networks, namely the Goodwin model [4,5]. A single gene, x, periodically inhibits itself through a self-negative feedback loop, leading to oscillations. Specifically, the gene x synthesizes a messenger RNA, X, using a template of the existing DNA in order to copy the genetic information. This process of synthesizing RNA from DNA or vice versa is known as *transcription*. Then, a sequence of nucleotides in X is converted into an enzyme, Y, and this process is referred to as *translation*. Consequently, Y catalyzes the production of a metabolite, Z, which causes the inhibition of the expression of X. We note that during transcription, the concentration of X increases, while during the production of Z, the concentration of X decreases. The process is cyclic and leads to the oscillations in the concentration of Y (*cf.* Figure 7.1b).

In principle, to generate the sustained oscillations, the Goodwin oscillator required an unrealistically large Hill coefficient value ($n > 8$) [6]. Nonetheless, it is the first model to position negative feedback loops as a critical element to obtain sustained oscillations. Goodwin's theoretical model was leveraged when a negative feedback loop was discovered in the circadian oscillators of *Drosophila melanogaster* and *Neurospora crassa* [6–8], and undoubtedly, his model laid the groundwork for other models to follow suit. A more detailed introduction to biological oscillators and their types can be found in Shitiri et al. [9].

7.2.2 Inhibitor-Based Oscillators and Synchronization

Inhibitors, that is, molecules that inhibit the chemical process responsible for generating a molecule or itself, are the main drivers for inhibitor-based oscillators. Inhibitor-based oscillators are based on a naturally occurring phenomenon, wherein the coupling of two feedback signals, excitatory and delayed inhibitory, results in oscillations [10].

Consider an inhibitor molecule, type X, which is produced through some chemical process by molecules of type x [11]. As shown in Figure 7.2, the excitatory feedback is formed when the type x molecules chemically react to generate type X molecules. As the concentration of X begins to increase, type X molecules begin to inhibit the type x from the further generation of X, forming the delayed inhibitory feedback. Simultaneously, the

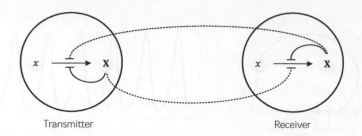

Transmitter   Receiver

FIGURE 7.2   Illustration of the transmitter and the receiver having identical inhibitor-based oscillators and the synchronization process through the diffusion of the inhibitors in the environment (dashed lines).

type X molecules disperse into the environment, resulting in a decrease in their concentration around x. Below a certain concertation threshold, H, they can no longer inhibit x, and therefore, x can then restart the production of type X molecules. The attainment of the threshold H marks the completion of one oscillation cycle. The quick release and slow dispersion give the oscillation a relaxation oscillator attribute. These processes keep repeating, forming the oscillations.

Synchronization through inhibitor-based oscillators is as follows. A transmitter emits the inhibitors as spikes when it pulses. The inhibitors propagate independently to the receiver, with the same oscillator, located at a distance d from the transmitter. Assuming that a spike of N molecules is released at time $t = 0$, the impulse response has concentration distribution $C(d, t)$ from the location of release and follows a Gaussian distribution [11]:

$$C(d,t) = \begin{cases} 0, & t \leq 0 \\ \dfrac{N}{(4\pi Dt)^{p/2}} e^{-\frac{d^2}{4Dt}}, & t > 0 \end{cases} \tag{7.1}$$

where D is the diffusion coefficient of the molecules, and p is the dimensionality of the space [12]. When the receiver receives the inhibitors, they cause the receiver from pulsing for a short time [11]. In particular, the receiver can pulse before the inhibitory molecules arrive (i.e., at a time similar to when the transmitter pulsed), and the receiver cannot pulse for some time after, since the molecules from the transmitter are then already inhibiting the receiver. Synchronization is attained when the nanomachines are either pulsing with the same oscillation period (T) at the same time, *in-phase*, or are pulsing alternating halfway out of phase, *anti-phase*.

Assuming initially that the nanomachines are oscillating at different time instances, there exists a *phase difference* θ between the transmitter and the receiver, given by $\theta = t - t_i$, where t_i is the time of the most recent pulse at the transmitter and t is the time of the most recent pulse at the receiver. θ lies in the range $(0, T)$. If the transmitter sends out L pulses with constant T and θ values, then the net effect of the pulses on the inhibitor concentration at the receiver can be expressed as [13]:

$$\Gamma(t) = \sum_{j=1}^{K} \sum_{i=1}^{L} C(d_j,(i-1)T+\theta_j+t), \tag{7.2}$$

where $\Gamma(t)$ is the inhibitor concentration at the receiver at time t, K is the number of nanomachines, d_j is the distance between the nanomachines, and C is defined by (7.2). The time period $(i-1)T+\theta_j$ represents the time elapsed from the i-th most recent pulse of the j-th nanomachine to time $t=0$.

7.2.2.1 Convergence Analysis

When the nanomachines oscillate, performing an infinite number of pulses ($L=\infty$) according to fixed T and θs, a steady state is achieved. Note that when a pulse occurs at the receiver, the concentration $\Gamma(t)$ is equal to the concentration threshold H. Then, from (7.2), we obtain H as [13]:

$$H = \sum_{j=1}^{K} \sum_{i=1}^{\infty} C(d_j,(i-1)T+\theta_j), \tag{7.3}$$

For a single nanomachine, where $K=1$, $d_1=0$, and $\theta_1=0$, the period T can be expressed as [13]:

$$T = \frac{1}{4\pi D}\left(\frac{N}{H}\sum_{i=1}^{\infty}\frac{1}{i^{p/2}}\right)^{2/p}. \tag{7.4}$$

The summation in (7.4) is a Riemann Zeta series, which diverges to ∞ for $p=1$ or $p=2$ but converges for $p=3$. That implies that in one-dimensional (1-D) and two-dimensional (2-D) spaces, T is infinitely long for the concentration to decrease below H, whereas in three-dimensional (3-D) space, T is finite for the concentration to decrease below H. For the oscillator to work in 1-D and 2-D spaces, other chemical processes would be required to reduce the concentration of the molecules. For example, the exponential decay of molecules throughout the environment may prevent excessive accumulation of concentration and cause the oscillations to converge to a stable period in 1-D or 2-D spaces or closed environments [13].

7.2.3 Inducer-Based Oscillators and Synchronization

Unlike inhibitors, *inducers* are molecules that promote the chemical process responsible for generating a molecule or itself and are the main drivers for inducer-based oscillators. Inducer-based oscillators are also based on a naturally occurring population-based phenomenon, *quorum sensing*. Quorum sensing is a phenomenon by which a population of bacteria species performs coordinated behavior through the emission and reception of the inducers [14].

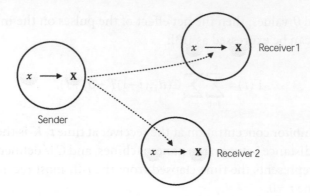

FIGURE 7.3 Illustration of a network of inducer-based oscillators and the synchronization process through the diffusion of the inducers in the environment (dashed lines).

Consider an inducer molecule, type X, which is produced through some chemical process by a molecule of type x. As illustrated in Figure 7.3, X is then released into the environment by a sender, and they propagate via diffusion to nearby nanomachines (receivers 1 and 2). When received, X triggers the nanomachines to generate and release more X molecules into the environment. This domino-like effect causes an increase in the concentration of X among the network. When the concentration of X reaches an upper concentration threshold, the nanomachines stop the production of X molecules. However, the type X molecules continue to disperse into the environment, away from the nanomachines that generated them, and as a result, their concentration around the network decreases. When the concentration of X reaches a lower threshold concentration, the cycle is restarted, resulting in oscillations in the concentration of X.

One way in which synchronization is achieved is when the total concentration of X in the network reaches the upper concentration threshold. At this point, the network can collectively perform a common task [15]. In particular, global synchronization is attainable through inducer-based oscillators. Another way to utilize inducer-based oscillators is to form a cluster of nanomachines that act as the synchronization component with the sole purpose of conveying timing information to the communication nanomachines within a nanonetwork [16].

In a network of K nanomachines, the concentration of the inducers at an arbitrarily chosen nanomachine at a time t is the sum of the inducers that all nanomachines released and is given by [17]

$$\Gamma(t) = \sum_{j=1}^{K} C\big(d_j, (t - t_j)\big), \tag{7.5}$$

where t_j is the release time by the j-th nanomachine relative to the current nanomachine and C is defined by (7.1).

7.2.4 Joint Synchronization and Detection

Synchronization may be performed to allow coordinated actions, as described earlier, or to synchronize to the transmitter clock and detect the incoming symbols by choosing the correct sampling time. Here, we discuss the latter one.

Considering the fact that the concentration of the molecules at a receiver depends heavily on the diffusion coefficients, instead of emitting only the information molecules, a transmitter emits pilot molecules at the same time. The pilot molecules have a diffusion coefficient that is larger than that of the information molecules; this allows the pilot molecules to propagate faster. It is thus possible to estimate the starting of the symbol duration based on the high concentration level of pilot molecules that are expected to arrive in advance, owing to higher diffusion coefficient [18]. As the pilot molecules perform the synchronization, they are referred to as the synchronization molecules. In particular, the receiver estimates the start of a symbol based on the high concentration level of the received synchronization molecules.

Consider type x molecules for information transmission and type y molecules for synchronization purpose. Under the assumptions of a point transmitter and a spherical receiver, the fraction of type m, $m \in \{x, y\}$ molecules hitting the receiver at time t is given by [19]

$$F_m(t) = \frac{r}{d+r} erfc\left(\frac{d}{\sqrt{4D_m t}}\right), \tag{7.6}$$

where r is the radius of the receiver, d is the distance between the transmitter and the receiver, D_m is the diffusion coefficient of type m molecules, and $erfc(.)$ is the complementary error function.

Let $N_{T,m}(k,t)$ be the number of type m molecules released for the $s[k]$-th symbol. Then, the total number of molecules received for the k-th symbol is given as

$$N_{R,m}(k,t) = \underbrace{s[k]N_{T,m}(k)F_m(t)}_{\text{desired signal}} + \underbrace{\sum_{i=1}^{k-1} a[i]N_{T,m}(i)F_m(t-iT_s) + n_b}_{\text{intersymbol interference}}, \tag{7.7}$$

where $N_{R,m}(k,t)$ is the total number of received molecules, and n_b is the Brownian noise following a Gaussian distribution $n_r \sim \mathcal{N}\left(0, N_{T,m}(k)F_m(t)(1-F_m(t))\right)$ [20]. Using (7.7), the receiver can obtain $\hat{t}_{\text{sync,peak}}(k)$ as the estimated peak time of the received synchronization molecule for the k-th symbol. Then, the receiver obtains $\hat{t}_{\text{info,peak}}(k) = \hat{t}_{\text{sync,peak}}(k)$, where $\hat{t}_{\text{info,peak}}(k)$ is the estimated start time for the k-th symbol. Once the $\hat{t}_{\text{info,peak}}(k)$ is obtained, the receiver can perform the symbol detection.

Joint synchronization and symbol detection have the advantage of handling nonequal symbol durations and emitting frequency. In addition, the constraint of maintaining a fixed clock frequency at the receiver is relaxed [18].

7.3 TRADITIONAL SYNCHRONIZATION APPROACHES

Maximum likelihood (ML) estimation techniques are well defined for conventional communication systems. Several challenges in MC systems push the need to derive suitable ML estimators. Mainly, two statistical channel models are significant in MC. One is the free-diffusion channel in which the unknown propagation delay, τ, is modeled as a Gaussian distribution $\tau \sim \mathcal{N}(\mu, \sigma^2)$, having a mean μ and variance σ^2. The other is an assisted-diffusion channel in which the unknown propagation delay is modeled as an inverse Gaussian distribution $\tau \sim IG(\mu, \lambda)$, having a mean $\mu > 0$ and shape parameter $\lambda > 0$. Here, we treat the free-diffusion case. First, we lay down the clock model.

7.3.1 Clock Model

Let us consider two nanomachines, A and B, in the environment. Each nanomachine has a clock that provides timing information. In practice, the clocks drift away from each other owing to perturbations caused by *clock skews* (α) and *clock offsets* (β). Clock skews refer to the differences in the frequencies of the clock signals, while clock offset refers to the advance or delay in the clock times. In ideal situations, $\alpha = 1$ and $\beta = 0$. At any time t, the clock reading $Clk(t)$ at a nanomachine is given by

$$Clk_i(t) = t\alpha_i + \beta_i, \qquad i \in \{A, B\}. \tag{7.8}$$

Clearly, the presence of perturbations can hinder the performance of an MC system.

7.3.2 ML-Based Timing Synchronization Under Gaussian Distribution Delay Model

The goal of ML estimation is to estimate the clock skew and clock offset and use that information to update the local clocks, thereby achieving timing synchronization. Let us consider the case where nanomachine A wants to synchronize with B. Using a two-way message exchange, A and B exchange time-stamped messages with each other [21]. As shown in Figure 7.4, nanomachine A sends a synchronization message to B at the time $T_{1,i}$. Nanomachine B replies to A with the time stamps $T_{2,i}$ and $T_{3,i}$. Then, A records the reception time

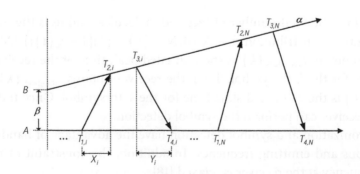

FIGURE 7.4 Two-way message exchange method for MC clock synchronization.

of B's reply as $T_{4,i}$. After N rounds of message exchange, nanomachine A obtains a set a time stamps $\{T_{1,i}, T_{2,i}, T_{3,i}, T_{4,i}\}_{i=1}^{N}$. From (7.6), we obtain the relationship among the time stamps as

$$T_{2,i} = (T_{1,i} + X_i)\alpha_0 + \beta_0, \tag{7.9}$$

$$T_{3,i} = (T_{4,i} - Y_i)\alpha_0 + \beta_0, \tag{7.10}$$

where α_0 and β_0 are the relative clock skew and clock offsets, respectively. X_i and Y_i are the random propagation delays from A to B and B to A, respectively. As stated earlier, the unknown propagation delay, $\tau \in \{X_i, Y_i\}$, is modeled as a Gaussian distribution $\tau \sim \mathcal{N}(\mu, \sigma^2)$, whose probability density function can be expressed as

$$f(\tau; \mu, \sigma) = \frac{1}{\sigma\sqrt{2\pi}} \exp\left[-\frac{(\tau - \mu)^2}{2\sigma^2} \right]. \tag{7.11}$$

7.3.2.1 ML Estimator

The terms X_i, Y_i, α_0, and β_0 are unknown and need to be estimated through ML estimation. From (7.9) and (7.10), X_i and Y_i can be derived as

$$X_i = \frac{T_{2,i}}{\alpha_0} - T_{1,i} - \frac{\beta_0}{\alpha_0} \tag{7.12}$$

$$Y_i = T_{4,i} - \frac{T_{3,i}}{\alpha_0} + \frac{\beta_0}{\alpha_0}. \tag{7.13}$$

The likelihood function can be derived from the joint probability density functions of X_i and Y_i $\prod_{i=1}^{N} p(X_i)p(Y_i)$, where $p(X_i)$ and $p(Y_i)$ are defined by (7.11). Taking a different approach, a new random variable, W_i, is constructed as

$$W_i = X_i - Y_i$$

$$= \frac{1}{\alpha_0}(T_{2,i} + T_{3,i}) - T_{1,i} - T_{4,i} - \frac{2\beta_0}{\alpha_0}. \tag{7.14}$$

From (7.14), the unknown variable μ is eliminated, and that makes the derivation of the ML estimators easier. The log-likelihood function for α_0 and β_0 can be expressed as

$$\ln \Lambda(\alpha_0, \beta_0) = \ln \prod_{i=1}^{N} p(W_i)$$

$$= \sum_{i=1}^{N} \ln p(W_i), \tag{7.15}$$

where $p(W_i)$ is the probability density function of W_i and follows a Gaussian distribution with zero mean and variance $\sigma_w^2 = 2\sigma^2$. To obtain the ML estimates of α_0 and β_0, (7.15) is set to 0 and the partial derivative is taken. Solving the partial derivative yields the estimates of α_0 and β_0 as

$$\hat{\alpha}_0 = \frac{\sum_{i=1}^{N}\left(T_{2,i}-T_{3,i}\right)^2 - N\sum_{i=1}^{N}\left(T_{2,i}-T_{3,i}\right)^2}{\sum_{i=1}^{N}\left(T_{2,i}-T_{3,i}\right)\sum_{i=1}^{N}\left(T_{1,i}-T_{4,i}\right) - N\sum_{i=1}^{N}\left(T_{2,i}-T_{3,i}\right)\left(T_{1,i}-T_{4,i}\right)}$$

(7.16)

$$\hat{\beta}_0 = \frac{\sum_{i=1}^{N}\left(T_{1,i}-T_{4,i}\right)\sum_{i=1}^{N}\left(T_{2,i}-T_{3,i}\right)^2 - \sum_{i=1}^{N}\left(T_{2,i}-T_{3,i}\right)^2\sum_{i=1}^{N}\left(T_{2,i}-T_{3,i}\right)\left(T_{1,i}-T_{4,i}\right)}{2\left(\sum_{i=1}^{N}\left(T_{2,i}-T_{3,i}\right)\sum_{i=1}^{N}\left(T_{1,i}-T_{4,i}\right) - N\sum_{i=1}^{N}\left(T_{2,i}-T_{3,i}\right)\left(T_{1,i}-T_{4,i}\right)\right)},$$

where $\hat{\alpha}_0$ and $\hat{\beta}_0$ are the estimates of α_0 and β_0, respectively. The steps leading to (7.16) can be found in Appendix A.

7.3.3 ML-Based Joint Synchronization and Detection

Under the assumption that two different types of molecules represent different data symbols, joint synchronization and detection can be attained [22]. Type x represents symbol "0," and type y represents symbol "1." Both types of molecules have identical diffusion coefficient [23].

Let $s[k] \in \{0,1\}$ and $t_s[k] \in \mathbb{R}[k]$ denote the symbol of the k-th symbol duration and the start of the k-th symbol duration, respectively. $\mathbb{R}[k]$ is a random variable in $[T^{\min}, T^{\max}]$, where $T^{\min}$ and $T^{\max}$ are the minimum and maximum possible symbol durations, respectively. Assuming that the symbols are transmitted at nonequal intervals and that the $t_s[k]$ is unknown to the receiver, the goal is to estimate $t_s[k]$ and subsequently detect the transmitted $s[k]$. Let $r_x(t_n)$ and $r_y(t_n)$ denote the received signal for type x and type y molecules, respectively. $t_n = (n-1)\Delta t, n = 1,2,3...$, is the discretized time for observing the molecules arriving at the receiver. Considering a *symbol-by-symbol* joint synchronization and detection, the ML problem can be formulated as

$$\left[\hat{t}_s[k], \hat{s}[k]\right] = \underset{\forall t \in \mathbb{R}[k], s \in \{0,1\}}{\arg\max} \Lambda_x(t,s)\Lambda_y(t,s),$$

(7.17)

where $\Lambda_x(t,s)$ and $\Lambda_y(t,s)$ are the likelihood functions, and they can be expressed as

$$\Lambda_m(t,s) = \prod_{t_n \in \mathbb{R}^{ow}[k]} f_P\left(r_m(t_n), \bar{r}_m(t_n) \mid t_s[k]=t, s[k]=s\right),$$

(7.18)

where $m \in \{x,y\}$ and $f_P(\psi,\lambda) = \frac{\lambda^\psi e^{-\lambda}}{\psi!}$ is the probability mass function of a Poisson random variable with mean λ. $\mathbb{R}^{ow}[k]$ is the observation window, and $\bar{r}_m(t_n)$ are the expected molecules. The log-likelihood function is given by

$$\left[t_s[k], \hat{s}^{ml}[k]\right] = \underset{\forall t \in \mathbb{R}[k], s \in \{0,1\}}{\arg\max} \ln \Lambda_x(t,s) + \ln \Lambda_y(t,s)$$

$$= \underset{\forall t \in \mathbb{R}[k], s \in \{0,1\}}{\arg\max} \sum_{t_n \in \mathbb{R}^{ow}[k]} \left[r_x(t_n) \ln(\overline{r}_x(t_n)) - r_x(t_n) - \ln(\overline{r}_x(t_n)!) \right. \tag{7.19}$$

$$\left. + r_y(t_n) \ln(\overline{r}_y(t_n)) - r_y(t_n) - \ln(\overline{r}_y(t_n)!) \right].$$

The function in (7.19) does not lend itself to a closed-form solution, and therefore, a one-dimensional search can be used to find the optimal ML solution [22].

7.3.4 Nondecision-Directed ML Estimator

Under the assumption that M different molecules represent different data symbols and the molecules have different diffusion coefficients, a blind synchronization ML estimator can be derived [24]. Considering additive Brownian noises $n_b \sim \mathcal{N}(0, s(t))$ and residual noises $n_r \sim \mathcal{N}(\eta, \kappa)$, a symbol $s(t)$ transmitted at a time t will be corrupted by n_b and n_r. Then, the received symbol, $r(t)$, is given as

$$r(t) = s(t) + n_b + n_r \sim \mathcal{N}(s(t) + \eta, s(t) + \kappa), \tag{7.20}$$

where η and κ are the mean and variance, respectively, of the Gaussian distribution, and their values depend on the propagation environment and system design parameters [20].

If there are M parallel bank of chemical receptors, each for the M different molecule types, then a correct detection will occur only if a received symbol molecule type matches the predetermined receptor type; else, the received symbol is treated as a noise. Considering these points, the probability density function of the received symbol $r(t) = \{r_m(t), m = 1, ..., M\}$, given that the symbol $s_i(t)$ is transmitted and the propagation delay is τ, is

$$p(r(t) \mid s_i(t), \tau) = p(r_1(t), r_2(t), ..., r_M(t) \mid s_i(t-\tau))$$

$$= \prod_{m=1}^{M} p(r_m(t) \mid s_i(t-\tau))$$

$$= p(r_i(t) \mid s_i(t-\tau)) \prod_{m=1, m \neq i}^{M} p(r_m(t)), \tag{7.21}$$

where $p(r_i(t) \mid s_i(t-\tau))$ and $p(r_m(t))$ are defined by the Gaussian distributions $\mathcal{N}(s(t)+\eta, s(t)+\kappa)$ and $\mathcal{N}(\eta, \kappa,)$, respectively.

For nondecision-directed estimation, the receiver treats the data as a random variable. Here, the estimation is obtained by averaging the likelihood function over the probability of the information symbols [1]. Then, we can express the nondecision directed likelihood function as

$$\ln \Lambda(\tau) = \sum_{i=1}^{M} \ln \left(p(r_1(t), r_2(t), ..., r_M(t) \mid s_i(t-\tau)) \right) p(s_i), \tag{7.22}$$

where $p(s_i)$ is the probability that the i-th symbol was transmitted. For equiprobable symbols, the $\ln \Lambda(\tau)$ can be expressed as

$$\ln \Lambda(\tau) = -\sum_{i=1}^{M}\left[r_i^2(t)*\left(\frac{1}{s_i(-t)+\kappa} \right) - 2r_i^2(t)*\left(\frac{s_i(-t)+\eta}{s_i(-t)+\kappa} \right) \right], \qquad (7.23)$$

where $*$ represents the convolution operation. Then, the estimate of the τ is the value of τ that maximizes (7.23) as

$$\hat{\tau} = \arg \max_{\tau} \ln \Lambda(\tau). \qquad (7.24)$$

7.4 SUMMARY

In this chapter, we delved into the problem of synchronization in MC system and networks. We emphasized that the purpose of synchronization is twofold—to perform coordinate actions among a network of nanomachines and to provide correct sampling time for data detection. The naturally available biological oscillators can be valid tools to achieve the coordinated actions, as the oscillations can be used to approximate the clock signals. We have also seen that extending the conventional synchronization concepts tuned to adapt to the unique challenges in MC can be valid tools for symbol synchronization and data detection.

Although the concepts presented in this chapter lay down the groundwork of synchronization for MC, there are still many open problems for researchers to address. One of the open challenges is the tradeoff between complexity and accuracy. The level of accuracy may differ from application to application, depending on the timing precision requirements constrained by the complexity that the nanomachines can handle. It, therefore, creates a caveat necessitating application-oriented synchronization techniques.

Other challenges include unidirectional flowing environments such as blood streams. Such environments put a constraint on using mechanisms that require feedbacks. On that note, two studies have recently surfaced. One is the work by Hsu et al. [25] and more recently the work by Luo et al. [26]. However, both works involve complex estimation techniques, causing complexity issues. Inducer-based oscillators may not be applicable as well, since the inducers will be continually flowing away from the nanomachines and the upper threshold concentration may never be attained. Therefore, there remains an open issue for low-complexity synchronization schemes specific to flowing environments.

APPENDIX A

The log-likelihood function described in (7.15) can be rewritten as

$$\ln \Lambda(\alpha_0, \beta_0) = \sum_{i=1}^{N}\left(-\ln\left(\sigma_w \sqrt{2\pi} \right) - \frac{W_i^2}{2\sigma_w^2} \right). \qquad (7.25)$$

Taking the partial derivative of (7.25) with respect to α_0 yields the following expressions.

$$\frac{\partial \ln \Lambda(\alpha_0, \beta_0)}{\partial \alpha_0} = -\frac{\sum_{i=1}^{N} \left(2W_i \times \frac{\partial W_i}{\partial \alpha_0} \right)}{2\sigma_w^2}$$

$$= \frac{\sum_{i=1}^{N} \left(\frac{1}{\alpha_0} (T_{2,i} + T_{3,i}) - T_{1,i} - T_{4,i} - \frac{2\beta_0}{\alpha_0} \right) (T_{2,i} + T_{3,i} - 2\beta_0)}{\alpha_0^2 \sigma_w^2}$$

$$= \frac{1}{\alpha_0^2 \sigma_w^2} \left(\frac{1}{\alpha_0} \sum_i^N (T_{2,i} + T_{3,i})^2 - \sum_i^N (T_{2,i} + T_{3,i})(T_{1,i} + T_{4,i}) \right.$$

$$\left. -\frac{4\beta_0}{\alpha_0} \sum_i^N (T_{2,i} + T_{3,i}) + 2\beta_0 \sum_i^N (T_{1,i} + T_{4,i}) + \frac{4\beta_0 N}{\alpha_0} \right) \qquad (7.26)$$

Taking the partial derivative of (7.26) with respect to β_0 yields the following expressions

$$\frac{\partial \ln \Lambda(\alpha_0, \beta_0)}{\partial \beta_0} = -\frac{\sum_{i=1}^{N} \left(2W_i \times \frac{\partial W_i}{\partial \beta_0} \right)}{2\sigma_w^2}$$

$$= \frac{2 \left(\frac{1}{\alpha_0} \sum_{i=1}^{N} (T_{2,i} + T_{3,i}) \sum_i^N (T_{1,i} + T_{4,i}) - \frac{2\beta_0 N}{\alpha_0} \right)}{\alpha_0 \sigma_w^2}. \qquad (7.27)$$

Let $A, B, C,$ and D denote the summands as

$$A = \sum_{i=1}^{N} (T_{2,i} + T_{3,i})$$

$$B = \sum_{i=1}^{N} (T_{1,i} + T_{4,i})$$

$$C = \sum_{i=1}^{N} (T_{2,i} + T_{3,i})^2$$

$$D = \sum_{i=1}^{N} (T_{2,i} + T_{3,i})(T_{1,i} + T_{4,i}). \qquad (7.28)$$

Substituting (7.28) in (7.26) and (7.27) and setting them to 0, we have

$$\frac{C}{\alpha_0^2} - D - \frac{4\beta_0}{\alpha_0} \times A + 2\beta_0 \times B + \frac{4\beta_0^2 N}{\alpha_0} = 0$$

$$\frac{A}{\alpha_0^2} - B - \frac{2\beta_0^2 N}{\alpha_0} = 0 \qquad (7.29)$$

On further simplification, it yields

$$C - D\alpha_0 - 4\beta_0 A + 2\alpha_0 \beta_0 B + 4\beta_0^2 N = 0$$

$$A - B\alpha_0 - 2\beta_0^2 N = 0$$

(7.30)

The set of equations in (7.30) can be viewed as a quadratic equation set with two variables α_0 and β_0. The solution can be obtained as

$$\hat{\alpha}_0 = \frac{A^2 - CN}{AB - DN}$$

(7.31)

$$\hat{\beta}_0 = \frac{BC - AD}{2(AB - DN)}.$$

BIBLIOGRAPHICAL NOTES

The first work that proposed the concept of bio-inspired synchronization for MC networks was the inducer-based oscillator paper by Abadal and Akyildiz [15]. Recently, Lin et al. [27] have carried out a study considering the complete architecture of inducer-based oscillators. In the same study, they considered synchronization using different types of inducers for each nanomachine.

The first work extending the concepts of traditional synchronization techniques was the 2013 paper by ShahMohammadian et al. [24]. Besides deriving the ML estimator, they also derived the Cramer-Rao lower bound for the proposed ML estimator. The paper by Lin et al. [21] was the basis of Section 7.3.2, and a detailed analysis of the Cramer-Rao lower bound on the ML estimators of the clock skew and the clock offset can be found in the paper. Other ML estimations under the Gaussian distributed delay model include the works of Luo et al. [28], Lin et al. [29], and Yang et al. [30]. For articles covering the ML estimators of clock parameters under inverse Gaussian distribution delay model, the readers are directed to references [25,31–33].

Other related works not included in this chapter include the following. Extending the concept of a phase-locked loop (PLL) to MC, Lo et al. [34] proposed the MC version called the molecular PLL (MPLL). Similarly, a suboptimal low-complexity scheme based on linear filters was also studied by Jamali et al. [22]. In the same paper, they have carried out studies on a suboptimal low-complexity scheme based on peak observations and threshold triggering. A two-way message without time stamps was investigated by Shitiri and Cho [35].

REFERENCES

1. J.G. Proakis and M. Salehi, *Digital Communications*, 5th ed., McGraw-Hill, Boston, MA, 2008.
2. A. Goldsmith, *Wireless Communications*, Cambridge University Press, Cambridge, UK, 2005.
3. W.O. Friesen and G.D. Block, What is a biological oscillator? *Am. J. Physiol. Integr. Comp. Physiol.* 246 (1984), pp. R847–R853.

4. B.C. Goodwin, *Temporal Organization in Cells: A Dynamic Theory of Cellular Control Processes*, Academic Press, London, UK, 1963.
5. B.C. Goodwin, Oscillatory behavior in enzymatic control processes, *Adv. Enzyme Regul.* 3 (1965), pp. 425–437.
6. J.S. Griffith, Mathematics of cellular control processes I. Negative feedback to one gene, *J. Theor. Biol.* 20 (1968), pp. 202–208.
7. P.E. Hardin, J.C. Hall and M. Rosbash, Circadian oscillations in period gene mRNA levels are transcriptionally regulated, *Proc. Natl. Acad. Sci. U.S.A.* 89 (1992), pp. 11711–11715.
8. B. Aronson, K. Johnson, J. Loros and J. Dunlap, Negative feedback defining a circadian clock: Autoregulation of the clock gene frequency, *Science* 263 (1994), pp. 1578–1584.
9. E. Shitiri, A. Vasilakos and H.-S. Cho, Biological oscillators in nanonetworks—Opportunities and challenges, *Sensors* 18 (2018), p. 1544.
10. U. Alon, *An Introduction to Systems Biology*, Chapman and Hall/CRC, New York, 2006.
11. M.J. Moore and T. Nakano, Synchronization of inhibitory molecular spike oscillators, in *Conference on Bio-Inspired Models of Networks, Information, and Computing Systems*, Springer, Berlin, Germany, 2012, pp. 183–195.
12. H.C. Berg, *Random Walks in Biology*, Princeton University Press, Princeton, NJ, 1993.
13. M.J. Moore and T. Nakano, Oscillation and synchronization of molecular machines by the diffusion of inhibitory molecules, *IEEE Trans. Nanotechnol.* 12 (2013), pp. 601–608.
14. M.B. Miller and B.L. Bassler, Quorum sensing in bacteria, *Annu. Rev. Microbiol.* 55 (2001), pp. 165–199.
15. S. Abadal and I.F. Akyildiz, Bio-Inspired synchronization for nanocommunication networks, in *2011 IEEE Global Telecommunications Conference—GLOBECOM 2011*, 2011, pp. 1–5.
16. Ö.U. Akgül and B. Canberk, An interference-free and simultaneous molecular transmission model for multi-user nanonetworks, *Nano Commun. Netw.* 5 (2014), pp. 83–96.
17. F. Li, L. Lin, C. Yang and M. Ma, Evaluation of molecular oscillation for nanonetworks based on quorum sensing, in *2015 1st Workshop on Nanotechnology in Instrumentation and Measurement (NANOFIM)*, 2015, pp. 233–237.
18. M. Mukherjee, H.B. Yilmaz and B.B. Bhowmik, Joint synchronization and symbol detection for diffusion-based molecular communication systems, CoRR abs/1804.0 (2018), pp. 1–4.
19. H.B. Yilmaz, A.C. Heren, T. Tugcu and C.-B. Chae, Three-dimensional channel characteristics for molecular communications with an absorbing receiver, *IEEE Commun. Lett.* 18 (2014), pp. 929–932.
20. H. ShahMohammadian, G.G. Messier and S. Magierowski, Optimum receiver for molecule shift keying modulation in diffusion-based molecular communication channels, *Nano Commun. Netw.* 3 (2012), pp. 183–195.
21. L. Lin, C. Yang, M. Ma, S. Ma and H. Yan, A clock synchronization method for molecular nanomachines in bionanosensor networks, *IEEE Sens. J.* 16 (2016), pp. 7194–7203.
22. V. Jamali, A. Ahmadzadeh and R. Schober, Symbol synchronization for diffusion-based molecular communications, *IEEE Trans. Nanobioscience* 16 (2017), pp. 873–887.
23. N. Farsad, H.B. Yilmaz, A. Eckford, C.-B. Chae and W. Guo, A comprehensive survey of recent advancements in molecular communication, *IEEE Commun. Surv. Tutorials* 18 (2016), pp. 1887–1919.
24. H. ShahMohammadian, G.G. Messier and S. Magierowski, Blind synchronization in diffusion-based molecular communication channels, *IEEE Commun. Lett.* 17 (2013), pp. 2156–2159.
25. B. Hsu, P. Chou, C. Lee and P. Yeh, Training-based synchronization for quantity-based modulation in inverse Gaussian channels, in *2017 IEEE International Conference on Communications (ICC)*, 2017, pp. 1–6.
26. Z. Luo, L. Lin, W. Guo, S. Wang, F. Liu and H. Yan, One symbol blind synchronization in SIMO molecular communication systems, *IEEE Wirel. Commun. Lett.* 7 (2018), pp. 530–533.

27. L. Lin, F. Li, M. Ma, S. Member and H. Yan, Synchronization of bio-nanomachines based on molecular diffusion, *IEEE Sens. J.* 16 (2016), pp. 7267–7277.

28. Z. Luo, L. Lin and M. Ma, Offset estimation for clock synchronization in mobile molecular communication system, in *2016 IEEE Wireless Communications and Networking Conference*, 2016, pp. 1–6.

29. L. Lin, C. Yang and M. Ma, Offset and skew estimation for clock synchronization in molecular communication systems, in *Proceedings of the 9th EAI International Conference on Bio-inspired Information and Communications Technologies (formerly BIONETICS)*, 2016.

30. C. Yang, L. Lin, F. Li, S. Ma and M. Ma, Reference broadcast synchronization scheme for nanomachines, in *2016 IEEE International Instrumentation and Measurement Technology Conference Proceedings*, 2016, pp. 1–5.

31. L. Lin, C. Yang and M. Ma, Maximum-likelihood estimator of clock offset between nanomachines in bionanosensor networks, *Sensors* 15 (2015), pp. 30827–30838.

32. L. Lin, C. Yang, M. Ma and S. Ma, Diffusion-based clock synchronization for molecular communication under inverse Gaussian distribution, *IEEE Sens. J.* 15 (2015), pp. 4866–4874.

33. L. Lin, J. Zhang, M. Ma and H. Yan, Time synchronization for molecular communication with drift, *IEEE Commun. Lett.* 21 (2017), pp. 476–479.

34. C. Lo, Y.-J. Liang and K.-C. Chen, A phase locked loop for molecular communications and computations, *IEEE J. Sel. Areas Commun.* 32 (2014), pp. 2381–2391.

35. E. Shitiri and H.-S. Cho, Achieving in-phase synchronization in a diffusion-based nano-network with unknown propagation delay, in *Proceedings of the 4th ACM International Conference on Nanoscale Computing and Communication—NanoCom'17*, 2017, pp. 1–6.

CHAPTER 8

Multiple Access Control Strategies for Nanoscale Communications and Networking

Fabrizio Granelli, Cristina Costa, and Riccardo Bassoli

CONTENTS

8.1 INTRODUCTION

Nowadays, communication technologies are breaking the boundaries of common wireless and wired channels toward innovative areas, aiming to go beyond traditional mediums to novel channels to extend the opportunities and increase the range of communicating devices. In this framework, nanoscale and molecular communications represent an interesting subject of investigation.

Nanocommunications imply the exchange of information at the nanoscale level, on the basis of any wired or wireless interconnection of nanomachines in a nanonetwork. Communications are of extremely low range (tens of millimeters) and high frequency (in the terahertz band).

Molecular communications represent an emerging communication paradigm for bio-nanomachines (e.g., artificial cells and genetically engineered cells) to perform coordinated actions in an aqueous environment.

As it can be seen by the definitions given previously, both communication paradigms share similar features, as the operation at the molecular level implies nanomachines or the usage of nanolevel communication concepts. Nevertheless, while the theoretical and technical description of the technologies for enabling nanoscale communications is provided in other chapters of the book, this chapter focuses on the management of the communications resources and on the access to those resources: the medium access control (MAC).

The MAC is a key functionality for obtaining good levels of performance from the physical transmission medium. Indeed, the MAC is responsible for the following:

- Frame delimiting and recognition

- Addressing of destination stations (both as individual stations and as groups of stations)

- Conveyance of source-station addressing information

- Transparent data transfer of higher-level protocol data units (PDUs) or of equivalent information through the physical channel

- Protection against errors, generally by means of generating and checking frame check sequences

- Control of access to the physical transmission medium

The chapter focuses mainly on the last item of the above list, that is, access control to the physical transmission medium.

The channel access control mechanisms provided by the MAC layer are also known as multiple access protocol. This makes it possible for several stations connected to the same physical medium to share it. The multiple access protocol may detect or avoid data packet collisions if a packet mode contention-based channel access method is used or may reserve resources to establish a logical channel if a circuit-switched or channelization-based channel access method is used. The channel access control mechanism relies on a physical layer multiplex scheme.

The most widespread multiple access protocol is the contention-based carrier sense multiple access (CSMA)/collision detection (CD) protocol used in Ethernet networks. This mechanism is utilized only within a network collision domain, for example, an Ethernet bus network or a hub-based star topology network. An Ethernet network may be divided into several collision domains, interconnected by bridges and switches.

Effective MAC strategies are well-known to networking and communication experts and are highly dependent on the properties of the communication channel and on the characteristics of communication in such environment. Therefore, designing an effective MAC strategy requires knowing the characteristics and services provided by the physical communication medium.

The purpose of this chapter is to illustrate the design issues and existing proposals for MAC in bio- and nanocommunication environments. The structure of the chapter is as follows. Section 8.2 reviews the existing well-known approaches to MAC, outlining their advantages and disadvantages, as well as their expected performance. This section provides a vision of the existing approaches in today's networks, in order to better understand the background knowledge on MAC.

The following sections address two technological scenarios and the respective MAC solutions existing in the literature: nanoscale communications and molecular communications. Section 8.3 introduces nanoscale communications in the terahertz band and surveys the available MAC strategies for such scenario, while Section 8.4 introduces the concept of bio-nanocommunications while presenting proposed techniques for MAC, with their pros and cons.

Section 8.5 lists the main simulators for nano- and bio-communications, which are currently available in the research community, useful to experiment with the solutions described in the previous sections. Finally, Section 8.6 briefly discusses the integration between nano- and molecular communications within the Internet, in the so-called Internet of Bio-Nano Things (IoBNT), and Section 8.7 provides a summary and conclusions.

8.2 TRADITIONAL MULTIPLE ACCESS CONTROL STRATEGIES

In general, multiple access control strategies can be clustered into three major categories:

- Channel partitioning
- Random access
- Rotation

Moreover, duplexity in the physical communication can be represented as a further dimension to consider, since the design of a proper MAC scheme clearly depends on the possibility for devices to transmit and receive at the same time (full-duplex mode) against the need for either transmitting or receiving (half-duplex mode). In addition to that, duplexity can impact the complexity of the communication devices: it is possible to obtain full-duplex communication by using two half-duplex channels (one in each direction), but this requires to fully duplicate the transmission/reception chains.

The following paragraphs qualitatively describe the performance limitation of the different approaches to MAC, as well as the major parameters of the system impacting the MAC performance.

In *channel partitioning schemes*, MAC is able to guarantee a constant performance level by partitioning the channel resources. Since partitioning is performed during system design, such scheme is appropriate for constant bit-rate data flows. Once a new data flow requests access to communication resources, admission control is required to check the availability of a proper partition of the channel and to allocate it to the flow for its entire duration. The only relevant parameter in this case is represented by the channel setup delay, that is, the delay to grant access to the actual communication resources. Moreover, additional resources (or a separate signaling channel) might be requested to enable control signaling during the channel setup phase.

In *random access schemes*, access control is fully distributed, with the possibility of having collisions (i.e., overlapping communications) due to lack of coordination among the devices. Regardless of this serious inconvenience, random access is extremely popular in wired and wireless communications, starting from the historical ALOHA and slotted-ALOHA MAC strategies up to carrier sense multiple access (CSMA) and its variants (CSMA/CD, CSMA/ CA). Indeed, in cases where the signal propagation time is smaller than the transmission time, the CSMA is able to obtain very high utilization of the transmission medium. An additional advantage of random access is that no separate signaling channel is required.

In *rotation schemes*, MAC tries to obtain the most stable performance with respect to the two strategies described previously, that is, good performance at both low and high loads. Decentralized rotation-based MAC schemes imply the exchange or "rotation" of a token (describing the device allowed to transmit at a given time); thus, an additional (small) signaling overhead is required. However, the token is often transmitted using the same communication resources at the actual data frames (in-band signaling). Rotation-based MAC can also be implemented in a centralized architecture, where a specific device acting as the master periodically polls the slave devices. However, this architecture is proven to provide lower performance with respect to decentralized solutions owing to the additional communication delay between master and slave devices.

From these considerations, the major parameters influencing MAC performance are the following:

- Number of communicating devices
- Traffic model(s)
- Type of signaling (in-band and out-of-band)
- Transmission and propagation time
- Frame format (header and data length and the presence of a signaling sequence to initiate transmission)

As a consequence, based on the specific scenario parameters, it is possible to understand the approaches on resource allocation that might be appropriate.

8.3 NANODEVICES COMMUNICATING IN THE TERAHERTZ BAND

A nanomachine is an integrated device, with sizes ranging from one to a few hundred nanometers, that is able to pursuit simple tasks such as sensing, plain computation, communication, and local actuation. A nanonetwork is formed through communication between nanomachines, in order to perform more complicated and collaborative tasks in a distributed manner, such as drug delivery, health monitoring, and military and industrial applications. Networked nanomachines can carry out macroscale objectives and cover a larger area varying from meters to kilometers, while a single nanomachine can act only in limited workspace and nanoscale targets.

This section analyzes the existing MAC solutions for a scenario where nanodevices are in the range of one to a few hundreds of nanometers. In such situation, communication between nanodevices can be implemented in the terahertz band, to build wireless networks for applications in medical, industrial, biomedical, military, and environmental fields.

The introduction of graphene nanoantennas made it possible to support the terahertz frequency band at nanoscale, achieving bitrates in the order of terabits/s over a transmission range of around tens of millimeters [1].

Indeed, the Terahertz band provides wireless communication with a huge bandwidth (0.1–10 THz), limited by the following impairments [2]:

- High propagation loss, with resulting limited transmission range

- Molecular absorption in the terahertz band, which introduces noise and high path loss

The unique medium characteristics of nanoscale communications in the terahertz medium require to revise traditional assumptions in MAC design. Even wireless sensor network (WSN) MAC protocols do not fully comply with its requirements.

The molecular absorption is the main phenomenon affecting transmission, and it is the process where part of the electromagnetic (EM) energy of the wave is converted into kinetic energy internal to the vibrating molecules [3]. The impact of the molecular absorption depends on the distance.

For short distances, less than 1 m, where we can assume a low number of molecules, the terahertz band behaves as a single transmission window that is several terahertz wide. When considering wider distances, of a few meters, the molecular absorption becomes more significant, but still, various transmission windows tens or hundreds of gigahertz wide are available, depending on the type and concentration of molecules found in the channel. Indeed, the molecular composition of the transmission medium (e.g., the presence of water vapor molecules) affects the number and size of the transmission windows, as well the transmission distance. Since type of application (e.g., intrabody and space) often imposes a transmission medium of choice, there is a close relationship between it and the MAC options that can be adopted.

Owing to its size, nanodevices have limited processing, memory, and energy resources available. This requires the development of efficient modulation and medium sharing schemes that maintain the simplicity necessary for nanoscale devices. It has been demonstrated that physical data rate can reach hundreds of Gbps and a few Tbps, even with low-complexity modulations [3]. Besides low computational power, the amount of memory available on the nanonode is limited. This limitation may affect the MAC implementation in various way. As an immediate consequence, the size of the transmission queue is limited, till the size of just one packet waiting for transmission. This means that if a packet is not delivered before the subsequent one arrives, one of the two should be dropped. This places a strong constraint in terms of delivery timeout. Allowing operations that take more time than the frame data production rate may lead to lost data, even before transmission [4]. For overcoming these limitations, a possibility is to exploit the hierarchical nature of the nanosensor networks and shift the complexity of the MAC protocol toward more resourceful nodes (e.g., nanocontrollers) [5], thus maximizing the usage of resource of each node.

Another important limitation that should be taken into account is the nature of energy source. Nanodevices are based on energy-harvesting systems and have a limited energy-storage capacity. This represent a major constraint of nanodevice communication, since it is not feasible to generate a high-power carrier signal in the nanoscale at terahertz frequencies. Since energy is mostly based on harvesting, its availability nature has temporal fluctuations, owing to charge—discharge cycles and environmental factors. A nanosensor can complete only a very few tasks in a single harvest consumption cycle. The charge available after harvesting may be enough for the transmission of just a frame, and recharging may be long (with respect the transmission time). Besides, the harvesting source may not always be available or may have varying nature.

Harvesting-aware solutions cannot be designed as traditional energy-aware protocols. Solutions that are harvesting-aware cannot be designed as traditional energy-aware protocols, since the latter are based on the fact that the total available energy is limited and they aim at minimizing its general consumption [6]. Rather than simply minimizing the energy consumption, it is possible to obtain better results considering systems that are able to optimize the harvesting-consumption process by capturing its temporal characteristics [5]. Wang introduced the concept of critical packet transmission ratio (CTR), which is the allowable ratio between the transmission time and the energy-harvesting time, below which a nanosensor can harvest more energy that the consumed one. Jointly optimizing the energy-harvesting and energy-consumption processes may lead toward a perpetual data transmission.

A consequence of the limited energy is that when transmitting, there is no assurance that the receiver has the required energy to receive the packet and process it. Considering that a nanonode may be able to send just a few packets before having to recharge, a missed transmission is a high cost for the transmitter. A possible approach is to include in the MAC design the probability that the receiver is able to receive and process the frame [6]. Also, packet retransmissions, due to errors or collisions, are costly. For every transmission missed, the time before the next retransmission can be very high, owing to the time that the node may need to recharge. This introduces unavoidable delays that may render retransmission useless. Therefore, it should be considered to avoid to heavily rely on retransmissions and use with error-control protocols and controlling the access to the channel.

For the reasons given previously, heavy signaling protocols in general are not appropriate: they can easily become too energy- and processing-consuming, at both ends of the transmission chain. For example, if we consider a handshake exchange, there is no guarantee that both the transmitter and the receiver will have enough energy to successfully complete the packet transmission in one round. Nodes might have to wait to harvest the required energy, and MAC should take waiting times into account. The handshake process, as well as MAC protocols that involve heavy signaling, can limit the real potential of the terahertz network.

The density and distance between nodes also affect the MAC protocol design. Depending on the density and transmission duty cycle, devices may not need to aggressively contend for the channel for two reasons: the very large bandwidth available and the very short transmission time that minimizes the collision probability.

In scenarios with very high number of nanosensors instead, even with a high transmission bandwidth, MAC protocols have to regulate the access to the channel and coordinate and synchronize the transmissions among them. Traditional MAC protocols that address this issue are too demanding, and more low-complexity solutions should be adopted.

Finally, application requirements and assumptions should be considered, since they may affect and influence the MAC design. For example, many scenarios envisage delay-tolerant applications, where it is possible to assume that the energy-harvesting rate is lower than the consumption rate. In these cases, the delay in packet transmission and propagation is of the order of picoseconds, and all delay is dependent on the energy-harvesting limitations.

Another factor is the packet size. The amount of the information that each node has to transmit is expected to be encoded in a few bits. This poses some limits (e.g., the handshake approaches may not be efficient), but for example, it is more probable that the receiver has enough resources for receiving and processing the packet. Other factors that depend on the application requirements that may affect the MAC design include, if the application is loss sensible, its duty cycle, the delay tolerance, packets rate generation (e.g., constant rate), fixed position (e.g., grid-aware approaches [7]), and moving nodes.

MAC protocol solutions available in the literature for nanonetworks in the terahertz band can be classified as [2] (i) physical-layer-aware schemes, (ii) energy-aware schemes, (iii) receiver-initiated protocols, and (iv) optimization-based schemes.

8.3.1 Physical-Layer-Aware MAC Schemes

Physical-layer-aware schemes are MAC protocols that are designed considering the specific features of the physical channel in nanocommunications. The CSMA approaches can be classified into this area, as handling collisions requires knowledge of the communication characteristics.

Recently, a MAC protocol derived from an existing Zigbee MAC protocol scheme has been proposed [8]. The slotted CSMA/certification authority (CA) mechanism based on the superframe structure fairly provides communication chance for each node and makes a reasonable usage of the available energy, like in beacon-enabled Zigbee networks.

Next, it is important to notice that a scheme has been proposed that exploits the concept of timing channel [9], that is, the logical communication channel in which information is encoded in the timing of transmissions, called timing channel for nanonetworks (TCN).

This allows low rate communications in an energy efficient and reliable manner. Timing channels are logical channels in which the information to be transmitted is encoded in silence between subsequent events. Therefore, the information is actually not transmitted, as it is contained in the duration of the silence interval in which the transmitting front end can be turned off.

This approach is suitable for nanocommunications, since by exploiting graphene-based antennas, it is possible to transmit pulses whose duration can be lowered up to 1 ps. Such clock accuracy enables to reduce the duration of silence periods encoding the information to a few hundreds of microseconds, thus making transmission delays introduced by timing channels tolerable for several application scenarios (e.g., health and environmental monitoring). Moreover, despite their considerable sensing capabilities, nanomachines are also expected to have limited transmission capabilities; thus, low-rate timing channel communications are well suited to transmit small-size data, and the fact that the information is encoded in the interval between packets allows to save energy. This encoding scheme is not suitable in applications where delay in the order of several hundreds of milliseconds is not tolerable. Furthermore, if compared with traditional transmission schemes that exploit packets to transmit the information, this approach offers lower achievable transmission rates.

Let's consider the transmitter and the receiver agree on the communication parameters and channel coding scheme during the handshake phase, which occurs before the real data transmission takes place [10]. The Data and Coding Scheme (DCS) is selected by the receiver according to the channel quality. The quality of the channel can be evaluated and predicted from the measured noise intensity and the pulse shape, in order to achieve a target Packet Error Rate (PER). Finally, the data transmission process will take place, where the data packet is transmitted at the specified symbol rate β.

- *Advantages*: Reduce/control nanonetwork interference and maximize probability of decoding received information

- *Disadvantages*: Coordination among nodes required and might require relevant computational resources

8.3.2 Energy-Aware MAC Schemes

Energy-aware schemes are MAC strategies designed for the purpose of saving energy and thus providing long lifetime of nanonodes.

An energy- and spectrum-aware MAC protocol exploits a hierarchical network architecture [11], where a more powerful nanocontroller coordinates the nanosensor channel access and controls the synchronization among these nanodevices by using a Time-division multiple access (TDMA) approach. The behavior of the MAC is based on the CTR parameter, which represents the highest acceptable ratio between the transmission time and the energy-harvesting time. Nanosensors can recharge their batteries in sleeping and transmission time slots, enabling to balance between the consumed energy and the harvested energy.

A new energy-efficient wireless nanosensor network MAC protocol (EEWNSN-MAC) for mobile multihop wireless nanonetworks has been proposed [12], which takes advantage of the clustering mechanism and TDMA scheduling scheme to alleviate the mobility effects and transmission collisions. Experiments outline good performance in terms of total consumed energy per sent/received packet on the network packet loss ratio (PLR) and scalability.

- *Advantages*: Different time slots length, reliable, energy-aware scheduling, and potentially high throughput
- *Disadvantages*: Centralized network topology and controller-to-nanodevice communication

8.3.3 Receiver-Initiated MAC Protocols

Received-initiated schemes are MAC strategies based on the receiver coordination of the access to the communication channel.

A coordinated energy consumption scheduling (CECS) with the distributed receiver-initiated harvesting-aware MAC protocols (DRIH-MAC) has been studied [13]. Such protocol is based on the usage of ready-to-receive packet sent by potential receivers in order to poll for data packets. Experiments showed that RIH-MAC maximizes energy efficiency, minimizes the collision probability, and achieves a high ratio of packet delivery, but in several cases, this might lead to relevant energy consumption.

- *Advantages*: Support energy harvesting, limited collision probability, high packet delivery ratio, and scalable
- *Disadvantages*: Hidden terminal problem and waste of energy for the lookup table

8.3.4 Optimization-Based MAC Protocols

Optimization-based approaches are MAC strategies designed to optimize one performance parameter.

A MAC scheme has been designed for Internet of Nano Things (IoNT) that enables nanodevices to make optimal transmission decisions locally based on their incoming traffic rate, virtual debts, and channel-sensing results [14]. This approach not only leads to high network throughput but also guarantees that the memory of each device is empty before the next packet arrives, thus addressing the fundamental challenge imposed by the extremely limited memory of nanodevices.

- *Advantages*: Support energy harvesting and high packet delivery ratio
- *Disadvantages*: Complexity

8.4 NANO-BIO COMMUNICATIONS

The main characteristics of biomolecular nanocommunications (MCs) is the use of molecules to transport information instead of electromagnetic waves [15]. In classical communications, the carrier is employed to transport the message. In MCs, the carriers are particular molecules that can carry chemosignals or molecular structures containing information. Molecular carriers enhance the propagation capabilities of single molecules to provide more reliable communications and allow the setup of multiple independent channels by using the same medium [16].

In MCs, the medium can be wet or dry and the conditions significantly affect the propagation. For example, the speed of the medium (faster than the speed of the molecules) can influence the communication between nanomachines.

Nano-bio communication paradigm consists of intrabody nano-/micronetworks (nanonodes, nanosensors, nanorouters, and nanogateways) Thus, it is very likely acceptable to depict the network architecture of nano-bio communications as in Figure 8.1a.

There are five main types of nano-bio communication paradigms [15,17] (Figure 8.1b):

1. *Diffusion-based*: Molecules are immersed in a fluid (the medium), and they move because of free diffusion. Frames' information is encoded either in the kind of molecules, which are diffused, in the release time of the molecules or in the intensity of their concentration.

2. *Wired active*: Molecules move in predefined microtubules to send information. In particular, the flow of molecules can be self-propelled or generated by another system.

3. *Wireless active*: There are two subcategories: bacteria-based and catalytic nanomotor-based. The former employs bacteria—carrying DNA-based messages—which react to specific attractants to be led toward the sink. The latter uses nanorods (e.g., platinum and gold), which can propel themselves by catalyzing the free chemical energy of the environment.

4. *Physical-contact-based*: The communication happens via contacts, which can be gap junctions or synapses (also known as neuron-like communications).

5. *Förster resonance energy transfer (FRET)-based*: The communication employs specific molecules called fluorophores, which can be excited by either optical or chemical stimuli, so that when they come back to the normal state, they release visible-light photons.

Side by side, another classification of nano-bio communications is possible according to the way in which molecules propagate [18]. First, walkway-based, which includes communications that follow predefined paths (e.g., wired active and wireless active catalytic nanomotor). Second, flow-based, which groups communications where molecules are released in a fluid and guided by currents or flows (e.g., FRET-based and wireless active bacteria-based). Finally, diffusion-based communications, where molecules move because of free diffusion (as discussed previously).

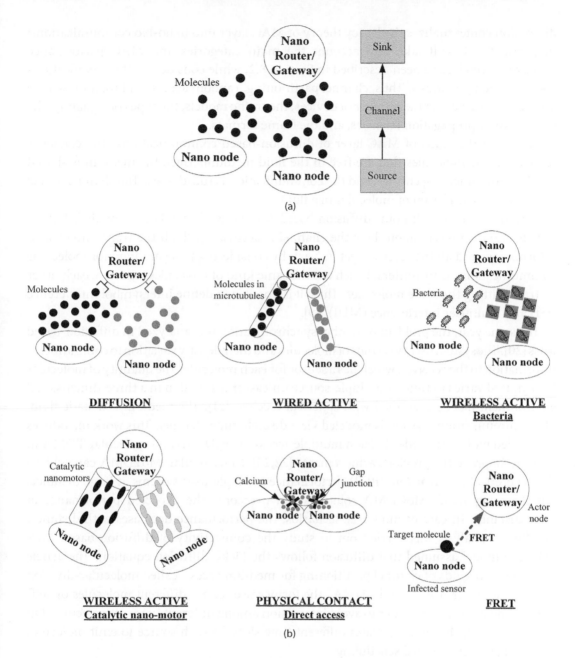

FIGURE 8.1 (a) Network architecture for the molecular communications and the Internet of Bio-nano Things. (b) Different molecular communication paradigms.

The main role of legacy MAC layer and its characteristics were described in Section 8.2. The remainder of this section tries to classify and overview the main up-to-date results in MAC layer protocols for nano-bio molecular communications. The heterogeneous nature of communication's physical layer and of the propagation paradigms (as previously discussed) implies more complex classification of MAC approaches than the legacy one proposed for the existing electromagnetic-based communications. The subsequent discussion contextualizes the legacy theory of MAC layer into nano-bio communications: in particular, it will take into account the main categories and MAC performance indicators, which have been described in Section 8.2, while considering the classifications just presented previously. Thus, channel partitioning, random access, and rotation will be analyzed by surveying the number of devices, the traffic models, the types of signaling, the transmission/propagation latencies, and the frame formats.

Let's start the study of MAC layer of diffusion-based communications. The carrier is represented by molecules that are free in the fluid medium. Thus, the medium is shared by the communicating entities, and the communication is full-duplex. The channel model consists of free diffusion of molecules in a fluid.

A multipoint-to-multipoint diffusion-based MC network has been provided, but no specification is present about how the channel is accessed [18]: all the transmitters send information, and all the receivers get it. Next, by considering this paradigm for molecular communications, transmitters, which use the same kind of molecules, interfere each other by transmitting the same molecules. Thus, it is possible to define intersymbol interference (ISI) and multiuser interference (MUI) [19].

The same year, the problem of symbol synchronization was analyzed in diffusion-based MCs [20]. The system uses two kinds of molecules and their density variation to communicate information to the receiver, which has a sensor for each molecule. The quantity of molecules transmitted varies in time. A multiple-source unicast transmission in a three-dimensional environment has been studied with spherical receiver [21]. The medium is a static fluid. The communications system is modeled via a discrete time channel. This work introduces a so-called molecular-code-division multiple access (MoCDMA) and molecular TDMA in order to improve the previous achievements [22,23]. In particular, MoTDMA can achieve the theoretical bit error rate (BER) performance in single-user systems at large distance. On the other hand, MoCDMA solutions can overcome the expected performance at large distance. In case of multiuser scenarios, the performance decreases with distance. Finally, an analysis was carried out to study the connectivity of diffusion-based MCs [24]. The model assumed that diffusion follows the Fick's diffusion equation. The article describes two kinds of channel partitioning for medium access, called molecular-division multiple access (MDMA) and TDMA. The former uses either different molecules or self-identifying molecules in order to avoid interaction among different types of molecules. On the other hand, the latter allocates different time slots for each source to emit molecules according to a centralized scheduling.

Microtubule-based MCs deploy molecules to transmit information. The medium can be shared, and the communication is half-duplex. In this case, no broadcasting is available. A study on in-cell microtubule propagation [25] was performed via molecular motors, which were transmitted along protein filaments from source to sink. The authors used a dynamic topology, where transmitter and receivers were dynamically connected via microtubules.

Bacteria-based wireless active MC is a communication paradigm that supports full-duplex communications, broadcasting, and medium sharing. In this type of bio-nano networks, bacteria carry information via their DNA molecules, called plasmids. Sensor nanomachines reveal the flow of bacteria [26]. Bacteria also respond and adapt to the environment by exchanging DNA plasmids among them. Furthermore, bacteria, which are transporting messages, carry information encoded in DNA plasmids from source. Next, they move in the environment following chemical gradient released by destination, to which the information is delivered [27]. For example, bacterium *Escherichia coli* is capable of sensing at least 12 different attractants: its attractant processing allows precise navigation, and the propulsion is represented by its flagella. The bacterial system for molecular communications considers time division reception of bacteria at receivers [28]. The MAC techniques mainly supported by this kind of communications are either time-division approach or code-division approach (i.e., deployment of different kinds of bacteria for each user). Finally, an approach called amplitude-division multiple access (ADMA) was proposed, in which sources transmit to a single receiver in a star topology [29]. The receiver gets the sum of the amplitudes of sources' signals and detects the information of each source by knowing that each one has a constant amplitude for transmission. The experiment in this work was performed with bacteria *E. coli*.

Molecular communications based on catalytic nanomotors use nanorods as carriers and support full duplex communications without sharing the medium, thus without multicasting transmissions. In this context, single-source unicast communications were studied by research community [16].

Physical-contact-based MCs transmit molecules carrying information. They support full duplexing, but in general, no medium is shared, and no broadcasting is supported. Nevertheless, short-range communications can use calcium signaling to provide communication among several nanomachines [16]. In direct access, nanomachines are physically connected, and calcium signals pass through the gates, On the other hand, in indirect access, they releases diffuse information molecules in the medium (and they fall in the first "diffusion" communication paradigm). These two techniques can support multicast and broadcast transmissions. A neuronal-time-division multiple access (TDMA) approach based on a noise-aware evolutionary multiobjective optimization algorithm (EMOA) was developed [23]. This solution is applicable in nanocommunication between neurons. In particular, the algorithm performs single-bit TDMA scheduling in order to decide which neighboring neurons and thus which multiplex neuronal signals to activate. Side by side, long-range communication using pheromones [16] is the paradigm that the majority of biological systems do. Animals use molecular messages (pheromones) to communicate with members of the same species. Pheromones can carry information, and they can be decoded only by specific receivers. The main difference with short range is that the channel cannot be considered deterministic. The transmission process consists of releasing selected pheromones to the medium (liquid or gas). Next, the propagation via diffusion permits the information to reach the specific receiver. A neural TDMA (NTDMA) scheduling algorithm to trigger the activation of neurons was proposed in order to avoid collisions in transmitting a signal from source to sink [30]. The nanocommunications are divided in timeslots assigned to each neuron to transmit. This NTDMA algorithm exploits a genetic-based algorithm (GBA) to multiplex neural signals and to solve the multiobjective optimization problem of scheduling.

Regarding molecular communications' paradigms, which allow multiple access to transmission medium, biological systems normally avoid interference by using different kinds of molecules [31]. If interference occurs, methods to manage the access to the channel are required. Currently, very few techniques have been developed so far for MAC operations [31], which are in line with the ones described previously. For example, wireless bionanosensors can present recurrent collisions in the order of femtoseconds, while protocols to avoid that have still to be researched [32].

8.5 BIO- AND NANOCOMMUNICATION SIMULATORS

Owing to the complexity and cost of deploying solutions at MAC level in actual testbeds, most of the existing works on MAC strategies for nanoscale and molecular communications are performed through the use of network- or system-level simulation platforms.

Few simulators are available to study bio- and nanocommunications. The most interesting approaches are listed as follows:

- TeraSim [33]: An open-source network simulation platform for terahertz communication networks is presented. TeraSim is built as an extension for Network Simulator 3 (ns-3), which is one of the most widely used teaching and education network simulation software. The simulator has been developed by considering two major types of application scenarios, namely nanoscale communication networks (average transmission range usually below 1 m) and macroscale communication networks (distances larger than 1 m). The simulator consists of a common channel module, separate physical and link layers for each scenario, and two assisting modules, namely terahertz antenna module and energy-harvesting module, originally designed for the macroscale and nanoscale scenarios, respectively.

- TouchCom [34]: It focuses on the controllable and trackable properties of message carriers. A potential therapeutic application of TouchCom is the targeted delivery of drugs, which utilizes physically transient nanobots as vehicles to deliver drug particles. Meanwhile, the IEEE 1906.1 group, an IEEE standard working group for developing a common framework for nanoscale and molecular communications, has proposed a new standard to the IEEE Communications Society Standards Development Board, which has also been approved recently. This paper maps the paradigm of TouchCom for drug-delivery systems onto the 1906.1 framework. It also describes the nanoscale simulation package for the ns-3 platform that implements the IEEE 1906.1 standard to build a TouchCom example module.

- A modular and easily upgradable simulation platform [35] was intended for wireless nanosensor networks based on electromagnetic communication in the terahertz band. The paper is based on the well-known ns-3 open-source network simulator.

8.6 THE BIG PICTURE: THE INTERNET OF BIO-NANO THINGS

Nowadays, for a technology to be useful and usable, it requires to be integrated in the Internet. Indeed, any communication technology might fit the design paradigms of the current Internet and can unleash its full potential.

The interconnection of "things" (interconnected machines and objects with embedded computing capabilities) to the Internet allowed to extend the Internet to novel application domains, leading to the Internet of Things (IoT). While research and development continue for general IoT devices, there are many application domains where very tiny, concealable, and nonintrusive things are needed. In this framework, the surveyed nanoscale communication paradigm might further extend the scope of the IoT toward new areas, leading to two novel concepts: the Internet of Nano Things (IoNT) and the Internet of Bio-Nano Things (IoBNT) [36,37]. Especially, this paradigm consists of intrabody nano-/micronetworks connected via nano-/microinterfaces to external gateways, which connect to the Internet via common telecommunications paradigms (e.g., electromagnetic waves and Ethernet).

One of the relevant issues related to integrating nanocommunications within a network is related to routing. Indeed, communication in nanonetworks still poses a nontrivial challenge, owing to the constraint of processing, storage, energy, and communication range capabilities of nanonodes. In particular, short communication range in the terahertz band severely limits communication in nanonetworks. Hence, multihop communication among nanonodes is currently regarded as a viable solution for nanonetwork realization. Three routing protocols are studied: controlled flooding, coordinate/routing for nanonetworks, and hierarchical ad hoc on demand distance vector [38]. In the paper, the performance of the three protocols is evaluated with respect to energy consumption and network delay against transmission range and network density.

The concept of IoNT [36] represents an extension of the IoT concept based on the interconnection of nanoscale devices, which can then be integrated with biological technologies, leading to the IoBNT. The paradigm of the IoBNT is introduced by stemming from synthetic biology and nanotechnology tools that allow the engineering of biological embedded computing devices [37]. Based on biological cells, and their functionalities in the biochemical domain, Bio-nano things promise to enable applications such as intrabody sensing and actuation networks and environmental control of toxic agents and pollution. The IoBNT stands as a paradigm shifting concept for communication and network engineering, where novel challenges are faced to develop efficient and safe techniques for the exchange of information, interaction, and networking within the biochemical domain, while enabling an interface to the electrical domain of the Internet.

From the point of view of the architecture, a possible paradigm to enable the integration of nano-bio communications within an IP-based network infrastructure is the one presented in Figure 8.2, where the use of proper gateway devices is employed in order to act as middleware or "virtual bridges" toward the remaining devices of the network.

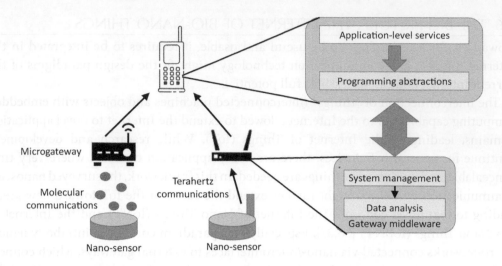

FIGURE 8.2 Integration of nano-bio communications within the Internet of Nano Things. Molecular and terahertz communications are integrated through the use of proper gateway devices.

8.7 SUMMARY

MAC in nanoscale communication networks represents an open challenge for the communications community. Indeed, both in communications in the terahertz band and in molecular communications, research is still ongoing on the identification of the most suitable strategies for accessing and sharing the communication channel.

This chapter provided the current state of the art and some considerations on the most interesting schemes, as well as some hints on performance analysis and possible integration issues within the emerging paradigms of the IoBNT.

REFERENCES

1. Jornet, J. M. and Akyildiz, I. F. (2013). Graphene-based plasmonic nano-antenna for terahertz band communication in nanonetworks. *IEEE Journal on Selected Areas in Communications*, 31(12):685–694.
2. Alsheikh, R., Akkari, N., and Fadel, E. (2016). MAC protocols for wireless nano-sensor networks: Performance analysis and design guidelines. In *Sixth International Conference on Digital Information Processing and Communications (ICDIPC)*, IEEE, December, pp. 129–134.
3. Jornet, J. and Akyildiz, I. (2014). Femtosecond-long pulsebased modulation for terahertz band communication in nanonetworks. *IEEE Transactions on Communications*, 62:1742–1754.
4. Akyildiz, I. F. and Jornet, J. M. (2010). Electromagnetic wireless nanosensor networks. *Nano Communication Networks*, 1(1):3–19.
5. Wang, P., Jornet, J. M., Malik, M. G. A., Adra, N. A., and Akyildiz, I. F. (2013). Energy and spectrum-aware mac protocol for perpetual wireless nanosensor networks in the terahertz band. *Ad Hoc Networks*, 11:2541–2555.
6. Mohrehkesh, S. and Weigle, M. (2014a). RIH-MAC: Receiver-initiated harvesting-aware mac for nanonetworks. *IEEE Transactions on Molecular, Biological and Multi-Scale Communications*, Vol. 1.

7. Alsheikh, R., Akkari, N., and Fadel, E. (2016). Grid based energy-aware mac protocol for wireless nanosensor network. In *2016 8th IFIP International Conference on New Technologies, Mobility and Security (NTMS)*, Larnaca, Cyprus, November 21–23, pp. 1–5.

8. Lee, S. J., Choi, H., and Kim, S. (2018). Slotted CSMA/CA based energy efficient MAC protocol design in nanonetworks. *International Journal of Wireless & Mobile Networks*, 10(1):1–12.

9. D'Oro, S., Galluccio, L., Morabito, G., Palazzo, S., D'Oro, S., Galluccio, L., and Palazzo, S. (2015). A timing channel-based MAC protocol for energy-efficient nanonetworks. *Nano Communication Networks*, 6(2):39–50.

10. Jornet, J. M., Capdevila Pujol, J., and Solé-Pareta, J. N. (2012). PHLAME: A physical layer aware MAC protocol for electromagnetic nanonetworks in the terahertz band. *Nano Communication Networks*, 3(1):74–81.

11. Wang, P., Jornet, J. M., Malik, M. G. A., Akkari, N., and Akyildiz, I. F. (2013). Energy and spectrum-aware MAC protocol for perpetual wireless nanosensor networks in the terahertz band. *Ad Hoc Networks*, 11(8):2541–2555.

12. Rikhtegar, N., Keshtgari, M., and Ronaghi, Z. (2017). EEWNSN: Energy E-cient wireless nano sensor network MAC protocol for communications in the terahertz band. *Wireless Personal Communications*, 97(1):521–537.

13. Mohrehkesh, S. and Weigle, M. (2014). RIH-MAC: Receiver-initiated harvesting-aware mac for nanonetworks. In *Proceedings of the 1st ACM International Conference on Nanoscale Computing and Communication, NANOCOM 2014*, Atlanta, GA.

14. Akkari, N., Wang, P., Jornet, J. M., Fadel, E., Elrefaei, L., Malik, M. G. A., Almasri, S., and Akyildiz, I. F. (2016). Distributed timely throughput optimal scheduling for the internet of nano-things. *IEEE Internet of Things Journal*, 3(6):1202–1212.

15. Akan, O. B., Ramezani, H., Khan, T., Abbasi, N. A., and Kuscu, M. (2017). Fundamentals of molecular information and communication science. *Proceedings of the IEEE*, 105(2):306–318.

16. Akyildiz, I. F., Brunetti, F., and Blázquez, C. (2008). Nanonetworks: A new communication paradigm. *Computer Networks*, 52(12):2260–2279.

17. Darchini, K. and Alfa, A. S. (2013). Molecular communication via microtubules and physical contact in nanonetworks: A survey. *Nano Communication Networks*, 4:73–85.

18. Llatser, I., Cabellos-Aparicio, A., and Alarcon, E. (2012). Networking challenges and principles in diffusion-based molecular communication. *IEEE Wireless Communications*, 19(5):36–41.

19. Dinc, E. and Akan, O. B. (2017). Theoretical limits on multiuser molecular communication in internet of nano-bio things. *IEEE Transactions on NanoBioscience*, 16(4):266–270.

20. Jamali, V., Ahmadzadeh, A., and Schober, R. (2017). Symbol synchronization for diffusion-based molecular communications. *IEEE Transactions on NanoBioscience*, 16(8):873–887.

21. Korte, S., Damrath, M., Damrath, M., and Hoeher, P. A. (2017). Multiple channel access techniques for diffusion-based molecular communications. In *11th International ITG Conference on Systems, Communications and Coding (SCC 2017)*, Hamburg, Germany, pp. 1–6.

22. Giné, L. P. and Akyildiz, I. F. (2009). Molecular communication options for long range nanonetworks. *Computer Networks*, 53(16):2753–2766.

23. Suzuki, J., Phan, D. H., and Budiman, H. (2014). A nonparametric stochastic optimizer for TDMA-based neuronal signaling. *IEEE Transactions on NanoBioscience*, 13(3):244–254.

24. Arifler, D. (2017). Connectivity properties of free diffusion-based molecular nanoscale communication networks. *IEEE Transactions on Communications*, 65(4):1686–1695.

25. Moore, M. J., Enomoto, A., Suda, T., Kayasuga, A., and Oiwa, K. (2008). Molecular communication: Unicast communication on a microtubule topology. In *2008 IEEE International Conference on Systems, Man and Cybernetics*, pp. 18–23.

26. Balasubramaniam, S., Lyamin, N., Kleyko, D., Skurnik, M., Vinel, A., and Koucheryavy, Y. (2014). Exploiting bacterial properties for multihop nanonetworks. *IEEE Communications Magazine*, 52(7):184–191.

27. Unluturk, B. D., Balasubramaniam, S., and Akyildiz, I. F. (2016). The impact of social behavior on the attenuation and delay of bacterial nanonetworks. *IEEE Transactions on NanoBioscience*, 15(8):959–969.

28. Qiu, S., Haselmayr, W., Li, B., Zhao, C., and Guo, W. (2017). Bacterial relay for energy-efficient molecular communications. *IEEE Transactions on NanoBioscience*, 16(7):555–562.

29. Krishnaswamy, B., Jian, Y., Austin, C. M., Perdomo, J. E., Patel, S. C., Hammer, B. K., Forest, C. R., and Sivakumar, R. (2017). Adma: Amplitude-division multiple access for bacterial communication networks. *IEEE Transactions on Molecular, Biological and Multi-Scale Communications*, 3(3):134–149.

30. Ghasempour, A. (2015). Using a genetic-based algorithm to solve the scheduling optimization problem for long-range molecular communications in nanonetworks. In *2015 IEEE 26th Annual International Symposium on Personal, Indoor, and Mobile Radio Communications (PIMRC)*, pp. 1825–1829.

31. Nakano, T., Moore, M. J., Wei, F., Vasilakos, A. V., and Shuai, J. (2012). Molecular communication and networking: Opportunities and challenges. *IEEE Transactions on NanoBioscience*, 11(2):135–148.

32. Islam, N., Misra, S., Mahapatro, J., and Rodrigues, J. J. P. C. (2013). Catastrophic collision in bio-nanosensor networks: Does it really matter? In *2013 IEEE 15th International Conference on e-Health Networking, Applications and Services (Healthcom 2013)*, pp. 371–376.

33. Hossain, Z., Xia, Q., and Jornet, J. M. (2018). Terasim: An ns-3 extension to simulate terahertz-band communication networks. *Nano-communication Networks*, 17:36–44.

34. Zhou, Y., Chen, Y., Murch, R. D., Wang, R., and Zhang, Q. (2018). Simulation framework for touchable communication on NS3Sim. *Nano Communication Networks*, 16:26–36.

35. Piro, G., Grieco, L. A., Boggia, G., and Camarda, P. (2013). Simulating wireless nano sensor networks in the NS-3 platform. In *Proceedings—27th International Conference on Advanced Information Networking and Applications Workshops, WAINA 2013*, Barcelona, Spain, pp. 67–74.

36. Balasubramaniam, S. and Kangasharju, J. (2013). Realizing the internet of nano things: Challenges, solutions, and applications. *Computer*, 46(2):62–68.

37. Akyildiz, I. F., Pierobon, M., Balasubramaniam, S., and Koucheryavy, Y. (2015). The internet of bio-nano things. *IEEE Communications Magazine*, 53(3):32–40.

38. Abuali, N., Aleyadeh, S., Djebbar, F., Alomainy, A., Ali Almaazmi, M. M., and Al Ghaithi, S. (2018). Performance evaluation of routing protocols in electromagnetic nanonetworks. *IEEE Access*, 6:35908–35914.

CHAPTER 9

Media Access Control for Nanoscale Communications and Networking

Sergi Abadal, Josep Solé-Pareta, Eduard Alarcón, and Albert Cabellos-Aparicio

CONTENTS

DIFFERENT FORMS OF NANOSCALE communications and networking are a reality, thanks to recent advances in nanotechnology, integrated circuit design, antenna miniaturization, and synthetic biology. As first nanonetwork systems start to appear, one of the main challenges from the communications perspective is the development of effective and efficient means for medium access control, where this chapter sets the focus. Existing techniques cannot be applied directly in nanoscale scenarios because of the extreme differences with respect to traditional contexts. Here, these differences are reviewed for three representative nanonetwork paradigms spanning several transmission ranges, communication mechanisms, energy policies, or mobility characteristics. Then, their impact on the design of appropriate shared medium management procedures is analyzed through a brief survey of protocols proposed in the literature.

9.1 INTRODUCTION

The design of the medium access control (MAC) layer has been a key research issue since the creation of the first computer networks. Basically, the MAC protocol defines mechanisms to ensure that all nodes can access a shared medium in a reliable manner. This is mandatory because two or more simultaneous accesses to the same channel generally *collide* and cannot be received correctly, resulting in a waste of resources. Thus, the MAC mechanisms need to determine how to avoid them and/or how to recover from them, in order to guarantee a successful transmission. These decisions play a decisive role in determining the performance of any network.

Wireless communications are particularly prone to collisions owing to several factors, namely (i) the explosion of multiuser wireless networks in terms of applications and adoption, (ii) the traditional use of rather broad beams, and (iii) the employment of air as the fundamental shared medium for communication. Since the inception of the ALOHA system [1], considered one of the first wireless packet data networks with a MAC protocol implementation, MAC design has evolved to embrace the constant scaling and growing sophistication of wireless networks. In short, huge efforts have been directed toward reducing the cost of shared medium management, while still addressing multiple challenges associated with node mobility, disjoint transmission ranges, and intermittent failures, so as to be applicable to a wide range of scenarios such as local area networks (LANs), mobile networks, and wireless sensor networks (WSNs) [2–4].

Nanonetworks are the last frontier in communications [5] and hold great promise for the implementation of systems capable of reaching unprecedented locations, reacting to nanoscale phenomena in a biocompatible way, conveying information wirelessly at terabit-per-second speeds, and/or being perpetual in nature through energy harvesting. These features come at the cost of outstanding challenges at all communication levels, and the MAC layer is no exception. In particular, nanonetworks either (i) exacerbate the problems found in traditional wireless applications, owing to the extreme downscaling of their components or (ii) create completely new challenges, owing to the use of novel paradigm-shifting approaches for communication.

This chapter aims to review the problem of MAC for nanoscale communication networks. In essence, we provide a brief overview of the main characteristics of different nanonetwork contexts and analyze how they affect the design of appropriate methods at the MAC level. We set the focus on three representative scenarios, summarized in Table 9.1:

- In Section 9.2, we examine electromagnetic (EM) nanonetworks in the terahertz (THz) band, seen as extremly downscaled versions of personal area networks (PANs) or WSNs [5]. Distinctive features are massive node densities, the use of energy harvesting, limited computing capabilities, and irregular and intermittent connectivity.

TABLE 9.1   Summary of the Analyzed Scenarios

	THz Nanonets	Monolithic Nanonets	Molecular Nanonets
Section	Section 9.2	Section 9.3	Section 9.4
Carrier	EM waves	EM waves	Molecules
Range	μm–km	μm–cm	μm
Node density	Medium-high	Medium-high	Extreme
Energy	Depends on scale	Guaranteed	Constrained
Computation	Limited	Ample	Very limited
Mobility	Yes	No	Limited
Connectivity	Intermittent	Steady	Intermittent
Other features	Deafness problem	Knowledge on traffic, environment	Extreme delays, jitter

- In Section 9.3, we analyze EM nanonetworks located within computing packages. The small size of the front-end components allows their massive integration on chips to improve the performance of monolithic computing systems [6]. Distinctive features are static and known distribution, energy awareness, and unique knowledge of the potential traffic patterns.

- In Section 9.4, we delve into the biocompatible molecular communications scenario [7]. The distinctive feature is the use of molecules instead of EM waves to convey the information. This entails a new set of challenges, for which we have a completely different set of tools to leverage. In this case, we need to overcome extremely large delays, minimal and unconventional computing capabilities, and scarcity of communication resources.

We finally summarize the main conclusions of the analysis and provide a brief outlook of future perspectives in Section 9.5.

9.2 THE MAC LAYER IN TERAHERTZ-BAND NANONETWORKS

Progress in THz technology has recently opened the door to efficient signal sources [8], antennas [9], and circuits for imaging or communications [10,11]. In the latter case, the use of this frequency band becomes extremely attractive owing to the abundance of bandwidth and the potential for low area and power footprints. However, it is challenging, given the large propagation losses or the lack of appropriate protocols, among other reasons [12].

The range of applications for wireless communications in the THz band is very wide, starting from macroscale examples such as cellular networks beyond 5G, high-speed data kiosks, high-throughput wireless LANs and PANs, and secure wireless communications for the military [5]. Further ahead, miniaturization of antennas and transceivers, together with ultra-efficiency energy harvesting methods, will enable the micro- and nanoscale paradigms such as wireless nanosensor networks (WNSNs) [13] and the Internet of Nano-Things [14]. In this context, intrabody applications show great promise.

In next sections, we highlight the main characteristics of these communication scenarios and review the existing approaches at the MAC level.

9.2.1 Context Analysis

As outlined previously, we can distinguish between two main scenarios for wireless communications in the THz band: macroscale and microscale. Although we are particularly interested in the latter, we will review the particularities in both cases.

Propagation in the THz range is dominated by spreading losses due to the constant shrinkage of antennas. This affects both microscale networks with short transmission ranges and simple antennas and macroscale networks, where directive antenna arrays can compensate for high losses due to relatively long propagation paths. Although the use of directive antennas provides an opportunity to exploit beam-multiplexing methods, it also creates the link-level phenomenon denoted as the *deafness problem* [15]. This refers to the situation where two nodes, which are equipped with directive antennas, cannot communicate properly, because their beam is not pointing to each other (see the left inset of Figure 9.1). The effects of the deafness problem can appear at both scales and to varying extents, depending on the directivity of the antennas.

Another particularity of THz propagation is the existence of the frequency-selective attenuation peaks due to molecular absorption or particle scattering [16]. These effects mainly depend on the composition of the medium and scale exponentially with the distance between transmitter and receiver, thus becoming significant at the macroscale [17]. The main consequence at the link level is that, depending on the transmission distance, certain frequency windows may not be available. For instance, in the example presented in the right inset of Figure 9.1, the short-range link has a much wider band available than the long-range link. Therefore, THz networks may need to account for methods that estimate the distance and point to the right frequencies to establish a link.

Besides spreading losses and a frequency-selective channel, THz signals suffer from severe blockage effects. Materials such as bricks and wood amd even the human body, which did not suppose a problem for microwaves, become obstacles for THz waves [18,19]. This is a problematic issue at the macroscale, where devices will most likely be surrounded by such obstacles, and affects the link layer design deeply (see the middle inset of Figure 9.1

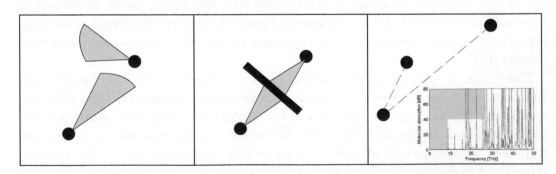

FIGURE 9.1 Link-level impairments inherent to THz communications. From left to right: deafness problem, signal blocking, and distance-dependent molecular absorption effects.

for an example). Link establishment and tracking will be required to ensure that signals are not blocked as transmitters or receivers change their positions.

As we move down to the microscale, other impairments further challenge the reliable transmission of data and complicate the MAC protocols. In this case, problems are related to high density of nodes and the scarcity of resources. Let us consider the example of WNSNs for intrabody applications, mainly composed by nanosensor motes with energy-harvesting means and limited computing capabilities [20]. Harvesting is used to gather energy from the environment and power up the transceiver and antenna for short intervals of time. This leads to intermittent connectivity and very short windows of opportunity to transmit information. Besides, harvesting severely limits (i) the complexity of the employed protocols, as the energy used for computations is also restricted, and (ii) the transmission range of the nanosensor mote. To compensate for these effects, WNSNs are expected to scale to thousands or even millions of nodes in an attempt to increase the likelihood of success. However, this also implies that the protocol should scale as well.

9.2.2 Existing Approaches and Outlook

Although it is not strictly a MAC protocol, one of the first efforts worth nothing is the distance-aware multi-carrier (DAMC) modulation [21]. DAMC selects the transmission windows depending on the distance between nodes to avoid the undesired attenuation peaks at the channel response. Once the transmission window is chosen, the MAC protocol comes into play. For macroscale networks in the millimeter-wave (mmWave) and THz bands, Han et al. [25] provide a comprehensive review of different techniques as well as a taxonomy of the existing MAC protocols in [27]. The work highlights a preference of the research community and the standardization bodies [1] for centralized approaches, some of which we detail next.

Most MAC protocols aim to overcome the deafness problem, which is pervasive at many scales. Different approaches are used; however, they generally boil down to a discovery service that uses omnidirectional antennas in lower bands, space sweeping, or side-lobe information to establish a connection, agreeing on the frequency channel and/or the beam direction [22–24]. In an extended definition, some authors argue that the handshaking can have different meanings at the macro- and microscale: Xiao et al. propose that the negotiation be triggered by the receiver, indicating the amount of time for which it will remain facing a given direction (macroscale, deafness-bound) or the amount of time for which it will be active, accepting packets (microscale, energy-bound) [15]. The interested reader will find deeper explanations of macroscale THz protocols in [25].

As we move down to the nanoscale, explicit MAC protocols are scarcer, owing to the extreme requirements of the scenario. One of the first proposals, physical layer aware MAC protocol for electromagnetic (PHLAME) [26], is an opportunistic very-low-complexity MAC solution based on the Time Spread On-Off Keying (TS-OOK) modulation. TS-OOK is a pulse-based modulation, where pulses are very spread over time, leaving long gaps of silence in between. PHLAME works over TS-OOK by assigning co-prime pulse repetition rates among the different nodes; this way, the probability of collision between pulses is minimized, without requiring stringent computation capabilities. A similar strategy is followed in [27], which uses pseudo-random codes and an adaptive approach to maximize throughput.

Other works on microscale MAC protocols attempt to circumvent the limitations of the scenario with different assumptions. In [59], the authors assume the existence of a nano-controller or gateway, a bit more powerful that the plain nodes, which acts as a centralized scheduler. The proposal by Mohrehkesh et al., instead, is fully distributed and takes the receiver-initiated approach mentioned earlier, with the particularity that control signals carry information about the amount of remaining energy to optimize the lifetime of the network [29]. These are the foundations of more theoretical works that evaluate the scalability of nanonetworks taking into consideration the harvesting limitations of the scenario, to validate the approach of "perpetual networks" [30,31].

On a final note, it is worth highlighting that graphene-based plasmonic THz antennas are expected to be key in future wireless networks (see Chapter 1 for more details). Their miniaturization and tunability properties make them particularly suitable for the microscale scenario and, as discussed in recent works [32,33], could be leveraged at the MAC layer. The reason is that, thanks to the aforementioned properties, joint frequency-beam reconfigurability is achievable with a simple change of voltage. The work in [33] leverages this property in an antenna device that is able to reconfigure the beam direction on a per-pulse basis, which would be key in massive beam-multiplexing protocols, even in the microscale. Hosseininejad et al. generalize the approach by creating programmable antennas that expose the frequency-beam reconfigurability to the MAC layer [32]. The device, schematically represented in Figure 9.2, allows to simplify macroscale protocols (e.g., no need for multiple radios for service discovery) and port to microscale environments.

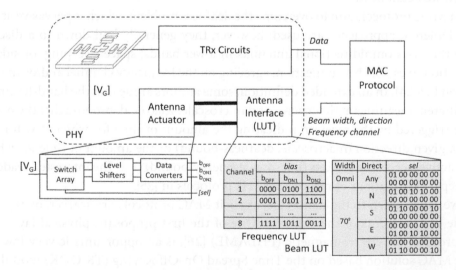

FIGURE 9.2 MAC-oriented programmable graphene-based antenna at THz frequencies. Voltages required to tune the antenna elements for a certain frequency and beam direction are stored in look-up tables. Voltages are then applied, thanks to an array of bias level shifters and switch arrays.

9.3 THE MAC LAYER IN NANONETWORKS FOR MONOLITHIC SYSTEMS

The continued downscaling of the wireless radio frequency (RF) front ends is currently enabling their massive on-chip integration. Initially, integration was leveraged to improve the performance and reduce the cost of wireless systems. However, current capabilities have led to the conception of the wireless on-chip communication paradigm, where multiple antennas and transceivers are placed within the same chip to build a wireless network within the computing system. This approach is a promising candidate to complement the existing network-on-chips (NoCs) in the pathway to overcoming the performance and efficiency bottlenecks of massively parallel processors [34].

In general, wireless on-chip communications can be applied to any monolithic system within a common computing package. This definition applies not only to Besides multiprocessor chips but also to data center racks [35], multichip modules [36], and embedded systems such as those found in latest reconfigurable metasurface prototypes [37]. In these systems, the benefits of wireless communications are low latency, natural broadcast capabilities, complementary metal–oxide semiconductor (CMOS) technology compatibility, and system-level flexibility given by the lack of path infrastructure between the interconnected entities. In next sections, we describe the main challenges and the existing approaches at the MAC level of design.

9.3.1 Context Analysis

A wireless on-chip network, illustrated in Figure 9.3, consists in the co-integration of wireless links with the processing cores, memory, and existing wired network. The processors and memory act as clients of the network, preparing and sending messages through either the wired network or the wireless links.

From the communications perspective, the chip scenario does not present as many issues as traditional wireless networks but is overall more demanding in terms of performance and reliability [6]. In broad strokes, the node density reaches levels commensurate to those of massive WSN or machine-to-machine (M2M) networks for 5G and beyond [38].

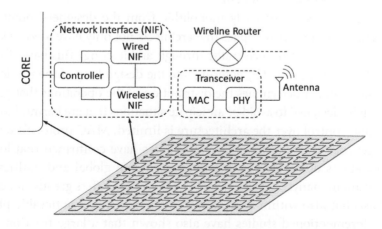

FIGURE 9.3   Schematic diagram of a multiprocessor chip with a wireless on-chip network.

The throughput demands also lead to strong resemblances with M2M networks, including the use of mmWave technologies. Finally, the on-chip networking scenario shares with mission-critical WSNs the need for latency and reliability guarantees, although with much more restrictive deadlines and power budget [39].

Such a distinctive combination of requirements would be unsolvable in many scenarios, owing to the problems related to the unknown topology, intermittent nodes, and blockage, to name a few. However, we will see next that the unique characteristics of the on-chip scenario virtually eliminate most of these issues. We enumerate the most important ones next; more details can be found in related works [6].

9.3.1.1 Physical Constraints

The physical landscape of the on-chip scenario is generally confined and dense but (unlike other wireless scenarios) static and known, because the chip layout is decided and fixed at the time of design [40]. Thus, the on-chip channel becomes quasi-deterministic and can be modeled with high accuracy by exploiting the previous knowledge of the scenario. As a result, problems such as hidden terminal can be eliminated, distributed protocols can be greatly simplified, and collision detection methods (generally restricted to wired networks) can be developed.

Being integrated within a monolithic computing system, the wireless on-chip networks are guaranteed energy availability. However, heat dissipation problems and efficiency requirements suggest that the on-chip networks should be very resource-aware. Since latency demands are also strict, such awareness implies that MAC designers should work toward streamlining the protocols and minimizing overhead, while finding optimal schedules to avoid collisions. This needs to be accomplished in a dense network, with potentially thousands of nodes within a single hop distance, which practically discards approaches solely based on multihop or multiplexing and hinders acknowledging. It is worth noting that the underlying wired network within the chip could also be leveraged to simplify the wireless protocol to some extent by, for instance, providing a synchronized clock signal or a token passing network [6].

9.3.1.2 Traffic in the Monolithic System

Most computing systems are basically monolithic from the designers' point of view and often a proprietary solution. This implies that architects not only have *a priori* knowledge on the traffic, but also can control such traffic through architecting. This basically implies that MAC protocols can be optimized by entering into the design loop of the whole architecture. For instance, the compiler determines the distribution of the operations that generate traffic and, as such, can be designed to avoid or anticipate situations that are harmful to the network.

Assuming that control over the architecture is limited, MAC protocols need to handle highly heterogeneous traffic profiles. Several studies have confirmed that local and unicast communications dominate but that the presence of global and multicast flows can become significant in manycore processors [41,42]. Traffic differs greatly not only on a per-application basis but also within applications, often showing a noticeable phased behavior [43]. The aforementioned studies have also shown that a large fraction of the traffic is generated in bursts and by a rather small subset of processors. These spatiotemporal characteristics are generally detrimental to performance and to adaptive approaches that can react to packet bursts or changes in the number of active processors or even anticipate those situations by leveraging the monolithic nature of the system.

9.3.1.3 Performance Requirements

Computing systems are not *best effort*. Each clock cycle of latency in communication lags the whole execution, whereas each erroneous bit or lost packet may lead the application to crash. This has three main consequences: (i) latency comes first (even before throughput [44]), and therefore, MAC protocols should emphasize latency sensitivity. (ii) The MAC protocol shall be stable to avoid latency peaks that could harm execution, even in the presence of bursty traffic, and provide quality of service (QoS) features to adapt to traffic flows that are latency-critical. (iii) Communications should have a reliability close to that of the wired network, unless specified in an approximate computing framework [45]. To achieve that, the MAC layer should be able to detect and remediate any error, distinguishing between collisions and regular bit flips caused by thermal noise or interferences.

9.3.2 Existing Approaches and Outlook

We have seen that the multiprocessor scenario is driven by latency and reliability, yet with a strong emphasis on energy efficiency. Besides the need for simplicity, these requirements are considered the main driver of a first wave proposals that relied on time and frequency multiplexing [34]. Multiplexing techniques, however, do not scale well, owing to their inherent rigidity and the implementation complexity as we increase the number of required channels. In response to this problem, another set of works instead proposed the use of different variants of the well-known token passing protocol [46–48], but the token round-trip time remains as a scalability concern. Variants of ALOHA and Carrier Sense Multiple Access (CSMA) protocols have been rarely evaluated [49,50] despite of their low latency and simplicity, arguably due to the relatively low performance in terms of throughput.

A recent scalability analysis confirms the points described in previously [6], clearly pointing toward CSMA-like protocols for low latency and token passing as a potential solution for high throughput. The work suggests that a sweet spot may exist between those two alternatives, in protocols that combine both approaches. The challenge is to find a graceful balance between performance and reconfigurability. A few proposals go in this direction by implementing both approaches and connecting them to a controller that selects the best strategy according to the load [50,47]. Although not stated in the aforementioned articles, the controller can take advantage of knowledge on the application to better drive the decisions and improve the performance further.

Another source of optimization resides in the development of custom variants of the CSMA protocol leveraging collision detection. As in wired networks, collision detection could be used in this scenario to reduce the impact of collisions on performance. The work in [51] explores different ways to actually implement collision detection, agreeing that a helper node [52] different than the transmitter should perform this action. Collisions could then be notified by using a sort of jamming signal as soon as they are detected, so that the transmitters would stop sending and start backing off as soon as possible.

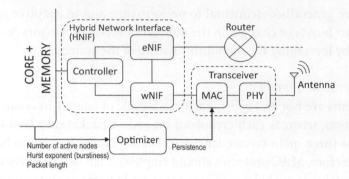

FIGURE 9.4 Schematic of a persistence optimizer in a wireless on-chip network. (Barcia, A., *Persistence Optimization for MAC Protocols in Wireless Networks on Chip*, PhD thesis, UPC-BarcelonaTech, 2018.)

Other ways to further optimize performance in CSMA protocols take advantage of the monolithic nature of the system. Recent work in [53] argues that the persistence of any CSMA variant, including that of [51], can be optimized with previous knowledge on the traffic. The work follows well-stated intuitions for other scenarios [54], and, as shown in Figure 9.4, it relates the number of active nodes, the burstiness of traffic, and the average packet length with the persistence that optimizes either latency or throughput. A generalization of this notion would be the application of machine learning methods that would automatically detect the features of traffic and tweak the protocol accordingly. A step further would consist of developing algorithms that would anticipate slots where no one would transmit, avoiding unnecessary collisions, without having to resort to a random backoff. A first approximation to this approach is presented in [41], where the authors predict up to 80% collisions in several parallel computing benchmarks. However, this strategy needs to be examined deeper before determining its validity, as the number of parameters that can affect such predictions is huge.

9.4 THE MAC LAYER IN MOLECULAR NANONETWORKS

Molecular communication is a novel communication paradigm that is based on the physical transport of molecules [55,7]. As shown in Figure 9.5, the molecules are released by the transmitter in a controlled manner; then, they propagate through the medium via diffusion, by directed flows, or using bacteria or molecular motors, until impacting the receiver [56]. Information is encoded either within the individual molecules or using the density of molecules as a magnitude.

The main distinctive trait of molecular communication is its biocompatibility, since all the elements of communication mimic or directly leverage existing biological processes to take place [7]. As a result, molecular nanonetworks have shown an unparalleled potential for the implementation of the Internet of Nano-Things [57] within biological systems. This approach is also referred to as the Internet of Bio-Nano Things [58] and targets intrabody applications such as disease detection and monitoring with unprecedented accuracy, intelligent drug delivery, and repair of damaged natural communication links in the body and in other scenarios for environmental control or cleaning [58].

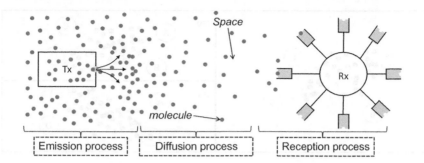

FIGURE 9.5 Representation of a molecular communication process: emission, diffusion, and reception.

Given the high node density of nanonetworks, access to the shared medium is an issue for molecular communication networks. Next sections detail the main characteristics of the scenario and the approaches at the MAC level of design.

9.4.1 Context Analysis

Broadly speaking, molecular communications encompass a variety of propagation and information encoding mechanisms. In the former, the transport means (e.g., regular diffusion, diffusive flows, and molecular motors) determine the speed of propagation and are generally orders of magnitude slower than in conventional EM networks [56]. In the latter, it is generally considered that information will be encoded using the density of molecules as a signal, requiring at least hundreds of units to encode a binary symbol. These two characteristics suggest that molecular communications will be subject to very high propagation delays, jitter, and loss rate, all leading to very low throughput levels.

Another important characteristic in the molecular communications scenario is the scarcity of resources for communication and computing [7]. Molecules will be created, captured, and released by bio-inspired nanomachines with minimal computing capabilities—possibly limited operations doable through gene regulation or molecular computing techniques. This restricts the amount of implementable protocols even further than in EM nanonetworks, but it also offers an advantage: in the same way that different biological processes use different types of molecules, nanomachines can use different molecular carriers to distinguish between *colliding* messages [59]. Last but not least, and since molecules are used both for computing and communicating, there is an inherent trade-off between the complexity of the protocol and the number of molecules that will be available for transmission.

9.4.2 Existing Approaches and Outlook

Molecular communication is a technique that, albeit highly ICT-centric, is heavily based on biological processes. As such, the associated communication protocols will most likely differ greatly from those found in conventional networks. In fact, some already existing complex processes such as quorum sensing [60] can be already seen as protocols by themselves, without having to build from scratch using the new molecular communication paradigm. Quorum sensing methods can be actually used for synchronization [61] and signal amplification [62] at the link level.

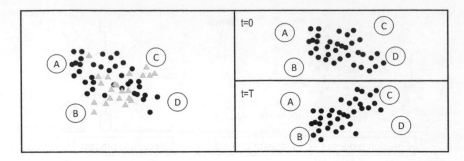

FIGURE 9.6 Mutiplexing techniques in molecular communications. Left: molecular division multiple access (black dots connect nodes A and D; gray triangles connect nodes B and C). Right: time division multiple access (A–D and B–C connections are active at $t = 0$ and $t = T$, respectively).

This opportunistic and bio-inspired line of work is followed in works proposing the use of different molecule types to allow multiple concurrent transmissions on the same channel [59]. This approach, represented in the left chart of Figure 9.6, would be useful to communicate different coexisting subsystems, each with its own molecule type. Applied to more generic frameworks, however, this molecular multiplexing scheme does not scale well, as it requires nanomachines to be equipped with selective ligand receptors and many different types of molecules, which may not be available.

Time-division multiplexing is another approach that has been proposed for molecular communications [63]. The main idea, represented in the right chart of Figure 9.6, is to have predefined temporal windows of activity in an attempt to minimize cross-talk interference and ensure proper multiplexing of communications. It is acknowledged that this requires either a previous effort to build the schedule or a synchronization mechanism that allows to negotiate the time-slot distribution among neighboring transmitters. This, however, is simpler and more adequate for such high-delay networks than more conventional CSMA approaches, which would instead require methods to verify that the channel is idle and to perform random backoffs [59]. Even if those methods would be available, collisions would not be completely avoided and would add extra delay to an already-very-slow process.

9.5 SUMMARY

Nanonetworks in the THz band, within monolithic systems, and those based on molecular communications are three distinctive paradigms of nanoscale networking. We have seen that the three alternatives push performance and cost to the limit for different reasons, suggesting the use of simple and effective MAC protocols tailored to the particularities of each scenario. In THz band nanonetworks, propagation of THz waves leads to deafness problems, blocking, and distance-dependent attenuation that need to be solved via intelligent service discovery means, possibly enabled by graphene-based programmable antennas. In nanonetworks within monolithic systems, MAC mechanisms need to be simple yet fast owing to the stringent latency requirements of the scenario, and to this end, they can leverage knowledge on the static environment and on the traffic patterns to streamline the protocols. Last but not least, molecular nanonetworks are based on completely different communication mechanisms, and therefore, MAC protocols need a profound rethinking. Resources are extremely limited, and moreover, there is a strong trade-off between communication and computation, which ends up reducing the number of alternatives. However, the existence of multiple molecule types can help in building scalable MAC strategies capable of overcoming the expected massive density of nanomachines in this context.

REFERENCES

1. N. Abramson. The ALOHA system: Another alternative for computer communication. In *AFIPS Fall Joint Computer Conference*, Vol. 37, pp. 281–285, 1970.
2. 802.15.3c: Part 15.3—Wireless Medium Access Control (MAC) and Physical Layer (PHY) Specifications for High Rate Wireless Personal Area Networks (WPANs): Amendment 2—Millimeter-wave-based Alternative Physical Layer Extension, 2009.
3. I. Akyildiz, W. Su, Y. Sankarasubramaniam, and E. Cayirci. Wireless sensor networks: A survey. *Computer Networks*, 38(4):393–422, 2002.
4. B. Crow, I. Widjaja, L. Kim, and P. Sakai. IEEE 802.11 wireless local area networks. *IEEE Communications Magazine*, 35(9):116–126, 1997.
5. I. F. Akyildiz, J. M. Jornet, and C. Han. Terahertz band: Next frontier for wireless communications. *Physical Communication*, 12:16–32, 2014.
6. S. Abadal, A. Mestres, J. Torrellas, E. Alarcon, and A. Cabellos-Aparicio. Medium access control in wireless network-on-chip: A context analysis. *IEEE Communications Magazine*, 56(6):172–178, 2018.
7. T. Nakano, M. J. Moore, F. Wei, A. V. Vasilakos, and J. Shuai. Molecular communication and networking: Opportunities and challenges. *IEEE Transactions on NanoBioscience*, 11(2):135–148, 2012.
8. R. A. Lewis. A review of terahertz sources. *Journal of Physics D: Applied Physics*, 47(37), 2014.
9. S. W. Qu, H. Yi, B. J. Chen, K. B. Ng, and C. H. Chan. Terahertz reflecting and transmitting metasurfaces. *Proceedings of the IEEE*, 105(6):1166–1184, 2017.
10. S. Kang, S. V. Thyagarajan, and A. M. Niknejad. A 240 GHz fully integrated wideband QPSK transmitter in 65 nm CMOS. *IEEE Journal of Solid-State Circuits*, 50(10):2256–2267, 2015.
11. H. Rucker, B. Heinemann, and A. Fox. Half-terahertz SiGe BiCMOS technology. In *Proceedings of the SiRF'12*, pp. 133–136, 2012.
12. I. Akyildiz, J. Jornet, and C. Han. TeraNets: Ultra-broadband communication networks in the terahertz band. *IEEE Wireless Communications*, 21(4):130–135, 2014.
13. I. F. Akyildiz and J. M. Jornet. Electromagnetic wireless nanosensor networks. *Nano Communication Networks (Elsevier) Journal*, 1(1):3–19, 2010.
14. I. F. Akyildiz and J. M. Jornet. The Internet of nano-things. *IEEE Wireless Communications*, 17(6):58–63, 2010.
15. Q. Xiao, Z. Hossain, M. Medley, and J. M. Jornet. A link-layer synchronization and medium access control protocol for terahertz-band communication networks. In *Proceedings of the GLOBECOM'15*, 2015.
16. J. Kokkoniemi, J. Lehtomaki, K. Umebayashi, and M. Juntti. Frequency and time domain channel models for nanonetworks in terahertz band. *IEEE Transactions on Antennas and Propagation*, 63(2):678–691, 2015.
17. I. Llatser, A. Mestres, S. Abadal, E. Alarcon, H. Lee, and A. Cabellos-Aparicio. Time-and frequency-domain analysis of molecular absorption in short-range terahertz communications. *IEEE Antennas and Wireless Propagation Letters*, 14:350–353, 2015.
18. K. Guan, B. Ai, A. Fricke, D. He, Z. Zhong, D. W. Matolak, and T. Kurner. Excess propagation loss of semi-closed obstacles for inter/intra-device communications in the millimeter-wave range. *Journal of Infrared, Millimeter, and Terahertz Waves*, 1–15, 2016.

19. R. Piesiewicz, T. Kleine-Ostmann, N. Krumbholz, D. Mittleman, M. Koch, J. Schoebel, and T. Kurner. Short-range ultra-broadband terahertz communications: Concepts and perspectives. *IEEE Antennas and Propagation Magazine*, 49(6):24–39, 2007.

20. C. Liaskos and A. Tsioliaridou. A promise of realizable, ultra-scalable communications at nano-scale: A multi-modal nano-machine architecture. *IEEE Transactions on Computers*, 64(5):1282–1295, 2015.

21. C. Han and I. F. Akyildiz. Distance-aware multi-carrier (DAMC) modulation in Terahertz Band communication. In *Proceedings of the ICC'14*, pp. 5461–5467, 2014.

22. C. Han, W. Tong, and X.-W. Yao. MA-ADM: A memory-assisted angular-division-multiplexing MAC protocol in Terahertz communication networks. *Nano Communication Networks*, 13:51–59, 2017.

23. Q. Xia and J. M. Jornet. Leveraging antenna side-lobe information for expedited neighbor discovery in directional terahertz communication networks. In *Proceedings of the VTC-Spring'18*. IEEE, 2018.

24. X.-W. Yao and J. M. Jornet. TAB-MAC: Assisted beamforming MAC protocol for Terahertz communication networks. *Nano Communication Networks*, 9:36–42, 2016.

25. C. Han, X. Zhang, and X. Wang. On medium access control schemes for wireless networks in the millimeter-wave and Terahertz bands. *Nano Communication Networks*, 19:67–80, 2018.

26. J. M. Jornet, J. Capdevila-Pujol, and J. Sole-Pareta. PHLAME: A physical layer aware MAC protocol for electromagnetic nanonetworks in the terahertz band. *Nano Communication Networks*, 3(1):74–81, 2012.

27. H. Mabed and J. Bourgeois. A flexible medium access control protocol for dense terahertz nanonetworks. In *Proceedings of the NANOCOM'18*, 2018.

28. P. Wang, J. M. Jornet, M. G. Abbas Malik, N. Akkari, and I. F. Akyildiz. Energy and spectrum-aware MAC protocol for perpetual wireless nanosensor networks in the Terahertz Band. *Ad Hoc Networks*, 11(8):2541–2555, 2013.

29. S. Mohrehkesh, M. C. Weigle, and S. K. Das. DRIH-MAC: A distributed receiver-initiated harvesting-aware MAC for nanonetworks. *IEEE Transactions on Molecular, Biological, and Multi-scale Communications*, 1(1):97–110, 2015.

30. R. Cid-Fuentes, S. Abadal, A. Cabellos-Aparicio, and E. Alarcon. Scalability of network capacity in nanonetworks powered by energy harvesting. In *Proceedings of the NANOCOM'15*, 2015.

31. X.-W. Yao, C.-C. Wang, W.-L. Wang, and J. M. Jornet. On the achievable throughput of energy-harvesting nanonetworks in the terahertz band. *IEEE Sensors Journal*, 18(2):902–912, 2018.

32. S. E. Hosseininejad, S. Abadal, M. Neshat, R. Faraji-Dana, M. C. Lemme, C. Suessmeier, P. Haring Boljvar, E. Alarcon, and A. Cabellos-Aparicio. MAC-oriented programmable terahertz PHY via graphene-based Yagi-Uda antennas. In *Proceedings of the WCNC'18*, 2018.

33. J. Lin and M. A. Weitnauer. Pulse-level beam-switching for terahertz networks. *Wireless Networks*, 7:1–16, 2018.

34. S. Deb, A. Ganguly, P. P. Pande, B. Belzer, and D. Heo. Wireless NoC as interconnection backbone for multicore chips: Promises and challenges. *IEEE Journal on Emerging and Selected Topics in Circuits and Systems*, 2(2):228–239, 2012.

35. J.-Y. Shin, E. G. Sirer, H. Weatherspoon, and D. Kirovski. On the feasibility of completely wireless datacenters. *IEEE/ACM Transactions on Networking*, 21(5):1666–1679, 2012.

36. X. Timoneda, A. Cabellos-Aparicio, D. Manessis, E. Alarcon, and S. Abadal. Channel characterization for chip-scale wireless communications within computing packages. In *Proceedings of the NOCS'18*, 2018.

37. A. C. Tasolamprou, M. S. Mirmoosa, O. Tsilipakos, A. Pitilakis, F. Liu, S. Abadal, A. Cabellos-Aparicio et al. Intercell wireless communication in software-defined metasurfaces. In *Proceedings of the ISCAS'18*, 2018.

38. Y. Niu, Y. Li, D. Jin, L. Su, and A. V. Vasilakos. A survey of millimeter wave communications (mmWave) for 5G: Opportunities and challenges. *Wireless Networks (Springer)*, 21(8):2657–2676, 2015.

39. P. Suriyachai, U. Roedig, and A. Scott. A survey of MAC protocols for mission-critical applications in wireless sensor networks. *IEEE Communications Surveys & Tutorials*, 14(2):240–264, 2012.

40. D. Matolak, A. Kodi, S. Kaya, D. DiTomaso, S. Laha, and W. Rayess. Wireless networks-on-chips: Architecture, wireless channel, and devices. *IEEE Wireless Communications*, 19(5), 2012.

41. S. Abadal, R. Martjnez, J. Sole-Pareta, E. Alarcon, and A. Cabellos-Aparicio. Characterization and modeling of multicast communication in cache-coherent manycore processors. *Computers and Electrical Engineering*, 51:168–183, 2016.

42. V. Soteriou, H. Wang, and L. Peh. A statistical traffic model for on-chip interconnection networks. In *Proceedings of MASCOTS'06*, pp. 104–116, 2006.

43. T. Sherwood, E. Perelman, G. Hamerly, S. Sair, and B. Calder. Discovering and exploiting program phases. *IEEE Micro*, 23(6):84–93, 2003.

44. D. Sanchez, G. Michelogiannakis, and C. Kozyrakis. An analysis of on-chip interconnection networks for large-scale chip multiprocessors. *ACM Transactions on Architecture and Code Optimization*, 7(1):Article 4, 2010.

45. T. Moreau, J. San Miguel, M. Wyse, J. Bornholt, A. Armin, L. Ceze, N. Enright Jerger, and A. Sampson. A taxonomy of approximate computing techniques. *IEEE Embedded Systems Letters*, 10(1):2–5, 2016.

46. S. Deb, K. Chang, X. Yu, S. P. Sah, M. Cosic, P. P. Pande, B. Belzer, and D. Heo. Design of an energy efficient CMOS compatible NoC architecture with millimeter-wave wireless interconnects. *IEEE Transactions on Computers*, 62(12):2382–2396, 2013.

47. N. Mansoor, S. Shamim, and A. Ganguly. A demand-aware predictive dynamic bandwidth allocation mechanism for wireless network-on-chip. In *Proceedings of the SLIP'16*, 2016.

48. M. Palesi, M. Collotta, A. Mineo, and V. Catania. An efficient radio access control mechanism for wireless network-on-chip architectures. *Journal of Low Power Electronics and Applications*, 5(2):38–56, 2015.

49. P. Dai, J. Chen, Y. Zhao, and Y.-H. Lai. A study of a wire-wireless hybrid NoC architecture with an energy-proportional multicast scheme for energy efficiency. *Computers and Electrical Engineering*, 45:402–416, 2015.

50. N. Mansoor and A. Ganguly. Reconfigurable wireless network-on-chip with a dynamic medium access mechanism. In *Proceedings of the NoCS'15*, page Article 13, 2015.

51. A. Mestres, S. Abadal, J. Torrellas, E. Alarcon, and A. Cabellos-Aparicio. A MAC protocol for reliable broadcast communications in wireless network-on-chip. In *Proceedings of the NoCArc'16*, pp. 21–26, 2016.

52. J. J. Garcia-Luna-Aceves and M. M. Carvalho. Collaborative collision detection with half-duplex radios. In *Proceedings of the WCNC'18*, 2018.

53. A. Barcia. *Persistence Optimization for MAC Protocols in Wireless Networks on Chip*. PhD thesis, UPC-BarcelonaTech, 2018.

54. R. Bruno, M. Conti, and E. Gregori. Optimization of efficiency and energy consumption in p-Persistent CSMA-based wireless LANs. *IEEE Transactions on Mobile Computing*, 1(1):10–31, 2002.

55. I. F. Akyildiz, F. Fekri, R. Sivakumar, C. R. Forest, and B. K. Hammer. MoNaCo: fundamentals of molecular nano-communication networks. *IEEE Wireless Communications*, 19(5):12–18, 2012.

56. M. Pierobon and I. F. Akyildiz. A physical end-to-end model for molecular communication in nanonetworks. *IEEE Journal on Selected Areas in Communications (JSAC)*, 28(4):602–611, 2010.

57. S. Balasubramaniam and J. Kangasharju. Realizing the internet of nano things: Challenges, solutions, and applications. *Computer*, 46:62–68, 2013.

58. I. F. Akyildiz, M. Pierobon, S. Balasubramaniam, and Y. Koucheryavy. The internet of bio-nano things. *IEEE Communications Magazine*, 53(3):32–40, 2015.

59. I. Llatser, A. Cabellos-Aparicio, and E. Alarcon. Networking challenges and principles in diffusion-based molecular communication. *IEEE Wireless Communications*, 19(5):36–41, 2012.

60. S. Abadal and I. F. Akyildiz. Automata modeling of Quorum Sensing for nanocommunication networks. *Nano Communication Networks*, 2(1):74–83, 2011.

61. S. Abadal and I. Akyildiz. Bio-inspired synchronization for nanocommunication networks. In *Proceedings of the GLOBECOM'11*, 2011.

62. S. Abadal, I. Llatser, E. Alarcon, and A. Cabellos-Aparicio. Cooperative signal amplification for molecular communication in nanonetworks. *Wireless Networks*, 20(6):1611–1626, 2014.

63. T. Nakano, T. Suda, Y. Okaie, M. J. Moore, and A. V. Vasilakos. Molecular communication among biological nanomachines: A layered architecture and research issues. *IEEE Transactions on NanoBioscience*, 13(3):169–197, 2014.

Signal Processing for Nanoscale Communication and Networking

Hossein Moosavi and Francis M. Bui

CONTENTS

10.1 INTRODUCTION

When interconnected nanoscale devices are able to collaboratively execute complex tasks in a distributed manner, they lead to the concept of nanonetworks [1,2]. Nanonetworks are envisioned to have all features analogous to wired and wireless communication networks. They must be able to collect, encode, transport, receive, decode, and deliver information to the appropriate application, which entails the traditional concepts of communication and information theory. A major challenge in nanonetworks involves interfacing between nanoscale components and between nanoscale and macroscale networks.

10.1.1 Nanocommunication Paradigms

Interfacing between nanoscale components can be categorized into two main paradigms: molecular communication and electromagnetic nanocommunications. In molecular communication, the transmitter makes a molecular concentration. More specifically, the transmitter emits molecules, which should be detected by the receiver. Molecular communication is emerging to be very promising networking paradigm in nanonetworks owing to the dimensional similarities of miniature human-designed entities with biological structures [3–5].

Molecular communication can be engineered in two ways [4]: first, an entirely artificial device could be designed to communicate, using signaling molecules, and second, the molecular communication capabilities, which occur ubiquitously at all levels of biological systems, including molecule, cell, tissue, and organ levels, could be engineered to transport artificial information.

Electromagnetic nanocommunication (aka, terahertz [THz] communications) is an alternative to molecular communication at the nanoscale. It utilizes electromagnetic waves to carry messages following the similar philosophy as in current wireless communication systems. The IEEE 1906.1-2015 [2] has recently become a recommended practice for the electromagnetic nanocommunications and molecular communication framework standard. Nonetheless, electromagnetic nanocommunication requires addressing a number of new challenges not covered by current wireless communication paradigms [6]. This primarily suggests unexplored frequency bands to be used, owing to the size of nanocomponents and their antennas. Moreover, nanonetworking requires a large set of functions to be performed, including propagation modeling, capacity analysis, modulation schemes, access control, and addressing, where traditional solutions may not be applicable owing to the limitations in terms of communication capabilities assigned to nanocomponents.

For intrabody communication, advantages of molecular communication over THz communications are propagation gain and energy consumption. More specifically, molecular energy attenuates at a lower rate than THz wave signals, and molecular processes consumes significantly less energy than THz waves. These advantages particularly stand out when considering the lossy nature of human body as the signal-propagation medium and the limited output power of nanotransceivers [1]. On the other hand, the major drawbacks of molecular communication are the propagation speed and delay of molecules. These characteristics prevent molecular communication in nanonetworks used for delay-sensitive applications, which require effective signal propagation and negligible delays. That said, among biological molecular communications, neuron-based communication does not have the issues of particle-propagation speed and delay. Furthermore, in terms of reliability, capacity, data storage, and energy consumption, the neural communications have superior characteristics over other intrabody nanoscale communications. These collectively make neural communication a promising candidate for engineering nanoscale communications and networking solutions.

10.1.2 Neural Communication

Neurons are communication entities within the neural network and are special among the cells of the body in their abundance and ability to propagate information rapidly within the network. The neural nanonetwork uses neural signaling to communicate. In neural signaling, a transmitting neuron sends a message to propagate as reaction diffusion waves up to distances of over a few meters. The emitted message propagates through the neural infrastructure and is ultimately received by receiving neurons.

During the signaling process, the temporal sequences of electrical pulses called action potentials (APs) or spikes from the neural body travel through neural fibers and reach a synapse. The synapse is the site of functional apposition between two cells, where a transmitting neuron converts the electrical signal into a chemical/molecular signal, which in turn is released into the synaptic cleft to propagate and eventually bind to the receptors located on the membrane of the receiving neuron. In principle, the synaptic transmission is the molecular transmission where molecules diffuse slowly. Nevertheless, the synaptic width is approximately 20 nm [7], which is notably less relative to widths of channels in other types of molecular communication. Hence, the propagation speed and delay are rather tolerable in synaptic channel within the neural communication.

Major types of noise that affect the signaling in neural communication include molecules in the synaptic cleft, induced electromagnetic fields, and shape deviation of APs.

10.1.3 Contributions and Organization

This chapter aims to provide a literature review of the theoretical frameworks for the neural communication in man-made nanonetworks. More specifically, we give an overview and identify the deficiencies of the existing solutions in the following areas.

- Neural compartments relevant to the signaling aspects of neurons

- Communication system models for neural communication

The conclusion will be the motivation and justification for novel research efforts in signal processing for neural communication.

The remainder of the chapter is organized as follows. Section 10.2 reviews the latest developments in computational neuromodeling. An overview of system models for neural communication is given in Section 10.3. Section 10.4 concludes the chapter.

10.2 LATEST DEVELOPMENTS IN NEUROMODELING

Spiking neurons are fundamental computational and signaling units of the nervous system. Thus, modeling of the neural hardware and processes is fundamental.

10.2.1 Computational Modeling

Computational modeling is to characterize what nervous system actually does and how it functions or operates. Most computational models are intended to analytically describe and reproduce real-world behavior of cellular compartments or predict the response to specific input signals.

In classical von Neumann architecture, central processing units of a computing system are physically separated from the main memory areas. However, colocalized memory and computation are the main characteristics of both biological and artificial neural processing systems. Both memory storage and complex nonlinear operations are simultaneously carried out by the synapses in a neural network. Distributing memory with processing when building processor architectures allows to address the need for increased computational power [8].

Mimicking of the functionalities of biological neurons and synapses in nanoscale devices has been extensively studied in the literature [9–14]. Electronic multineuron computing platforms have been built at the network and system levels capable of performing pattern analysis and machine learning operations [15–17]. Although remarkable, these works are limited in terms of both precision and bandwidth requirements. Such limitations can be overcome by imitating biological neural processing systems more closely. More specifically, physics of the silicon medium can be exploited by designing electronic circuits biased in the subthreshold regime to reproduce the biophysics of real synapses, neurons, and other neural structures [14,18–20].

Brain-inspired neuromorphic computing architectures allow for building nanoscale cognitive systems with colocalized memory and computation resources. The system can learn the statistics of varying input signals and their internal states and then behave accordingly. Such architectures provide state- and context-dependent information processing and support models of cortical networks and deep neural networks [8]. The reconfigurable on-line learning spiking (ROLLS) neuromorphic processor introduced in [21] is an example of the neuromorphic multineuron chip that follows this approach. Here, a spiking neural network is implemented using neuromorphic circuits biased in the subthreshold regime. The slow analog circuits directly mimic the physics of a real neural system for adaptation and learning, while the fast digital logic circuits provide an asynchronous communication protocol based on the address-event representation (AER) as well as the capability to configure the parameters of the neural elements in the chip.

10.2.2 Challenges and Issues

Although the current neuromorphic computing systems are promising, they have yet to be matured to the point that they can implement specialized autonomous small-scale cognitive agents.

10.3 SYSTEM MODELS FOR NEURAL COMMUNICATION

Communication modeling of neural signaling is to obtain quantitative estimates for properties such as ionic and chemical signals transmission, noise, and interference. Neural communication models are typically developed to resemble models of conventional wireless communication systems.

Neuron-to-neuron communication in the nervous system happens through two communication paradigms, namely electrochemical and molecular. The electrochemical communication is referred to as the intraneuronal communication because of the pulse transmission within the cell, while the molecular synaptic communication is referred to as the interneuronal communication because of the particle transmission between the cells.

Neural communication processes are modeled in the literature primarily from a neuroscientific point of view [22–24]. The purpose of such models is to characterize the functional relationship between sensory stimuli and neural spike trains. Some works base their representations on analytical expressions from validated mathematical models [25,26]. There are also efforts at examining neural communications from an engineering and fundamental communication theory perspective, with a view to construct body area nanonetworks [27–30].

10.3.1 System Model for Intraneural Communications

The electrochemical neuronal channel within the cell is considered as the preceding channel in the overall pathway of neural communication. A neuron generates APs in response to either chemical inputs normally collected by dendrites or the current induced in the soma. The neuronal response is then emitted from the soma, which serves as a transmitter, to propagate down through the axon, which serves as a communication channel connecting the soma and the presynaptic terminal of the same cell. The presynaptic terminal serves as a receiver of the electrochemical neural response (Figure 10.1).

The soma within the electrochemical communication pathway presents a nonlinear filtering and spike-generation mechanism. Depending on the type of the neuron, soma differently processes the sensory input and performs either as a low-pass or as a band-pass filter. The instant spiking rate of the neurons stems from their nonlinearity that transforms the filtered soma output into a spiking sequence.

10.3.2 System Model for Interneural Communications

In synapse, the electrochemical signals are transduced into molecular signals that are communicated following the law of diffusion. The diffusion-based molecular communication is limited to the transmission of particles/neurotransmitters between two neurons, that is, the pre- and postsynaptic neurons. In the molecular communication pathway, the presynaptic terminal within the axon of the presynaptic neuron acts as the molecular transmitter,

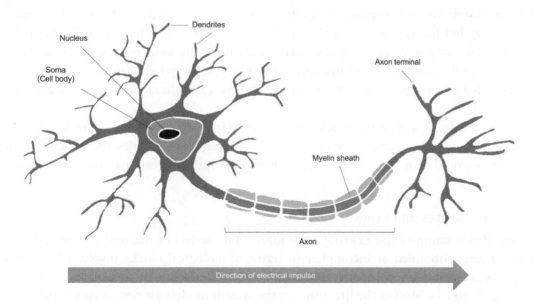

FIGURE 10.1   Intraneuronal communication. (From Kazilek, Neuron anatomy, Kazilek, 3 May 2011, https://askabiologist.asu.edu/neuron-anatomy.)

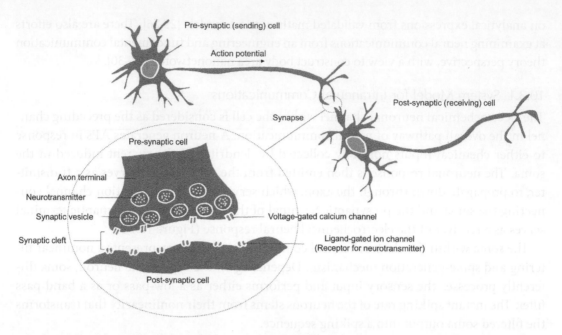

FIGURE 10.2 Interneuronal communication. (From The Synapse, Khan Academy, https://www. khanacademy.org/science/biology/human-biology/neuron-nervous-system/a/the-synapse.)

the synaptic cleft acts as the molecular channel, and the postsynaptic terminal at the next interconnected neuron acts as the molecular receiver. We refer to the synaptic neuronal transmission as the interneuronal communication (Figure 10.2).

The interneural communication is explored with the aim to characterize the spiking propagation between neurons and develop a framework in terms of transfer functions. Assuming that the synapse is highly reliable (neurons in the spinal cord, for example, make highly reliable synapses), the sequential chemical and ionic processes can be modeled as signals, and the remaining communication pathway between two neurons can be modeled as an input—output system deterministically characterized by a frequency response [30].

This model maps neuronal blocks to electrical circuitry equivalents, similar to classical communication systems. Such circuital representation enables the measurement of interneuronal communication efficacy through stimulus—response analysis for each process.

10.3.3 Challenges and Issues

Major shortcoming of the existing communication models of neurons is that they are either overly simplified or incomplete in terms of biological blocks involved in signaling processes. Such models, however, streamline the development of more accurate and advanced models. Most of the literature on the system models for neural communication is focused on the estimation of stimuli from neural responses, which leave the rest of the synaptic communication pathway as a black box [22].

A few works have studied the detection of spikes at the receiving neurons in the presence of axonal or synaptic noise [28,31,32]. One notable direction toward more realistic neurospike communication models is to investigate the performance of the neural signaling under different stochastic impairments such as axonal shot noise, synaptic noise, glial cells, and random vesicle release. The goal is to detect the spike trains optimally at the receiving neuron.

10.4 SUMMARY

Neural communication is a promising nominee for applications that involve nanoscale networking and communications, owing to its effectiveness and superior characteristics over other relevant paradigms. Investigating the performance of neural communication systems from the signal processing and communication theoretic perspectives allows us to understand the limitations of neurospike communication and helps to design artificial neural systems.

The tremendous progress in nanotechnologies in recent years, paralleled by the remarkable progress in both experimental and theoretical neuroscience, has led to considerable breakthroughs in neuromorphic processing systems. Further research and development efforts are required for such systems to be able to show the richness of behaviors seen in biological systems such as robustness, learning, and cognitive abilities.

A neurospike communication system model can be based on a neuroscientific theory, analytical expressions from validated mathematical models, or communication engineering tools and abstractions. To this end, the information about chemical and ionic behavior can be represented as signals, and then, stimulus—response analysis can be carried out to study the contribution of each stage of the neural communication pathway as an input—output system characterized by a frequency response.

Realistic neurospike communication models that take into account different stochastic impairments are useful for the design of effective communication techniques for nanonetworks and intrabody communications.

REFERENCES

1. Akyildiz, I. F., F. Brunetti, and C. Blázquez. "Nanonetworks: A new communication paradigm," *Computer Networks*, vol. 52, no. 12, pp. 2260–2279, 2008.
2. "IEEE recommended practice for nanoscale and molecular communication framework," in *IEEE Std 1906.1-2015*, pp. 1–64, January 11, 2016.
3. Atakan, B. *Molecular Communications and Nanonetworks*. New York: Springer-Verlag, 2016.
4. Nakano, T., A. W. Eckford, and T. Haraguchi. *Molecular Communication*. Cambridge, UK: Cambridge University Press, 2013.
5. Pierobon, M., and I. F. Akyildiz. "Fundamentals of diffusion-based molecular communication in nanonetworks," *Foundations and Trends® in Networking*, vol. 8, no. 1–2, pp. 1–147, 2014.
6. Galluccio, L., O. Akan, S. Balasubramaniam, and R. Sivakumar. "Wireless communications at the nanoscale [guest editorial]," *IEEE Wireless Communications*, vol. 19, no. 5, pp. 10–11, 2012.
7. Ribrault, C., K. Sekimoto, and A. Triller. "From the stochasticity of molecular processes to the variability of synaptic transmission," *Nature Reviews Neuroscience*, vol. 12, no. 7, p. 375, 2011.

8. Indiveri, G., and S. Liu. "Memory and information processing in neuromorphic systems," *Proceedings of the IEEE*, vol. 103, no. 8, pp. 1379–1397, 2015.

9. Kim, S. et al. "Experimental demonstration of a second-order memristor and its ability to bio-realistically implement synaptic plasticity," *Nano Letters*, vol. 15, no. 3, pp. 2203–2211, 2015.

10. Saighi, S. et al. "Plasticity in memristive devices," *Frontiers in Neuroscience*, vol. 9, p. 51, 2015.

11. Suri, M. et al. "Bio-inspired stochastic computing using binary CBRAM synapses," *IEEE Transactions on Electron Devices*, vol. 60, no. 7, pp. 2402–2409, 2013.

12. Yu, S. et al. "A low energy oxide-based electronic synaptic device for neuromorphic visual systems with tolerance to device variation," *Advanced Materials*, vol. 25, no. 12, pp. 1774–1779, 2013.

13. Kim, K. et al. "A functional hybrid memristor crossbar-array/CMOS system for data storage and neuromorphic applications," *Nano Letters*, vol. 12, no. 1, pp. 389–395, 2011.

14. Jo, S. H. et al. "Nanoscale memristor device as synapse in neuromorphic systems," *Nano Letters*, vol. 10, no. 4, pp. 1297–1301, 2010.

15. Sengupta, A., Y. Shim, and K. Roy. "Proposal for an all-spin artificial neural network: Emulating neural and synaptic functionalities through domain wall motion in ferromagnets," *IEEE Transactions on Biomedical Circuits and Systems*, vol. 10, no. 6, pp. 1152–1160, 2016.

16. Merolla, P. A. et al. "A million spiking-neuron integrated circuit with a scalable communication network and interface," *Science*, vol. 345, no. 6197, pp. 668–673, 2014.

17. Furber, S., F. Galluppi, S. Temple, and L. Plana. "The SpiNNaker project," *Proceedings of the IEEE*, vol. 102, no. 5, pp. 652–665, 2014.

18. Chicca, E., F. Stefanini, C. Bartolozzi, and G. Indiveri. "Neuromorphic electronic circuits for building autonomous cognitive systems," *Proceedings of the IEEE*, vol. 102, no. 9, pp. 1367–1388, 2014.

19. Rajendran, B. et al. "Specifications of nanoscale devices and circuits for neuromorphic computational systems," *IEEE Transactions on Electron Devices*, vol. 60, no. 1, pp. 246–253, 2013.

20. Benjamin, B. V. et al. "Neurogrid: A mixed-analog-digital multichip system for large-scale neural simulations," *Proceedings of the IEEE*, vol. 102, no. 5, pp. 699–716, 2014.

21. Qiao, N. et al. "A re-configurable on-line learning spiking neuromorphic processor comprising 256 neurons and 128k synapses," *Frontiers in Neuroscience*, vol. 9, p. 141, 2015.

22. Dayan, P., and L. F. Abbott. *Theoretical Neuroscience: Computational and Mathematical Modeling of Neural Systems*. Cambridge, MA: MIT Press, 2001.

23. Hodgkin, A. L., and A. F. Huxley. "A quantitative description of membrane current and its application to conduction and excitation in nerve," *The Journal of Physiology*, vol. 4, no. 117, pp. 500–544, 1952.

24. Nadkarni, S., and P. Jung. "Dressed neurons: Modeling neural-glial interactions," *Physical Biology*, vol. 1, no. 1, p. 35, 2004.

25. Paninski, L., J. W. Pillow, and E. P. Simoncelli. "Maximum likelihood estimation of a stochastic integrate-and-fire neural encoding model," *Neural Computation*, vol. 16, no. 12, pp. 2533–2561, 2004.

26. Izhikevich, E. "Simple model of spiking neurons," *IEEE Transactions on Neural Networks*, vol. 14, no. 6, pp. 1569–1572, 2003.

27. Galluccio, L., S. Palazzo, and G. E. Santagati. "Modeling signal propagation in nanomachine-to-neuron communications," *Nano Communications Networks*, vol. 2, pp. 213–222, 2011.

28. Balevi, E., and O. Akan, "A physical channel model for nanoscale neuro-spike communications," *IEEE Transactions on Communications*, vol. 61, no. 3, pp. 1178–1187, 2013.

29. Song, D., R. Chan, V. Marmarelis, R. Hampson, S. Deadwyler, and T. Berger. "Nonlinear dynamic modeling of spike train transformations for hippocampal-cortical prostheses," *IEEE Transactions on Biomedical Engineering*, vol. 54, no. 6, pp. 1053–1066, 2007.

30. Veletić, M., P. A. Floor, Z. Babić, and I. Balasingham. "Peer-to-peer communication in neuronal nano-network," *IEEE Transactions on Communications*, vol. 64, no. 3, pp. 1153–1166, 2016.

31. Manwani, A., and C. Koch. "Detecting and estimating signals over noisy and unreliable synapses: Information-theoretic analysis," *Neural Computation*, vol. 13, no. 1, pp. 1–33, 2001.

32. Maham, B. "A communication theoretic analysis of synaptic channels under axonal noise," *IEEE Communications Letters*, vol. 19, no. 11, pp. 1901–1904, 2015.

33. Kazilek. "Neuron anatomy," Kazilek, May 3, 2011, https://askabiologist.asu.edu/neuron-anatomy.

34. "The synapse," Khan Academy, https://www.khanacademy.org/science/biology/human-biology/neuron-nervous-system/a/the-synapse.

III

Molecular Nanoscale Communication and Networking of Bio-Inspired Information and Communications Technologies

CHAPTER 11

Communication Between Living and Nonliving Systems

Uche K. Chude-Okonkwo, B. T. Maharaj,
Reza Malekian, and A. V. Vasilakos

CONTENTS

11.1 INTRODUCTION

It is often common to categorize objects/structures into living systems and nonliving systems. Each of these categories of systems is differentiated from the other by certain unique characteristics. A biophysical characteristic that is peculiar to living systems is their ability to maintain stable ordered states, far from thermodynamic equilibrium [1]. On the other hand, nonliving systems if isolated or placed in a uniform environment usually cease all motions very quickly, such that no macroscopically observable events occur, thereby maintaining permanent equilibrium.

From biological perspective, the ordered state of living systems is implicit on their highly organized, coordinated structure that is a product of evolution-influenced progressive arrangement. The evolution theory of living systems indicates the progression of living systems from molecules through unicellular organisms to multicellular organisms. Hence, to maintain a stable ordered state in a living system, and yet preserve life process far from equilibrium, there is constantly the need to exchange entropy (information and energy) along and across the different levels of the living system's structure and the environment. In particular, the exchange of information (communication) between a living system and its environment (which includes other systems) enables it to self-organize and replicate (if necessary). As soon as a living system becomes incapable of exchanging entropy, it meets death.

The breakdown in the communication among living systems, for instance, humans, is implicated in various conflicts that lead to unhealthy and unproductive associations. On the other hand, the breakdown in communication among the fundamental units that make up a living system has been implicated in almost all diseases associated with the organism. To address the challenges posed by the breakdown in communication among humans, various man-made devices, such as cellphones, real-time electronic language interpreter, computers, and satellites, have been developed and are in use. These devices have tremendously impacted the way we live. On the other hand, man-made devices have also been developed to interface with living systems at the fundamental level of the system to probe and obtain information about how well the communication among the components of a living system behaves.

The discussion in this article primarily focuses on the fundamentals of establishing communication between living and man-made (nonliving) systems. The living-to-nonliving communication systems find application in fields that cut across areas such as medical, agriculture, environmental management/control, and human security. First, we explore the basics and characteristics of communication among living system. Second, we discuss the communication basics and potentials of communication between living systems and nonliving systems. Third, we present some exemplary application concepts for the living-to-nonliving system communication. Finally, we highlight issues such as toxicity and biocompatibility, which arise from the introduction of man-made nonliving system into the microenvironment of the cells.

11.2 COMMUNICATION BETWEEN LIVING SYSTEMS

In general, communication involves the conveyance of information or meaningful message from one system or group of systems to another by using mutually understood functions and axioms. Effective communication among living systems at the various levels of their existence is essential to their abilities to establish relationships, spread knowledge, and

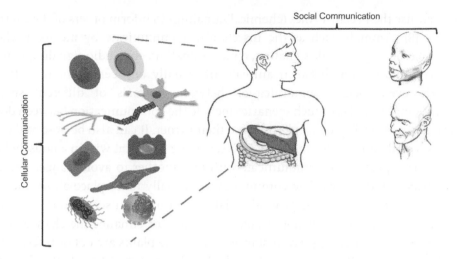

FIGURE 11.1   Illustration of social and cellular communication among living systems.

form collaborative structure to enhance productivity and survival. We categorize communication among the living units in an organism and that among organisms into two major groups, namely social communication and cellular communication, as illustrated in Figure 11.1.

11.2.1 Social Communication

Social communication is employed here to define the form of communication that occurs among multicellular organism such as animals and plants. The social communication platform enables the organisms to establish relationships, spread knowledge, enhance cohesive organization, increase productivity, and improve their survival probability at all levels of their existence. The breakdown in communication brings about ineffective interactions, which result is uncertainty, misunderstanding, confusion, argument/conflict, and ultimately poor productivity, as well as, sometimes, death [2].

To engage in social communication, the organisms are naturally equipped with transmit/receive faculties such as auditory systems, olfactory systems, visual systems, chemoreceptor systems, and mechanoreceptor systems. With the above transmit–receive systems, they equivalently exchange information by means of acoustic signaling, optical signaling, chemical signaling, and mechanical signaling. For instance, unaided communication among humans is achieved by means of syntactically organized system of signals, such as sounds, intonations or pitch, and gestures or written symbols that communicate thoughts or feelings. Among other animals, the exchange of information is often achieved by making special sounds, sending out optical signals, emitting/receiving chemical signals (pheromone communication) [3,4], and tactile communication [5]. These signaling modes basically enable them to organize themselves to achieve tasks, to search/reach for food, to mate with one another, and to avoid dangers/preys. For examples, monkeys use calls to warn one another of danger [6]. Often, birds such as the peacocks can make imposing feather display to communicate a territorial warning to other competing peacocks [7]. Some other animals may use their scent/urines (chemical signaling) to inform others of their territorial boundaries [8]. Animals such as the insects also communicate by means of chemical signaling (pheromones), tactile signaling, optical signaling, and audio signaling [9,10].

Social communications between animals are usually between animals of the same species. However, social communication between animals of different species is not uncommon. Examples of such scenarios include the communication between domesticated animals such as dog, cat, and horse and their owner. It can also be observed between two domesticated animals, for instance, between a dog and a cat within a household.

Like animals, plants also communicate with one another to avoid danger and locate/absorb nutrients and energy. The communication usually takes place above the ground or inside the ground. The above-ground social communication scenario is often used by plants to exchange information about potential danger or enhance the chance of species to compete for sunlight energy. For instance, when some plants are cut or infested by herbivores, they emit volatile organic compounds (VOCs) to warn other plants of impending danger [11,12]. The underground social communication scenario is often used by plants to obtain information about nutrient source. For instance, plants can use their roots to locate water [13]. There is also experimental suggestion that plants communicate to recognize their relatives [14]; this capability can modify their growth behavior. Even when the above-ground and below-ground communication signals are blocked, some plants can sense their neighbors and identify their relatives by means of uncharted communication channels used by seeds and seedling to sense neighbors and identify relatives [15]. It has also been found that besides communicating with other plants, plants can communicate with fungus and animals. For instance, a carnivorous plant has evolved ultrasound reflector features that can hijack bat communication systems of echolocation to its advantage [16]. More also, by employing a network of fungi, plants are able to communicate over a long distance [17–19]; such network is related to a neural network [20].

11.2.2 Cellular Communication

In this chapter, cellular communication is employed to define the form of communication that occurs among unicellular organisms. We shall only concentrate on the animal cells, plant cells, bacteria, virus, and fungi. The social communications between each type of the unicellular organism enable the organisms to enhance their survival probability over time.

11.2.2.1 Animal Cell-to-Cell Communications

Animal cells are fundamental units of life and form complex communication network in the body of the organism. There are trillions of cells in an adult human, with each cell size ranging from 1 to 100 microns. These cells are grouped into various types, which include skin cells, blood cells, nerve cells, fat cells, muscle cells, and bone cells. The cells in an organism do not live in isolation but interact. Their survival depends on receiving and processing information from each other and the extracellular environment. The information may reflect the availability of nutrients in the microenvironment, the changes in the working condition of a cell, the changes in the environmental conditions, the need to reproduce/grow, or the need to undergo extinction (apoptosis).

At the basic level, the exchange of information between cells essentially occurs by means of chemical signaling and electric impulses. Many animal cells employ various kinds of chemical signaling methods to communicate, which are dependent on their proximity to one another. These signaling are grouped into four types, namely autocrine, juxtacrine, paracrine, and endocrine signaling. *Autocrine signaling* occurs when a cell responds to its own biochemical signaling molecules that it produced. Some examples of this type of signaling include lipophilic and prostaglandins binding to membrane receptors. *Juxtacrine signaling* occurs between adjacent cells that are in contact; hence, this type of signaling is often referred to as a contact-dependent signaling. This type of signaling plays a very significant role in controlling cell fate and embryonic development [21]. *Paracrine signaling* involves signaling between cells that are within the same vicinity. An example of this is the histamine hormones, which are released as local responses to stress and injury [22]. Endocrine signaling is the most common type of cell signaling and involves sending a signal throughout the whole body by secreting hormones into the bloodstream or sap of the organism. Examples include adrenal signaling, thyroid signaling, and pancreatic signaling [23,24].

On the other hand, nerve cells called neurons communicate by means of a combination of electrical and chemical signals. Within the neuron, electrical signals driven by the movement of charged molecules across the cell membrane allow rapid propagation of electric pulses from one end of the cell to the other. Communication between neurons occurs at tiny gaps called synapses, where specialized parts of the two cells (i.e., the presynaptic and postsynaptic neurons) come within nanometers of one another to allow for chemical transmission [25].

11.2.2.2 Plant Cell-to-Cell Communications

Like the animal cells, the plant cells communicate among themselves to coordinate both developmental and environmental responses across diverse cell types. The ability of plant cells to make these cellular communications enables the plant to make the astonishing social communications discussed earlier. The exchange of information between the plant cells is conducted using specific mechanisms. Short-distance communications between cells in close proximity often occur through the symplastic and apoplastic pathways. Communication through the symplastic pathways is mediated by plant-specific structures called plasmodesmata. The plasmodesmata are specialized plasma membrane pores that traverse the cell walls of adjacent cells, thus connecting their cytoplasms [26]. Communication through the apoplastic pathways employs the extracellular space between cell walls as a route to the targeted cells through their plasma membranes by various mechanisms such as polar auxin transport [27].

11.2.2.3 Bacterium/Virus/Fungus-to-Bacterium/Virus/Fungus Communications

In single-cell primitive organisms such as bacteria, viruses, and fungi, communication between organisms has been known to occur by means of chemical signaling. Bacteria can communicate by quorum signaling/sensing, especially when they are in high density [28]. This form of collaborative communication provides the group of bacteria a way to adapt to the environment.

When temperate viruses infect and become dormant in their host cells, they usually decide between whether to undergo lytic cycles (replicate by using their host cell's protein synthesis mechanism), thereby killing the host cell, or to undergo lysogenic cycle (become latent in the host cell) and keep the host viable. It is shown in [29] that these viruses use a small-molecule communication system to coordinate lysis–lysogeny decisions.

For the purposes of basic biological functions such as growth, mating, and morphogenesis, fungi are known to engage in communication within species and across kingdoms. Such communication mechanisms are controlled by a variety of messenger molecules, including small peptides, alcohols, lipids, and volatile compounds [30].

11.3 COMMUNICATION BETWEEN LIVING AND NONLIVING SYSTEMS

We have discussed various forms of communication between living systems, from social communication to cellular communication, and established that these communication mechanisms are very crucial for the survival/harmonious existence of living systems. Different factors may affect effective communications at the social network level and the cellular network level. In the social network level, issues such as language/literacy barrier, distance, location, disability, and information organization are some of the factors that militate against effective communication. In the attempts to address these odds in social communication, man has, over many years, designed and developed many nonliving communication technologies and devices. Such devices and technologies include phones, computers, telepresence, hearing-aid devices, visually impaired aid devices, and the Internet. These devices basically act as the interface between two humans in an effort to bridge the communication gaps.

Similarly, the communication challenges at the level of cellular networks are very crucial and will, in many cases, interfere significantly with the organism's social communication capability. The factors that affect effective cellular communication come from many forms of impairments and perturbations to the effective working mechanisms of the body cells, tissues, and organs. This could be as the results of physical injuries, interaction with pathogens and toxins, inherent system degradation, and adverse environmental conditions. The impairments in the cellular communication typically disrupt the hitherto nonequilibrium state of the living host system, which may push the system toward the equilibrium state and eventual death. Typically, the breakdown in the cellular communication leads to the manifestation of disease symptoms [31]. For instance, breakdown in the cellular communication has been implicated in many diseases such as Alzheimer [32], diabetes [33], Parkinson [34], cancer [35], and Huntington [36]. If effective actions are not taken to normalize the breakdown in communication or where such actions are not available, the anomaly in cellular communication may lead to the death of the organism.

Ideally, when breakdown in communication occurs among cells, complex cell-to-cell communications that involve specific cells in the organism are triggered to correct the problem. For instance, when organisms encounter injuries, their body employs cellular signaling pathways that involve complex cell-to-cell communication to heal wounds [37]. The correction mechanisms for cellular communication breakdown are usually fine-tuned over the period of evolution. In this sense, the vast cellular network of the organism has been equipped with memory and the know-how required for addressing many possible communication issues over the course of evolution. However, in the case where such memory or/and the know-how for addressing certain communication anomalies is inaccessible, the cellular communication network is faced with serious challenges. In some cases, the pace of correcting the breakdown may be too slow, or the specific biological systems tasked with the repair of the communication breakdown may not have sufficient resources required to address the problem. In this case, artificially synthesized resources need to be introduced to address the communication problem—this is the fundamental goal of medicine and healthcare services. This goal fundamentally requires the establishment of communication between living systems (in this case, body cells) and artificially engineered systems (basically, nonliving systems). We shall proceed to discuss the basics, techniques, and applications of artificially engineered communication between the living and nonliving systems.

11.3.1 Fundamentals of Living Systems-to-Nonliving Systems Communication

As we have stated earlier, the fundamental goal of medicine is to provide the additional resources required to correct cells-to-cells communication anomalies in the body. The additional resources include artificially synthesized systems and functions that can basically communicate directly with the affected body cells in the cellular network to repair, replace, or/and reconfigure the network desirably. To communicate with a system implies being able to change the state of one or more fundamental components of the systems by using any form of information carrying signals such as electromagnetic, electrical, biological, chemical, and mechanical signals. Hence, establishing communication between a nonliving system and living systems implies using information signals produced by man-made systems to artificially change the state of a cell or group of cells in a living system, as desired, and *vice versa*, as depicted in Figure 11.2.

11.3.2 Macroscale Communication Between Living and Nonliving Systems

To be able to correct anomalies in cellular communications, one has to first obtain a detailed understanding of the architecture, material composition, and working principles of the cells or biological unit of interest in both healthy and unhealthy conditions. This enables us to know the set of the candidate signaling functions that can communicate with the cells of interest to achieve the desired objectives. Many contemporary man-made communication devices are typically based on semiconductor technology, where elements such as silicon, gold, boron, and phosphorous are used. Hence, the signaling functions associated with such devices include electric, magnetic, electromagnetic, and thermal signaling functions, which are fundamentally energized by the interchange of electrons in semiconductors. Since the biological systems such as cells are fundamentally made up of atoms, the signaling function from the made-made devices can interact with the material components on the subatomic level. In that case, and depending on the desired task, these signaling functions may be used directly as communication signals to interact with certain cells of interest. The man-made devices in this case are usually of macroscale dimension; hence, this form of living-to-nonliving systems communication is termed *macroscale cell-to-man-made devices communication* here. For instance, let us consider the medical application where the task at hand is to destroy some unhealthy cells in the body. Let us assume that these cells

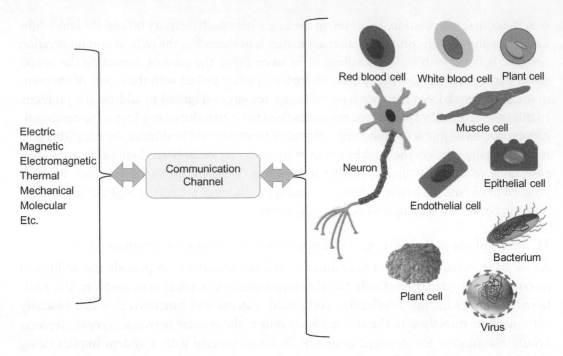

FIGURE 11.2 Illustration of communication between living and nonliving systems as being fundamentally based on the communication between man-made devices and the fundamental units of life.

can desirably respond (be destroyed) to certain frequency and intensity of electromagnetic signals, which can significantly disrupt the biological mechanism and maybe the physical structure of the targeted cells, thereby destroying them. Then, we can build systems that can respond to specific information signal to generate the required frequency and intensity of the electromagnetic wave and direct that to the targeted cells. This form of cell-to-man-made system communication is exemplified in the radiation therapy used in cancer treatment [38]. Magnetic signaling [39], electric signaling [40], and thermal signaling [41] to cells have been used as therapeutic modalities. Man-made devices other than those build on semiconductor technology can also communicate with the cells. A typical example is the use of injection pumps to send specific biomolecular signals (drugs) to act as extracellular chemical signals, which inform the defective cells to directly carry out the desired chemical, physical, and biological modifications to achieve therapeutic results. The source that introduces the drug particles represents a transmitter and the targeted defective cells, the receiver. This can be regarded as a communication between the nonliving and living systems. Even in plants, man-made signaling sources for electromagnetic [42–44] and magnetic [45–47] fields have been employed to communicate with plant cells to treat plant disease and alter the growth and development of plants for the improvement of food production.

The nonliving system-to-living system communications highlighted earlier are basically on-way systems, which imply that the communication is unidirectional (from devices to cells). This is often referred to as simplex communication in the communication engineering parlance. Simplex communication can also occur from the cells to man-made devices. For instance, the specialized sensors can read biosignal information that originates from the variation in the characteristics of certain cells in the body. For instance, defects in the endothelial cells can result in the thickening of the blood vessels, which resultantly affects the blood pressure. This variation in blood pressure (biosignal) can be detected using specialized sensors in a form of communication setting. Hence, the biosignals are usually associated with various pathological conditions such as blood pressure, sugar level, pulse rate, body temperature variation, electrical activity of the brain, electrical activity of the muscles, and electrical activity of the heart. Such simplex communication can also be regarded as a communication between cells and man-made systems.

11.3.3 Nanoscale Communication Between Living and Nonliving Systems

Fundamental biological units such as cells and unicellular organisms are naturally equipped with reception/transmission facilities that have been fine-tuned over the period of evolution to enable them to communicate with one another by means of high molecular signaling specificity in specialized microenvironmental conditions. The molecular signaling can be through protein exchange or ion exchange. Other forms of signaling other than molecular signaling, such as the combination of molecular signaling and electric signaling in neuronal communication and of "nonphysical" signaling and "noncontact" signaling [48], have also been observed between cells. Hence, the possibility of engineering communication between living cells and man-made nonliving systems at the nanoscale level is conceivable if there is the know-how to design such nanosystems. The advent of nanotechnology offers us exceptional concepts, tools, and technologies to achieve the objective of designing nanosystems that can help us realize communications at the nanoscale level [49,50]. Essentially, the man-made nanosystems can establish communications with cells by means of the exact or replica signaling format that the specific cells use in communicating with other cells to achieve specified tasks. This type of communication is generally termed nanocommunication [51,52].

11.3.3.1 Autonomous Nanoscale Communication Between Living and Nonliving Systems
Presently, extensive research is going on in the field of nanocommunication [31,53,54]. These research efforts build on the tremendous state of the art in communication engineering and the gain of nanotechnology to propose different approaches to information exchange between man-made nanosystems and different components of a living system. Hence, these artificial nanosystems must be able to transmit and receive molecular signals and electric signals (in the case of neuronal communication). Depending on the design and task, they must also be able to operate in any of the modes typical of cellular communications such as juxtacrine, paracrine, and endocrine signaling modes. More also, when deployed to the targeted microenvironment in the body, they must seamlessly form part of the cellular network of the body and work collaboratively with the entire network. Since they will usually work in the remote parts of the body, they should, like typical natural cells work *autonomously*.

Autonomous communication between cells and man-made nanosystems can be used *in vivo* to monitor local changes in pH [55]; the health of a cell; the protein expression in a cell; and changes in the local concentration of metabolites or biochemicals, such as glucose [56], glutamate [57], and dopamine [58]. The capability of achieving autonomous nano scale communication can provide us with the ability to address the challenges of diseases such as cancer [59], diabetes, Alzheimer's disease and AIDS at the cellular level. Extensive research in molecular biology and synthetic biology has provided us with a bunch of candidate artificially synthesized nanosystems that can be used to communicate with the cells. Examples include (i) lipid-based vesicles such as liposome (and its different versions, namely nisosomes, transfersomes, and ethosomes), (ii) polymer-based particles such as nanosphere and nanocapsules, (iii) conjugates such as antibody–nanoparticle conjugates and polymer–protein conjugates, and (iv) polymer-based self-assembly vesicles such as micelles, DNA-origami capsules, and dendrimer [31]. Other advances in scientific research can also be incorporated to achieve cell-to-man-made nanosystem communications. For instance, a bioelectronic device with ion channels that control the flow of ions across a supported lipid layer is developed in [60], which can be used as a communication nanosystem. In [61], the concept of harnessing the information interconversion between electronic and ionic modalities based on redox reaction [62] to carry electronic information to engineered bacterial cells, in order to control transcription from a simple synthetic gene circuit, is proposed. The idea of relayed communication at the nanoscale level has also been explored. In this sense, where there is biochemical message discrepancy between the nanosystems and the targeted receiver cells, an artificially synthesized nanosystem that acts as a translator relays the information to the cell in a format understandable by the cell [63]. Mobility attributes can be integrated in the nanosystems by engineering them to respond mechanically to changes in the concentration of specific biochemical in the microenvironment [64]. Other ideas such as the development of genetic circuit components and the use of genetically modified biological systems such as modified cells [65], viruses [66], bacteria [67], and bacteriophage [68] as synthesized nanosystems can be adopted. In a more advanced autonomous systems, the artificially nanosystems can be designed to be controlled by specific natural neural activities [69].

11.3.3.2 Interfaced Nano-to-Macroscale Communication Between Living and Nonliving Systems

In many scenarios, there may be obvious incompatibility between some man-made communication devices signaling and the signaling formats of the biological systems with which man-made device intends to communicate. In some other scenarios, there may be the need to monitor the activities of the nanosystems or the communication process from a remote location or to connect the *in vivo* nanonetwork to different networks of diverse signaling, such as in the Internet of Bio-Nano Things (IoBNT) [70,71]. The major issue here is how to design the communication system to realize effective information exchange between man-made communication devices and the living cells, taking the incompatibility of signaling formats into consideration.

To overcome this barrier, a device or a system that acts as an interface between the biological systems and the man-made communication devices is required. Such interface device must be structurally viable and can emit/transmit, sense/receive, and transduce/

manipulate various forms of signals into the forms that are desirable at the receiving end. Depending on the application/task at hand, the location of the targeted cells, and the architecture/working principles of the cells of interest, the interfacing unit can be located on or within the body system, in proximity to the targeted microenvironment/cells. These factors help in making the best choice of the candidate signaling format and, of course, in optimizing the design of the interface unit. For instance, the intended application and the working principles of the targeted cells/man-made nanosystems define the candidate signaling formats that the interface unit will employ in the communication setup. This in turn influences the potential design approaches and architectures for the interface unit.

Hence, the choice of the architecture/operation of the interface unit depends on the nature of the input and output signals of the cells and the proximity/nature of the medium through which the signals will propagate. For instance, communication among neuronal cells is basically by means of the combination of molecular signaling and electric signaling. Hence, interfacing of the neural systems with many man-made systems is commonly achieved by electrical stimulation [72]. In some instances, the molecular aspect of the neuronal signaling can be mediated by the use of synthetic ion pump to control signaling by mean of ions, neurotransmitters, and other molecules [73]. The possibility of bioelectrochemical transduction is discussed in [61], where an electrode can be energized to modulate the activation or deactivation of naturally occurring electrochemical redox-based mediators, which in turn trigger the biological response in cells.

In some examples, the use of optical communication to interface made-made systems and cells can be achieved with a man-made nanosystem that can respond to light from an external light source at a specific wavelength [74]. Also, luminescence phenomenon can be employed in the design of an interface unit. Such approach converts chemical signals to optical signals [75]. By using temperature-sensitive materials to design the interface unit, thermal signaling from external thermal sources can be used to produce molecular signaling [76]. It is also possible to convert magnetic signaling to molecular signaling. For instance, by coating magnetic nanoparticles with specific ligands that enable them to bind to receptors on a cell's surface, they can respond to externally applied magnetic field and pull on the particles, so that they deliver nanoscale forces at the ligand–receptor bond [76]. In another instance, an interface unit can be integrated with gold nanoparticles, which will respond to the concentration of DNA molecules in the presence of magnetic signals from an external source [77].

In some design, multiple types of signaling can be employed on one interface unit. For instance, in [71], an interface unit is designed to interface biological systems with contemporary electrical transceiver. The unit is modeled by employing biological concepts, such as the responsiveness of certain biomolecules to thermal and light stimuli, and the bioluminescence phenomenon of some biochemical reactions. It converts electrical signal to heat, which in turn is used to generate molecular signal. It is also capable of converting molecular signal to optical signal, which is used to generate electromagnetic signal.

Some other proposals for addressing the interface issue have been presented in [78–80]. In [78], the issue of establishing interfaces that interconnect a molecular communication environment and an environment external to the molecular communication environment is examined. The use of silicon nanowire field effect transistors as molecular antennas to realize bio-cyber interfaces between molecular and conventional electronic communication networks is proposed in [79]. In [80], different CMOS bioelectronics platforms are presented for interfacing electronic systems to biological systems.

11.4 ILLUSTRATIVE APPLICATION EXAMPLES OF LIVING SYSTEMS-TO-NONLIVING SYSTEMS COMMUNICATIONS

The ability to enable communication between man-made systems and the fundamental units of living systems can enable a wide range of applications such as medical, environmental management/security, and energy/food production applications. In medical application, the know-how to communicate with the cells and identify complex biological interaction and cellular stress responses can be used for molecular-level disease diagnosis, therapy, and monitoring of disease/therapeutic progress in animals. More also, by being able to communicate with plant cells, we are availed with a high degree of freedom to provide molecular-level therapy to diseased plants, as well as improve the efficiency of food production from plants. In terms of environmental management, the ability to build the communication network of bacteria to sense and destroy harmful agents in the environment [81,82] and the use of artificially synthesized nanosystems to eliminate bacteria/fungi in the contaminated environment can be employed in cleaning up water, soil, and atmosphere pollution, as well as improving manufacturing methods to reduce pollution. Further application in alternative power sources can be obtained from the ability to build controlled communication network of genetically engineered bacteria [83,84]. We present exemplary scenarios where basic cell-to-nonliving communication systems, supported by contemporary research results, can be employed to address the existing medical and bioprocess challenges.

11.4.1 Example 1

Figure 11.3 illustrates the communication between sets of man-made nonliving system and living system, the natural cells. In Figure 11.3a, following [63], an artificially synthesized nanotransmitter is intended to send molecular signal to sets of living cells to respond as desired. However, in this scenario, the natural cells are unable to relate to the message molecules' format. Hence, an artificially synthesized interface nanosystem is used as a molecular message translator to convey the acceptable message format to the natural cells. In Figure 11.3b, a similar idea to that in Figure 11.3a is shown; however, in this case, an externally generated signaling from a contemporary electronic device is used to elicit response in the natural cell of a living system through an artificially synthesized nanosystem that functions as a translator.

The above ideas can be employed to deploy targeted delivery of drugs, which ensures the minimization of side effects of conventional drug-delivery methods such as toxicity. The treatments of cancer by using chemotherapy can benefit from these ideas. For instance, the idea in Figure 11.3a is related to the proposal in [59], where a set of nanosystems acting as an interface system employs enzymatic reaction process to translate/activate prodrug molecules to active drugs, to which cancer cells respond. The method described in Figure 11.3b can be employed to remotely trigger the release of chemotherapy drugs from nanocarriers that have accumulated at the disease site [85].

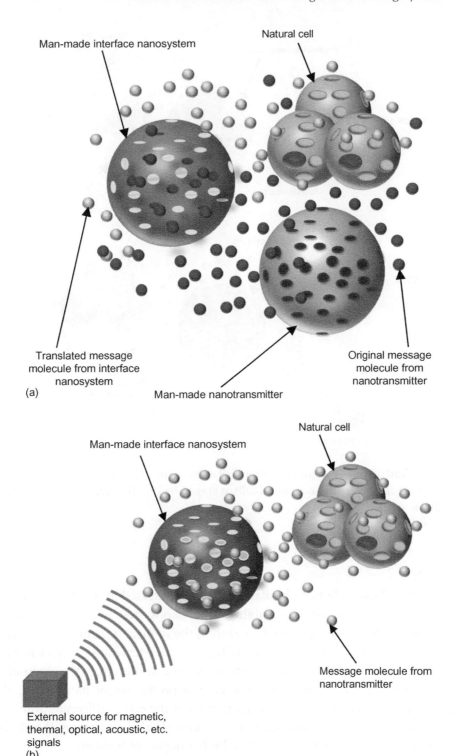

FIGURE 11.3 (a) Illustration of communication between man-made nanosystems and natural cells within the microenvironment of the cells. (b) Illustration of communication between man-made macrosystems and natural cells, with a man-made nanosystem acting as the interface system.

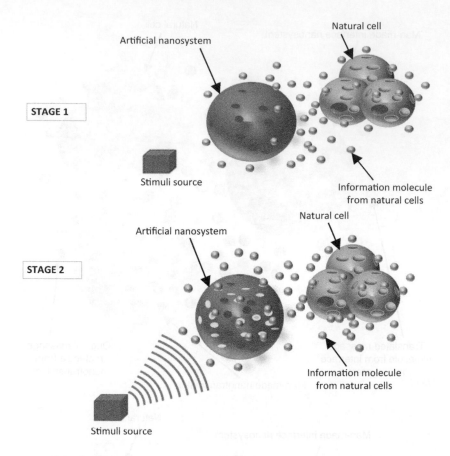

FIGURE 11.4 Illustration of communication between man-made nanosystem with stimuli-controlled surface characteristic and natural cells at Stage 1 before stimulus is applies and at Stage 2 after stimulus is applied.

11.4.2 Example 2

In this example, following [64], molecular information released by a natural cell is sensed by a nanosystem, where the sensing capability of the nanosystem is mediated by the external stimuli. As illustrated in Figure 11.4, in *Stage 1*, the information molecules released by the cells cannot be communicated to the artificial nanosystem, since it does not express the complementary reception surface proteins. In *Stage 2*, stimuli from the external source trigger the expression of the complementary receptor protein by the nanosystem, thereby enabling it to receive the molecular information from the cells. Different stimuli can be used to express different surface characteristics on a set of the nanosystems, thereby modulating the reception of the information with a higher degree of freedom.

The above ideas can be employed to relay information about the state of a cell in diagnostic scenarios. For instance, the interaction between the stimuli-induced surface proteins and the molecules released by the cells can give bioluminescence, electric, or chemical response that can be used to identify the state of the cells.

11.4.3 Example 3

In this example, the application of communication between living system and nonliving system in bioprocess is illustrated. In the conversion of waste products to biofuels, it is required that microbes break down a particular component of the waste called lignocellulose biomass. Hence, exploiting a way to communicate and artificially modulate the population of the lignocellulose-degrading bacteria for efficient biofuel production is important. The population of these bacteria can be modulated by many different factors such as environment (pH, temperature, salinity, etc.), resource, and toxin production. For instance, different species of bacteria prefer different pH values [86]. Hence, the modification of any of the bacteria population-dependent factors can influence the production of the biofuel.

In *Stage* 1 of Figure 11.5, there are few lignocellulose-degrading microbes to degrade the lignocellulose in the waste, and owing to the competition between lignocellulose-degrading bacteria and lignocellulose-nondegrading bacteria, as wells the presence of toxins from the lignocellulose-nondegrading bacteria in the waste, the population of the lignocellulose-degrading microbes refuses to grow. In *Stage* 2a of Figure 11.5, the external stimulus is used to change the environmental conditions or toxin factor in favor of the lignocellulose-degrading microbes, hence enabling their population to increase to encourage more biofuel production. In *Stage* 2b, in addition to the objective in *Stage* 2a, additional artificially synthesized nanosystems are introduced to the waste environment to sense the level of the different factors militating against efficient biofuel production. This can be achieved by incorporating the idea that is introduced in Example 2.

11.5 TOXICITY AND BIOCOMPATIBILITY ISSUES WITH THE COMMUNICATING MAN-MADE NONLIVING SYSTEMS

The introduction of artificially fabricated nanosystems and devices into the body of a living organism to communicate with the fundamental units of the organism may sometime give rise to toxicity- and biocompatibility-related issues. These issues are associated with how the nanosystems interact with the body biomolecules in a way that may alter their structure and functions, thereby creating adverse effects in the organism. Hence, these nanosystems and devices must be designed with optimal physical, electrical, and chemical characteristics to minimize toxicity and bioincompatibility [31]. The physical, electrical, and chemical characteristics that are currently known to influence the level of toxicity from artificial nanosystems and devices introduced into the body include the fabrication material, size, shape, surface chemistry, and charge [87,88]. The nanoparticle materials are implicated in the sterility and bioincompatibility of particles.

One of the nanosystem-synthesizing methods that will offer very low toxicity and good biocompatibility performance is the use of unmodified cells [89] such as erythrocytes, leukocytes, and stem cells circulating in the blood system, as natural nanosystems show more promising capabilities compared with artificially synthesized nanosystems. Indeed, nature favors simple designs to improve efficiency while minimizing resource usage. Hence, the introduction of artificially synthesized nanosystems into the natural cellular networks in any part of the body has to be critically analyzed and must take the resultant potential of energy/material imbalance into consideration [90].

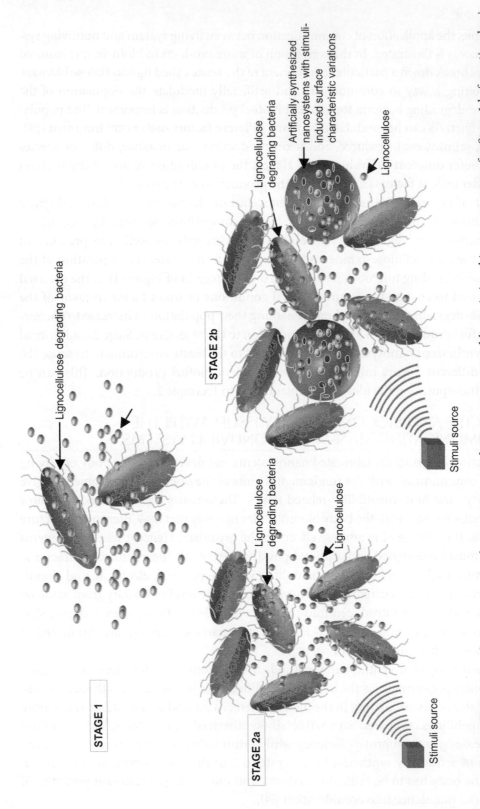

FIGURE 11.5 Illustration of communication between man-made macrosystems and bacteria in a bioprocess system for Stage 1 before stimulus is applied, Stage 2a after stimulus is applied, and Stage 2b after stimulus is applied from a macrosystem on interface nanosystem with stimuli-controlled surface characteristic.

11.6 SUMMARY

Communication between living entities is very vital to helping them establish/improve relationships, spreading knowledge, and building cohesive organization in order to increase their probability of surviving. Typically, information exchange occurs at every level of the existence of the organism, across cells, through tissues/organs, and between organisms. In this chapter, we methodically explored the fundamentals, objectives, and impacts of information exchange among living systems at both the organism level and the cellular level. We also discussed the basics and need to reconfigure the communication among organisms and cells by using artificial resources in the event of communication breakdown, which typically results in diseases. This forms the basics of modern medicine. Within the context of simplified communication engineering milieu, we explored the idea of using artificially fabricated nonliving systems to enhance effective communication between living systems and among their fundamental units, the cells.

REFERENCES

1. B. R. Frieden and R. A. Gatenby, "Information dynamics in living systems: Prokaryotes, eukaryotes, and cancer," *PLoS One*, vol. 6, p. e22085, 2011.
2. B. J. King, A. L. Gilmore-Bykovskyi, R. A. Roiland, B. E. Polnaszek, B. J. Bowers, and A. J. Kind, "The consequences of poor communication during transitions from hospital to skilled nursing facility: A qualitative study," *Journal of the American Geriatrics Society*, vol. 61, pp. 1095–1102, 2013.
3. K.-E. Kaissling, "Pheromone reception in insects," in *The Example of Silk Moths in Neurobiology of Chemical Communication*, C. Mucignat-Caretta (ed.), Boca Raton, FL: CRC Press/Taylor & Francis Group, 2014, pp. 99–146.
4. A. Bigiani, C. Mucignat-Caretta, G. Montani, and R. Tirindelli, "Pheromone reception in mammals," in *Reviews of Physiology, Biochemistry and Pharmacology*, Offermanns, S(Ed.), Springer, Berlin Heildelberg, Germany: 2005, pp. 1–35.
5. J. H. Kirman, "Tactile communication of speech: A review and an analysis," *Psychological Bulletin*, vol. 80, p. 54, 1973.
6. C. Coye, K. Ouattara, K. Zuberbühler, and A. Lemasson, "Suffixation influences receivers' behaviour in non-human primates," *Proceedings of the Royal Society B: Biological Sciences*, vol. 282, p. 20150265, 2015.
7. J. L. Yorzinski, G. L. Patricelli, S. Bykau, and M. L. Platt, "Selective attention in peacocks during assessment of rival males," *Journal of Experimental Biology*, vol. 220, pp. 1146–1153, 2017.
8. L. M. Gosling and S. C. Roberts, "Scent-marking by male mammals: Cheat-proof signals to competitors and mates," in *Advances in the Study of Behavior*, P.J.B Slater, J.S. Rosenblatt, C.T. Snowdon, and T.J. Roper (Eds), vol. 30, Elsevier, 2001, pp. 169–217.
9. B. Leonhardt, "Insect communication codes," *Analytical Chemistry*, vol. 57, pp. 1240A–1249A, 1985.
10. J. Billen, "Signal variety and communication in social insects," in *Proceedings of the Section Experimental and Applied Entomology-Netherlands Entomological Society*, 2006, p. 9.
11. H. Ueda, Y. Kikuta, and K. Matsuda, "Plant communication: Mediated by individual or blended VOCs?" *Plant Signaling & Behavior*, vol. 7, pp. 222–226, 2012.
12. R. Karban, L. H. Yang, and K. F. Edwards, "Volatile communication between plants that affects herbivory: A meta-analysis," *Ecology Letters*, vol. 17, pp. 44–52, 2014.
13. M. Gagliano, M. Grimonprez, M. Depczynski, and M. Renton, "Tuned in: Plant roots use sound to locate water," *Oecologia*, vol. 184, pp. 151–160, 2017.
14. S. A. Dudley and A. L. File, "Kin recognition in an annual plant," *Biology Letters*, vol. 3, pp. 435–438, 2007.

15. M. Gagliano, M. Renton, N. Duvdevani, M. Timmins, and S. Mancuso, "Out of sight but not out of mind: Alternative means of communication in plants," *PLoS One*, vol. 7, p. e37382, 2012.

16. M. G. Schöner, C. R. Schöner, R. Simon, T. U. Grafe, S. J. Puechmaille, L. L. Ji et al., "Bats are acoustically attracted to mutualistic carnivorous plants," *Current Biology*, vol. 25, pp. 1911–1916, 2015.

17. M. A. Gorzelak, A. K. Asay, B. J. Pickles, and S. W. Simard, "Inter-plant communication-through mycorrhizal networks mediates complex adaptive behaviour in plant communities," *AoB Plants*, vol. 7, 2015, pp. 1–13.

18. S. C. Jung, A. Martinez-Medina, J. A. Lopez-Raez, and M. J. Pozo, "Mycorrhiza-induced resistance and priming of plant defenses," *Journal of Chemical Ecology*, vol. 38, pp. 651–664, 2012.

19. S. W. Simard, D. A. Perry, M. D. Jones, D. D. Myrold, D. M. Durall, and R. Molina, "Net transfer of carbon between ectomycorrhizal tree species in the field," *Nature*, vol. 388, p. 579, 1997.

20. S. W. Simard, "Mycorrhizal networks facilitate tree communication, learning, and memory," in *Memory and Learning in Plants*, F. Baluska and M. Gagliano (eds.), Boston, MA: Springer, 2018, pp. 191–213.

21. N. Perrimon, C. Pitsouli, and B.-Z. Shilo, "Signaling mechanisms controlling cell fate and embryonic patterning," *Cold Spring Harbor Perspectives in Biology*, vol. 4, p. a005975, 2012.

22. L. Kennedy, K. Hodges, F. Meng, G. Alpini, and H. Francis, "Histamine and histamine receptor regulation of gastrointestinal cancers," *Translational Gastrointestinal Cancer*, vol. 1, p. 215, 2012.

23. J. D. Veldhuis and M. L. Johnson, "A novel general biophysical model for simulating episodic endocrine gland signaling," *American Journal of Physiology-Endocrinology and Metabolism*, vol. 255, pp. E749–E759, 1988.

24. B. Kleine and W. G. Rossmanith, *Hormones and the Endocrine System: Textbook of Endocrinology*. Cham, Switzerland: Springer, 2016.

25. D. M. Lovinger, "Communication networks in the brain: Neurons, receptors, neurotransmitters, and alcohol," *Alcohol Research & Health*, Vol. 31, no. 3, 2008, pp. 196–214.

26. I. Sevilem, S. Miyashima, and Y. Helariutta, "Cell-to-cell communication via plasmodesmata in vascular plants," *Cell Adhesion & Migration*, vol. 7, pp. 27–32, 2013.

27. H. S. Robert and J. Friml, "Auxin and other signals on the move in plants," *Nature Chemical Biology*, vol. 5, p. 325, 2009.

28. M. B. Miller and B. L. Bassler, "Quorum sensing in bacteria," *Annual Reviews in Microbiology*, vol. 55, pp. 165–199, 2001.

29. Z. Erez, I. Steinberger-Levy, M. Shamir, S. Doron, A. Stokar-Avihail, Y. Peleg et al., "Communication between viruses guides lysis–lysogeny decisions," *Nature*, vol. 541, p. 488, 2017.

30. F. Cottier and F. A. Mühlschlegel, "Communication in fungi," *International Journal of Microbiology*, vol. 2012, 2011, pp. 1–9.

31. U. A. Chude-Okonkwo, R. Malekian, B. T. Maharaj, and A. V. Vasilakos, "Molecular communication and nanonetwork for targeted drug delivery: A survey," *IEEE Communications Surveys & Tutorials*, vol. 19, pp. 3046–3096, 2017.

32. G. A. Garden and A. R. La Spada, "Intercellular (mis) communication in neurodegenerative disease," *Neuron*, vol. 73, pp. 886–901, 2012.

33. R. K. Benninger and D. W. Piston, "Cellular communication and heterogeneity in pancreatic islet insulin secretion dynamics," *Trends in Endocrinology & Metabolism*, vol. 25, pp. 399–406, 2014.

34. P. Gómez-Suaga, J. M. Bravo-San Pedro, R. A. González-Polo, J. M. Fuentes, and M. Niso-Santano, "ER–mitochondria signaling in Parkinson's disease," *Cell Death & Disease*, vol. 9, p. 337, 2018.

35. M. H. Oktay, Y.-F. Lee, A. Harney, D. Farrell, N. Z. Kuhn, S. A. Morris et al., "Cell-to-cell communication in cancer: Workshop report," *NPJ Breast Cancer*, vol. 1, p. 15022, 2015.

36. S. Humbert and F. Saudou, "Huntington's disease: Intracellular signaling pathways and neuronal death," *Journal de la Societe de biologie*, vol. 199, pp. 247–251, 2005.
37. H. P. Ehrlich, "A snapshot of direct cell–cell communications in wound healing and scarring," *Advances in Wound Care*, vol. 2, pp. 113–121, 2013.
38. R. Baskar, K. A. Lee, R. Yeo, and K.-W. Yeoh, "Cancer and radiation therapy: Current advances and future directions," *International Journal of Medical Sciences*, vol. 9, p. 193, 2012.
39. Z. Wang, P.-L. Che, J. Du, B. Ha, and K. J. Yarema, "Static magnetic field exposure reproduces cellular effects of the Parkinson's disease drug candidate ZM241385," *PLoS One*, vol. 5, p. e13883, 2010.
40. K. D. Swanson, E. Lok, and E. T. Wong, "An overview of alternating electric fields therapy (NovoTTF Therapy) for the treatment of malignant glioma," *Current Neurology and Neuroscience Reports*, vol. 16, p. 8, 2016.
41. E. S. Glazer and S. A. Curley, "The ongoing history of thermal therapy for cancer," *Surgical Oncology Clinics*, vol. 20, pp. 229–235, 2011.
42. M. N. Halgamuge, "Weak radiofrequency radiation exposure from mobile phone radiation on plants," *Electromagnetic Biology and Medicine*, vol. 36, pp. 213–235, 2017.
43. M.-L. Soran, M. Stan, Ü. Niinemets, and L. Copolovici, "Influence of microwave frequency electromagnetic radiation on terpene emission and content in aromatic plants," *Journal of Plant Physiology*, vol. 171, pp. 1436–1443, 2014.
44. B. Gustavino, G. Carboni, R. Petrillo, G. Paoluzzi, E. Santovetti, and M. Rizzoni, "Exposure to 915 MHz radiation induces micronuclei in vicia faba root tips," *Mutagenesis*, vol. 31, pp. 187–192, 2015.
45. S. Pietruszewski and E. Martínez, "Magnetic field as a method of improving the quality of sowing material: A review," *International Agrophysics*, vol. 29, pp. 377–389, 2015.
46. J. Li, Y. Yi, X. Cheng, D. Zhang, and M. Irfan, "Study on the effect of magnetic field treatment of newly isolated Paenibacillus sp," *Botanical Studies*, vol. 56, p. 2, 2015.
47. E. Fu, "The effects of magnetic fields on plant growth and health," *Young Scientists Journal*, vol. 5, p. 38, 2012.
48. V. V. Chaban, T. Cho, C. B. Reid, and K. C. Norris, "Physically disconnected non-diffusible cell-to-cell communication between neuroblastoma SH-SY5Y and DRG primary sensory neurons," *American Journal of Translational Research*, vol. 5, p. 69, 2013.
49. U. A. Chude-Okonkwo, R. Malekian, and B. Maharaj, "Diffusion-controlled interface kinetics-inclusive system-theoretic propagation models for molecular communication systems," *EURASIP Journal on Advances in Signal Processing*, vol. 2015, p. 1, 2015.
50. T. Nakano, T. Suda, Y. Okaie, M. J. Moore, and A. V. Vasilakos, "Molecular communication among biological nanomachines: A layered architecture and research issues," *IEEE Transactions on Nanobioscience*, vol. 13, pp. 169–197, 2014.
51. T. Nakano, T. Suda, M. Moore, R. Egashira, A. Enomoto, and K. Arima, "Molecular communication for nanomachines using intercellular calcium signaling," in *5th IEEE Conference on Nanotechnology*, 2005, pp. 478–481.
52. I. F. Akyildiz, F. Brunetti, and C. Blázquez, "Nanonetworks: A new communication paradigm," *Computer Networks*, vol. 52, pp. 2260–2279, 2008.
53. T. Nakano, M. J. Moore, F. Wei, A. V. Vasilakos, and J. Shuai, "Molecular communication and networking: Opportunities and challenges," *IEEE Transactions on Nanobioscience*, vol. 11, pp. 135–148, 2012.
54. N. Farsad, H. B. Yilmaz, A. Eckford, C.-B. Chae, and W. Guo, "A comprehensive survey of recent advancements in molecular communication," in *IEEE Communications Surveys & Tutorials*, pp. 1887–1919, 2016.
55. A. Chanu and S. Martel, "MRI controlled magnetoelastic nano biosensor for in-vivo pH monitoring: A preliminary approach," in *Nanotechnology, 2007. IEEE-NANO 2007. 7th IEEE Conference on*, pp. 166–170, 2007.

56. W. Zhang, Y. Du, and M. L. Wang, "Noninvasive glucose monitoring using saliva nano-bio-sensor," *Sensing and Bio-Sensing Research*, vol. 4, pp. 23–29, 2015.

57. G. Xiao, Y. Song, S. Zhang, L. Yang, S. Xu, Y. Zhang et al., "A high-sensitive nano-modified biosensor for dynamic monitoring of glutamate and neural spike covariation from rat cortex to hippocampal sub-regions," *Journal of Neuroscience Methods*, vol. 291, pp. 122–130, 2017.

58. I. Gualandi, D. Tonelli, F. Mariani, E. Scavetta, M. Marzocchi, and B. Fraboni, "Selective detection of dopamine with an all PEDOT: PSS organic electrochemical transistor," *Scientific Reports*, vol. 6, p. 35419, 2016.

59. U. A. Okonkwo, R. Malekian, and B. S. Maharaj, "Molecular communication model for targeted drug delivery in multiple disease sites with diversely expressed enzymes," in *IEEE Transactions on Nanobioscience*, pp. 230–245, 2016.

60. Z. Hemmatian, S. Keene, E. Josberger, T. Miyake, C. Arboleda, J. Soto-Rodríguez et al., "Electronic control of H+ current in a bioprotonic device with gramicidin A and alamethicin," *Nature Communications*, vol. 7, p. 12981, 2016.

61. T. Tschirhart, E. Kim, R. McKay, H. Ueda, H.-C. Wu, A. E. Pottash et al., "Electronic control of gene expression and cell behaviour in escherichia coli through redox signalling," *Nature Communications*, vol. 8, p. 14030, 2017.

62. E. Kim, W. T. Leverage, Y. Liu, I. M. White, W. E. Bentley, and G. F. Payne, "Redox-capacitor to connect electrochemistry to redox-biology," *Analyst*, vol. 139, pp. 32–43, 2014.

63. R. Lentini, S. P. Santero, F. Chizzolini, D. Cecchi, J. Fontana, M. Marchioretto et al., "Integrating artificial with natural cells to translate chemical messages that direct E. coli behaviour," *Nature Communications*, vol. 5, p. 4012, 2014.

64. J. L. Terrell, H.-C. Wu, C.-Y. Tsao, N. B. Barber, M. D. Servinsky, G. F. Payne et al., "Nano-guided cell networks as conveyors of molecular communication," *Nature Communications*, vol. 6, p. 8500, 2015.

65. S. Tan, T. Wu, D. Zhang, and Z. Zhang, "Cell or cell membrane-based drug delivery systems," *Theranostics*, vol. 5, p. 863, 2015.

66. D. Lockney, S. Franzen, and S. Lommel, "Viruses as nanomaterials for drug delivery," in *Biomedical Nanotechnology: Methods and Protocols*, S. J. Hurst (ed.), Totowa, NJ: Humana Press, 2011, pp. 207–221.

67. L. Steidler, "Live genetically modified bacteria as drug delivery tools: At the doorstep of a new pharmacology?" *Expert Opinion on Biological Therapy*, vol. 4, pp. 439–441, 2004.

68. I. Yacoby, H. Bar, and I. Benhar, "Targeted drug-carrying bacteriophages as antibacterial nanomedicines," *Antimicrobial Agents and Chemotherapy*, vol. 51, pp. 2156–2163, 2007.

69. S. Arnon, N. Dahan, A. Koren, O. Radiano, M. Ronen, T. Yannay et al., "Thought-controlled nanoscale robots in a living host," *PLoS One*, vol. 11, p. e0161227, 2016.

70. I. Akyildiz, M. Pierobon, S. Balasubramaniam, and Y. Koucheryavy, "The internet of bio-nano things," *IEEE Communications Magazine*, vol. 53, pp. 32–40, 2015.

71. U. A. Chude-Okonkwo, R. Malekian, and B. Maharaj, "Biologically inspired bio-cyber interface architecture and model for internet of bio-nano things applications," *IEEE Transactions on Communications*, vol. 64, pp. 3444–3455, 2016.

72. T. Kawano, T. Harimoto, A. Ishihara, K. Takei, T. Kawashima, S. Usui et al., "Electrical interfacing between neurons and electronics via vertically integrated sub-4 μm-diameter silicon probe arrays fabricated by vapor–liquid–solid growth," *Biosensors and Bioelectronics*, vol. 25, pp. 1809–1815, 2010.

73. D. T. Simon, K. C. Larsson, M. Berggren, and A. Richter-Dahlfors, "Precise neurotransmitter-mediated communication with neurons in vitro and in vivo using organic electronics," *Journal of Biomechanical Science and Engineering*, vol. 5, pp. 208–217, 2010.

74. Y. I. Wu, D. Frey, O. I. Lungu, A. Jaehrig, I. Schlichting, B. Kuhlman et al., "A genetically encoded photoactivatable rac controls the motility of living cells," *Nature*, vol. 461, p. 104, 2009.

75. T. Ozawa, H. Yoshimura, and S. B. Kim, "Advances in fluorescence and bioluminescence imaging," *Analytical Chemistry*, vol. 85, pp. 590–609, 2012.

76. D. Needham, G. Anyarambhatla, G. Kong, and M. W. Dewhirst, "A new temperature-sensitive liposome for use with mild hyperthermia: Characterization and testing in a human tumor xenograft model," *Cancer Research*, vol. 60, pp. 1197–1201, 2000.

77. K. Hamad-Schifferli, J. J. Schwartz, A. T. Santos, S. Zhang, and J. M. Jacobson, "Remote electronic control of DNA hybridization through inductive coupling to an attached metal nanocrystal antenna," *Nature*, vol. 415, p. 152, 2002.

78. T. Nakano, S. Kobayashi, T. Suda, Y. Okaie, Y. Hiraoka, and T. Haraguchi, "Externally controllable molecular communication," *IEEE Journal on Selected Areas in Communications*, vol. 32, pp. 2417–2431, 2014.

79. M. Kuscu and O. B. Akan, "Modeling and analysis of SiNW BioFET as molecular antenna for bio-cyber interfaces towards the internet of bio-nano things," in *Internet of Things (WF-IoT), 2015 IEEE 2nd World Forum on*, pp. 669–674, 2015.

80. D. L. Bellin, S. B. Warren, J. K. Rosenstein, and K. L. Shepard, "Interfacing CMOS electronics to biological systems: From single molecules to cellular communities," in *Biomedical Circuits and Systems Conference (BioCAS), 2014 IEEE*, pp. 476–479, 2014.

81. V. de Lorenzo, "Cleaning up behind us: The potential of genetically modified bacteria to break down toxic pollutants in the environment," *EMBO Reports*, vol. 2, pp. 357–359, 2001.

82. K. Pathakoti, M. Manubolu, and H.-M. Hwang, "Nanotechnology applications for environmental industry," in *Handbook of Nanomaterials for Industrial Applications*, C. M. Hussain (ed.), Amsterdam, the Netherlands: Elsevier, 2018, pp. 894–907.

83. R. Kumar and P. Kumar, "Future microbial applications for bioenergy production: A perspective," *Frontiers in Microbiology*, vol. 8, p. 450, 2017.

84. S. K. Srivastava, P. Piwek, S. R. Ayakar, A. Bonakdarpour, D. P. Wilkinson, and V. G. Yadav, "A biogenic photovoltaic material," *Small*, vol. 14, p. 1800729, 2018.

85. M. Huo, Y. Chen, and J. Shi, "Triggered-release drug delivery nanosystems for cancer therapy by intravenous injection: Where are we now?" *Expert Opinion on Drug Delivery*, vol. 13, pp. 1195–1198, 2016.

86. C. Ratzke and J. Gore, "Modifying and reacting to the environmental pH can drive bacterial interactions," *PLoS Biology*, vol. 16, p. e2004248, 2018.

87. Y. S. Choi, M. Y. Lee, A. E. David, and Y. S. Park, "Nanoparticles for gene delivery: Therapeutic and toxic effects," *Molecular & Cellular Toxicology*, vol. 10, pp. 1–8, 2014.

88. C. Buzea, I. I. Pacheco, and K. Robbie, "Nanomaterials and nanoparticles: Sources and toxicity," *Biointerphases*, vol. 2, pp. MR17–MR71, 2007.

89. Y. Su, Z. Xie, G. B. Kim, C. Dong, and J. Yang, "Design strategies and applications of circulating cell-mediated drug delivery systems," *ACS Biomaterials Science & Engineering*, vol. 1, pp. 201–217, 2015.

90. S. B. Laughlin and T. J. Sejnowski, "Communication in neuronal networks," *Science*, vol. 301, pp. 1870–1874, 2003.

CHAPTER 12

Molecular Communication and Cellular Signaling from an Information-Theory Perspective

Kevin R. Pilkiewicz, Pratip Rana, Michael L. Mayo, and Preetam Ghosh

CONTENTS

12.1 INTRODUCTION

Traveling to a foreign country can be a daunting experience if you do not speak the local language. Even the most rudimentary of social interactions can become tribulation. Imagine, however, that instead of traveling to another country, you sojourned to another planet, and suppose this planet was inhabited by a race of sentient alien creatures who communicated through complex ululations of a balloon-like throat sac. This would be a troublesome situation, indeed, for not only you would lack an understanding of the language, but also you would not even possess the capacity to converse in it if you did, as you lack the organ required to produce its phonemes. This may seem like a far-fetched scenario, but it is essentially the situation in which scientists in the field of nanocommunications now find themselves as they attempt to engineer devices and constructs that can bridge the communication gap between the synthetic and organic worlds.

We find ourselves increasingly surrounded by an interconnected web of electronic smart devices, each of which speaks its own language, yet all of which are able to communicate with each other, because their languages utilize the same medium of electromagnetic signals. Cells, the "smart" devices that constitute our bodies, principally communicate in a fundamentally different way—through the transport and binding of molecules.

Figure 12.1 schematically illustrates the processes of gene transcription and translation, which represent just one example of molecular communication at work. The transcription of a gene begins when a molecule of the enzyme RNA polymerase diffuses through the cell nucleus and binds to an initiation site along the contour of a DNA segment. The double helical structure of DNA is then temporarily unzipped, and RNA polymerase slides along the unwound DNA, reencoding its genetic data, stored in quaternary by a sequence of nucleobases, into a signaling molecule called the messenger RNA (mRNA). The mRNA molecule must then diffuse out of the nucleus and through the cytoplasm to an organelle called a ribosome, whereby the information stored in the mRNA is decoded and translated into an amino acid sequence. This sequence is assembled by another class of signaling molecules called transfer RNAs (tRNAs), and thus, a protein is born.

The preceding description, already dripping with biological jargon, is still but a gross oversimplification of the actual process of cellular protein production, whose stage is cluttered with dozens of additional molecular actors [1]. Nature has contrived a way for all these biomolecules to act in concert so efficiently and consistently, despite their noisy environment and principally diffusive transport, and this is an astounding achievement in

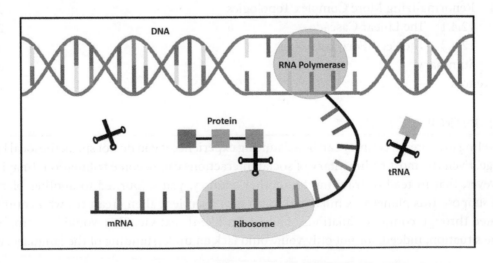

FIGURE 12.1   A schematic representation of gene transcription and translation.

robust communication. It is an accomplishment that scientists must tease apart in order to build devices that can interface with nature in its native tongue.

When it comes to human languages, even grammatically disparate tongues can be related to one another, because certain fundamental linguistic concepts such as nouns, verbs, and adjectives (things, actions, and descriptors) are universal features of human communication. To better translate the molecular language of biology into the signal-processing language of electronics and computer science, it would be helpful if there were a similarly universal set of concepts to which one could appeal. Fortunately, such a framework has existed for more than half a century in the form of the information theory devised principally by Claude Shannon [2].

For a discrete random variable X whose value x has probability $p(x)$, Shannon [2] defined the information in a measurement of X to be $-\log p(x)$. The negative sign, coupled with the fact that $0 \le p(x) \le 1$ for all $x \in X$, ensures that the information is always nonnegative and that the less-likely values of the variable contain more information. The sense in this latter property can be understood intuitively in the context of password cracking. If an individual uses a common password such as their birthdate or their favorite sports team, it will require little knowledge of the person to break into their data. On the other hand, if the person uses something less common such as their mother's maiden name or the name of their first pet, a greater amount of information will be required.

The logarithm in the definition makes the information in statistically independent measurements additive, because if for two random variables X and Y, $p(x, y) = p(x)p(y)$, then $-\log p(x, y) = -\log p(x) - \log p(y)$. It should be noted that the base of the logarithm here is arbitrary, and although information is a formally dimensionless quantity, different choices of this base are referred to as different "units," the most common being the base-2 unit of bits. (It is trivial to show that for a random variable with two equally likely states, such as the flipping of a fair coin, the information in each measurement is exactly 1 bit, consistent with our understanding of the bit in computer science as a binary choice of zero or one.)

Rather than concerning ourselves with the information content of individual measurements of the variable X, it is typically more useful to consider its average information, known as the Shannon entropy $H(X)$:

$$H(X) \equiv -\sum_{x \in X} p(x)\log p(x) \tag{12.1}$$

The term entropy is used because, like the thermodynamic quantity of the same name, the Shannon entropy is essentially a measure of randomness, quantifying the amount of uncertainty in a random variable. To ensure that this quantity remains finite, $p(x)\log p(x)$ is defined to be zero when $p(x) = 0$.

To understand the relevance of Shannon's information theory to the study of communication, consider the transmission of a time-varying signal $S(t)$ across some sort of channel. At the other end of the channel, a time-varying response $R(t)$ will be measured by an observer, and the hope is that the observer can determine the information content of the original signal from this response. The issue, of course, is that most channels suffer from some sort of noise that will degrade the quality of the signal as it is transmitted. One can thus think of the response as the sum of the original signal plus some function $\eta(t)$, representing the distortion caused by the noise (the noise is seldom additive like this, but one can always at least formally define the response in this way with an appropriate definition of $\eta(t)$). Because the deterministic processes causing the noise typically occur over much smaller length and time scales than the signal transmission process of interest and are effectively unmeasurable, the noise is usually well-approximated as a stochastic process, meaning that $R(t)$ is, at every point in time, a random variable whose average information content can be described by the Shannon theory.

One usually wants to know how much information is shared between the signal and the response, and that is most aptly measured by the mutual information between two random variables, $MI(X;Y) \equiv H(X) + H(Y) - H(X,Y)$. Figure 12.2 illustrates with a Venn diagram that the mutual information does indeed correspond to the average amount of information shared between the variables X and Y. The figure also demonstrates why the mutual information can equivalently be defined as $H(X) - H(X|Y)$, where the conditional entropy $H(X|Y)$ can be defined as in Equation (12.1), as long as one understands that $\sum_{x|y} = \sum_x \sum_y p(y)$. This form for the mutual information is particularly useful for signaling applications, as the starting point of most analyses is a model for the probability of observing an output response R, given a known input signal S, which can be used to directly compute the entropy $H(R|S)$.

The remaining entropy needed to compute the mutual information is $H(R)$, which can be computed from the conditional probability $p(R|S)$ if the signal probability $p(S)$ is known, but this is often not the case. In fact, since the signal can usually be controlled by the experimenter, ascribing a likelihood to any particular signal is meaningless. This wrinkle is traditionally ironed out by using the signal probability function that maximizes the mutual information of the signal and the response. This maximal information is called the

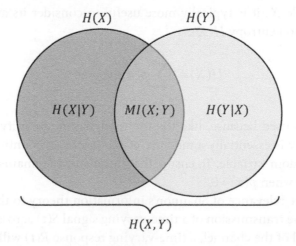

FIGURE 12.2 A Venn diagram summary of how the mutual information can be constructed from different combinations of the Shannon entropies.

channel capacity and represents the optimal performance possible for a given communication channel.

An astute reader will already notice several problems with applying information theory to signaling systems, as described. For starters, in both electronic and molecular communications, the signals of interest can take on a continuum of values, but the Shannon entropy is formally defined only for discrete random variables. Second, there is no systematic method for finding the function $P(S)$ that maximizes the mutual information, which makes the channel capacity very difficult to compute. In biological applications, the channel capacity may not even be that useful, as the signal probability is often a meaningful physical quantity in these contexts. For example, in the case of gene transcription and translation, one is often interested in studying how the dynamic expression level of one protein varies in response to changes in a different protein expression level. In this situation, both the signal and the response are represented by effectively stochastic cellular processes, neither of which can be controlled directly by the experimenter.

The objectives of this chapter are to provide some resolution to these issues and to illustrate through several (relatively) simple examples how information theory may be implemented properly to describe molecular communication processes. The treatment presented here will be far from exhaustive and will focus more on results than on the math required to derive them. This is because the most difficult aspect of information theory is not acquiring the results themselves, which can often be done by straightforward numerical methods, but correctly interpreting them, which involves a surprising amount of subtlety and a high-level understanding of what metrics such as mutual information actually mean.

12.2 INFORMATION THEORY AND MOLECULAR TRANSPORT

The first order of business in attempting to couch molecular transport and bind in the language of signal processing and information theory is to identify the nature of the signal. In electronics, the signal is an electromagnetic wave, and information is encoded into some combination of its time-varying amplitude and frequency. In biological systems, the signal is usually composed of molecules such as RNA polymerase and mRNA that are synthesized and released into the cytosol and then propagate to a binding site, where they trigger another biological process, whose purpose is often to synthesize and release a different signaling molecule.

The type of information encoded in these molecular signals can vary. For example, the nucleobase sequence of mRNA encodes the amino acid sequence of a protein. At some level, however, most cellular processes occur in response to dynamic changes in the cell's external environment. For example, the bacterium *Escherichia coli* requires different proteins to metabolize different sugars, and the expression levels of these various proteins are modulated appropriately, depending on the relative concentration of different sugars present in the bacterial cell's surroundings. The sensitivity of this response is achieved through a complicated cascade of molecular signaling processes [3], but the important point is that information about the external state of the cell is encoded in the frequency at which these processes occur [4]. Thus, for a single molecular communication process, one can define the signal as the set of times at which signaling molecules are produced, and the response to this signal can be defined as the set of times at which these molecules bind to their target and thereby trigger the next cellular process in the cascade to commence.

12.2.1 Molecular Transport in One Dimension

For the purpose of this chapter, the discussion shall be restricted to facilitated molecular transport in one dimension. Facilitated means here that the molecules are transported toward their destination faster than they would be under purely diffusive motion but not as fast as they would under ballistic motion. These different transport regimes are quantified in Figure 12.3, using the temporal scaling of the mean-squared displacement (MSD)—the average squared distance traveled by a molecule over time. The MSD is used because the mean displacement will always be zero, owing to the assumed isotropy of the system. Pure diffusion corresponds to linear scaling of the MSD with time, whereas ballistic motion scales quadratically. Facilitated or "superdiffusive" motion falls somewhere in between these two extremes.

It may seem like a crass oversimplification to limit the discussion to one-dimensional (1D) transport, since the interior of a cell is a three-dimensional volume, but in fact, there are many cellular transport processes that are pseudo-1D. As just one example, because cellular vesicles are too large to diffuse through the crowded cellular environment with any efficiency, special motor proteins called kinesins bind to them and quite literally walk them along cytoskeletal filaments known as microtubules [5]. Despite their large aspect ratio, microtubules are of course not truly 1D. The kinesin proteins also sometimes detach from the tubule, diffuse in three dimensions for some distance, and then reattach further down the line, but despite these caveats, the transport can still be modeled quite reasonably as a 1D, facilitated process.

With this understood, the biological transport considered here and throughout the remainder of this chapter will be modeled as a 1D drift–diffusion process characterized by a drift speed v and a diffusion constant D, such that the stochastic distance traveled by a molecule in a time interval of length Δt is normally distributed as $N(v\Delta t, 2D\Delta t)$. Note

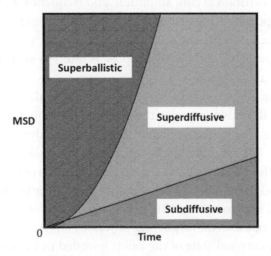

FIGURE 12.3 The three classes of transport, demarcated by the scaling regimes of the mean-squared displacement versus time.

that the drift speed determines the mean distance traveled, whereas the diffusion constant controls the width of the fluctuations about that mean.

The information in this system is encoded in the pattern of molecular emissions into the channel, and this information must be decoded from the pattern of molecular absorptions at the end of the channel, where molecular binding is assumed to occur. For a molecule released into one end of a channel of length ℓ at time τ, the conditional probability $p(t|\tau)$ that it is absorbed at the other end of the channel at some later time t can be shown to equal the inverse Gaussian distribution $IG(\mu, \lambda, t - \tau)$ [6]:

$$p(t|\tau) = IG(\mu, \lambda, t - \tau) = \left(\frac{\lambda}{2\pi(t-\tau)^3} \right)^{1/2} \exp\left[\frac{-\lambda(t-\tau-\mu)^2}{2\mu^2(t-\tau)} \right], \qquad (12.2)$$

where the parameter $\mu \equiv \ell/v$ is the time required to traverse the channel in the absence of diffusion, and $\lambda \equiv \ell^2/2D$ is the average time required to traverse the channel in the absence of drift. Figure 12.4 plots this distribution as a function of its time interval argument for $\mu = 1$ and four different values of λ. As λ grows larger, diffusion becomes weaker and $t - \tau$ approaches μ, reducing Equation (12.2) to a standard Gaussian distribution $N(\mu, \mu^3/\lambda)$. This Gaussian approximation is plotted in Figure 12.4 as dashed curve for each value of λ to illustrate this trend. This simplification will be used extensively in the sections that follow, in order to considerably simplify the math of subsequent derivations.

For similar reasons, the release time τ will likewise be chosen to be a Gaussian random variable, with mean value $\bar{\tau}$ and variance σ^2. If one assumes that the emission into the channel is equally likely to occur at any instant in time, the probability that it gets released after a finite time τ should formally be described by an exponential distribution, not by a Gaussian distribution, but, as will be demonstrated numerically later, this choice ends up making little quantitative difference to the value of the mutual information, while it greatly reduces the complexity of its analytic evaluation.

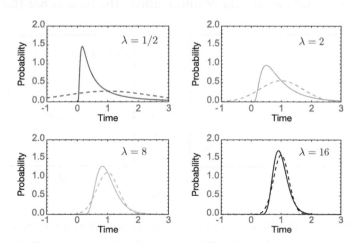

FIGURE 12.4 Comparison of the inverse Gaussian distribution with its Gaussian approximation for $\mu = 1$ and increasing values of λ.

12.2.2 Information in the Continuum Limit

In the simplest case of a single molecule traversing the 1D drift–diffusion channel outlined in the previous subsection, the information content of the signal is encoded entirely in the timing of the molecule's emission. However, the amount of information in this signal is difficult to interpret by using the formalism of the Shannon entropy, because the emission time τ is a continuous rather than a discrete random variable.

One approach to solving this problem is to divide the domain of the continuous distribution $p(\tau)$ into discrete bins of some width $\Delta\tau$ and assign to each bin a probability mass equal to the integral of the continuous distribution over that bin. For the Gaussian distribution chosen for $p(\tau)$, this leads to a Shannon entropy equal to

$$H(\tau) = -\sum_{n=0}^{\infty} \Delta\mathrm{erf}\left(\frac{n\Delta\tau}{\sigma\sqrt{2}}\right) \log\left[\frac{1}{2}\Delta\mathrm{erf}\left(\frac{n\Delta\tau}{\sigma\sqrt{2}}\right)\right], \tag{12.3}$$

where $\mathrm{erf}(x)$ is the standard error function and $\Delta\mathrm{erf}(nx) \equiv \mathrm{erf}((n+1)x) - \mathrm{erf}(nx)$. Because the error function asymptotes toward unity for large values of its argument, the terms in this sum will eventually approach $0\log 0 \equiv 0$. As the bin width $\Delta\tau$ shrinks, the number of terms making a significant contribution to the entropy grows, resulting in $H(\tau)$ diverging logarithmically, as plotted in Figure 12.5a.

It is somewhat troubling to find that the information content of the signal is dependent on an arbitrary discretization choice. While in certain applications, there may exist a maximum precision with which a continuous random variable can be measured, setting a natural scale for the bin width, it would be preferable to look for a metric whose value does not depend on the resolution of the measuring device or diverge in the continuum limit.

Irksome though they may be, the results described previously are actually not unreasonable. If a variable can take on any value from some interval of the real line, then knowing that value with absolute certainty, that is, to an infinite number of significant digits, understandably requires an infinite amount of information. The issue is not that the Shannon

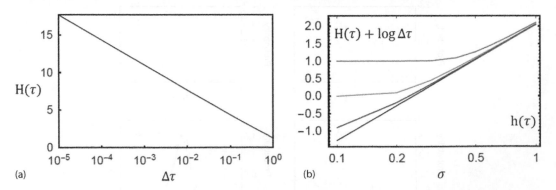

FIGURE 12.5 (a) Plot of the Shannon entropy for different discretizations of a normally distributed continuous random variable τ. (b) A comparison of the differential entropy of the variable τ and its Shannon entropy, corrected by the addition of $\log\Delta\tau$, as a function of the variance of the distribution σ for different values of $\Delta\tau$. In both panels, note the log scale on the abscissa.

entropy cannot be used to characterize continuous random variables, but merely that what it reveals about them is not particularly useful.

An alternative approach would be to simply take the sum in Equation (12.1), replace it with an integral, and swap the probability mass function with a probability distribution function. This immediately smacks of mathematical chicanery, as a continuous distribution function, unlike a discrete mass function, is not dimensionless and has no business under a logarithm. Proceeding nonetheless, as physicists so often do, one can define what is known as the differential entropy $h(X)$:

$$h(X) \equiv -\int dx\, p(x) \log p(x).$$
(12.4)

The reason for naming this quantity so will become clear shortly.

Returning to the Gaussian signal distribution $p(\tau)$, one can capitalize on the niceties of Gaussian integration techniques to evaluate Equation (12.4) analytically. Technically, the variable τ is constrained to the interval $(0, \infty)$, but so long as the mean of the distribution is much larger than its standard deviation, the lower integration limit can be extended to $-\infty$ to make the integration much easier without accruing a significant error. However, the result,

$$h(\tau) = \frac{1}{2} \log(2\pi e \sigma^2),$$
(12.5)

where e is the Euler number, is immediately problematic, because it can become negative if the distribution $p(\tau)$ is too narrow $(\sigma < 1/\sqrt{2\pi e} \approx 0.242)$. While thermodynamic entropies can, in contrived circumstances, be negative, in the context of information theory, this would correspond to knowing less than nothing, which is nonsensical.

To iron out all these conceptual wrinkles, it is useful to establish a rigorous connection between the differential and Shannon entropies by invoking the mean value theorem of calculus. To wit, for each bin in the discretization scheme employed earlier, there must exist a value τ_i, such that the integral of $p(\tau)$ over that bin equals $p(\tau_i)\Delta\tau$. This fact makes it possible to rewrite the Shannon entropy $H(\tau)$ as follows:

$$H(\tau) = -\sum_{i=-\infty}^{\infty} p(\tau_i)\Delta\tau \log\left[p(\tau_i)\Delta\tau\right]$$

$$= -\sum_{i=-\infty}^{\infty} \Delta\tau\, p(\tau_i) \log p(\tau_i) - \log\Delta\tau.$$
(12.6)

To get the second equality, the algebraic properties of the logarithm were used in conjunction with the fact that summing $p(\tau_i)\Delta\tau$ over all i must equal unity, since this is equivalent to integrating the properly normalized $p(\tau)$ over its entire domain. The first term to the right of the second equal sign is now properly the Riemann sum of the function $p(\tau)\log p(\tau)$, meaning that in the limit $\Delta\tau \to 0$, it will rigorously approach the differential entropy $h(\tau)$ (assuming that the distribution function is sufficiently well-behaved). This connection between the Shannon and differential entropies is illustrated clearly in Figure 12.5b, wherein the differential entropy of the Gaussian distribution (Equation 12.5) is plotted as a function of its standard deviation σ, (bottom-most curve) and the sum of $\log \Delta \tau$ and the Shannon entropy (Equation 12.3) is plotted for values of $\Delta \tau$ equals to, from top to bottom, 1, 1/2, and 1/4, respectively. As $\Delta \tau$ gets smaller, the latter curves merge with the former, starting at smaller values of σ. This makes sense, since capturing the features of the distribution will require a finer binning when its unimodal peak is sharper.

Thus far, all that has been accomplished is separating out the component of the Shannon entropy that diverges as the bin width shrinks. The important point is that this term depends *only* on the bin width. If one were to consider, for example, the difference $H(X) - H(Y)$ for two continuous random variables X and Y, one would find that, in the limit of the discretized bins for both distributions $p(x)$ and $p(y)$ going to zero, it approaches the difference $h(X) - h(Y)$. So, although the absolute information content of both continuous random variables is formally infinite, the relative difference in their information content is typically finite and is thus a useful comparative metric. This, by the way, is the origin of the term "differential" entropy—it is a quantity that only has physical relevance as a difference, much like the potential energy of a mechanical system. In fact, at the quantum level, the potential energy of a field is formally infinite as well.

Fortunately, the mutual information, which is often the information theory metric of principal interest in the study of communication, is defined as just such a difference. For the single-particle, 1D drift–diffusion model of molecular communication being considered presently, the mutual information between the response and signal is $MI(t;\tau) = h(t) - h(t|\tau)$. As a consequence of approximating the conditional distribution $p(t|\tau)$ as a Gaussian and choosing a Gaussian signal distribution $p(\tau)$, both these two differential entropies can be evaluated analytically, assuming once again that their means are sufficiently larger than their standard deviations. For the conditional distribution, this is equivalent to the condition $\lambda \gg \mu$, which has already been assumed in order to replace the inverse Gaussian with its Gaussian approximation. The result of this analysis is the following expression for the mutual information:

$$MI(t;\tau) = \frac{1}{2}\log\left(1 + \frac{\lambda\sigma^2}{\mu^3}\right). \tag{12.7}$$

As required, this expression will always be nonnegative and will grow larger as the variance of the conditional distribution μ^3/λ, that is, the channel noise, grows smaller.

12.3 COMPARING MOLECULAR COMMUNICATION PARADIGMS

In the previous section, the discussion was limited to describing a molecular communication event in which a single molecule was transported across a 1D channel. In more biologically relevant scenarios, communication occurs through a continual sequence of molecular channel traversals with stochastically distributed emission and absorption times. The signal in the system can be thought to consist of the specific pattern of molecular emission times $\tau_1, \tau_2, \tau_3, \ldots$, and the response to that signal will be the corresponding sequence of absorption times $t_1, t_2, t_3, \ldots$, but there is some ambiguity in how these time variables ought to be indexed. If the signaling molecules are biologically indistinguishable, then these time arguments can be thought to be indexed, such that t_2, for example, is the time at which the second molecule arrives and τ_2 is the time at which the second molecule is released, but due to the randomness of the diffusive transport, these may not be the same molecule.

Another scenario to consider is the case in which the molecules are distinguishable. For example, if the system under consideration consists of an entire chromosome sending mRNA molecules to a ribosome, each mRNA will likely represent a different gene, and the signal will contain information about not merely the release times but also the release order of the distinct molecules. In this case, it makes more sense to index the time variables by molecule, so that t_2 and τ_2 refer to the same distinguishable particle.

In the subsections that follow, this latter situation will be considered exclusively, owing to it being a little simpler as well as leading to integrals that can be evaluated directly. The ante will be upped only modestly compared with the previous section by considering the addition of just one additional particle to the 1D drift–diffusion channel, but this marginal increase in complexity will permit an instructive comparison between three signaling paradigms that will reveal some of the subtleties inherent to the correct interpretation of information theory results.

12.3.1 Bocce Rules

In the Italian game of bocce, one player tosses a ball onto a field or court, and all players then compete to lob their balls as close to that first ball as possible. The absolute position of each ball on the field is irrelevant—only its position relative to the initial ball matters. This seems like a bit of a non sequitur, but one can envision a continual biological process, where the amount of time that elapses in between events is the relevant variable rather than the absolute times of their occurrence. For example, a cell at the steady state (homeostasis) will always be manufacturing proteins—what matters is how the frequency of expression of different genes varies in response to the cell's evolving external environment.

Within this paradigm, a two-particle signal would be described by only a single random variable—the time *difference* between the particle emissions $\Delta\tau$. Similarly, the response to this signal would be the time interval between particle absorptions Δt. Because the particles are assumed to be distinguishable if $\Delta\tau$ is defined to be greater than or equal to zero, then Δt can potentially be negative, indicating that the particles arrived in the reverse order of their release. It will further be assumed that the particles are of negligible size and can therefore pass through each other without interaction, thereby rendering their channel traversal times as independent random variables.

If a clock is started when the first particle is emitted, then the two particle emission times will be recorded as zero and $\Delta\tau$; their two arrival times will consequently be t and $t + \Delta t$ for some time t. Since t is the total traversal time of the first particle across the 1D drift–diffusion channel, it will be inverse-Gaussian-distributed. The total traversal time of the second molecule, $t + \Delta t - \Delta\tau$, will be distributed identically. This results in the conditional probability of measuring a response time interval Δt, given a known signal time interval $\Delta \tau$, as the following inverse Gaussian convolution:

$$p(\Delta t|\Delta \tau) = \int_0^\infty dt\, IG(\mu,\lambda,t)\, IG(\mu,\lambda,t+\Delta t - \Delta \tau). \tag{12.8}$$

The integration over t is performed because only the relative times matter in this paradigm. Because, by assumption, Δt can be negative in Equation (12.8), it is possible for the second inverse Gaussian to have a less-than-zero argument, which is normally undefined for this distribution. This can be resolved by analytically continuing the standard inverse Gaussian definition of Equation (12.1) to be zero when its time argument is negative.

As before, in the regime where $\lambda \gg \mu$, the inverse Gaussian distributions in Equation (12.8) can be approximated as standard Gaussians, resulting in $p(\Delta t|\Delta \tau) = N(\Delta \tau, 2\mu^3/\lambda)$. This result is a consequence of the convolution of two Gaussians being another Gaussian; however, it should be noted that because the traversal times of the two molecules are independent, identically distributed random variables, the mean and variance of the exact conditional distribution will be precisely those of its normal approximation. Indeed, this normal approximation is reasonably accurate, even when μ is modestly greater than λ, as demonstrated in Figure 12.6a, where the numerically integrated conditional distribution is compared with its Gaussian approximation for a range of values of its variance.

If the distribution of the signal variable $\Delta \tau$ is once again chosen to be of the form $N(\overline{\Delta \tau}, \sigma^2)$, where the mean emission time interval $\overline{\Delta \tau}$ is chosen to be much larger than the standard deviation σ, then the mutual information between the response and signal time intervals immediately follows from the previous result for the single-particle channel:

$$MI(\Delta t; \Delta \tau) = \frac{1}{2}\log\left(1 + \frac{\lambda \sigma^2}{2\mu^3}\right). \tag{12.9}$$

Note the factor of two that distinguishes this result from Equation (12.7). Figure 12.6b compares the above result with a more exact numerical evaluation of the mutual information that does not use a Gaussian approximation for the conditional distribution, and the difference is negligible over the range of parameters considered. The inset of the figure illustrates the impact of choosing a uniform or exponential distribution for the signal, instead of a Gaussian. For all the signal distributions considered, the differential entropy $h(\Delta \tau)$ was held fixed, so there is no difference in the amount of information they contain, only in how that information is distributed. As the inset of Figure 12.6(b) makes clear, this difference has little impact on the amount of information that can be transmitted across the channel, justifying the choice of a Gaussian signal for this system. As shall be demonstrated in the next subsection, this insensitivity is not universal.

While the signal in this two-particle system has the same information content as the signal in the one-particle case of the previous section, Equations (12.7) and (12.8) make it clear that the two-particle channel transmits less information overall. This is because each

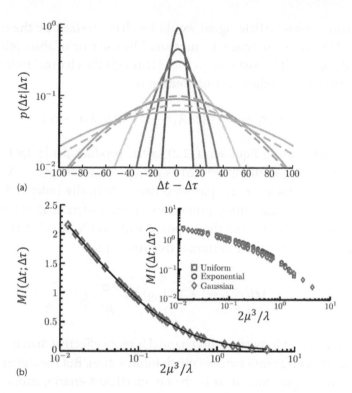

FIGURE 12.6 (a) Plots of the conditional distribution $p(\Delta t \,|\, \Delta \tau)$ as a function of $\Delta t - \Delta \tau$ for different values of its variance. The solid curves are a numerical evaluation of Equation (12.8), and the dashed curves are the corresponding Gaussian approximations. For all but the two largest choices of the variance, the Gaussian fits are practically indistinguishable from the exact distributions. (b) The mutual information of the stochastic time intervals Δt and $\Delta \tau$ plotted as a function of the conditional distribution variance. The diamonds use the exact conditional distribution (evaluated numerically), whereas the solid curve is the approximate analytic result of Equation (12.9). The inset demonstrates the lack of sensitivity of this result to the shape of the signal distribution. (Adapted from Rana, K.R.P. et al., *AIP Adv.*, 8, 055220, 2018.)

molecule undergoes the stochastically diffusive transport of the channel independently, so the overall impact of the noise is doubled in the two-particle scenario. Note that in making this comparison between systems, only differences in differential entropies have been considered—specifically, the difference in the signal differential entropies and the difference in the mutual information.

12.3.2 Billiard Rules

If bocce is typical of a game in which only relative distances matter, then billiards is typical of one in which absolute positions must be known. This is because billiards is played on a small table with fixed features (pockets or bumpers), so any shift in the absolute positions of the balls—even one that keeps their overall configuration fixed—will drastically affect the game. In biology, absolute times are important for processes that occur outside of homeostasis; for example, a sudden sharp change in a cell's environment may drastically change its internal chemical state.

In this scenario, a two-particle signal would be characterized by the pair of emission times (τ_1, τ_2), and the response would be measured by the pair of absorption times (t_1, t_2). The two molecules can still be assumed to travel through the channel independently, so in this case, the appropriate conditional distribution is

$$p(t_1, t_2 | \tau_1, \tau_2) = IG(\mu, \lambda, t_1 - \tau_1) IG(\mu, \lambda, t_2 - \tau_2), \qquad (12.10)$$

which can be rewritten using Equation (12.2) as the product $p(t_1|\tau_1)p(t_2|\tau_2)$. If the signal distribution $p(\tau_1, \tau_2)$ is similarly defined as $p(\tau_1)p(\tau_2)$, where $p(\tau) = N(\bar{\tau}, \sigma^2)$, then the two-particle system completely decouples into two statistically independent one-particle problems. Owing to the logarithmic properties of the differential entropy, the mutual information between the signal and the response will simply be $MI(t_1; \tau_1) + MI(t_2; \tau_2)$. If both emission times have the same variance, this sum equals

$$MI(t_1, t_2; \tau_1, \tau_2) = \log\left(1 + \frac{\lambda \sigma^2}{\mu^3}\right). \qquad (12.11)$$

At first blush, it is tempting to say that this channel is more effective than its single-molecule counterpart, since it can transmit twice as much information. But the signal in this case also contains twice as much information as in the one-particle scenario, and, as one will recall, each independent, continuous random variable formally contains an infinite amount of information. Thus, increasing the number of molecules in the channel results in a finite gain in the information transmitted at the cost of an infinite increase in the information needed to specify the signal.

Simply adding more molecules cannot increase the informational efficiency of the drift–diffusion channel, but Equation (12.11) suggests that increasing the variance of the signal may be another path for improving information transmission. Unfortunately, a simple calculation shows that the change in mutual information achieved by such an increase is exactly equal to the (now finite) change in the signal entropy. The only apparent way to achieve a net improvement in the performance of the channel is to reduce its noise by enhancing the facilitative degree of its transport, that is, by lowering μ and/or raising λ. This will increase the mutual information, without changing the information content of the signal. But could there be another way to achieve this?

12.3.3 Staggering the Signal

In the previous example, the emission of each molecule into the channel was assumed to be an identically distributed, statistically independent process. This requires both release times to be distributed about the same mean time, which will tend to make the interval between releases somewhat short. While this may be sensible for two particles, it will necessarily become unphysical as the number of molecules is increased. Assuming that a single biological process is responsible for the emissions and that it takes an average time $\bar{\tau}$ to produce one molecule, the mean emission time of the second molecule should logically be $2\bar{\tau}$. This suggests that a more reasonable signal distribution for the two-particle system might be slightly more complicated.

$$p(\tau_1,\tau_2) = \frac{1}{2\pi\sigma^2} \exp\left[\frac{-(\tau_1-\overline{\tau})^2}{2\sigma^2}\right] \exp\left[\frac{-(\tau_2-\tau_1-\overline{\tau})^2}{2\sigma^2}\right], \tag{12.12}$$

which is equal to $N(\overline{\tau},\sigma^2)N(\tau_1+\overline{\tau},\sigma^2)$. This distribution shall be referred to as the "staggered" signal, whereas the separable $N(\overline{\tau},\sigma^2)N(\overline{\tau},\sigma^2)$ of the previous subsection will be denoted the "symmetric" signal. The conditional distribution can remain unchanged from Equation (12.10).

When evaluating the differential entropy of the distribution in Equation (12.12), the substitution $\Delta\tau \equiv \tau_2 - \tau_1$ renders the two Gaussians independent once more, with no impact on the integration range $(-\infty,\infty)$. The staggered and symmetric signals thus contain the same amount of information; however, that information is distributed differently over the time domain. Figure 12.7 demonstrates this with side-by-side plots of the two distributions when both σ and $\overline{\tau}$ are unity. Note that this trick required at least two particles in the channel, because the only way to modulate the shape of a single Gaussian is to change its variance, but that would necessarily alter its differential entropy.

Staggering the signal may not affect its information content, but it does impact the information content of the response, as can be confirmed through an algebraically tedious calculation. This results in the following modification to the mutual information:

$$MI(t_1,t_2;\tau_1,\tau_2) = \frac{1}{2}\log\left[\left(1+\frac{\lambda\sigma^2}{\mu^3}\right)^2 + \frac{\lambda\sigma^2}{\mu^3}\right]. \tag{12.13}$$

Note that the above expression reduces to Equation (12.11) in the absence of the extra $\lambda\sigma^2/\mu^3$ term under the logarithm. This means that staggering the emission times of the molecules necessarily leads to an enhancement in information transmission, without a compensatory increase in the information content of the signal, which is precisely the result hoped for at the end of the previous subsection. In verbal communication, this result would be analogous to speaking more slowly in order to be better understood. Annunciating each syllable does not change the information content of the message being conveyed, but it does make it easier for a listener to discern your message in a noisy room.

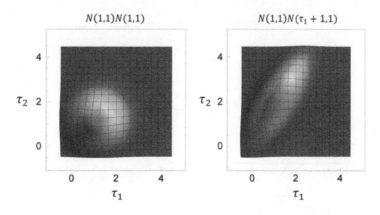

FIGURE 12.7 Comparison of the symmetric and staggered two-molecule signal distributions. Though they are distributed differently, there is no difference in their information content.

Thus far, the operating assumption has been that one is measuring the response in order to predict the signal. This is typically the case in biological applications, but an inversion of this paradigm leads to a puzzling result. Specifically, one finds that the increase in the mutual information achieved by staggering the signal exactly equals the change in the information content of the response. This is a little discomforting, because the mutual information is a symmetric quantity, that is, $MI(X; Y) = MI(Y; X)$. If one attempts to guess the molecular emission times from a measurement of their absorption times, one will be able to do so with more certainty if those release times were staggered. On the other hand, if one attempts to predict the absorption times, given a controlled set of release times, it will make no difference whether one staggers the release times or not!

There is, of course, not really a problem here, and one way to see this intuitively is to once again make an analogy to verbal communication. Speaking more slowly necessarily takes more time, and a longer time interval provides more opportunities for noise to distort the words, greatly increasing the range of possible messages that a listener could conceivably hear. As in the children's game of "telephone," wherein an initial message is successively whispered from one child to the next, it is much easier for the final participant to guess the initial message from the corrupted version that he receives (the usual objective of the game) than it is for the initial participant to predict how her message will be corrupted after successive miscommunications. In general, it is always easier to guess a lower-information quantity from the knowledge of a higher-information quantity than vice versa, and this is precisely due to the symmetry of the mutual information.

It seems as if one might be getting a "free lunch," as it were, by exploiting this symmetry, but the hidden cost comes in the act of measuring. The response contains more information than the signal, so a more sensitive device is needed to measure it accurately. In the verbal example, the person at the end of the telephone chain cannot rely on hearing something that makes logical sense, so she must divert a larger proportion of her brain's processing power toward listening to compensate for the larger space of possible messages.

The mutual information for the different molecular communication scenarios discussed thus far is plotted in Figure 12.8 versus the dimensionless parameter $\lambda \sigma^2 / \mu^3$.

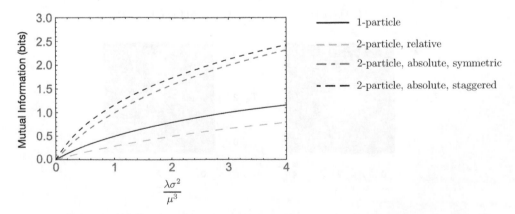

FIGURE 12.8 The mutual information (in bits) for four different molecular communication scenarios, plotted, versus the dimensionless parameter $\lambda \sigma^2 / \mu^3$.

The 1-particle channel, Equation (12.7), is plotted as a solid curve, and the three 2-particle channels–relative-time (Equation [12.9]), absolute-time, symmetric (Equation [12.11]), and absolute-time, staggered (Equation [12.13])–have their results plotted as variously dashed curves. If the reader has gleaned anything from this past section, it is that a comparative plot like this is effectively meaningless, and it is included here only to serve as a convenient summary of the results.

12.4 RENORMALIZING MORE COMPLEX TOPOLOGIES

The analysis of the previous section has emphasized how the communication performance of an isolated molecular channel can vary depending on how the signal is defined, with the focus being on how one can make fair and meaningful comparisons between different cases. In the world of biology, however, signaling processes seldom exist within this sort of vacuum. The regulation of gene transcription in *E. coli*, for example, involves hundreds of unique proteins called transcription factors that regulate the transcription and translation of thousands of functional proteins [8]. These signaling interactions can be represented as a complex network, with nodes representing proteins and directed bonds representing regulating interactions. Although the overall topology (connectivity) of this network is quite complicated, there are certain substructures or *motifs* that appear with an abnormally high frequency [9]. It is worth demonstrating how the basic tools developed in this chapter can be readily extended to these more sophisticated communication architectures and how, under suitable approximations, these motifs can be renormalized to a single-channel system with modified "effective" parameters.

12.4.1 The Linear Cascade

The simplest motif that arises in biological networks is the linear cascade, wherein a sequence of signaling processes is daisy-chained together, with the response of each subprocess serving as the signal for the next. The simplest nontrivial cascade consists of just three nodes, which may be labeled as S, R_1, and R_2, as shown in Figure 12.9a. Assuming that both the $S \rightarrow R_1$ and $R_1 \rightarrow R_2$ channels may be described as having 1D drift–diffusive transport and further assuming that once a molecule reaches the terminus of the first channel, the next signaling molecule in the cascade is instantaneously emitted into the second channel, the probability that the second molecule is absorbed at time t, given that the first molecule was emitted at time τ, is simply an inverse Gaussian convolution:

$$p(t|\tau) = \int_0^\infty dt' IG(\mu, \lambda, t - t') IG(\mu, \lambda, t' - \tau). \tag{12.14}$$

Note the similarity between this one-particle distribution and the two-particle distribution of Equation (12.8). The lower limit of the integral should formally be τ, since the molecule cannot arrive at the junction of the two channels earlier than the time of its original emission, but if the inverse Gaussian is analytically continued to be zero for negative values of its argument, the lower limit can be replaced with zero.

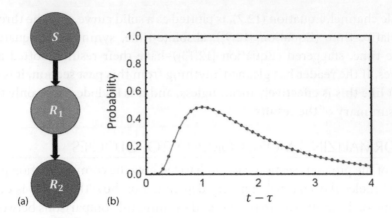

FIGURE 12.9 (a) A schematic representation of the two-channel linear cascade. (b) The conditional probability distribution $p(t|\tau)$ for the two-channel cascade. The data points are numerical evaluations of Equation (12.14), and the solid curve is a plot $IG(2\mu, 4\lambda, t-\tau)$ for $\mu = \lambda = 1$.

In the limit $\lambda \gg \mu$, wherein the inverse Gaussian can be replaced by its Gaussian approximation and the integration safely extended to negative infinity, Equation (12.14) actually becomes identical in form to the analogously approximated Equation (12.8) and reduces to $N(2\mu, 2\mu^3/\lambda)$. If one defines the scaled parameters $\mu' \equiv 2\mu$ and $\lambda' \equiv 4\lambda$, then this reduces to $N(\mu', \mu'^3/\lambda')$, which is just the Gaussian approximation to the conditional distribution for a molecule traversing a single drift–diffusion channel. In other words, the two-channel cascade is equivalent to a single channel with twice the length (recall that $\mu \equiv \ell/v$ and $\lambda \equiv \ell^2/2D$).

This result is perfectly sensible. In fact, it is so sensible that it ought not to depend on any kind of Gaussian approximations. Indeed, though the convolution integral of Equation (12.14) is difficult to evaluate analytically, Figure 12.9b demonstrates that its numerical value is identical to the function $IG(2\mu, 4\lambda, t-\tau)$, as expected. This result follows from the Markovity of the drift–diffusion process. By absorbing the incident molecule and then instantaneously emitting a new one, the junction effectively reinitializes the system and erases any influence that the first particle's trajectory might have had on the second, but the transport process has no memory in the first place, so this is no different from a single particle traversing the entire double-length channel uninterrupted. This logic can be extended to a cascade consisting of n daisy-chained drift–diffusion channels, implying that its conditional distribution must be equal to $IG(n\mu, n^2\lambda, t-\tau)$.

Rather than being instantaneous, each biological process producing a new signaling molecule should actually require a finite, stochastic amount of time, and accounting for these delays can complicate the statistics of the linear cascade considerably. One way to overcome this is to assume that the time it takes for each junction to produce and emit a new molecule is also inverse Gaussian-distributed, which effectively just increases the length of the cascade. For the simple two-channel cascade, for example, one can assume that the junction release time interval $t''-t'$ obeys the distribution $IG(\ell'/v, \ell'^2/2D, t''-t')$ for some effective "channel" length ℓ'. This effective length will serve as a tunable fitting parameter. If one defines a renormalized length $\tilde{\ell} \equiv \ell + \ell'/2$, then it can be shown with the same sort of reasoning as that used in the previous paragraph that the conditional distribution of the cascade channel must now equal $IG(2\tilde{\ell}/v, 4\tilde{\ell}^2/2D, t-\tau)$. This is exactly the same result as for the two-channel cascade with an instantaneous junction, only with the parameters μ and λ now defined in terms of the renormalized length scale.

12.4.2 The Feed-Forward Loop

Another topological motif that is overrepresented in biological networks is the feed-forward loop (FFL) [10]. Built up from the two-channel cascade considered previously, the standard FFL motif consists of both a direct path between nodes S and R_2 and an indirect path that goes through R_1 as an intermediary. Figure 12.10a provides a schematic representation of this topology.

Since the FFL consists of two parallel paths, this motif is most interesting to study in cases where signaling molecules are permitted to traverse both paths at once. As such, a two-particle model will be considered here, wherein each molecule is assumed to be equally likely to take either path. To restrict the number of random variables, the signal will be identified as the difference $\Delta\tau$ in the emission times of the two particles at S, and the response will be the difference Δt in the absorption times at R_2. Once again, the molecules will be assumed to be distinguishable from one another. There are clearly four cases that must be considered, the first of which is the one where both molecules take the direct path. This problem was solved in Section 12.3.1, and the conditional distribution, denoted as $p_{dd}(\Delta t | \Delta\tau)$ to indicate the direct–direct case, is given by Equation (12.8).

The second possibility is that both molecules start down the indirect path, which is assumed to be composed of two copies of the direct channel connected end to end by an instantaneous junction. There are in fact several similar ways of expressing the indirect–indirect conditional distribution, perhaps the simplest being

$$p_{ii}(\Delta t | \Delta\tau) = \int_0^\infty dt\, IG(2\mu, 4\lambda, t)\, IG(2\mu, 4\lambda, t+\Delta t - \Delta\tau), \qquad (12.15)$$

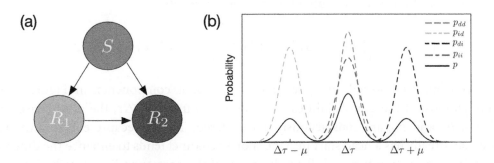

FIGURE 12.10 (a) A schematic representation of the feed-forward loop (FFL). (b) The total conditional probability distribution $p(t|\tau)$ for the FFL alongside its four contributions. The curves were generated by numerically evaluating Equations (12.8), (12.15), and (12.17) through (12.19) for $\mu = 5$ and $\lambda = 625$.

which combines the renormalized two-channel cascade result of the previous subsection with the two-particle result of Equation (12.8). As an alternative, the above distribution can also be expressed as a convolution of two direct–direct path distributions:

$$p_{ii}(\Delta t|\Delta\tau) = \int_{-\infty}^{\infty} d\Delta T \, p_{dd}(\Delta t|\Delta T) p_{dd}(\Delta T|\Delta\tau). \tag{12.16}$$

The interval ΔT represents the difference in the absorption times of the signaling molecules at R_1, which can be negative owing to the molecules potentially arriving in the opposite order of their emission. While both Equations (12.15) and (12.16) can be expanded into similar-looking convolutions involving four inverse Gaussian functions, no change of variables seems to transform one into the other, despite them having the same numerical value.

The remaining two cases to be considered are those in which one molecule takes the direct path while the other takes the indirect path. In the direct–indirect case, wherein the first molecule to be emitted takes the direct path, the conditional probability should take the following form:

$$p_{di}(\Delta t|\Delta\tau) = \int_{0}^{\infty} dt \, IG(\mu,\lambda,t) IG(2\mu,4\lambda,t+\Delta t-\Delta\tau). \tag{12.17}$$

In the indirect–direct case, an analogous expression is obtained:

$$p_{id}(\Delta t|\Delta\tau) = \int_{0}^{\infty} dt \, IG(\mu,\lambda,t+\Delta t-\Delta\tau) IG(2\mu,4\lambda,t). \tag{12.18}$$

Because the four emission possibilities are mutually exclusive, and because the choice of the emission paths is unbiased, the total conditional probability describing the transport of the FFL is just the weighted sum of the preceding results:

$$p(\Delta t|\Delta\tau) = \frac{1}{4}\left[p_{dd}(\Delta t|\Delta\tau) + p_{ii}(\Delta t|\Delta\tau) + p_{di}(\Delta t|\Delta\tau) + p_{id}(\Delta t|\Delta\tau) \right]. \tag{12.19}$$

This probability is plotted, along with its constituent components, in Figure 12.10b. Although most of these probabilities are centered around $\Delta t = \Delta\tau$, the mixed case probabilities, p_{di} and p_{id}, are conspicuously not centered. This is because emitting the molecule that already has a head start into the shorter channel tends to enhance the difference in the final absorption times, shifting the mean of the corresponding distribution to the right. Similarly, placing the first molecule in the longer channel tends to shrink this difference, since the particle with the later emission time can catch up along the shorter path. This results in the total probability being trimodal, approximately a sum of four Gaussian distributions with three different means.

Unfortunately, the logarithm of a sum does not generally simplify, so the mutual information $MI(\Delta t; \Delta \tau)$ for the two-particle FFL cannot be exactly evaluated, even in the Gaussian-approximation regime. However, it can be readily computed numerically, assuming a normally distributed $\Delta \tau$ as usual, and the result is plotted in Figure 12.11 as a function of the drift speed v for two different diffusion constants D. The solid curves are plots of Equation (12.9) for a single-channel system, with channel length equal to the average length of the direct and indirect paths. For small values of the drift, the fit is quantitative, though it gets progressively worse as v increases. So, for at least some parameter ranges, the two paths of the FFL can be effectively renormalized as a single average path.

To flesh out the region of parameter space where this approximate renormalization holds, one must first note that the sum of two Gaussians with the same mean and variances σ_1^2 and σ_2^2 can be well-approximated as a single Gaussian with an averaged variance $(1/2)(\sigma_1^2 + \sigma_2^2)$. This simplification reduces Equation (12.19) to a sum of only three approximately Gaussian terms, each of which has the same variance of $3\mu^3/\lambda$. Because the indirect path is twice the length of the direct path, this is precisely the variance that one would compute from two distinguishable particles propagating down a single channel whose length is the average of the two paths.

The two heteropath distributions in Equation (12.19) have means offset from that of the averaged homopath distribution by a temporal distance of μ, so it is conceivable for these distributions to have negligible overlap when μ is sufficiently large. So long as the standard deviation of the signal distribution is much less than μ, this lack of overlap will hold for the corresponding contributions to the total response distribution $p(\Delta t)$. In this special case, the logarithm of the sum of these contributions can be approximately separated into a sum of logs, leading to precisely the mutual information plotted in Figure 12.11. This renormalization is thus valid whenever $\sigma_S \ll \mu \ll \lambda$. It is the first of these inequalities that breaks down as the drift speed increases.

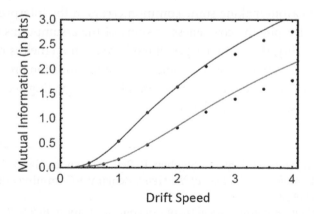

FIGURE 12.11   A plot of the FFL mutual information as a function of the channel drift speed for two different diffusion constants, $D = 0.02$ (bottom) and $D = 0.005$ (top). The data points come from numerical evaluations and the solid curves are plots of Equation (12.9) for a single channel of length of $\ell = 7.5$, the average of the direct and indirect channel lengths.

12.5 SUMMARY

The strength of the Shannon information theory is that it provides a common framework for quantitatively comparing disparate communication paradigms. However, this generality is also its greatest weakness and ironically limits how much information can be gleaned from these comparisons. For example, although mutual information is a quantitative metric of data transmission, its value is not terribly predictive. A common error is to assume that if a channel is capable of transmitting a single bit of information, then an observer of the channel output should be capable of making some binary determination about the input signal such as, "Is the signal above or below some threshold value?" The problem is that for a signal represented by a continuous random variable, an infinite number of such binary determinations are possible, and the 1 bit that gets transmitted will generally be some superposition of all of them. Information theory quantities are thus often most useful as comparative metrics, but, as has been demonstrated in this chapter, even this application is fraught with perils for the uninitiated.

The goal of this chapter has thus been principally to provide some level of initiation to those interested in using information in the study of biological systems. Indeed, though the Shannon theory has been around since the middle of the last century, its application to biological signaling is by comparison in its infancy. The current standard textbook in the field (Cover and Thomas's *Elements of Information Theory* [11]), while an excellent and comprehensive treatment, is written more with statisticians and computer scientists in mind, and a biologist attempting to learn the subject matter from that tome will consequently have difficulty in abstracting the worked examples to biological systems of interest, if they aren't immediately turned off by the rigorous and formal mathematics.

While a single chapter cannot hope to treat the entire breadth of applications that might interest biologists (e.g., this chapter does not even touch upon the use of information theory in understanding the collective behavior of animals or the self-assembly of supramolecular structures), it uses several exactly solvable examples to help the reader build a functional intuition about how to think about biological systems from an information theoretic perspective and how to avoid making some common errors in the interpretation of information theory results. Despite the coarseness of some of the assumptions made in the name of computational facility, the examples presented here are nonetheless reasonable baseline models for the study of certain classes of molecular communication processes, and the final section demonstrates how even much more complicated signaling scenarios can often be reduced to one of these simpler problems.

REFERENCES

1. Tania A. Baker, James D. Watson, Stephen P. Bell, Alexander Gann, Michael Levine, and Richard Losick. *Molecular Biology of the Gene*. Benjamin-Cummings Publishing Company, San Francisco, CA, 2003.
2. Claude E. Shannon. A mathematical theory of communication. *Bell System Technical Journal*, **27**(3), pp. 379–423, 1948.
3. David A. Dean, Jonathan Reizer, Hiroshi Nikaido, and Milton H. Saier, Jr. Regulation of the maltose transport system of *Escherichia coli* by the glucose-specific enzyme III of the phosphoenolpyruvate-sugar phosphotransferase system. *The Journal of Biological Chemistry*, **265**(34), pp. 21005–21010, 1990.

4. Xi Chen, Han Xu, Ping Yuan, Fang Fang, Mikael Huss et al. Integration of external signaling pathways with the core transcriptional network in embryonic stem cells. *Cell*, **133**(6), pp. 1106–1117, 2008.

5. Nobutaka Hirokawa, Yasuko Noda, Yosuke Tanaka, and Shinsuke Niwa. Kinesin superfamily motor proteins and intracellular transport. *Nature Reviews Molecular Cell Biology*, **10**, pp. 682–696, 2009.

6. Kothapalli V. Srinivas, Andrew W. Eckford, and Raviraj S. Adve. Molecular communication in fluid media: The additive inverse Gaussian noise channel. *IEEE Transactions on Information Theory*, **58**(7), pp. 4678–4692, 2012.

7. Pratip Rana, Kevin R. Pilkiewicz, Michael L. Mayo, and Preetam Ghosh. Benchmarking the communication fidelity of biomolecular signaling cascades featuring pseudo-one-dimensional transport. *AIP Advances*, **8**, pp. 055220, 2018.

8. Shai S. Shen-Orr, Ron Milo, Shmoolik Mangan, and Uri Alon. Network motifs in the transcriptional regulation network of *Escherichia coli*. *Nature Genetics*, **31**, pp. 64–68, 2002.

9. Ron Milo, Shai S. Shen-Orr, Shalev Itzkovitz, Nadav Kashtan, Dmitri Chklovskii, and Uri Alon. Network motifs: Simple building blocks of complex networks. *Science*, **298**(5594), pp. 824–827, 2002.

10. Shmoolik Mangan and Uri Alon. Structure and function of the feed-forward loop network motif. *PNAS*, **100**(21), pp. 11980–11985, 2003.

11. Thomas M. Cover and Joy A. Thomas. *Elements of Information Theory*. John Wiley and Sons, New York, 2012.

CHAPTER 13

Design and Applications of Optical Near-Field Antenna Networks for Nanoscale Biomolecular Information

Hongki Lee, Hyunwoong Lee, Gwiyeong Moon, and Donghyun Kim

CONTENTS

13.1 INTRODUCTION

Nanoantennas refer to nanoscale transducers that convert electric energy into electromagnetic (EM) wave propagation as a transmitting antenna (and vice versa as a receiving antenna) in a communication channel, with low energy loss, working at optical frequencies [1–3]. A schematic of a nanoantenna network operating in a transmitter–receiver link at different wavelengths is illustrated in Figure 13.1. Using an antenna, a receiver or a transmitter in a communication network may interact with a radiated EM wave,

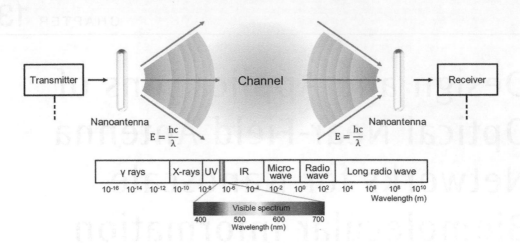

FIGURE 13.1 Schematic illustration of communication channel based on a nanoantenna network (*E*: energy, *h*: Plank constant, *c*: speed of light, and *λ*: wavelength). Below, the schematic are the spectral wavebands in which nanoantennas may operate.

which may be controlled or modulated on a nanometer scale for enhancement of the performance and efficiency in light emission and sensing [4,5].

The design and operating principle of nanoantennas are based largely on the impedance matching, a technique that has been applied to various types of antennas in communication networks at microwave and radio frequencies (RFs) [6,7]. The geometry and material configuration of an antenna are carefully chosen to meet required scattering parameters and frequency response. The dimension is highly dependent on the wavelength; that is, an antenna is sized to be commensurate with operating wavelengths. As a result, a nanoantenna system may become smaller (in principle, on a nanoscale) and thereby more complex. For nanoantennas, many researchers have been working to implement sophisticated nanoscale devices with sufficiently high yields [8–12]. With recent advancement in nanofabrication and optical technology, diverse implementations of optical nanoantennas have been demonstrated on a scale of a few nanometers in size [13–16].

The working principle of a nanoantenna depends on electron oscillation in a medium (like that of a traditional antenna). Despite many similarities with traditional macroscale microwave and RF antennas, nanoantennas have differences in the very small size, as well as resonance characteristics arising from metallic or nonmetallic nanostructures [17]. The properties of a metallic nanoantenna are heavily affected by surface plasmon (SP), which represents collective electron oscillations formed in a metal. An EM wave at or near plasma frequency can penetrate metal with partial energy absorption and create SP rather than being perfectly reflected. Plasma inside metal nanostructures can now be strongly coupled to an external EM wave. Under surface plasmon resonance (SPR), electron oscillation causes strong electric field enhancement [18,19]. SP may also confine an electric field to the vicinity of nanoscale structures and produce surface plasmon polaritons (SPPs) propagating at metal–dielectric interfaces and/or nonradiative localized surface plasmon (LSP) in nanostructures [20–23]. The size of field confinement in the near-field region can be reduced to a few nanometers, which is much smaller than Abbe's diffraction limit. This suggests that optical energy emitted or absorbed by a nanoantenna acting as a transmitter (or reciprocally a receiver) can be extremely localized within a subdiffraction-limited area [24–27].

An optical nanoantenna with such strongly localized light has great advantages in many research areas in which an optical signal needs to be controlled with high signal-to-noise ratio (SNR) and fine spatial precision [28–30]. A variety of designs, including Yagi–Uda antenna structure, have been considered for effective and efficient nanoscale communication [31,32]. In this chapter, we discuss theoretical backgrounds of a nanoantenna in Section 13.2. Performance of various nanoantenna designs and effects of design parameters on the channel characteristics are explored in Section 13.3. Unlike microwave and RF antennas, optical nanoantennas may be considered for coupled operation with fluorescent molecules and quantum dots for a broad range of utilities and improved performances [33,34]. In fact, the receiver and/or transmitter in a nanoantenna network can be any object, even on a quantum scale, that may absorb or excite photons. For this reason, nanoantenna have emerged as a platform in biomedical imaging and sensing applications: this is detailed in Section 13.4. The chapter is summarized in Section 13.5.

13.2 BACKGROUNDS OF AN OPTICAL NANOANTENNA

13.2.1 Near-Field and Far-Field Characteristics

EM energy is transmitted by nanoantenna in two different spatial regions: near- and far-field (Fraunhofer) regions, as illustrated in Figure 13.2. EM radiation from nanoantenna induces near-field distribution on a wavelength spatial scale, which undergoes transition from Fresnel region into far-field propagation. From the nature of nanoantennas, we focus more on the near-field characteristics; that is, in contrast to the far-field region, where the radiation power decreases by an inverse-square law, EM radiation in the near-field region diminishes by an inverse of the higher-order polynomials. In the near-field region, the distribution of an EM field is closely related to the electron charge distributions on the

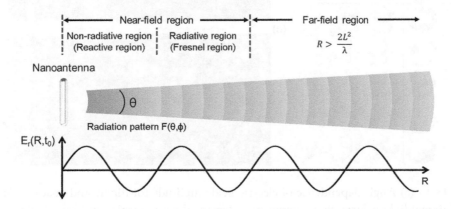

FIGURE 13.2 Schematic illustration of operating regions of a nanoantenna. The regions depend on the spatial distance away from an antenna (R: distance from antenna; θ and ϕ: polar and azimuthal angles, respectively, in the spherical coordinate system; t_0: observation time; L: antenna length; λ: wavelength; E_r: amplitude of an electric field; and F: radiation pattern).

antenna surface, and the intensity can be enhanced by using SP excitation of electrons. From the basic antenna theory, a nanoantenna has reactive characteristics to any energy fluctuation in an evanescent field, which rapidly extinguishes away from the near-field region. The impedance of a reactive nanoantenna consists of ohmic and radiation resistance determined by radiative damping and intrinsic properties of the material in different geometries. An evanescent field formed in the near field exists so tightly close to the nanoantenna surface that an EM field can be reduced to the nanoscale volume beyond the limit imposed by the far-field optics. In this sense, an optical nanoantenna can be made feasible to build nanoscale biomolecular network. EM fields being tightly localized in the near-field region implies significant enhancement in the interaction between an emitter/absorber and an EM field. In the far-field region, the EM radiation can be redirected with enhanced emission, using various nanoantenna structures. Even in the far-field region, by combining nanoscale transmitters and receivers with SP excitation, we can achieve directional confinement of an EM field and modulation of coupled emitters with increased power.

Note that both near-field distribution and far-field radiation are governed by electron oscillation in a nanoantenna. The resonant modes of electron concentration are dependent on the material dispersion and the geometry of a nanoantenna. For example, a spherical nanoantenna attains dipolar resonance easily by coupling to plane waves, owing to the symmetric geometry. Nanoantennas strongly coupled with an incident EM wave may exhibit multipolar resonances in addition to dipolar modes, as shown in Figure 13.3.

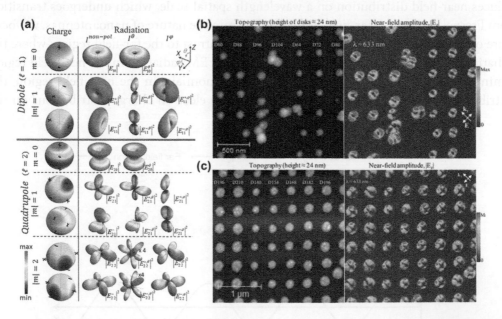

FIGURE 13.3 (a) Angle dependence of electric field amplitude for dipole and quadrupole modes under different light polarization component. (Reprinted with permission from Thollar, Z. et al., *ACS Photonics.*, 5, 2555–2560. Copyright 2017 American Chemical Society.) (b) The topographic image and near-field amplitude image of small-diameter (64–104 nm) gold nanodisks. (c) The topography and near-field amplitude distribution of gold nanodisks with a large diameter (140–210 nm). (Adapted from Habteyes, T.G. et al., *Opt. Express*, 21, 21607–21617, 2013. With permission.)

On the other hand, higher-order resonant modes in an antenna provide angular dependence of the radiation pattern. Optical nanoantennas can be formed as a phased array to produce diverse EM field distribution in the near-field and far-field regions by superposing resonant modes [35,36].

As described in Figure 13.2, nonradiative EM field can be localized in the vicinity of a nanoantenna, with a spatial distribution attributed to LSP. Electric dipole oscillation can be produced in the process of SP excitation and localization in a dimension much smaller than an optical wavelength, as shown in Figure 13.3a. Figure 13.3b and c shows that a nanoantenna can sustain complex field distribution arising from multipolar resonant SP modes and experience phase retardation as the size increases [37]. Interestingly, the intensity of localized EM fields can be enhanced, compared with that of an incident EM field. Inhomogeneous charge distribution on edged nanoantenna surface can further induce dramatic field enhancement, inducing additional multipolar oscillation [38,39]. Such localization of near-fields was explored in diverse applications, which will be discussed in the following sections.

13.2.2 Theoretical Backgrounds

A nanoantenna plays two major roles—as a transmitting antenna and as a receiving antenna—as described in Section 13.1. As a transmitting antenna, it converts the localized EM energy into far-field EM radiation, while as a receiving antenna, it receives and extracts EM energy from external EM fields. In some sense, the operation of optical antennas is analogous to that of conventional RF antennas, which transfer confined electrical signals into radiated propagating EM waves in the RF frequency. Even if their functionality is similar, optical and RF antennas show drastic differences in physics, material properties, and excitation features.

Many parameters of a nanoantenna such as impedance, directivity, and efficiency gain characterize and govern the performance. For one, impedance represents the complex ratio of the source voltage to the current, which is an indispensable factor when describing an antenna. In the antenna theory, the source is connected to the antenna through a transmission line. However, an optical antenna receives energy from an emitter, not the actual source current. To illustrate this point, the concept of local density of EM states (LDOS) and Green's function can be employed with a single emitter treated quantum mechanically; that is, the presence of a nanoantenna can be described via LDOS, while the Green's function describes the energy dissipation of a dipole that approximates a nanoantenna [40,41].

Directivity of a nanoantenna is the ability to concentrate the radiation in a certain direction and can be written by the expression:

$$D(\theta,\varphi) = \frac{4\pi \, p(\theta,\varphi)}{P_{rad}} \tag{13.1}$$

where P_{rad} is the total power radiated to the far field by the nanoantenna system and $p(\theta,\varphi)$ is the angular distribution of the radiated power. In particular, studies have been carried out to control the radiation pattern of an emitter or single molecule by using nanoantenna.

Metal-based structures were mainly used to direct the emission, while they were coupled with the single quantum system [42]. Also, spherical silicon nanoparticles were employed to achieve directional light scattering to simultaneously excite magnetic and electric dipole resonance [43].

The dissipative loss produced by a nanoantenna can be described by radiation efficiency (ε_{rad}) and is expressed as

$$\varepsilon_{rad} = \frac{P_{rad}}{P_{rad} + P_{loss}} \tag{13.2}$$

where P_{loss} is the total power loss incurred by a nanoantenna. If intrinsic quantum yield (η_i) of an emitter is considered to distinguish the dissipation of nanoantenna and transmitter, Equation 13.2 can be rewritten as:

$$\varepsilon_{rad} = \frac{P_{rad}/P_{rad}^o}{P_{rad}/P_{rad}^o + P_{loss}/P_{rad}^o + (1-\eta_i)/\eta_i} \tag{13.3}$$

where P_{rad}^o is the radiated power by an emitter in case there is no nanoantenna in the network. Intrinsic quantum yield can be assumed to be unity in Equation 13.3, because the estimation of radiation power loss of an emitter is practically difficult, if not impossible.

Directivity and radiation efficiency can be linked by antenna gain (G), which is one of the important characteristics of an antenna, and is written as:

$$G = \varepsilon_{rad} D \tag{13.4}$$

In other words, antenna gain for the case of a transmitting antenna represents the efficiency of converting incident power into a specific direction. In general, a radiation pattern is considered only in the known directions; therefore, relative gain is often used by defining it through the comparison of power gain in a certain direction with that of a reference antenna.

Note that efficiency of a quantum emitter can be significantly increased in a transmitting nanoantenna system in relation to the Purcell effect [44]. The Purcell effect is defined as the change of the spontaneous emission rate of a quantum emitter by its environment. Nanoantenna absorbs photons radiated by a quantum emitter and acts as an inhomogeneous environment. Equation 13.3 can then be rewritten as:

$$\varepsilon_{rad} = \frac{\Gamma_{rad}}{\Gamma_{rad} + P_{loss}/P_{rad}^o} \tag{13.5}$$

Γ_{rad} denotes the ratio of power radiated by the nanoantenna system to that of emitter radiation in the absence of a nanoantenna [40]. Based on Equation 13.5, a nanoantenna can be used to modify the spontaneous emission rate of a quantum emitter, on which many related studies have been conducted [5,45–49].

13.3 PERFORMANCE OF VARIOUS NANOANTENNA STRUCTURES

Optical nanoantennas aim to improve the radiation efficiency and modulate directivity in the course of signal propagation of light fields. This is similar to conventional antennas except that the propagation takes place over the nanoscale communication channel. In this way, the antenna networks can be utilized to transmit and receive signals in applications of nanoscale sensors and imaging systems.

The performance and effectiveness of an optical nanoantenna are affected by geometrical parameters used to define the structure, materials, and environment. In this section, we explore the effect of these parameters on the nanoantenna performance by considering nanoantenna structures that exploit localized and propagating light fields.

13.3.1 Overview of Metallic and Dielectric Nanoantenna

Material is one of the governing parameters of the antenna performance. For metallic nanoantennas, the optical response can be explained by the negative real part of dielectric constant derived from the effect of electron damping. In addition, the imaginary part of dielectric constant accounts for the absorption of radiation energy and can be minimized for efficient signal delivery. Incident EM radiation can be strongly localized near the nanoparticle or nanostructure or into highly directional scattering fields. In this context, metal nanoparticle or nanostructure can be considered as a nanoantenna by transferring the EM energy from near to far field (or vice versa). Effective cross-section much stronger than geometric cross-section can be accompanied.

Property of an emitter is related to not only the intrinsic nature of the emitter but also its environment [44,50]. When an emitter is placed close to metal nanoantenna, the radiative and nonradiative properties can be modified by weak or strong coupling. It gives rise to the modification of the decay rate of the emitter and enhanced interaction between emitter and far-field radiation.

Metallic nanoantenna can be roughly classified into monopole, dipole, bowtie, and Yagi–Uda nanoantennas [40]. The monopole antenna is of the simplest form, which serves as a receiver by confining strong EM field. It can be composed of vertically oriented nanorods and can be used for single-molecule imaging, sensing, or optoelectronic devices [51]. Dipole antenna, with a gap structure between two metal nanoparticles, usually represents nanodimer [40]. Dipole antenna has been widely used as excitation plasmonic waveguide [52], coupled to optical fibers or broadband wide-angle holography [53]. Bowtie antenna, where two metallic triangles are facing with a small gap between them, is analogous to classical conical RF antennas [54,55]. Bowtie antennas have been applied to surface-enhanced Raman scattering (SERS) [56,57], enhancement of fluorescence [56–58], and solar cell [56]. Yagi–Uda nanoantenna consists of feed and parasitic elements, which include reflector and directors, like RF Yagi–Uda antennas [40]. The directivity diagram can be controlled by adjusting the antenna dimension or parameters [42].

On the other hand, studies on dielectric materials have been actively conducted as materials of optical nanoantennas. Dielectric nanoantennas have drawn interests for several factors, for example, low dissipative loss compared with metallic materials, which have intrinsic loss at optical frequencies. A dielectric nanoantenna operates based on field and displacement currents induced by incident radiation, while metallic ones rely on oscillation of electrons. The Purcell factor is also lower than metal nanoantenna. Owing to high permittivity of dielectric material, electric and magnetic dipole resonances can be observed in visible and near-infrared (IR) spectral [62–65], creating Huygens element, using a single particle [66]. For example, spherical silicon nanoparticles with a diameter of 200 nm show the first two low-frequency Mie resonance, which correspond to electric and magnetic dipole moments [62]. These features can allow the possibility of an optical nanoantenna for the inducement of strong local magnetic field enhancement or unidirectional emission [63,67,68].

13.3.2 Nanorod Monopole Antenna

We now explore an optical monopole nanoantenna in more detail, using metallic nanorod as an antenna element. In the RF domain, antenna elements implemented by metallic rods resonate at about twice the length of the antenna. In the optical domain, much smaller spatial scale makes the performance of a nanoantenna extremely sensitive to the structural parameters. For nanoantennas based on sufficiently small particles, it was found that energy transfer in the EM response may generate LSP, which is determined largely by the shape and the material (dielectric constant) and less by the particle size [69]. In the case of nanorods, LSP resonance (LSPR) depends on the aspect ratios, and it was shown that resonance is achieved at longer wavelengths, compared with spherical particles, if the cross-section of nanorods is set to be constant and only axial length is changed.

13.3.3 Yagi–Uda Nanoantennas with Propagating Light Fields

Yagi–Uda designs have been extremely popular for optical nanoantennas; that is, arranging metal nanostructures at sufficiently narrow intervals can induce signal propagation by coupling plasmonic waves between the metals [70]. In this sense, the way that nanoantennas of this type propagate signals involves localization of SP, to some degree, near metallic nanostructures. Operating resonant wavelengths are configured by the geometrical parameters to achieve the desired antenna performance [71–75]. For example, optical nanoantennas using metal nanorods tend to have a longer resonant wavelength per antenna size than the conventional ones [76].

In this section, as an example of nanoantennas that employ propagating light fields, we examine Yagi–Uda designs, which are known to have high directivity in the RF regime, to transmit radiation from the nanoscale to the far field. Early studies of optical Yagi–Uda nanoantennas have been approached in the same way as RF antennas. Dipole antennas using nanorods were extended to Yagi–Uda designs [77]. Since then, continuous research has been carried out and has led to optical nanoantennas with strong directivity in nanoscale.

The widely known Yagi–Uda antennas with high directivity in conventional RF antennas consist of a linear array of metal bars. Each metal rod acts as a feed, reflector, and director. The fundamental principle is that EM waves from the feed induce currents in different metal rod arrays to cause phase-coherent emission to all antenna arrays. Yagi–Uda designs in nanoscale can be implemented with the same geometry scaled down from macroscopic antenna structure. For this purpose, resonant wavelengths with respect to the geometrical structure of nanorods should be determined by the measurement of transmission spectra. In this example, nanorods of 75-, 106-, and 125-nm length produced resonant wavelengths of about 610, 655, and 770 nm, respectively. A study conducted about a Yagi–Uda design employed 125-nm-long nanorods as reflectors, 106-nm-long nanorods as feeds, and 75-nm-long nanorods as directors [69]. Figure 13.4 shows all the elements of Yagi–Uda nanoantennas in two and five elements. The design using five elements in Figure 13.4c demonstrated the highest directivity.

Although most studies of nanoantennas based on Yagi–Uda designs investigated radiation directivity in the lateral plane, the emission properties were also explored in three dimensions (3D). Figure 13.5a shows the effect of coupling antenna arrays to dipole emitters. The directivity was calculated by increasing the number of Yagi–Uda antenna arrays stacked in 3D. Figure 13.5b presents a schematic of the actual operation with phase modulation [78].

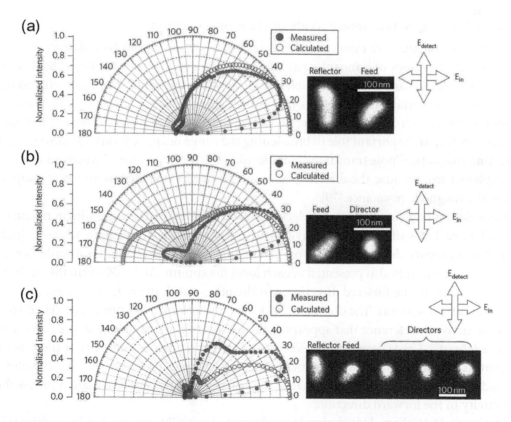

FIGURE 13.4   Measured (solid circles) and predicted (open circles) radiation patterns of Yagi–Uda nanoantennas. Two-element design with (a) feed-reflector, (b) feed-director, and (c) five-element Yagi–Uda antenna. The inset shows SEM images of the antennas. (Reprinted by permission from Macmillan Publishers Ltd. *Nat. Photonics*, Kosako, T. et al., 2010, copyright 2010.)

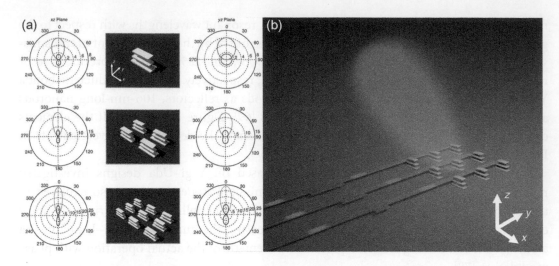

FIGURE 13.5   (a) The directivity of the transmitting Yagi–Uda array coupled to dipole emitters in various stacking scheme. (b) Conceptual illustration of 3 × 3 Yagi–Uda antenna array. (Reprinted by permission from Macmillan Publishers Ltd. *Nat. Commun.*, Dregely, D. et al., 2011, copyright 2011.)

13.3.4 Dielectric Nanoantennas Without Plasmon Excitation

Optical nanoantennas are mainly implemented in metal, which involves plasmon excitation. However, studies of nanoantennas using dielectric materials have been steadily on the rise. Dielectric nanoantennas exhibit low losses at the optical frequencies owing to the nature of the material. Nanoantennas may produce both electric and magnetic resonant modes based on high-k dielectric material. The unique feature of dielectric optical nanoantennas can play an important role in broadening the range of applications, for example, for detecting magnetic dipole transitions of molecules. Furthermore, the Huygens source can be implemented, because the dielectric nanosphere can be operated as an electric dipole under the magnetic resonance [79].

Consider an implementation of dielectric optical nanoantennas using silicon nanoparticles. Figure 13.6a shows the directivity with respect to the wavelength when a single dielectric nanoparticle is excited by an electric dipole source. 3D angular distribution of the radiated pattern is also presented at each local maximum. At $\lambda = 590$ nm, the radiated field is directed in the forward direction. On the other hand, the radiation is dominantly backward at $\lambda = 480$ nm. The dominant backward radiation at 480 nm is associated with the destructive interference that appears in the forward direction, owing to the phase difference (about 1.3 rad) between the electric and magnetic dipole moments of the sphere. In contrast, total electric and magnetic moment oscillate in phase at 590 nm. Therefore, the radiation pattern is identical to that of the Huygens source, and the main lobe has the directivity in the forward direction.

In Figure 13.6b, Yagi–Uda design is implemented with 70-nm spacing by adding silicon nanoparticles. A single reflector was constructed with a radius of 75 nm, and four directors were constructed with a radius of 70 nm to improve the performance of all-dielectric Yagi–Uda nanoantennas. The radius of each reflector and director was obtained

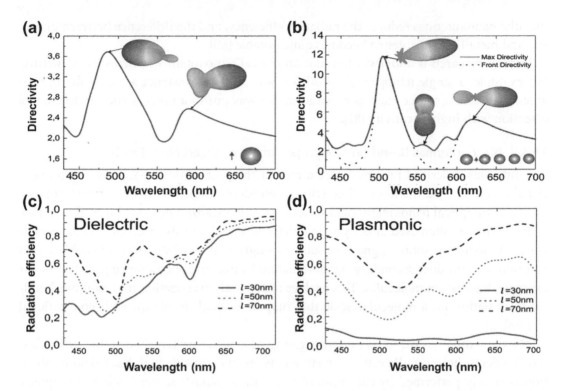

FIGURE 13.6 Wavelength dependence of the directivity for all-dielectric nanoantennas: (a) single particle and (b) Yagi–Uda array. Radiation efficiencies of (c) dielectric (Si) and (d) plasmonic (Ag) Yagi–Uda optical nanoantennas for an identical geometrical design. The spacing between antenna elements in the array is varied. (Adapted from Krasnok, A.E. et al., *Opt. Express*, 20, 20599–20604. With permission.)

by calculating the maximum constructive interference with directivity in the forward direction. The Yagi–Uda nanoantenna with optimal performance was found to consist of a reflector with a resonant radius for electric resonance at certain frequencies and a director with a radius that resonates with magnetic resonance. The strongest directivity was obtained at $\lambda = 500$ nm. The main lobe at this time showed a very narrow beam width of 40 degrees and negligible level of backscattering.

What is observed in Figure 13.6a and b is that the Yagi–Uda design presents significantly improved directivity over single particle-based antenna. The wavelength associated with the maximum directivity are not identical between nanoantennas based on a single nanoparticle and Yagi–Uda design. In fact, the wavelength is red-shifted by the introduction of nanoparticle arrays in the Yagi–Uda nanoantenna. The shift is due to the interaction between dielectric nanoparticles and does not exactly coincide the electric and magnetic resonances of a single nanosphere.

Figure 13.6c and d compare the radiation efficiency of Si-based dielectric and Ag nanoparticle-based metallic Yagi–Uda nanoantennas. When the spacing between nanoparticles of a nanoantenna was sufficiently large at 70 nm, the radiation efficiency was measured to be comparable in the two cases. As the spacing decreased, the radiation efficiency turned significantly lower for metallic nanoantenna, because much higher loss incurred in metallic nanoantennas reduces the radiation efficiency, and the difference between dielectric and metallic nanoantenna becomes quite notable [80].

Further research is underway to tailor an optical nanoantenna in dielectric structure. For example, a single nanoparticle of silicon was used to construct a highly directional antenna, where a source such as a quantum dot was put in a notch structure to control direction with high directivity [81].

13.3.5 Nanoantennas Based on Nonpropagating Localized Near Fields

The antennas described previously are aimed at the propagation of signals like conventional antennas, and therefore, directivity is considered to be an important performance measure. In optical nanoantennas, however, uniquely localized near field near the surface of an antenna is often used directly as an enhanced signal. In this case, optical nanoantennas act to focus the source signal to a subwavelength scale, and the resulting field can act as a source to produce secondary signal radiation by exciting photoemitting materials, for example, fluorescent molecules. This enables functional investigation to obtain biochemical information on a molecular scale that may be difficult to obtain with conventional technology.

The nanoantennas based on nonpropagating near fields are largely metallic, with plasmonic effects; however, dielectric nanoantennas are also feasible [82]. Near-field localization is largely performed by excitation of LSP with a nanostructure, such as nanoposts, nanoholes, nanoparticles, and arrays of periodic or random nanostructures arranged with enough distance to guarantee little plasmonic coupling. Absence of a gap in the nanostructures produces relatively weak enhancement in the near field as compared with a structure with a nanoscale gap. Localized near field is formed at a specific position of a nanoantenna [83,84]. For example, a structure may induce well-known lightning rod effect and plasmonic hybridization to localize the field at vertices and edges with a subwavelength-scale light volume.

Most of these nanoantennas are fabricated by e-beam lithography (EBL) and/or using focused ion beam (FIB), reactive-ion etching (RIE), and nanoimprints. These techniques allow fabrication of nanostructures at the cost of equipment and consumption of time. For this reason, annealing is sometimes employed for fabrication, because annealing is cheap and fast without lithography, despite low degrees of freedom in the type of structures that may be fabricated. Although annealing-based nanoantennas are irregular in shape, statistical control of the structure is feasible. The irregular nanostructures, often called nanoislands, produce localized yet irregular distribution of near fields. In the recent years, it was reported that light source can be switched for almost complete surface coverage of localized near fields on the surface of the entire substrate [85].

With a nanogap, very strong fields can be formed in the gap to implement a dipole nanoantenna [86,87]. For example, a dimer structure between two nanoposts at very narrow intervals gives rise to asymmetrically enhanced near fields [88]. Furthermore, nanogaps of a few nanometers in size were developed by combining atomic layer deposition (ALD) and simple tape peeling [89,90]. Smaller gaps help to increase the sensitivity when applied to detection modalities such as SERS and create strong field gradient for optical trapping [89].

13.4 BIOMEDICAL APPLICATIONS OF NANOANTENNA NETWORKS

Growing interests have been paid to optical nanoantennas in biomedical engineering applications for acquiring molecular information. In this section, we focus on the applications in biology and related areas.

First, optical nanoantennas were applied to Raman spectroscopy. SERS continues to be one of the most popular applications of optical nanoantennas, which is useful for analyzing materials with extremely high sensitivity. Nanoantennas can enhance Raman signals by LSPR on the surface. It is known that the intensity of Raman signal is increased by 10^{10} times, or higher, to a degree that can be useful in many applications beyond simple substance analysis [88]. The best-known example is the detection of cancer [91–93]: Figure 13.7a shows in vivo multiplex detection in xenograft tumor. As a result, SERS spectra from a tumor site in Figure 13.7b showed peaks at 1120, 1175, and 1650 cm^{-1} from SERS nanotags corresponding to Cy5, MGITC, and Rh6G, respectively. Signal intensity was not detectable 6 hours after injection, and nanotags were cleared fast from the body owing to the lack of specific binding. SERS can also trace the mechanism leading to the death of cancer cells. In addition, it was applied to food safety by detecting pesticides on fruit surfaces and *E. coli* O157:H7 [94].

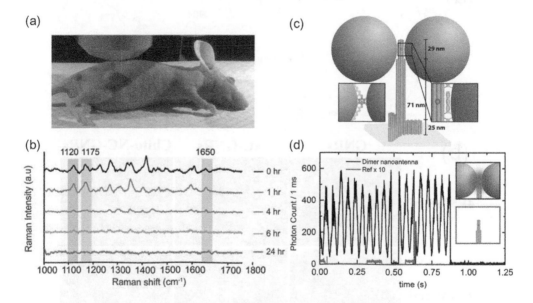

FIGURE 13.7   In vivo multiplex detection in xenograft tumor: (a) image of a tumor-bearing mouse from test group used in experiment. (b) SERS spectra from tumor site in a representative control mouse. (Reprinted by permission from Macmillan Publishers Ltd. *Sci. Rep.*, Dinish, U.S. et al., 2014, copyright 2014.) (c) Schematic illustration of the DNA origami pillar (gray) employed to build the optical nanoantenna with two 100-nm Au nanoparticles, together with a top view (lower-left inset). The lower-right inset describes the "zipper" binding strategy to incorporate the Au nanoparticles to the origami structure. (d) Single-molecule fluorescence transients for a dimer nanoantenna (black line) and for a DNA origami structure without nanoparticles (red line), obtained using 10 times more excitation intensity. (Reprinted with permission from Puchkova, A. et al., *Nano Lett.*, 15, 8354–8359. Copyright 2010 American Chemical Society.)

Nanoantennas can be used for biomolecular detection and imaging [95]. Many studies have been carried out to achieve intensity enhancement of fluorescence of quantum dots by modifying the spontaneous emission rate of an emitter creating highly enhanced EM fields [45,47,48,96–99]. Figure 13.7c presents a schematic of nanoantenna networks using DNA origami, where two colloidal gold nanoparticles form a dimer, with DNA origami between them. Greatly enhanced EM energy confined in the gap between nanoparticles improves the fluorescent signatures of DNA origami, as shown in Figure 13.7d. The extreme localization achieved by such an optical nanoantenna allows discrimination of the signal associated with a molecule or an emitter when located in or near localized fields.

By selective acquisition and amplification of the signal emitted by optical nanoantennas, the sensitivity of molecular detection can be improved. Biomolecular detection and cell imaging using diverse nanoantenna structures have been conducted by localizing EM fields [100–103]. Nanoantennas have led to great advances in single-molecule experiments. The signal intensity of a single molecule was unprecedently amplified by a nanoantenna [5,58,104,105], while the emission was controlled [42,106]. With enhanced field

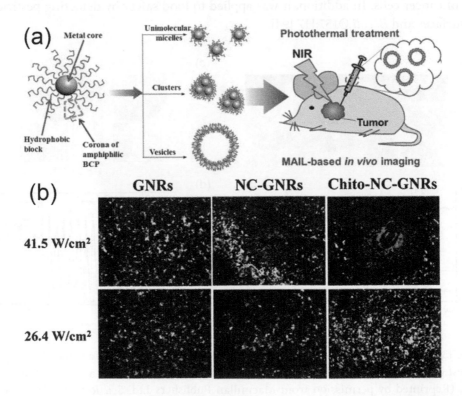

FIGURE 13.8 (a) Amphiphilic plasmonic micelle-like nanoparticle antenna developed for bioimaging and photothermal therapy of cancer. (Reprinted with permission from He, J. et al., *J. Am. Chem. Soc.*, 135, 7974–7984, 2013. Copyright 2013 American Chemical Society.) (b) Flourescence images after selective near-infrared photothermal therapy on SCC7 cancer cells with gold nanorod (GNR) or GNR-loaded nanocarriers irradiated by a laser at 780 nm with two different power densities. (Reprinted with permission from Choi, W.I. et al., *ACS Nano*, 5, 1995–2003, 2011. Copyright 2011 American Chemical Society.)

localization, nanoantenna network can also be applied to fluorescence correlation spectroscopy (FCS) to study biomolecular dynamics and to analyze diffusion characteristics inside and outside of a cell and on the cell membrane [107–116]. Extreme localization of light in the near field helps to reveal the nature of molecular interactions [117–119].

Nanoantenna was also used as a tool for cancer therapy. For example, amphiphilic plasmonic micelle-like nanoparticles, based on self-assembly of gold nanoparticles with block copolymers, were used for photothermal therapy of tumor in mice, as shown in Figure 13.8a [120]. Also, a hyperthermia system for photothermal therapy was developed by using gold nanorods as functional nanocarriers [121]. Figure 13.8b shows photothermolysis effect from cancer cell (SCC7) by selective near-IR photothermal therapy. Although not detailed here, nanoantennas even find applications for transmitting thermal energy and providing a stimulus to biological molecules, cells, and tissues.

13.5 SUMMARY

In this chapter, we presented near-field and far-field characteristics of nanoantenna networks and showed that signal propagation can be performed using diverse architectures. Network characteristics can be modulated by varying antenna parameters, including the geometry of metallic and dielectric nanoantennas. Intensive research has been underway in this direction. Many studies in the past were conducted based on plasmon excitation in metallic nanoantennas, a feature that distinguishes nanoantenna network from the counterparts in the microwave and RF regimes. We also discussed dielectric nanoantenna without plasmon excitation, which bears an advantage of low loss and high k. The capability of a nanoantenna converting EM fields for signal delivery in the nanoscale environment has been widely useful in various biomedical engineering applications; for example, selective acquisition and amplification of Raman and fluorescence signal were discussed. Without any doubt, the growing interests in nanoscale antenna networks can be extrapolated to the future. The properties and performance of nanoantennas are expected to be investigated in wider areas of research as a key component of nanotechnology and to find more applications for delivery of nanoscale biomolecular information.

REFERENCES

1. Giannini, V., Fernández-Domínguez, A. I., Heck, S. C., and Maier, S. A. 2011. Plasmonic nanoantennas: Fundamentals and their use in controlling the radiative properties of nano-emitters. *Chemical Reviews* 111:3888–3912.
2. Fitzgerald, J. M., Azadi, S., and Giannini V. 2017. Quantum plasmonic nanoantennas. *Physical Review B* 95:235414.
3. He, S., Cui, Y., Ye, Y., Zhang, P., and Jin, Y. Optical nano-antennas and metamaterials. *Materials Today* 12:16–24.
4. Eggleston, M. S., Messer, K., Zhang, L., Yablonovitch, E., and Wu, M. C. 2015. Optical antenna enhanced spontaneous emission. *Proceedings of the National Academy of Sciences of the United States of America* 112:1704–1709.
5. Yrofymchuk, K., Reisch, A., Didier, P. et al. 2017. Giant light-harvesting nanoantenna for single-molecule detection in ambient light. *Nature Photonics* 11:657–663.
6. Caron, W. N. 1989. *Antenna Impedance Matching*. Newington, CT: American Radio Relay League.

7. Shinki, Y., Shibata, K., Mansour, M., and Kanaya, H. 2017. Impedance matching antenna-integrated high-efficiency energy harvesting circuit. *Sensors* 17:E1763.

8. Wang, Q., Liu, L., Wang, Y. et al. 2015. Tunable optical nanoantennas incorporating bowtie nanoantenna arrays with stimuli-responsive polymer. *Scientific Reports* 5:18567.

9. Dipalo, M., Messina, G. C. Amin, H. et al. 2015. 3D plasmonic nanoantennas integrated with MEA biosensors. *Nanoscale* 7:3703–3711.

10. Law, S., Yu, L., Rosenberg, A., and Wasserman, D. 2013. All-semiconductor plasmonic nanoantennas for infrared sensing. *Nano Letters* 13:4569–4574.

11. Hensenk, M., Heilpern, T., Gray, S. K., and Pfeiffer, W. 2018. Strong coupling and entanglement of quantum emitters embedded in a nanoantenna-enhanced plasmonic cavity. *ACS Photonics* 5:240–248.

12. Tong, L., Pakizeh, T., Feuz, L., and Dmitriev, A. 2013. Highly directional bottom-up 3D nanoantenna for visible light. *Scientific Reports* 3:2311.

13. Wissert, M. D., Schell, A. W., Ilin, K. S., Siegel, M., and Eisler, H.-J. 2009. Nanoengineering and characterization of gold dipole nanoantennas with enhanced integrated scattering properties. *Nanotechnology* 20:425203.

14. Pfeiffer, M., Lindfors, K., Zhang, H. et al. 2014. Eleven nanometer alignment precision of a plasmonic nanoantenna with a self-assembled GaAs quantum dot. *Nano Letters* 14:197–201.

15. Zhang, J., Irannejad, M., and Cui, B. 2015. Bowtie nanoantenna with single-digit nanometer gap for surface-enhanced Raman scattering (SERS). *Plasmonics* 10:831–837.

16. Sundaramurthy, A. and Kino, G. S. 2006. Toward nanometer-scale optical photolithography: Utilizing the near-field of bowtie optical nanoantennas. *Nano Letters* 6:355–360.

17. Novotny, L. and Van Hulst, N. 2011. Antennas for light. *Nature Photonics* 5:83–90.

18. Lee, H., Kim, C., and Kim, D., 2015. Sub-10 nm near-field localization by plasmonic metal nanoaperture arrays with ultrashort light pulses. *Scientific Reports* 5:17584.

19. Nien, L.-W., Lin, S.-C., Chao, B.-K. et al. 2013. Giant electric field enhancement and localized surface plasmon resonance by optimizing contour bowtie nanoantennas. *Journal of Physical Chemistry C* 117:25004–25011.

20. Ives, M., Autry, T. M., Cundiff, S. T. and Nardin, G. 2016. Direct imaging of surface plasmon polariton dispersion in gold and silver thin films. *Journal of the Optical Society of America B* 33:C17–C21.

21. Beane, G., Yu, K., Devkota, T. et al. 2017. Surface plasmon polariton interference in gold nanoplates. *Journal of Physical Chemistry Letters* 8:4935–4941.

22. Kuisma, M., Sakko, A., Rossi, T. P. et al. 2015. Localized surface plasmon resonance in silver nanoparticles: Atomistic first-principles time-dependent density-functional theory calculations. *Physical Review B* 91:115431.

23. Hira, T., Homma, T., Uchiyama, T., Kuwamura, K., and Saiki, T. 2013. Switching of localized surface plasmon resonance of gold nanoparticles on a GeSbTe film mediated by nanoscale phase change and modification of surface morphology. *Applied Physics Letters* 103:231101.

24. Guo, K., Lozano, G., Verschuuren, M. A., and Rivas, J. G. 2015. Control of the external photoluminescent quantum yield of emitters coupled to nanoantenna phased arrays. *Journal of Applied Physics* 118:073103.

25. Lin, L. and Zheng, Y. 2015. Optimizing plasmonic nanoantennas via coordinated multiple coupling. *Scientific Reports* 5:14788.

26. Wang, C.-M. and Feng, D.-Y. 2014. Omnidirectional thermal emitter based on plasmonic nanoantenna arrays. *Optics Express* 2:1313–1318.

27. Gurlek, G., Sandoghdar, V., and Martín-Cano, D. 2018. Manipulation of quenching in nanoantenna–emitter systems enabled by external detuned cavities: A path to enhance strong-coupling. *ACS Photonics* 5:456–461.

28. Lechago, S., García-Meca, C., Sánchez-Losilla, N., Griol, A., and Martí, J. 2018. High signal-to-noise ratio ultra-compact labon-a-chip microflow cytometer enabled by silicon optical antennas. *Optics Express* 26:25645–25656.
29. Yardimci, N. T. and Jarrahi, M. 2017. High sensitivity terahertz detection through large-area plasmonic nano-antenna arrays. *Scientific Reports* 7:42667.
30. Xi, Z. Wei, L. Adam, A. J. L. Urbach, H. P., and Du, L. 2016. Accurate feeding of nanoantenna by singular optics for nanoscale translational and rotational displacement sensing. *Physical Review Letters* 117:113903.
31. Feichtner, T., Selig, O., and Hecht, B. 2017. Plasmonic nanoantenna design and fabrication based on evolutionary optimization. *Optics Express* 25:10828–10842.
32. Ho, J., Fu, Y. H., Dong, Z. et al. 2018. Highly directive hybrid metal–dielectric Yagi-Uda nano-antennas. *ACS Nano* 12:8616–8624.
33. Sun, S., Li, R., Li, M. et al. 2018. Hybrid mushroom nanoantenna for fluorescence enhancement by matching the stokes shift of the emitter. *Journal of Physical Chemistry C* 122:14771–14780.
34. Lyamkina, A. A., Schraml, K., Regler, A. et al. 2016. Monolithically integrated single quantum dots coupled to bowtie nanoantennas. *Optics Express* 24:28936–28944.
35. Schmidt, F. P., Ditlbacher, H., Hofer, F., Krenn, J. R., and Hohenester, U. 2014. Morphing a plasmonic nanodisk into a nanotriangle. *Nano Letters* 14:4810–4815.
36. Raza, S., Kadkhodazadeh, S., Christensen, T. et al. 2015. Multipole plasmons and their disappearance in few-nanometre silver nanoparticles. *Nature Communications* 6:8788.
37. Davis, T. J., Vernon, K. C., and Gómez, D. E. 2009. Effect of retardation on localized surface plasmon resonances in a metallic nanorod. *Optics Express* 17:23655.
38. Urbieta, M., Barbry, M., Zhang, Y., Koval, P., Sánchez-Portal, D., Zabala, N., and Aizpurua, J. 2018. Atomic-scale lightning rod effect in plasmonic picocavities: A classical view to a quantum effect. *ACS Nano* 12:585–595.
39. Bellido, E. P., Zhang, Y., Manjavacas, A., Nordlander, P., and Botton, G. A. Plasmonic coupling of multipolar edge modes and the formation of gap modes. *ACS Photonics* 4:1558–1565.
40. Bharadwaj, P., Deutsch, B., and Novotny, L. 2009. Optical antennas. *Advances in Optics and Photonics* 1:438–483.
41. Krasnok, A. E., Maksymov, I. S., Denisyuk, A. I. et al. 2013. Optical nanoantennas. *Physics-Uspekhi* 56:539–564.
42. Curto, A. G., Volpe, G., Taminiau, T. H., Kreuzer, M. P., Quidant, R., and van Hulst, N. F. 2010. Unidirectional emission of a quantum dot coupled to a nanoantenna. *Science* 329:930–933.
43. Fu, Y. H., Kuznetsov, A. I., Miroshnichenko, A. E., Yu, Y. F., and Luk'yanchuk, B. 2013. Directional visible light scattering by silicon nanoparticles. *Nature Communications* 4:1527.
44. Purcell, E. M. 1995. Spontaneous emission probabilities at radio frequencies. *Physical Review* 69:681.
45. Akselrod, G. M., Argyropoulos, C., Hoang, T. B. et al. 2014. Probing the mechanisms of large Purcell enhancement in plasmonic nanoantennas. *Nature Photonics* 8:835–840.
46. Krasnok, A., Glybovski, S., Petrov, M. et al. 2016. Demonstration of the enhanced purcell factor in all-dielectric structures. *Applied Physics Letters* 108:211105.
47. Mohammadi, A., Sandoghdar, V., and Agio, M. 2008. Gold nanorods and nanospheroids for enhancing spontaneous emission. *New Journal of Physics* 10:105015.
48. Liaw, J. W., Chen, J. H., Chen, C. S., and Kuo, M. K. 2009. Purcell effect of nanoshell dimer on single molecule's fluorescence. *Optics Express* 17:13532–13540.
49. Greffet, J. J., Laroche, M., and Marquier, F. 2010. Impedance of a nanoantenna and a single quantum emitter. *Physical Review Letters* 105:117701.
50. Krasnok, A. E., Slobozhanyuk, A. P., Simovski, C. R. et al. 2015. An antenna model for the Purcell effect. *Scientific Reports* 5:12956.

51. Yu, B., Goodman, S., Abdelaziz, A., and O'Carroll, D. M. 2012. Light-management in ultra-thin polythiophene films using plasmonic monopole nanoantennas. *Applied Physics Letters* 101:151106.

52. Alu, A. and Engheta, N. 2008. Tuning the scattering response of optical nanoantennas with nanocircuit loads. *Nature Photonics* 2:307–310.

53. Yifat, Y., Eitan, M., Iluz, Z., Hanein, Y., Boag, A., and Scheuer, J. 2014. Highly efficient and broadband wide-angle holography using patch-dipole nanoantenna reflectarrays. *Nano Letters* 14:2485–2490.

54. Fromm, D. P., Sundaramurthy, A., Schuck, P. J., Kino, G., and Moerner, W. E. 2004. Gap-dependent optical coupling of single "bowtie" nanoantennas resonant in the visible. *Nano Letters* 4:957–961.

55. Stutzman, W. L. and Thiele, G. A. 2013. *Antenna Theory and Design*. Hoboken, NJ: John Wiley & Sons.

56. Jäckel, F., Kinkhabwala, A. A., and Moerner, W. E. 2007. Gold bowtie nanoantennas for surface-enhanced Raman scattering under controlled electrochemical potential. *Chemical Physics Letters* 446:339–343.

57. Fromm, D. P., Sundaramurthy, A., Kinkhabwala, A., Schuck, P. J., Kino, G. S., and Moerner, W. E. 2006. Exploring the chemical enhancement for surface-enhanced Raman scattering with Au bowtie nanoantennas. *Journal of Chemical Physics* 124:061101.

58. Kinkhabwala, A., Yu, Z., Fan, S., Avlasevich, Y., Müllen, K., and Moerner, W. E. 2009. Large single-molecule fluorescence enhancements produced by a bowtie nanoantenna. *Nature Photonics* 3:654–657.

59. Lu, G., Li, W., Zhang, T. et al 2012. Plasmonic-enhanced molecular fluorescence within isolatedbowtie nano-apertures. *ACS Nano* 6:1438–1448.

60. Taminiau, T. H., Moerland, R. J., Segerink, F. B., Kuipers, L., and van Hulst, N. F. 2007../4 resonance of an optical monopole antenna probed by single molecule fluorescence. *NanoLetters* 7:28–33.

61. Sabaawi, A. M., Tsimenidis, C. C., and Sharif, B. S. 2011. Infra-red nano-antennas for solarenergy collection. In *Antennas and Propagation Conference* 1–4.

62. Evlyukhin, A. B., Novikov, S. M., Zywietz, U. et al. 2012. Demonstration of magnetic dipole resonances of dielectric nanospheres in the visible region. *Nano Letters* 12:3749–3755.

63. Staude, I., Miroshnichenko, A. E., Decker, M. et al. 2013. Tailoring directional scattering through magnetic and electric resonances in subwavelength silicon nanodisks. *ACS Nano* 7:7824–7832.

64. Cao, L., Fan, P., Barnard, E. S., Brown, A. M., and Brongersma, M. L. 2010. Tuning the color of silicon nanostructures. *Nano Letters* 10:2649–2654.

65. Seo, K., Wober, M., Steinvurzel, P. et al. 2011. Multicolored vertical silicon nanowires. *Nano Letters* 11:1851–1856.

66. Markov, G. T. and Sazonov, D. M. 1975. *Antennas*. Moscow, Russia: Izdatel Energiia.

67. Filter, R., Mühlig, S., Eichelkraut, T., Rockstuhl, C., and Lederer, F. 2012. Controlling the dynamics of quantum mechanical systems sustaining dipole-forbidden transitions via optical nanoantennas. *Physical Review B* 86:035404.

68. Bakker, R. M., Permyakov, D., Yu, Y. F. et al. 2015. Magnetic and electric hotspots with silicon nanodimers. *Nano Letters* 15:2137–2142.

69. Kosako, T., Kadoya, Y., and Hofmann, H. F. 2010. Directional control of light by a nano-optical Yagi–Uda antenna. *Nature Photonics* 4:312–315.

70. Jain, P. K. and El-Sayed, M. A. 2010. Plasmonic coupling in noble metal nanostructures. *Chemical Physics Letters* 487:153–164.

71. Li, J., Gu, Y., Zhou, F., Li, Z. Y., and Gong, Q. 2009. A designer approach to plasmonic nanostructures: Tuning their resonance from visible to near-infrared. *Journal of Modern Optics* 56:1396–1402.

72. Najafabadi, A. F. and Pakizeh, T. 2017. Analytical chiroptics of 2D and 3D nanoantennas. *ACS Photonics* 4:1447–1452.

73. Hofmann, H. F., Kosako, T., and Kadoya, Y. 2007. Design parameters for a nano-optical Yagi–Uda antenna. *New Journal of Physics* 9:217.

74. De Arquer, F. P. G., Volski, V., Verellen, N., Vandenbosch, G. A., and Moshchalkov, V. V. 2011. Engineering the input impedance of optical nano dipole antennas: materials, geometry and excitation effect. *IEEE Transactions on Antennas and Propagation* 59:3144–153.

75. Dorfmuller, J., Dregely, D., Esslinger, M. et al. 2011. Near-field dynamics of optical Yagi-Uda nanoantennas. *Nano Letters* 11:2819–2824.

76. Novotny, L. 2007. Effective wavelength scaling for optical antennas. *Physical Review Letters* 98:266802.

77. Li, J., Salandrino, A., and Engheta, N. 2007. Shaping light beams in the nanometer scale: A Yagi-Uda nanoantenna in the optical domain. *Physical Review B* 76:245403.

78. Dregely, D., Taubert, R., Dorfmüller, J., Vogelgesang, R., Kern, K., and Giessen, H. 2011. 3D optical Yagi–Uda nanoantenna array. *Nature Communications* 2:267.

79. Evlyukhin, A. B., Reinhardt, C., Seidel, A., Luk'yanchuk, B. S., and Chichkov, B. N. 2010. Optical response features of Si-nanoparticle arrays. *Physical Review B* 82:045404.

80. Krasnok, A. E., Miroshnichenko, A. E., Belov, P. A., and Kivshar, Y. S. 2012. All-dielectric optical nanoantennas. *Optics Express* 20:20599–20604.

81. Krasnok, A. E., Simovski, C. R., Belov, P. A., and Kivshar, Y. S. 2014. Superdirective dielectric nanoantennas. *Nanoscale* 6:7354–7361.

82. Hu, S., Khater, M., Salas-Montiel, R., Kratschmer, E., Engelmann, S., Green, W. M., and Weiss, S. M. 2018. Experimental realization of deep-subwavelength confinement in dielectric optical resonators. *Science Advances* 4:eaat2355.

83. Kim, K., Yajima, J., Oh, Y. et al. 2012. Nanoscale localization sampling based on nanoantenna arrays for super-resolution imaging of fluorescent monomers on sliding microtubules. *Small* 8:892–900.

84. Kim, K., Oh, Y., Lee, W., and Kim, D. 2010. Plasmonics-based spatially activated light microscopy for super-resolution imaging of molecular fluorescence. *Optics Letters* 35:3501–3503.

85. Son, T., Moon, G., Lee, H., and Kim, D. 2018. Metallic 3D random nanocomposite islands for near-field spatial light switching. *Advanced Optical Materials* 6:1701219.

86. Lee, C., Sim, E., and Kim, D. 2017. Effect of nanogap-based light-matter colocalization on the surface plasmon resonance detection. *Journal of Lightwave Technology* 35:4721–4727.

87. Oh, Y., Lee, W., Kim, Y., and Kim, D. 2014. Self-aligned colocalization of 3D plasmonic nanogap arrays for ultra-sensitive surface plasmon resonance detection. *Biosensors and Bioelectronics* 51: 401–407.

88. Lim, D. K., Jeon, K. S., Kim, H. M., Nam, J. M., and Suh, Y. D. 2010. Nanogap-engineerable Raman-active nanodumbbells for single-molecule detection. *Nature Materials* 9:60–67.

89. Chen, X., Park, H. R., Pelton, M. et al. 2013. Atomic layer lithography of wafer-scale nanogap arrays for extreme confinement of electromagnetic waves. *Nature Communications* 4:2361.

90. Yoo, D., Nguyen, N.C., Martin-Moreno, L. et al. 2016. High-throughput fabrication of resonant metamaterials with ultrasmall coaxial apertures via atomic layer lithography. *Nano Letters* 16:2040–2046.

91. Pallaoro, A., Hoonejani, M. R., Braun, G. B., Meinhart, C. D., and Moskovits, M. 2015. Rapid identification by surface-enhanced Raman spectroscopy of cancer cells at low concentrations flowing in a microfluidic channel. *ACS Nano* 9:4328–4336.

92. Lee, S., Chon, H., Lee, J. et al. 2014. Rapid and sensitive phenotypic marker detection on breast cancer cells using surface-enhanced Raman scattering (SERS) imaging. *Biosensors and Bioelectronics* 51:238–243.

93. Harmsen, S., Huang, R., Wall, M. A. et al. 2015. Surface-enhanced resonance Raman scattering nanostars for high-precision cancer imaging. *Science Translational Medicine* 7:271ra7.

94. Cho, I. H., Bhandari, P., Patel, P., and Irudayaraj, J. 2015. Membrane filter-assisted surface enhanced Raman spectroscopy for the rapid detection of *E. coli* O157: H7 in ground beef. *Biosensors and Bioelectronics* 64:171–176.

95. Choi, J. R., Kim, K., Oh, Y. et al. 2014. Extraordinary transmission-based plasmonic nanoarrays for axially super-resolved cell imaging. *Advanced Optical Materials* 2:48–55.

96. Cang, H., Liu, Y., Wang, Y., Yin, X., and Zhang, X. 2013. Giant suppression of photobleaching for single molecule detection via the purcell effect. *Nano Letters* 13:5949–5953.

97. Muskens, O. L., Giannini, V., Sanchez-Gil, J. A., and Gomez Rivas, J. 2007. Strong enhancement of the radiative decay rate of emitters by single plasmonic nanoantennas. *Nano Letters* 7:2871–2875.

98. Yuan, H., Khatua, S., Zijlstra, P., Yorulmaz, M., and Orrit, M. 2013. Thousand-fold enhancement of single-molecule fluorescence near a single gold nanorod. *Angewandte Chemie International Edition* 52:1217–1221.

99. Zhang, J., Fu, Y., Chowdhury, M. H., and Lakowicz, J. R. 2007. Metal-enhanced single-molecule fluorescence on silver particle monomer and dimer: Coupling effect between metal particles. *Nano Letters* 7:2101–2107.

100. Lee, W., Kinosita, Y., Oh, Y. et al. 2015. Three-dimensional superlocalization imaging of gliding Mycoplasma mobile by extraordinary light transmission through arrayed nanoholes. *ACS Nano* 9:10896–10908.

101. Chung, K., Lee, J., Lee, J. E. et al. 2013. A simple and efficient strategy for the sensitivity enhancement of DNA hybridization based on the coupling between propagating and localized surface plasmons. *Sensors and Actuators B: Chemical* 176:1074–1080.

102. Oh, Y., Son, T., Kim, S. et al. 2014. Surface plasmon-enhanced nanoscopy of intracellular cytoskeletal actin filaments using random nanodot arrays. *Optics Express* 22:27695–27706.

103. Bezryadina, A., Zhao, J., Xia, Y., Zhang, X., and Liu, Z. 2018. High spatiotemporal resolution imaging with localized plasmonic structured illumination microscopy. *ACS Nano* 12:8248–8254.

104. Kühn, S., Håkanson, U., Rogobete, L., and Sandoghdar, V. 2006. Enhancement of single-molecule fluorescence using a gold nanoparticle as an optical nanoantenna. *Physical Review Letters* 97:017402.

105. Punj, D., Mivelle, M., Moparthi, S. B. et al. 2013. A plasmonic "antenna-in-box" platform for enhanced single-molecule analysis at micromolar concentrations. *Nature Nanotechnology* 8:512–516.

106. Taminiau, T. H., Stefani, F. D., Segerink, F. B., and Van Hulst, N. F. 2008. Optical antennas direct single-molecule emission. *Nature Photonics* 2:234–237.

107. Vukojević, V., Heidkamp, M., Ming, Y., Johansson, B., Terenius, L., and Rigler, R. 2008. Quantitative single-molecule imaging by confocal laser scanning microscopy. *Proceedings of the National Academy of Sciences* 13:0809250105.

108. Hui, Y. Y., Zhang, B., Chang, Y. C. et al. 2010. Two-photon fluorescence correlation spectroscopy of lipid-encapsulated fluorescent nanodiamonds in living cells. *Optics Express* 18:5896–5905.

109. Wohland, T., Shi, X., Sankaran, J., and Stelzer, E. H. 2010. Single plane illumination fluorescence correlation spectroscopy (SPIM-FCS) probes inhomogeneous three-dimensional environments. *Optics Express* 18:10627–10641.

110. Hassler, K., Leutenegger, M., Rigler, P. et al. 2005. Total internal reflection fluorescence correlation spectroscopy (TIR-FCS) with low background and high count-rate per molecule. *Optics Express* 13:7415–7423.

111. Edel, J. B., Wu, M., Baird, B., and Craighead, H. G. 2005. High spatial resolution observation of single-molecule dynamics in living cell membranes. *Biophysical Journal* 88:L43–L45.

112. Block, S., Aćimović, S. S., Odebo Länk, N., Käll, M., and Höök, F. 2018. Antenna-enhanced fluorescence correlation spectroscopy resolves calcium-mediated lipid–lipid interactions. *ACS Nano* 12:3272–3279.

113. Langguth, L., Szuba, A., Mann, S. A., Garnett, E. C., Koenderink, G. H., and Koenderink, A. F. 2017. Nano-antenna enhanced two-focus fluorescence correlation spectroscopy. *Scientific Reports* 7:5985.

114. Kelly, C. V., Wakefield, D. L., Holowka, D. A., Craighead, H. G., and Baird, B. A. 2014. Near-field fluorescence cross-correlation spectroscopy on planar membranes. *ACS Nano* 8:7392–7404.

115. Regmi, R., Winkler, P. M., Flauraud, V. et al 2017. Planar optical nanoantennas resolve cholesterol-dependent nanoscale heterogeneities in the plasma membrane of living cells. *Nano Letters* 17:6295–6302.

116. Winkler, P. M., Regmi, R., Flauraud, V. et al. 2017. Transient nanoscopic phase separation in biological lipid membranes resolved by planar plasmonic antennas. *ACS Nano* 11:7241–7250.

117. Serov, A., Rao, R., Gösch, M. et al. 2004. High light field confinement for fluorescent correlation spectroscopy using a solid immersion lens. *Biosensors and Bioelectronics* 20:431–435.

118. Regmi, R., Al Balushi, A. A., Rigneault, H., Gordon, R., and Wenger, J. 2015. Nanoscale volume confinement and fluorescence enhancement with double nanohole aperture. *Scientific Reports* 5:15852.

119. Winkler, P. M., Regmi, R., Flauraud, V. et al. 2017. Optical antenna-based fluorescence correlation spectroscopy to probe the nanoscale dynamics of biological membranes. *The Journal of Physical Chemistry Letters* 9:110–119.

120. He, J., Huang, X., Li, Y.-C. et al. 2013. Self-assembly of amphiphilic plasmonic micelle-like nanoparticles in selective solvents. *Journal of the American Chemical Society* 135:7974–7984.

121. Choi, W. I., Kim, J. Y., Kang, C., Byeon, C. C., Kim, Y. H. and Tae, G. 2011. Tumor regression in vivo by photothermal therapy based on gold-nanorod-loaded, functional nanocarriers. *ACS Nano* 5:1995–2003.

122. Thollar, Z., Wadell, C., Matsukata, T., Yamamoto, N., and Sannomiya, T. 2017. Three-dimensional multipole rotation in spherical silver nanoparticles observed by cathodoluminescence. *ACS Photonics* 5(7):2555–2560.

123. Habteyes, T. G., Dhuey, S., Kiesow, K. I., and Vold, A. 2013. Probe-sample optical interaction: Size and wavelength dependency in localized plasmon near-field imaging. *Optics Express* 21:21607–21617.

124. Dinish, U. S., Balasundaram, G., Chang, Y. T., and Olivo, M. 2014. Actively targeted in vivo multiplex detection of intrinsic cancer biomarkers using biocompatible SERS nanotags. *Scientific Reports* 4:4075.

125. Puchkova, A., Vietz, C., Pibiri, E. et al, 2015. DNA origami nanoantennas with over 5000-fold fluorescence enhancement and single-molecule detection at 25 μM. *Nano Letters* 15:8354–8359.

CHAPTER 14

Basis of Pharmaceutical Formulation

Shahnaz Usman, Karyman Ahmed Fawzy,
Rayisa Beevi, and Anab Usman

CONTENTS

14.1 INTRODUCTION

An appropriate pharmaceutical formulation enables optimal performance of active pharmaceutical ingredients (API). In practical sciences, we design delivery systems and excipients in such a way that they work as a unit. From dry powders and multiphasic emulsions to encapsulation systems, all serve the same purpose, that is, getting the right amount of API to the right place at the right time.

Drug product formulation is the core of practical sciences. Delivery of APIs from their products has become the focus of many studies lately. Initially, industries used to focus on new chemical entities (NCEs) for drug-delivery systems, but now, they consider a "go-to approach" strategy (GTM strategy) for both NCE commercialization and modification and rechannelization of the existing APIs. The APIs biologically fall into one or a combination of four numbers of categories [1]:

1. *Psychopharmacological agents*: Compounds acting on the central nervous system (CNS) that comprises the brain and the spinal cord and is responsible for the control of thought processes, emotions, senses, and motor functions. Antidepressants and antipsychotics are the examples of this category.

2. *Pharmacodynamics agents*: Compounds that interact with the normal dynamic processes of the body, for example, circulation of blood. Antianginals and vasodilators are the common example of this class.

3. *Chemotherapeutic agents*: Compounds that are selectively more toxic for disease-causing microorganisms than for the host. The main examples of this category are antibiotics and antiviral.

4. *Agents acting on metabolic diseases and endocrine function*: This group includes drugs for the treatment of diabetes, arthritis, inflammation, atherosclerosis, and hormones.

APIs based on their sources can be divided into two major categories [1]:

1. *Chemical synthetic drugs*: They are further divided into organic synthetic drugs and inorganic synthetic drugs. (i) *Inorganic synthetic drugs* are based on inorganic compounds (10%). For example: Aluminum hydroxide and magnesium trisilicate are used for the treatment of gastric and duodenal ulcers. (ii) *Organic synthetic drugs* mainly comprise drugs prepared by basic organic chemical raw materials. They are obtained by a series of organic chemical reactions. Aspirin, chloramphenicol, and caffeine are the good examples.

2. *Natural chemical drugs*: Based on their sources, they are divided into two subcategories: (i) *biochemical drugs*, which include antibiotics obtained from microbial fermentation, and (ii) *plant chemical drugs*, which comprise a large number of compounds obtained from plants by extraction of their different parts. Quinine and aspirin used for the treatment of diseases are the examples of drugs obtained from plants. Medicines derived from plants and fungi are mostly used for their mind-altering or psychotropic properties [2]. On the other hand, a large number of semi-synthetic antibiotics are also available, which are obtained by the combined processes of biosynthesis and chemical synthesis.

On the basis of marketing, the medicines are segmented into proprietary (innovative) and generic drugs. A pharmaceutical company **is responsible for** exposing and developing a new chemical moiety as a therapeutic agent. After discovery, the company submits a file for getting patent to reserve their right for making and selling the drugs. Tylenol® is an example of a brand name that is used for the relief of pain, whereas its generic name is acetaminophen [3].

1. As per the World Health Organization (WHO), **generic medicine** is a drug that is used in place of an innovator product that can be manufactured by any pharmaceutical company other than innovator company after the expiry date of its patent [4].

The major emphasis of pharmaceutical companies is to produce safe and effective drug-dosage form, so that the general public could be able to take their medicines in an appropriate and accurate dose. Practically, it was believed that the pharmaceutical formulation is a data-ambitious science, and systematic proficient approach was used to prepare it. One of the most surprising characteristics of drugs is the ability to produce their actions and effects on the body in different assortments. Owing to this quality, the drugs can be used in the treatment of various conditions in different selective ways [5].

Thus, different types of dosage forms show their inimitable physical and pharmaceutical characteristics. These variations create challenges in formulation for manufacturing pharmacists. In this regard, the basic focusing areas of study are the design and development of formulation and the manufacturing, stability, and effectiveness of pharmaceutical dosage forms. In the appropriate fabrication of pharmaceutical products, the physical, chemical, and biologic properties of all APIs and excipients play a vital role. They must be well-matched with one another to produce a drug product that is stable, effective, good-looking, easy to use, and safe. Thus, the preparation of pharmaceutical dosage form should be appropriate with all the quality control parameters, and packaging containers should be of good qualities to help maintain the product stability. The containers should then be labeled correctly and appropriately and should be stored in a suitable condition to uphold the product for its utmost shelf-life [5].

14.2 DRUG-DOSAGE FORMS AND DELIVERY SYSTEMS

A drug dosage is often referred to as "pharmaceutical," and it is defined as a form or means by which the drug substances are delivered to the site of action to produce the desired therapeutic effect. Crude drug and drug substances are developed into dosage forms in order to optimize drug safety, stability, and effectiveness [5].

14.2.1 Importance of Developing Drug-Dosage Forms

Dosage forms help in delivering the prescribed or measured amount of drug to the site of action within the body. Hence, the development of dosage forms helps in administering an accurate dose of potent and low-dosage drugs to patients. Drug-dosage forms help in protecting the drug substance from atmospheric conditions such as humidity and oxygen (e.g., formulating coated tablets or packing the drug substance in sealed ampoules). Also, preparing dosage forms like enteric coated tablets protect the drug substances from the destructive effects of gastric acid after oral administration. Dosage forms such as coated tablets, capsules, and flavored syrups help in masking the unpleasant taste and odor of drugs and hence in increasing the patient palatability. Dosage forms such as suspensions are needed to provide liquid preparations of drugs that are unstable or insoluble in the desired vehicle. Also, dosage forms are needed to provide controlled and sustained release medications, for example, controlled-release suspensions, capsules, and tablets. The preparation of topical dosage forms such as creams, ointments, and transdermal patches, as well as ear, ophthalmic, and nasal preparations helps in providing optimal topical drug action. The availability of various types of dosage forms facilitates the placement of the drug substance in the desired part of the body. For example, injections are used to place the drug directly into the bloodstream, whereas vaginal or rectal suppositories are inserted into body cavities to provide a local or systemic effect. Also, the development of dosage forms such as inhalation aerosols and inhalants helps in providing optimal drug action through inhalation therapy [6].

14.3 DRUG-DELIVERY SYSTEMS

Drug delivery is defined as a process or method by which the drug substance is delivered or administered into the body in order to achieve a desired therapeutic effect. The term *drug-delivery system* may refer to a device that is used to deliver the drug substances, for example, nebulizers, syringes, teaspoon, and intravenous (IV) fluid infusion pumps. It can also refer to a certain design feature of a dosage form that can affect drug delivery, such as enteric coating. Enteric coating affects the release of medication from the dosage form, as it resists the breakdown by gastric fluids and releases the medication only in the intestine. Drug-delivery systems differ in their rate of delivery, site of action, and amount of active ingredients delivered. For instance, the drug nitroglycerin has three common delivery systems, which are sublingual tablets (placed under the tongue), ointment (applied directly on the skin surface), and transdermal patches. The sublingual tablets show the fastest onset of action but deliver the API for short period of time, that is, 30 minutes. On the other hand, transdermal patches and ointments act slowly, with longer duration of action, that is, 24 and 12 hours, respectively [7].

14.3.1 Types of Dosage Forms

Pharmaceutical formulations or dosage forms can be classified according to the route of administration or according to the physical forms. The physical forms of formulations help to predict the release pattern of drug molecules to the sites of action within the body. According to the physical properties the pharmaceutical dosage forms are classified as the following: [8]

Gaseous dosage forms

- Medicinal gases

- Aero dispersions

Liquid dosage forms

Semisolid dosage forms

- Unshaped (without specific physical shape)

- Shaped

Solid dosage forms

14.3.2 Route of Administration

Route of administration can be defined as the path by which a drug substance is introduced into the body. Routes of administration are broadly divided into enteral routes (oral, sublingual, buccal, and rectal), parenteral injections (IV, intramuscular [IM], and subcutaneous [SC]), topical, inhalation, vaginal, ophthalmic, otic, and intranasal routes of administration. Special routes such as intra-arterial routes of administration in cancer chemotherapy and intrathecal routes in spinal anesthesia or CNS infections provide better therapeutic outcomes to patients [9]. The choice of the route of administration depends mainly on two factors: drug properties and therapeutic concerns. Drug properties refer to the physicochemical characteristics of the drug (e.g., molecular size, lipid solubility, and ionization status), as well as the plasma concentration. Therapeutic concerns include the desired onset of action, duration of drug action, site of action, and patient compliance. In cases of emergency or acute conditions, the route of administration must provide fast onset of action by allowing fast absorption of the drug substance [9]. Parenteral injections such as IV, IM, and SC injections are often the route of choice in emergency conditions. In addition, other routes of administration can be considered in acute conditions; for example, inhalation route is used to deliver bronchodilators in cases of acute asthma attack, and sublingual route is used to deliver nitroglycerin in cases of acute angina attack. Also, the rectal route is preferred in young children, especially if the patient is continuously vomiting or is unconscious. On the contrary, the onset of action is not of much concern when treating chronic illness, whereas the choice of a convenient route of administration and simple dosage regimen (e.g., once daily or weekly) is highly considered. The commonly used routes of administration in the treatment of chronic conditions are oral, topical, and transdermal routes. This, in turn, provides better patient compliance, especially if a controlled- or sustained-release formulation is used [9].

14.3.3 Solid Dosage Forms

14.3.3.1 Tablets

Tablets are defined as a unit dose solid preparation containing one or more active ingredients, with or without excipients, prepared by molding or compression. Tablets are the most commonly used dosage forms for drug administration. Tablets differ in shape, size, weight, thickness, hardness, dissolution, and disintegration characteristics depending on their method of manufacture and the intended use. The excipients used in tablet preparation are diluents or fillers, binders, disintegrating agents, lubricants, glidants, antiadherents, as well as flavors and coloring agents [10]:

- *Buccal and sublingual tablets*: Sublingual tablets are placed under the tongue to be absorbed by the oral mucosa. They are relatively small in size and dissolve promptly and rapidly. After absorption, the drug enters directly into the systemic circulation, by passing the portal circulation and hence avoiding first-pass metabolism. Buccal tablets are small oval or flat tablets that are inserted into the buccal cavity, where they are eroded or dissolved slowly. Drugs that are destroyed by gastric fluids or subjected to substantial presystemic metabolism can be formulated in such dosage forms [10].

- *Chewable tablets*: Chewable tablets are chewed before swallowing. Chewable tablets are made up of a creamy base of a specially colored and flavored mannitol. Chewable tablets are useful in administering large tablets to children and adults who have difficulty in swallowing [10].

- *Effervescent tablets*: Effervescent tablets are uncoated tablets prepared by compressing the effervescent granular salts, which release gas when dissolved in water. Effervescent tablets are made up of organic acids and sodium bicarbonate, along with the drug substance. The acid and the base react in the presence of water, liberating carbon dioxide gas, thus producing effervescence [10].

14.3.3.2 Capsules

Capsules are solid dosage forms in which the drug substance along with excipients are enclosed in a small shell of hard or soft gelatin. Gelatin is a protein substance obtained by partial hydrolysis of collagen present in animal bones, skin, and white connective tissues. Hard and soft gelatin capsules can be opaque, semi-transparent, transparent, colored, or imprinted with an identifying mark. Capsules are intended to mask the smell and unpleasant taste of the API. Some patients prefer capsules over tablets, as they are tasteless and easily swallowed. The most common excipients used in capsules are disintegrating agents, preservatives, solubilizers, and coloring agents. Flavoring agents are not commonly used in this dosage form, since the ingredients are enclosed in a gelatin shell [10,11]:

- *Hard gelatin capsules*: Hard gelatin capsules are also known as the dry-filled capsules, since they are used for dry powdered ingredients. Hard gelatin capsules are made up of two sections: capsule body and shorter cap. The cap is slipped over the open end of the capsule body, thus protecting the integrity of the drug formulation. The empty shells of hard gelatin capsules are made up of gelatin, water, and sugar. They can be clear, colorless, or colored using drug and cosmetic (D&C) or food, drug and cosmetic (FD&C) dyes. Also, titanium dioxide may be used to make the hard gelatin capsules opaque [10,11].

- *Soft gelatin capsules*: Soft gelatin capsules may be oval, round, or oblong in shape. The gelatin used in preparing soft gelatin capsules is made of polyhydric alcohol such as sorbitol and glycerin. The moisture content of soft gelatin capsules is relatively higher than that of hard gelatin capsules; thus, preservatives such as propylparaben and methylparaben may be incorporated to retard the microbial growth. Soft gelatin capsules are used to hermetically seal or encapsulate liquids, suspensions, pasty materials, or dry powders. For instance, oil-soluble drugs such as vitamins A and E are administered in soft gelatin capsules [10,11].

- *Enteric-coated capsules*: Enteric coated capsules resist breakdown by gastric fluids; instead, they break down in the alkaline medium of the intestine. This dosage form is used for drugs that are destroyed by gastric fluids or that may cause gastric irritation [10,11].

14.3.3.3 Lozenges or Pastilles

Lozenges are in a solid dosage form and are made up of a sugar base and gum. Lozenges contain the APIs, along with flavoring agents, and are intended to be dissolved slowly in the mouth. The gum provides strength and cohesiveness to the preparation, as well as facilitates the slow release of API from the preparation. Lozenges are most likely used to produce local effects; for example, they are used to relieve sore throat and treat fungal infection of mouth and throat. However, some lozenges may also be used to produce systemic effect. Pastilles are solid preparations intended to be dissolved slowly in the mouth, but they are softer than lozenges. The most common bases used in the preparation of pastilles are gelatin, glycerol, sugar, and acacia [12].

14.3.3.4 Pills

Pills are spherical masses made up of one or more active ingredients, along with inert excipients, and they are coated with gelatin, talc, or sugar. Pills are rarely prescribed and used today [5].

14.3.3.5 Implants or Pellets

Implants or pellets are solid dosage forms intended to be inserted under the skin by means of a minor surgery in order to provide continuous, prolonged, and controlled release of the medication. For example, Norplant is an implant used to prevent pregnancy, as it contains levonorgestrel. The use of such dosage form improves patient compliance but results in complications at the site of insertion; thus, they are not used widely [5].

14.3.3.6 Powders

Powders are one of the oldest dosage forms used in history. Powders can be defined as dry, finely divided particles of drugs or chemicals that can be used internally or externally. The main advantages of using powders are good chemical stability and flexibility in compounding. Powders also possess some disadvantages, for example, inaccuracy of dosing, especially with bulk powder, and the unsuitability of dispensing powders of unpleasant taste and hygroscopic drug substances. Powders can be classified according to division into prescribed dose, that is, bulk and divided powders, or according to composition, that is, simple and complex powders [5]:

- *Bulk and divided powders*: Medicated powders may be dispensed as bulk powders or divided into unit doses. Bulk powders are intended for nonpotent drugs that are required in large doses. Bulk powder is divided into oral powder (e.g., antacids, laxatives, and dietary nutrient supplements), dusting powder, dentifrices, and insufflations. Dusting powders are nonirritating preparations applied externally on the skin surface. Commercial dusting powders are available in either pressure aerosols or sifter-top containers. Insufflations refer to powders introduced into body cavities (e.g., nose, ears, and throat) by using a device known as insufflator. Divided powder is divided into unit doses, depending on the amount to be taken each time. Divided powder may be prepared using block-and-divide method or by weighing it individually (used for potent drugs) [5].

- *Simple and complex powders*: Simple powder contains only one active ingredient along with excipients, whereas complex powder contains more than one active ingredient [5].

14.3.3.7 Granules

Granules are solid, dry agglomerates of powders that can be used directly for their medicinal value, or they may be used in making tablets or filling capsules. Granules are prepared by adding a suitable amount of liquid binder to powder, forming a damp mass, which is then passed through a screen to form granule. Granules have irregular shapes; thus, they have better flow properties and are more stable than powders. Granules are commonly supplied as single-dose sachets and are more suitable for preparing solutions of drugs as compared with powders, because they are less likely to float on the liquid surface. Granules may be placed on the tongue and swallowed with water, or they may be dissolved in water before taking or dispensing, such as antibiotic suspensions. Effervescent granules contain citric acid or tartaric acid, and bicarbonate along with the medicinal agent which release carbon dioxide when dissolved in water. The carbon dioxide released helps in masking the unpleasant taste, as well as providing better absorption of the drug [5].

14.3.4 Liquid Dosage Form

Liquid dosage forms are essential pharmaceutical preparations made up of a mixture of medicinal agents along with excipients in a liquid medium. Liquid dosage forms are formulated by one of the following methods:

- Dissolving the active ingredients in aqueous or nonaqueous vehicles
- Suspending drug particles in an appropriate liquid medium
- Dispersing undissolved drug substance in two-phase system, that is, oil and water phases

Liquid dosage forms can be intended for oral use, injected, applied externally, or introduced into body cavities. The advantages of oral liquid dosage forms over solid dosage forms include ease of swallowing for youngsters and elderly, flexibility of dosing, and faster rate of absorption. Disadvantages are also associated with the use of oral liquid dosage forms, for example, shorter shelf-life as compared with other dosage forms, bulky, difficulty to measure an accurate dose, and requirement of special storage conditions. Liquid dosage forms are less stable and at greater risk of microbial contamination than solid dosage forms, and as a result, they require proper preservation [13].

14.3.4.1 Solutions

Solutions are defined as one-phase, homogenous, and transparent systems prepared by dissolving one or more therapeutic agents in a suitable vehicle or a mixture of miscible vehicles. On the basis of the vehicle used, solutions can be classified as aqueous solutions (e.g., gargles, mouthwashes, nasal washes, douches, and enemas), nonaqueous solutions (e.g., elixirs, spirits, and collodions), and viscid or sweet solutions (e.g., syrups and honeys). Also, on the basis of pharmaceutical use, solutions can be classified as oral, ophthalmic, otic, and topical solutions. According to the contents, solutions may be classified as syrups, elixirs, aromatic water, extracts, fluid extracts, spirits, tinctures, and irrigating solutions [13,14].

- *Syrups*: Syrups are aqueous preparations containing a large amount of dissolved sugar such as sucrose or sugar substitutes such as propylene glycol and sorbitol. Syrups may be classified as simple syrups, medicated syrups, and flavored syrups. Simple syrups refer to aqueous solution in which water alone is used to prepare the syrup. Medicated syrups refer to syrups that contain one or more medicinal agents, whereas flavored syrups do not contain any therapeutic agents, only flavoring agents. Hence, they are used as flavored or nonmedicated vehicles. Flavored vehicles are used in preparing medicated syrups or extemporaneous compounding of prescriptions. The excipients used in preparing syrups include purified water, sugar or sugar substitutes, flavoring agents, antimicrobial agents, coloring agents, solubilizing agents, stabilizers, and thickening agents. Syrups are particularly effective in administering pediatric medications, as they do not contain alcohol and are thus preferred in masking the unpleasant taste of medications. In addition, syrups are most commonly used for elderly patients who have difficulty in swallowing the commercially available solid preparations of the same drug [14].

- *Elixirs*: Elixirs are clear, sweetened, and usually flavored hydroalcoholic aqueous solutions intended for oral use. Medicated elixirs are used for their therapeutic effect, whereas nonmedicated elixirs are used as vehicles. Elixirs are less sweet and less viscous than syrups, because they contain lesser portion of sugar as compared with syrups. As a result, elixirs are less effective in masking the unpleasant taste of the active ingredients. The proportion of water to ethanol in elixirs varies widely depending on the solubility characteristics of the components in both water and ethanol. For instance, to prepare an elixir containing poorly-water-soluble ingredients, the amount of alcohol added would be greater than that in the elixirs containing ingredients having good water solubility. The main ingredients used in preparing elixirs are alcohol, water, flavoring agents, coloring agents, sweeteners, and adjunctive solvents. The most frequently used adjunctive solvents are propylene glycol and sorbitol. The most commonly used sweeteners in the preparation of elixirs are sucrose or sucrose syrup, glycerin, sorbitol, and artificial sweeteners such as saccharin. Saccharin is usually used to sweeten elixirs containing high content of alcohol, because it is more alcohol soluble as compared with sucrose. Also, it is required in smaller quantities to produce the same sweetness as sucrose. Flavoring agents are used to enhance the patient palatability, whereas coloring agents are added to improve the appearance. Antimicrobial agents are not used if the elixir contains 10% to 12% of alcohol, that is, self-preserving elixir. The most important limitation of elixirs is their alcohol content, especially for children and adult patients who prefer to avoid alcohol. It is very important to store elixirs in tight, light-resistant container protected from sunlight and excessive heat, owing to their alcohol content [14].

- *Aromatic water*: Aromatic water is a clear solution of water, containing large amount of volatile oils and other volatile and aromatic substances, which have pungent and pleasing smell. The volatile substances used in the preparation of aromatic water include chloroform, camphor, peppermint oil, anise oil, rose oil, orange flower oil, winter green oil, and spearmint oil. Aromatic water is used in flavoring and perfuming [5].

- *Fluidextracts*: Fluidextracts are liquid solutions prepared from extractives of vegetable drugs by percolation (i.e., extraction process). Fluidextracts are considered too potent, and hence, they are not safe to be self-administered. Fluidextracts have bitter taste, which makes them unpalatable, and therefore, sweeteners or flavoring agents are added before use, or they are used as drug source in formulating liquid dosage forms such as syrups [5].

- *Extracts*: Extracts are potent and concentrated dosage forms obtained by extracting animal and vegetable drugs by using a suitable solvent. The extract contains mainly the active constituents of the crude drug, since most of the inactive constituents have been removed during the extraction process. The main advantage of extracts is that it provides the medicinal activity of the bulky plant material in small quantities and in a suitable form that is physically stable. The most commonly used method of obtaining extracts is percolation. The extract is obtained by evaporating the solvent, and the extent of evaporation of the solvent determines the physical form of the extract. For instance, the extract may be semi-liquid, solid, or pilular extract and powdered extract; each physical form has its own use in pharmaceutical formulation and compounding [5].

- *Spirits*: Spirits are alcoholic or hydroalcoholic solutions containing aromatic or volatile substances. The alcohol concentration of spirits is usually 60% or higher. Consequently, the concentration of the aromatic and volatile substances in spirit is more than their corresponding aromatic water, as they are more soluble in alcohol than in water. A milky solution is formed when spirits are mixed with aqueous preparations or water owing to the separation of the volatile substances from the solution. Nonmedicated spirits, also known as flavoring spirits, are used as flavoring agents to impart their flavor in different pharmaceutical formulations such as cardamom spirit and orange spirit. Medicated spirits may be taken orally, inhaled, or applied externally. The orally taken spirits should be mixed with water in order to reduce its pungency. Aromatic spirits of ammonia are a common example of spirits that are inhaled. The methods used to prepare spirits include distillation, solution by maceration, and simple solution [5].

- *Tinctures*: Tinctures are alcoholic or hydroalcoholic liquid extracts containing nonvolatile substances prepared from vegetable materials or chemical origin. Tinctures can be prepared using solvents such as alcohol and glycerin. Alcohol is the most commonly used solvent, as it helps in extracting various substances such as resins, most alkaloids, fats, some volatile oils, and wax. The alcohol content of tinctures varies depending on the preparation, from 15% to 80%. Alcohol acts as self-preservative and hence gives the formulation a shelf-life of more than 5 years. Glycerin is used in the form of solution, that is, glycerin and water, to prepare tincture for nonalcoholic patients. Also, glycerin as a solvent is more effective than alcohol in extracting tannins. Generally, three methods are involved in the preparation of tincture: maceration, percolation, and simple solution. Maceration is used to prepare tinctures of benzoin, tincture of orange, and tincture of opium, whereas percolation is used to prepare tincture of belladonna, tincture of ginger, and tincture of digitalis. Simple solution method is used when the tincture is prepared from chemical origin such as thimerosal and iodine. Tinctures may be classified as medicated and nonmedicated. Medicated tincture such as tincture of opium is used as a narcotic and analgesic, whereas tincture of benzoin is used as a stimulant and in the treatment of chronic and acute bronchitis and laryngitis when inhaled after dilution. Tinctures should be stored in an airtight container and protected from sunlight and excessive heat [5].

- *Irrigating solutions*: Irrigating solutions refer to solutions used in the cleansing of body cavities, wounds, or tissues exposed during surgery. Irrigating solutions may be used topically, that is, applied to eyes and ears. Also, evacuation enemas and vaginal douches are examples of irrigating solutions. Evacuation enemas are used to cleanse the bowel, whereas vaginal douches are used in irrigation cleansing of the vagina [5].

14.3.4.2 Emulsions

An emulsion is a biphasic system made up of two immiscible liquid phases, in which one is dispersed as small droplets into the other with the help of an emulsifying agent. The small dispersed droplets are referred to as the internal phase or the dispersed phase, whereas the dispersion medium is referred to as the external phase or the continuous phase. Emulsions are classified as oil-in-water emulsion and water-in-oil emulsion. Oil-in-water emulsion is made up of an oleaginous dispersed phase and an aqueous continuous phase. On the other hand, water-in-oil emulsion is made up of an aqueous dispersed phase and an oleaginous continuous phase. According to the constituents of the emulsion, it may be administered orally (e.g., cod liver oil emulsion), parenterally (e.g., emulsions of fat-soluble vitamins), or applied externally (e.g., oily calamine lotion). In fact, many pharmaceutical preparations are emulsions, but they fall under other pharmaceutical categories, where they fit more appropriately. For example, emulsions include liniments, lotions, ointments and creams. The main ingredients of an emulsion are emulsifying agents, antioxidants, preservatives, and flavoring agents. The addition of an emulsifying agent to emulsions is necessary to maintain their stability. The different types of emulsifying agents used in the preparation of emulsions may be classified as the following:

1. Carbohydrate substances such as tragacanth, acacia, agar, and pectin

2. Protein substances such as egg yolk, gelatin, and casein

3. Alcohols such as cetyl alcohol, stearyl alcohol, and glyceryl monostearate

4. Wetting agents, which may be cationic (e.g., benzalkonium chloride), anionic (e.g., sodium lauryl sulfate), or nonionic (e.g., sorbitan esters and polyoxyethylene derivatives)

The main purpose of emulsification is to produce a homogenous and stable emulsion. Generally, there are three emulsification theories: surface tension theory, oriented-wedge theory, and interfacial film theory. According to the surface tension theory, emulsifying agents act as surfactants or wetting agents and hence lower the interfacial tension between the two immiscible phases. The oriented-wedge theory is based on the assumption that a monomolecular layer of the emulsifying agent is formed around the globules of the dispersed phase, thus stabilizing the emulsion. The interfacial film theory is based on the principle that a thin film of the emulsifying agent is formed at the interface between the two immiscible phases, thereby preventing coalescence of the internal phase [15].

14.3.4.3 Suspensions

A suspension is a dispersion system in which finely divided drug particles of 0.5 to 5.0 microns in size are dispersed uniformly in a suitable liquid vehicle in which the drug has minimum solubility. Suspensions are available into two forms: ready-to-use form and dry powder for reconstitution. The ready-to-use form refers to suspensions that are already prepared and ready to be used by the patient. On the other hand, dry powder for reconstitution refers to a mixture of drug powder along with a suspending agent, which is diluted and agitated with purified water before use. Suspensions are dispensed as dry powder for reconstitution when the drug substance is not stable in aqueous vehicles for long periods, for example, antibiotics. Suspensions provide a suitable dosage form for drug substances that are not chemically stable in solution but are stable when suspended. Oral suspensions mask the unpleasant taste of poor-water-soluble drugs (e.g., erythromycin estolate), and they are advantageous for youngsters and elderly who have difficulty in swallowing solid dosage forms. A good suspension should settle slowly without caking and should be easily redispersed upon shaking. In addition, a properly prepared suspension should pour easily from the container, especially if it is intended for external application [16,17].

The formulation of suspensions depends on whether the suspension is flocculated or deflocculated. Flocculation is a phenomenon in which the repulsive forces between the dispersed particles are decreased by reducing the zeta potential, thereby increasing the forces of attraction between the dispersed particles. Flocculation results in the formation of loose aggregates known as floccules, which are held together by weak intermolecular forces of attraction. Flocculated suspensions have faster sedimentation rate as compared with deflocculated suspensions, owing to the greater size of the sedimenting particles. The loose structure of the floccules permits the easy break-up of the aggregates; thus, they get readily redispersed upon gentle shaking. In deflocculated suspensions, the particles exist as separate entities throughout the dispersion medium, as the electrical forces of repulsion is stronger than the forces of attraction between the particles. In deflocculated suspensions, the individual particles have smaller particle size as compared with the flocs formed in flocculated suspensions. As a result, the sedimentation rate is slow, since the individual particles are settling separately. The individual particles become closely packed during sedimentation; hence, they overcome the electrical repulsive forces between particles, forming a hard cake that is difficult to redisperse upon shaking of the container [5]. There are three approaches for formulation of suspensions: incorporation of structured vehicle, use of controlled flocculation, or combination of both methods, that is, addition of flocculating agent in a structured vehicle. *The main ingredients used in the preparation of pharmaceutical suspensions include the following:*

- *Flocculating agents* are used to reduce the zeta potential of the dispersed particles, which results in the formation of floccules. The different types of flocculating agents are electrolytes, surfactants, and hydrophilic polymers. Electrolytes such as sodium salts of phosphates, acetates, and citrates reduce the electrical barrier between the dispersed particles, resulting in flocculation. Monovalent and divalent ions are most commonly used as compared with trivalent ions, as they are less toxic. Surfactants bring about flocculation by reducing the surface tension between the dispersed particles and the dispersion medium. The particles with low surface energy are attracted toward each other through weak bonds, forming loose aggregates. Both ionic and nonionic surfactants can be used in suspension formulation, and they should be used in optimum concentrations. Polymers such as tragacanth, starch, alginates, and carbomers possess long, branched chains, which are used to control flocculation. One part of the chain gets adsorbed onto the surface of the dispersed particles, whereas the other part remains projected in the dispersion medium. Floccules are formed when the latter parts bridge through the formation of weak bonds [14].

- *Viscosity enhancers or thickeners* are used to increase the viscosity of the external phase, thus reducing the sedimentation rate and rendering the suspension more stable. Viscosity enhancers or thickening agents are also known as suspending agents. Suspending agents function by forming a thin film around the dispersed particles, hence reducing the forces of attraction between the particles. The most commonly used suspending agents include cellulose derivatives (carboxy methyl cellulose (CMC) sodium), natural gums (tragacanth and acacia), carbomers, clays (bentonite), and sugars (fructose and glucose). The addition of suspending agents prevents caking or aggregation of the suspended particles. Moreover, an optimum amount of the suspending agent should be added to avoid having a very viscous suspension that is difficult to agitate and pour. A good suspension should be viscous enough to maintain the particles suspended, as long as possible. However, the viscosity should decrease upon shaking or agitation, providing good flow properties to readily pour the suspension from the container [14].

- *Wetting agents* such glycerin and alcohol are added to improve the dispersion of the particles throughout the dispersion medium. **Buffers** are added to stabilize the suspension at a particular pH. **Osmotic agents** are used to adjust the osmotic pressure of the preparation to the biological fluids. **Coloring agents** are added to suspensions to impart color and improve the appearance of the suspension. **Preservatives** are added to prevent microbial growth. The most commonly used preservatives in suspensions include alcohols (5%–10%), parabens (0.02%–0.2%), benzoates (0.1%–0.3%), and phenols (0.2%–0.5%). Water is the most commonly used external phase for oral suspensions; however, other solvents such as alcohol and glycerin may be used in the formulation. Oral suspensions are packed in wide-mouth containers having enough space above the liquid to ensure proper mixing [14].

14.3.4.4 Drops

Drops are potent liquid preparations intended to be used orally or applied externally. For instance, oil-soluble vitamins such as vitamin A and D are formulated as drops for oral administration. Externally used drops include eye, nasal, and otic drops. Eye drops are sterile liquid preparations, that is, solutions and suspensions, that are instilled into the eyes, for example, moxifloxacin eye drops. Eye drops should be buffered, isotonic with the lachrymal fluids, and free from any particulate matter to avoid eye irritation. Nasal drops are mainly aqueous solutions administered through the nose with the help of a dropper to produce a local or systemic effect. Otic drops are liquid preparations applied into the ears, either for ear cleaning and softening of wax or for the treatment of mild infections of the middle ear and the external ear [13].

14.3.4.5 Liniments

Liniments are liquid preparations that can be alcoholic or oily solutions or emulsions containing drug substances intended for external use, that is, rubbing on the skin. Alcohol-based liniments are mainly used when counterirritant, rubefacient, or penetration action is required. On the other hand, oleaginous liniments cause less skin irritation as compared with alcoholic liniments, and they are mainly used for massage purposes. Fixed oils such as sesame oil and almond oil, volatile substances such as turpentine, or a combination of both can be used as vehicles to prepare oleaginous liniments. Liniments should not be applied to bruised or broken skin, as they might cause excessive skin irritation. Liniments should be stored in airtight containers, and they should be labeled for external use only [13,14].

14.3.5 Semi-Solid Dosage Forms

Semi-solid dosage forms are pharmaceutical preparations having semi-solid consistency and are applied topically on the skin or mucous membrane, providing a protective or a therapeutic effect. Semi-solid preparations may be used nasally or applied rectally, vaginally, or nasally. Semi-solid preparations include ointments, creams, gels, pastes, plasters, and suppositories [5]. Semi-solid preparation maybe medicated, that is, containing a therapeutic agent, or nonmedicated, that is, used as a lubricant, protectant, or emollient. Semi-solid dosage forms may be designed to produce a local or systemic effect. A locally acting semi-solid product delivers the medications into the skin for the treatment of dermal disorders, whereas a systemically acting semi-solid preparation delivers the medication to the general systemic circulation through the skin. An ideal semi-solid dosage form should be elegant in appearance, smooth in texture, nondehydrating, and nonhygroscopic. Also, it should not be gritty, greasy, or staining. The physiological properties of a good semi-solid preparation are that it is miscible with skin secretions and nonsensitizing. In addition, semi-solid dosage forms should be easily applied with good drug release, and it should be easily washable. The main ingredients used in the preparation of a semi-solid dosage forms include the drug substance, bases, antimicrobial agents, antioxidants, gelling agents, emulsifiers, fragrance, and humectants [5].

14.3.5.1 Ointments

Ointments are semi-solid dosage forms meant for external application on the skin or mucus membranes. Medicated ointments are used for their medicinal value, and they primarily consist of a drug substance and a vehicle known as a base. On the other hand, nonmedicated ointments are used as protective barriers, emollients, and vehicles to incorporate drug substances. Ointment bases are classified into four categories: oleaginous bases or hydrocarbon base (e.g., white petrolatum), absorption bases or emulsion bases (e.g., lanolin), water-removable bases (e.g., hydrophilic ointment), and water-soluble bases or greaseless bases (e.g., polyethylene glycol). The selection of ointment bases depends on the following:

1. The required drug release rate from the ointment base

2. The required rate and extent of percutaneous absorption of the drug

3. The required extent of moisture occlusion from the skin

4. The drug stability in the ointment base

5. The effect of the drug substance on the consistency of the ointment base

6. The desirability of the ointment base to be removed easily on washing

7. The nature of the surface to which the ointment is applied

Ointments are prepared by three methods: incorporation method, fusion method, and ointment mill method. The incorporation method involves the mixing of the medicinal agent into the ointment base by using a glass slab and spatula or by using a mortar and a pestle. In the fusion method, all or some of the ingredients are combined by melting them together and then cooling them with constant stirring to form a homogenous mixture [18,19]. This method is used when solids or waxes of high melting point are to be added to oils or semi-solid constituents. Heat-sensitive ingredients are added last, when the temperature of the mixture is low enough to avoid their decomposition. In the ointment mill method, triple roller mills are used, and hence, it is used for large-scale production. Ointments should be stored in well-closed containers to avoid contamination and to protect them from heat. Also, ointments can be stored in jars and plastic or metal tubes [19].

14.3.5.2 Creams

Creams are semi-solid preparations intended for external application and consist of one or more APIs that are dispersed either in an oil-in-water emulsion or in a water-in-oil emulsion or other water-soluble base. Vanishing creams are aqueous cream, that is, oil-in-water type of emulsion, which contains large amount of water and stearic acid or other oily component. On the other hand, cold creams are a good example of oily creams, that is, water-in-oil emulsion, which is used for the cleansing and softening of the skin. The use of creams is often preferred as compared with ointments, as they are easier to spread and remove [20].

14.3.5.3 Gels

Gels are semi-solid preparations in which small or large molecules are dispersed in a liquid phase that is constrained in a 3D polymeric matrix upon the addition of a gelling agent. Gelling agents may be synthetic macromolecules such as cellulose derivatives and carbomers or natural gums, for example, acacia, tragacanth, and pectin. Gels are usually nongreasy and translucent. Generally, gels are used either for lubrication (e.g., lubrication of diagnostic equipment) or for medication (e.g., antiseptic and spermicidal gels). The main ingredients used in the formulation of gels are drug substance, gelling agent, water, solvents such as propylene glycol and alcohol, stabilizers such as edetate disodium, and antimicrobial agents such as propylparaben [20].

14.3.5.4 Pastes

Pastes are semi-solid dosage forms intended for external application on the skin. Generally, pastes are stiffer than ointments, as they contain greater portion of solid particles, that is, 25% of solid material. Pastes can be prepared by either direct mixing or using heat to soften the base prior to the addition of the solid material. Pastes are generally used to provide a protective coat over the intended area of application, as well as absorbing secretions such as serous secretions from the skin. For instance, zinc oxide paste is more effective in protecting the skin and in absorbing secretions as compared with zinc oxide ointment [5].

14.3.5.5 Plasters

Plasters are solid or semi-solid adhesive masses that adhere to the skin surface when spread on a backing material such as cotton, moleskin, linen, muslin, paper, and plastic. Plasters are mainly intended to adhere to the site of application for a long period of time. Medicated plasters are used to provide a therapeutic effect at the site of application; for example, salicylic acid plasters are used to remove corns from toes. On the other hand, nonmedicated plasters are used to provide protection and mechanical support at the site of application. They are also used to bring the medication in close contact with the skin [5].

14.3.5.6 Suppositories

Suppositories are semi-solid preparations intended for insertion into body cavities or orifices, that is, rectum, vagina, and urethra, where they melt, soften, or dissolve, producing a local or a systemic effect. Suppositories are available in various shapes, sizes, and weights. The size and shape of the suppository should allow its easy insertion in the intended body orifice, without causing unnecessary distention. Also, a suppository must be retained once inserted for an appropriate period of time [21]. Adult rectal suppositories are usually 32-mm long and cylindrical in shape, with one or both tapered ends. Adult rectal suppositories weight 2 g if prepared using cocoa butter as a base, whereas suppositories for children are half the weight and size of the adult suppositories. Vaginal suppositories are also known as pessaries, and they weigh about 5 g when cocoa butter is used as the base. Vaginal suppositories are usually molded in an oviform, globular, or cone shape. Suppositories can be prepared using different types of bases, for example, fatty bases (e.g., cocoa butter) and water-soluble and water-miscible bases (e.g., polyethylene glycols and glycerinated gelatin) [21]. Generally, suppositories are prepared using three methods: hand rolling and shaping, compression, and molding from melt. Molding is the most frequently used method for industrial and small-scale preparations of suppositories. Suppositories prepared using cocoa butter base are stored below 30°C (preferably refrigerated between 2°C and 8°C), and they are wrapped individually or separated in compartmented boxes to avoid contact and adhesion. Glycerinated gelatin and glycerin suppositories are stored at controlled temperature, that is, at 20°C–25°C, and they are packaged in tightly closed glass containers to avoid changes in the moisture content. Suppositories containing light-sensitive drugs are wrapped individually in opaque material, for example, metallic foil. Generally, suppositories are wrapped individually in either plastic or foil material [21].

14.3.6 Inhalation-Dosage Forms

Inhalers or aerosols are defined as solutions, emulsions, or suspensions of the therapeutically active drug, which are held under pressure in a mixture of inert propellants, that is, compressed or liquefied gas in an aerosol dispenser. The delivered physical form of the therapeutically active agents, for example, foams, fine mist, solid stream, and dry spray, depends on the product formulation and the valve type. Furthermore, the delivery of the aerosol contents is dependent on the actuator, the valve assembly, the container, and the propellant [22]. The advantages of pharmaceutical aerosols over other dosage forms include the following:

- Protection of the therapeutically active agents from the effect of atmospheric oxygen and moisture, hence enhancing drug stability.

- Drugs can be applied directly to the site of action, providing local drug effect.

- Application of aerosols is a rapid and clean process.

- The medication can be easily withdrawn from the package without contaminating or exposing the remaining material [22].

Aerosols consist of two components: the product concentrate and the propellant. The product concentrate refers to the medicinal agents of the aerosol combined with the required excipients. Different types of propellants can be used in the preparation of aerosols: liquefied gas or compressed gas such as nitrous oxide, nitrogen, and carbon dioxide. Liquefied gas propellants can also be employed as vehicles or solvents. Propellants having vapor pressure greater than the atmospheric pressure at 40°C result in the development of the required pressure to expel the product concentrate out of the container in the desired physical form. Nebulizers are devices used to deliver the medication in the form of liquid mist to the patient. Nebulizers are commonly used in the treatment of asthma and other respiratory disorders [22,23]. In a nebulizer, oxygen or air is passed through a liquid medication, turning it into vapors, which are inhaled by the patient. Generally, physicians prefer aerosols or inhalers over nebulizers in mild respiratory conditions, as they are portable and cheaper as compared with nebulizers [22–24].

14.3.7 Containers and Closures System

Pharmaceutical packaging is required to dispense the pharmaceutical preparation in a suitable form that is marketable and attractive for patients. Pharmaceutical packaging plays an important role in providing identification, presentation, information, protection, stability, and integrity for the drug product from the time of production until it is administered or consumed by the patient. Pharmaceutical packaging consists of four components: containers, closures, cartons, and boxes. Containers refer to a device or an article that is used to hold the pharmaceutical product either directly (immediate container) or indirectly [25]. Closures are defined as the devices that tightly pack the container in order to protect the drug substance against the entry of microorganisms, air, moisture, and particulate matter. Closures are part of the container–closure system; both the container and the closure should not physically or chemically interact with the drug substance. A cartons is the outer covering or packaging that provides protection against environmental and mechanical hazards, and it is usually made up of cardboard or wood pulp. Boxes are used to hold multiple products, and they are usually made up of thicker cardboard material [25]. The following are the ideal qualities of containers and closures:

1. They should not react chemically or physically with the stored content.

2. The content should be easily withdrawn from the containers, and they should be elegant in shape.

3. They should have sufficient mechanical strength to withstand wear and tear during handling, filling, closing, and transportation.

4. They should protect the drug substance against environmental factors and maintain drug stability.

5. The container should be inert or neutral toward the drug substance stored in it.

6. Containers used for sterile drug products should withstand high temperature and pressure used in the sterilization process.

7. They should provide correct identification and information about the pharmaceutical product.

8. Closures should be nontoxic in nature and inert toward the drug product.

9. Closures should be easily removed and replaced [25,26].

Thus, the following are the types of packaging:

- *Primary packaging*: Primary packaging refers to the packaging that is in direct contact with the pharmaceutical formulation. Primary packaging is aimed at protecting the drug substance from environmental, chemical, and mechanical factors. Examples of primary packaging include blister packaging and strip packaging [25,26].

- *Secondary packaging*: Secondary packaging refers to the external or consecutive packaging to the primary packaging. It is aimed at providing additional protection to pharmaceutical formulation from handling during transportation, for example, cartons and boxes [25,26].

- *Tertiary packaging*: Tertiary packaging is the outer package or covering of the secondary packaging. It is required for bulk handling and shipping of drug products. Barrels and containers are the examples of tertiary packaging [25,26].

In addition, the following are the types of packaging material:

- *Glass* is the most widely used packaging material, as it ensures visibility, inertness, moisture protection, ease of filling, and cleaning. The four types of glass used in pharmaceutical packaging are type-1 borosilicate glass, type -2 treated soda lime glass, type 3 regular soda lime glass, and type-4: Nanoparticle (NP) general-purpose soda lime glass and colored glass. The disadvantages of glass are that it is fragile, it is expensive compared with plastic, and some glass leach alkali into the content held in the container [26].

- *Plastic* is synthesized from synthetic polymers of high molecular weight that can be molded easily into different forms and shapes by applying pressure or using heat. The advantages of using plastic as a packaging material are that it is cheap, flexible, light in weight, resistant to breakage, poor conductor of heat, and available in different sizes and shapes. The common disadvantages of using plastic are leaching, sorption, permeation, chemical reactivity, and modification [26]. Plastics are classified into two main types: thermoplastic and thermosetting plastic. Thermoplastics soften on heating, forming a liquid that hardens upon cooling and can be molded, for example, polyvinyl chloride (PVC), acrylic, and polystyrene. Thermosetting plastics are flexible upon heating but cannot be reshaped or molded upon reheating, for example, urea, phenolic, epoxy resins, and polyesters [26].

- *Rubber* is mainly used to manufacture caps, closures, plungers, and vial wrappers. Rubber is available in two types: natural rubber and synthetic rubber. Synthetic rubber is made up of long chains of polymers in which isoprene units are linked together, and it is most commonly used in the preparation of pharmaceutical closures [26], for example, butyl rubber and nitrile rubber. Natural rubber is obtained from the juicy substance, which exudates from the stem of *Hevea brasiliensis* plant. The advantages of using rubber are that it has a long shelf-life, is available in different shapes and sizes, is impermeable to air and moisture, and is elastic. The main disadvantages of using rubber are leaching and sorption [26].

- *Metals* are mainly used to manufacture containers for both semi-solid preparations (e.g., metal collapsible tubes) and aerosols (pressurized dosage form). The most commonly used metals are aluminum, iron, lead, tin, and stainless steel. The advantages of using metals include impermeability to moisture, air, bacteria, and light. Moreover, metals are rigid, unbreakable, and ideal packaging material for aerosols. The disadvantages of using metals include high chemical reactivity and being expensive as compared with plastic and glass [26].

14.3.7.1 Types of Containers
Based on the method of closure, permeability of the container, and utility, containers are classified as follows:

- *Well-closed containers* protect the pharmaceutical formulation from getting contaminated by extraneous matter as well as prevent the loss of content under normal conditions of handling, shipping, and storage [25,26].

- *Airtight containers* protect the pharmaceutical formulation from getting contaminated by extraneous matter such as air, dust, and moisture. They prevent the loss of the content under normal conditions of handling, shipping, and storage. Airtight sealed containers are used for injectables, whereas airtight closed containers are used to store other pharmaceuticals. Ear drops, eye drops, and multiple dose vials are common examples of airtight containers [25,26].

- *Hermetically sealed containers* are impervious to air under normal conditions of handling, shipping, and storage, for example, glass ampoules [25,26].

- *Light-resistant containers* are used to protect photo-sensitive drug substances from the harmful effect of light [25,26].

- *Aerosol containers* are used to store aerosol products. They should have good mechanical strength to bear the pressure during the process of aerosol packing [25,26].

- *Children-proof containers* refer to containers that are difficult to be opened by children [25,26].

14.3.7.2 Closures

Closures play a very important role in preventing the contamination of drug substances and in facilitating the usability of the container by the patient. It is crucial while selecting the closure to ensure that it will appropriately seal the container and protect the product. Hermetic seal is considered to be the most stringent type of closures, as it is impervious to the entry of gases, microorganisms, and moisture. Hermetic sealing is achieved by thermal fusion of the container's material. It is used for single-dose sterile drug products, because once the seal is broken, it cannot be reused [10]. Another type of closures is rubber bung, which does not allow the passage of microorganisms but may allow the diffusion of some other substances, and it can be reused unlike the hermetic seal. Also, there are other types of closures that provide reasonable degree of protection to the drug product [10]. For example, a cap on a tablet or a capsule bottle provides an acceptable barrier to moisture and microorganisms entry. This type of closure will inevitably allow the exposure of the drug product to environmental contaminants, as it may be opened several times per day. Flexible packs such as sachets, blister packs, and ampoules are heat-sealed, and hence, they do not need a separate closure. However, separate closures such as bungs and caps may be required for some other types of containers. The caps may be made up of aluminum, thermoplastic, or thermosetting plastic, and they are retained over the container through a screw thread [10].

14.4 PREFORMULATION STUDIES

Preformulation studies are carried out when a newly synthesized drug shows promising pharmacological actions in animal models, and hence, it needs to be developed into a pharmaceutical formulation that can undergo drug evaluation in clinical trials. Preformulation studies are aimed at studying the physicochemical properties of the drug, which can influence the development of an effective dosage form, the choice of drug additives, bioavailability, and drug performance. An adequate and thorough understanding of the physicochemical properties of the new drug reduces the chances of any problems in the subsequent stages of drug development, reduces the cost, and decreases the time for the new drug to reach the market. It is crucial to understand and evaluate the intrinsic physical and chemical properties of every new drug before developing a new pharmaceutical formulation [27]. This strategy allows the research formulator to determine the appropriate formulation design, formulation ingredients or excipients, processing methodology, and the analytical methods required to develop an efficacious drug-delivery system. Generally, the physicochemical factors that must be studied include drug solubility, dissolution rate, partition coefficient, physical form, and drug stability [27]. The objective of conducting preformulation studies include the development of an effective, safe, stable, and elegant dosage form by establishing the kinetic rate profile of the drug as well as evaluating the drug compatibility with the different pharmaceutical ingredients and hence understanding the nature of the new drug. Preformulation studies provide strong scientific foundation of guidance to research formulators. In addition, it helps in conserving the resources needed in the evolution process and development of new drugs. It also enhances the public safety standards and quality of the dosage form [27].

14.4.1 Drug Solubility

Solubility is one of the most important physicochemical properties of a drug substance. For a drug substance to be administered by any route, it must possess an acceptable rate of solubility for systemic absorption and therapeutic effect. Relatively soluble or poorly soluble compounds result in slow, incomplete, or erratic drug absorption, hence producing minimal therapeutic response. As a result, solubility enhancement must be considered for such poorly soluble drug substances. The techniques used to improve drug solubility depend on the chemical nature of the drug substance as well as the consideration of the desired type of drug product. These techniques include chemical complexation, addition of cosolvent, micronization, preparation of solid dispersion, adjustment of PH, and preparation of salt or ester forms of the parent substance [27,28].

14.4.2 Dissolution Rate

Dissolution rate refers to the speed by which the drug substance dissolves into the biological fluids at the site of absorption. In many instances, when the drug is administered orally or intramuscularly, the dissolution rate is the rate-limiting step in the absorption process. The drug dissolution rate influences the onset of action, the intensity, and the duration of response, thus affecting the overall drug bioavailability from the dosage form. The drug dissolution rate can be increased by reducing the particle size of the drug and increasing the solubility of the drug substance in the diffusion layer. Moreover, preparing a highly-water-soluble salt form of the parent compound is considered the most effective technique in increasing the dissolution rate of the drug substance [27,28].

14.4.3 Partition Coefficient

Partition coefficient measures the lipophilic character of the drug substance, that is, the drug preference for the aqueous phase or the oil phase. For a drug molecule to elicit a pharmacological response, it must first penetrate the biological membrane, which is considered a lipid barrier for many drugs. Consequently, it is crucial to determine the partition coefficient of the drug substance, that is, the distribution of the drug between the water phase and the oil phase, since the ability of the drug molecule to penetrate the cell membrane depends on its partition coefficient [27,28].

14.4.4 Physical Form

The physical form of the drug substance is an important factor in the preformulation studies. For instance, a drug may exist in the crystalline or amorphous form. Also, polymorphic forms of the drug may exist, which usually possess different physicochemical characteristics such as solubility and melting point. The amorphous or noncrystalline form of the drug is always more soluble as compared with the crystalline form. It is crucial for the research formulator to evaluate the crystal structure, solvate form, and polymorphism of the drug substance, as any change in the crystal characteristics of the drug can affect the physical and chemical properties of the drug as well as the drug bioavailability. Evaluation of the crystal properties of the drug substance is significant when it comes to the preparation of tablet dosage forms, as it influences the compaction and flow properties.

The techniques used in the evaluation of the crystal properties include thermal analysis, hot-stage microscopy, X-ray diffraction, and IR spectroscopy. The crystalline or amorphous form along with particle size of the drug influences the drug dissolution rate, hence affecting the rate and extent of drug absorption. For instance, by reducing the particle size of a poorly soluble drug and increasing its surface area, both the dissolution rate and the biological absorption are enhanced [27,28].

14.4.5 Stability

Drug stability can be defined as the extent to which a drug product can retain the same characteristics and properties that it possessed at the time of manufacture throughout its shelf-life, that is, period of storage and use. The five types of stability that concerns research formulators are chemical, physical, microbiological, therapeutic, and toxicological stability. Chemical stability refers to the ability of the drug product to retain its chemical integrity as well as the labeled potency throughout its shelf-life. Physical stability can be defined as the ability of the drug product to retain its original physical characteristics such as uniformity, suspendability, dissolution, palatability, and appearance throughout its shelf-life. Microbiological stability is the ability of the drug product to resist the growth of microorganisms as well as to retain its sterility per the specified requirements [27,28]. For instance, antimicrobial agents are added to the pharmaceutical formulation to maintain its microbiological stability. Therapeutic stability refers to the ability of the drug to maintain its therapeutic efficacy; that is, it remains unchanged throughout its shelf-life. A drug product is toxicological stable when the toxicity of the formulation does not show a significant increase during its period of storage and use. Evaluation of both the physical and chemical stability of the drug substance is a crucial step in the preformulation studies. The different types of stability studies conducted in the preformulation phase include solid-state stability studies of the pure drug alone, stability studies in the presence of the expected excipients, and solution-phase stability. The initial investigations about the chemical structure of the drug molecule allow the research formulator to determine the degradation reactions that lead to drug instability, such as oxidation and hydrolysis reactions [27,28]. Oxidation is a chemical process that destroys many types of drugs, for example, alcohols, aldehydes, sugars, phenols, and alkaloids. In order to prevent the oxidative destruction of the above-mentioned types of drugs, antioxidants are added to the pharmaceutical formulation, hence protecting its potency. Hydrolysis is a destructive process in which drug molecules interact with water molecules, resulting in the breakdown of the drug product. Esters and other medicinal compounds containing groups such as lactones and amides are more susceptible to hydrolysis. These types of medicinal agents must be protected against moisture during formulation, processing, and packaging, thus preventing their decomposition. Also, during the preformulation studies, the stability of the drug is tested at different temperatures and relative humidity (RH) (e.g., 40°C at 75% RH and 30°C at 60% RH). Information about drug stability is very important for developing label instructions for use and storage, as well as for determining the expiration date of the drug product [27,28].

14.5 EXCIPIENTS AND INGREDIENTS IN PHARMACEUTICAL FORMULATIONS

Dosage forms are complex system that consist of different components along the active drug moiety. These components are referred to as excipients, and they play an important role in formulating any dosage form. Excipients had received only little attention during the initial years, but a change has been observed over the recent years, because nowadays, excipients are not only seen as an opportunity of cost saving but also offer new opportunities to introduce new dosage forms. Adding excipients to the formulation facilitate preparation process, patient compliance, and the properties of dosage form. There are various kinds of excipients such as coloring agents, flavoring agents, lubricants, and surfactants. Various properties of excipients are influenced by other excipients, which include absorption, bioavailability, solubility, safety, dissolution, and efficacy. They also assist in formulating a more patient-friendly product along with improving the product performance and pharmacological action [29,30].

The word *excipient* originated from the Latin word "excipere," which means to receive. An excipient may be defined as a pharmacologically inactive substance formulated along with the API [31,32]. According to the European pharmacopeia, "An excipient is any component, other than active substance(s), present in a medicinal product or used in the manufacture of the product. The intended function of an excipient is to act as a carrier (vehicle or basis) or as a component of the carrier of the active substance(s) and, in doing so, contribute to the product attributes such as stability, biopharmaceutical profile, appearance and patient acceptability and to the ease with which the product can be manufactured. Usually more than one excipient is used in the formulation of a medicinal product" [33].

In the drug development process, excipients play a vital role by helping in the formulation of a stable dosage form. The intended role of excipients in any formulation is to ensure that the required physiochemical and biopharmaceutical properties of any formulation are met. Certain excipients are multifunctional in nature. For example, a cellulose derivative called hydroxypropyl methyl cellulose, used as an excipient, serves many functions such as emulsifying agent, viscosity enhancer, coating agent, and even suspending agent. Excipients are an integral part of any formulation, and for a particular dosage form, specific excipients are to be selected. Care should be exercised while selecting the appropriate excipient, so as to avoid any unwanted reaction. Ideally, an excipient is expected to be inert in nature. Excipients require stabilization and standardization procedure similar to an active ingredient [34]. There is also a need of excipients in the formulation:

1. To enhance the stability of the formulation: In most cases, the API is found to be unstable when present in its pure form. As a result, the product fails to retain stability for a longer duration. Addition of excipients ensures that the stability of the product is maintained, hence increasing the shelf-life of the preparation.

2. To bulk up the formulation: In case of drugs that are highly potent, adding excipients aid in increasing the bulk, thereby forming an accurate dosage form.

3. Patient acceptance is increased.

4. Bioavailability of the API is increased: In certain cases, the drug substance may not be easily absorbed by the body. In such situations, the API is mixed or dissolved with a suitable excipient, which enhances the absorption of the drug, which in turn enhances bioavailability.

5. Excipients ensure the safety and effectiveness of the formulation throughout its storage and use by patient [35].

14.5.1 Classification of Excipients

14.5.1.1 According to Origin

This classification is based on the source from which the excipient is obtained. Based on origin, excipients may be classified into four categories: animal source (musk, lanolin, honey, beeswax, and lactose), plant source (starch, turmeric, peppermint, acacia, guar gum, and arginates), mineral source (calcium phosphate, silica, talc, kaolin, and paraffin), and synthetic source (boric acid, saccharin, lactic acid, povidone, and polysorbates) [35].

14.5.1.2 Based on Function

In this classification, we categorize excipients based on the function they perform when administered into a solid or liquid dosage form. In a liquid dosage form, they function as solvents, cosolvents, flavoring agents, emulsifying agents, sweetening agents, and antimicrobial agents. In a solid dosage form, they may act as diluents, lubricants, or binders [35].

14.5.1.3 Based on Therapeutic Value

There are certain excipients that possess a therapeutic value. They can be categorized as pH modifiers (citric acid), anesthetics (chloroform), carminative (anise water or cinnamon), laxatives (bentonite or xanthan gum), and nutrient sources (agar or lactose) [35].

14.5.2 Selection of Excipients

In most formulations, excipients are found in a greater proportion compared with the active ingredient, and hence, they are considered as an indispensible component of a formulation. It is of utmost importance to select a compound that satisfies the ideal qualities of an excipient. Otherwise, it may lead to some untoward happening such as nonuniform weight, friability problems, and improper hardness. The main considerations in selecting an excipient is its cost, availability, functionality, material consistency, regulatory acceptance, and the source. In addition, material properties (such as chemical, mechanical, and rheological properties), physiochemical properties, compatibility, pharmacokinetic attributes, permeation characteristics, stability, drug-delivery platform, intellectual property, and segmental absorption behavior are considered to determine the desired delivery platform and absorption challenges for the API. The risk associated with any interaction between the drug and the excipient can be determined and avoided with the help of excipient compatibility test. This will minimize the risks associated with the excipients. The route of administration is also seen to influence the selection of excipients, along with their characteristics. Normal variability and its possible impact on the process of formulation development can be understood with the help of the concept of quality by design (QbD) [36].

14.5.3 Ideal Characteristics of Excipients

Though excipients are generally considered as inert substances, they do have the tendency to react with other excipients, the packaging system, or even the drug. The impurities present in the excipients may also deteriorate the API, reducing its shelf-life. Hence, it is very important to choose an appropriate excipient. The ideal characteristics that must be possessed by an excipient are as follows:

1. It should be chemically stable.

2. It should be nonreactive.

3. It should have low equipment and process sensitivity.

4. It should have efficiency with regard to the indented use.

5. It should be inert to the human body.

6. It should be nontoxic.

7. It should be economical.

8. It should be acceptable with regard to organoleptic characteristics [37].

14.5.3.1 Diluents, Solvents, and Liquid Vehicles

14.5.3.1.1 Diluents Diluents, also known as fillers, are added to a formulation if the active ingredient is not present in sufficient quantity, so as to fill a capsule shell or make a compressible tablet. If direct compression is used, then it is more appropriate to use the term "filler-binder," as it will provide additional binding property by imparting strength to the formulation in addition to increasing the bulk. Diluents help to enhance compressibility, powder flow, and homogeneity by locking the active drug molecule within the granules. This will enhance the manufacturability of the formulation. Diluents are also seen to improve the quality and performance of the dosage form with respect to dissolution, friability, content uniformity, and stability. For a solid oral-dosage form, either single diluent or combinations of diluents are used. Addition of a mixture of diluents is employed if a single diluent cannot serve the purpose. For example, a formulation consisting of only lactose as a diluent will have a sharp granulation end point, whereas addition of microcrystalline cellulose with its high water-absorbing capacity is seen to have a more forgiving granulation end point. Combination of diluents may also be used for economic reasons, such as to partially replace an expensive diluent with an inexpensive one, without affecting the performance of the formulation. Diluents should be nonhygroscopic in nature, so as to ensure that they do not absorb moisture from the surrounding [38,39].

Added diluents should be inert in nature, so that they do not exhibit any pharmacological activity of their own and hence affect the performance of the drug product. They must also be compatible with the drug molecule and other added excipients, along with being compactable. The particle size of the diluents should be similar to the particle size of the active drug ingredient. Materials used as diluents are mainly categorized as cellulose materials, inorganic salts, and sugars [39]. The following are the types of diluents:

1. *Microcrystalline cellulose (MCC)*: MCC is partially polymerized cellulose derived from α-cellulose. It has multiple applications such as a diluent, binder, and disintegrant. When used as a diluent in tablet formulations, MCC is used in the range of 20%–90%. Even at low compression pressures, it shows good compatibility and can undergo plastic deformations. Different densities and particle sizes of MCC are available. Larger-particle-size and high-density MCC particles assist in the flow characteristics of blends but reduce compactability. As MCC is moisture-sensitive, it is important to control the moisture content while using it. It is sensitive to magnesium stearate owing to its ability to undergo plastic deformation, which can be avoided by blending it with colloidal silica.

2. *Dibasic calcium phosphate (DCP)*: DCPs are used as fillers in oral solid dosage forms in both anhydrous and dihydrated forms. But the anhydrous form is preferred over the dihydrate form owing to better intraparticular porosity, which makes better disintegration, better mean yield pressure, and better compressibility compared with dihydrate form. Excellent compaction and flow properties make DCP desirable to be used as a filler in the pharmaceutical industry. DCP may be abrasive on the tablet tooling, as it is an inorganic salt.

3. *Mannitol*: It is nonhygroscopic in nature and is a good choice for drugs that are moisture-sensitive. It is a preferred diluent in chewable tablets, as it provides a cooling sensation owing to its negative heat of solutions. Different polymeric forms of mannitol with different compression characteristics are available. Spray-dried grades are used for direct compression, while wet granulation employs crystalline grades. Compared with granulations made with other diluents, higher levels of lubricants are needed in mannitol-containing granulations.

4. *Lactose*: It is the oldest and most widely used diluent in oral solid dosage forms. Mainly four different forms of lactose are available: α-lactose monohydrate, anhydrous α-lactose, anhydrous β-lactose, and amorphous lactose [40].

14.5.3.1.2 Solvents and Liquid Vehicles   Solvents and liquid vehicles are required for solutions or suspensions indented for injections, oral administration, and application to the skin or for introduction into the body cavities. A solvent may be defined as any substance used to dissolve a solute, which results in a solution [41]. The solvent may be solid, liquid, or even gas but is usually liquid in nature. For preparing injections, sterile solvents are used.

Solvents are used in the purification, processing, and manufacturing of pharmaceutical products. An ideal solvent is expected to have the following characteristics:

1. It should be colorless, odorless, and tasteless.

2. It should be nontoxic and nonflammable.

3. The drying rate should be rapid.

4. It should be inert in nature.

5. It should not have any environmental impact.

6. Addition of small concentration of solvents should not result in an extremely viscous solution, which might create processing problems [42].

Solvents are mainly classified into three categories, depending on the force of interaction between them: (i) polar solvents, (ii) nonpolar solvents, and (iii) semi-polar solvents. Polar solvents are seen to dissolve polar compounds better and vice versa. Polar solvent consists of strong dipolar molecules, which contain hydrogen bonding. Examples of polar solvents include hydrogen peroxide and water. In nonpolar solvents, the molecules possess little or no dipolar characteristics, such as vegetable oil and mineral oil. Semi-polar solvents such as acetone do not possess molecules with strong dipolar molecules, but they lack hydrogen bonding. Water is the main solvent used in most pharmaceutical preparations. If the drug is not soluble in water, then alternatives are chosen [43].

To select the appropriate solvent, we must be aware of the route of administration, the properties of the drug molecule or mixture of drugs required to be formulated, and any ancillary effect desired by the formulator.

The most important factor to be considered is the route of administration, as the excipient must be accepted biologically. This factor limits the choices of solvents available for injections compared with other routes of drug administration, in addition of safety. If the intended preparation is to be given by the oral route, then taste and smell of the preparation should be adjusted appropriately. For external preparations, the factors to be considered are skin or tissue irritation and staining. The next factor to be considered is the chemical compatibility of drug components and excipients. Any solvent used should be chemically inert in nature, similar to diluents, though it is not possible to achieve the idea of universal compatibility in all cases, for example, if the solvent used is processed by using heat sterilization by autoclave. The most common example of reactive solvent is water. There are many drugs that have a shelf-life of less than a week in the aqueous solution. If no other solvents can be used for such drugs, then the problem of incompatibility is overcome by preparing the formulation extemporaneously.

Solubility is another important factor to be considered along with compatibility. The drug molecule should be either highly soluble or highly insoluble in the solvent of choice. For example, the desired preparation is a suspension; slight solubility will be a problem because temperature fluctuation of normal storage will induce crystal growth.

With regard to the desired ancillary effect, this factor mainly applies to topical applications and injections. This involves the use of a suitable vehicle to control the release rate of drug molecules [44].

14.5.3.2 Thickeners and Binders

14.5.3.2.1 Thickeners   Thickeners, otherwise known as viscosity-enhancing agents, are used in a pharmaceutical formulation to control the viscosity of the solution. It is mostly employed in liquid dosage forms, and it helps to control properties such as easy of pouring, palatability, and rate of sedimentation (in case of suspensions). Hydrophilic polymers are often used as viscosity-enhancing agents.

They should be compatible with other ingredients of the formulation, and proper measures should be taken to ensure that there is no interaction between the active drug molecule and the thickener added. It is a well-known fact that certain hydrophilic polymers interact with the preservatives, and hence, care should be taken to avoid any such interaction. Increasing the concentration of the preservative helps to reduce polymer–preservative interaction. They must be easy to filter and should be easily sterilized. They must remain stable, both physically and chemically, during the sterilization process [14].

The thickener added to any pharmaceutical preparation should be nontoxic, have high viscosity at negligible shear during storage and low viscosity and high shear rate during the pouring of medication, and produce a structured vehicle; in addition, the viscosity of the preparation should not be altered by temperature or on storage. Some of the viscosity-modifying agents used in pharmaceutical formulation include the following:

1. *Polyvinyl alcohol*: It is a water-soluble vinyl polymer, and three different grades of this polymer are available, which include high-viscosity polymers, medium-viscosity polymers, and low-viscosity polymer. These polymers differ from each other in their average molecular weights.

2. *Polyacrylic acid*: It is a water-soluble polymer of acrylate polymer. It is cross-linked with either allyl ethers or allyl sucrose of pentaerythritol.

3. *Hydroxypropylmethylcellulose (HPMC)*: It is a partially methylated and O-(2-hydroxypropylated) cellulose derivative. It is used in the concentration range of 0.45%–1.0% w/w [14].

14.5.3.2.2 Binders   The role of a binder is to provide cohesiveness to the drug and excipient combination during the process of formulation. Cohesiveness helps in the formation and flow of granules during the manufacturing process and helps to maintain tablet integrity upon compression from these granules. Materials used as binders may be sugars (sucrose, glucose, mannitol, and sorbitol), natural polymeric materials (starch, gelatin, or acacia), and synthetic polymers (methyl cellulose, ethyl cellulose, povidone, or hydroxypropyl cellulose). Synthetic polymers are preferred for use as binders over natural polymers, so as to reduce the potential microbial contamination from natural polymers and lot-to-lot variability [45].

Several factors influence the choice and amount of binders used in the formulation process. They are as follows:

1. Compatibility of the binder with the drug and other excipients

2. Cohesiveness

3. Processability

4. Effect of binder on properties of dosage form such as friability, dissolution, and disintegration [46]

The following are also types of binders:

1. *Sugars*: Glucose, sucrose, and mannitol are the sugars employed as binders. 2%–20% w/w concentration range of sucrose is used for dry granulation and 50%–67% w/w is used for wet granulation. Granulating solvent employed is either water or hydroalcoholic solvent. 5%–10% w/w concentration of liquid glucose is used during wet granulation. With higher concentration of binders, hard and brittle tablets are produced. They are preferred as binders in chewable tablets and bitter-tasting drugs owing to their pleasant taste.

2. *Natural polymers*

   Starch: It is a carbohydrate made up of linear amylase and branched amylopectin. Since starch is insoluble in cold water, it is prepared in the form of paste by mixing starch with hot water. 5%–10% w/w concentration of aqueous starch is used in formulations.

   Pregelatinized starch: It is the chemically modified form of starch. Pregelatinized starch is soluble in cold water, unlike starch.

3. *Synthetic polymers*

   Povidone: Povidone or polyvinylpyrrolidone (PVP) is one of the most widely used binders. It is a polymer of 1-vinyl pyrrolidone and is available in a wide range of molecular weights that provide different viscosities in solution. 0.5%–5% w/w concentration of PVP is used for its use as a binder. One of the major drawbacks of using PVP is its hygroscopisity. PVP can take up large amount of water, and hence, when used, it can affect the hardness, disintegration, and dissolution of tablets.

   Methylcellulose (MC): MC is substituted cellulose in which methyl ether substitutes 27%–32% of hydroxyl groups of cellulose. Though different grades with varying molecular weight (which imparts different viscosity to the solution) are available, low- and medium-viscosity grades are preferred for use as a binder. 1%–5% w/w concentration of MC is used [47,48].

14.5.3.3 Surface Active Agents

Surface active agents, otherwise known as surfactants, are molecules that consist of hydrophilic (polar) and a hydrophobic (nonpolar) parts. The hydrocarbon chain can interact with water weakly in an aqueous environment, while the polar or ionic group can interact strongly with water molecules. These interactions may be either via dipole or through ion–dipole interactions. The factor that renders surfactant soluble in water is this strong interaction with water molecules. A proper balance between the hydrophilic and hydrophobic portions is necessary to give surfactants their special features such as the ability to form micelles [49].

Surfactants can be classified mainly into four categories: anionic, cationic, zwitterionic, and nonionic. The main difference between cationic and anionic surfactants is that the anionic surfactant possesses a negative charge on its polar head group, while the cationic surfactant possesses a positive charge. In case of zwitterionic surfactants, depending on the environment in which they are placed, they have the potential to possess both the positive and negative groups on them. Nonionic surfactants, on the other hand, possess no charge on their groups [50].

Surfactants tend to modify the interfacial and bulk properties of the solvent. Surfactants exhibit a property of self-aggregation to form a colloid known as micelle. This process also decreases the overall free energy of the system. At a lower concentration, the molecules of the surfactant will be adsorbed preferentially at the air–water interface in which the hydrophobic tail is seen to be pointing away from the surface of water, thereby reducing the interfacial tension.

With increase in concentration, the adsorption at the air–water interface becomes stronger, thereby forming a condensed monolayer (Gibb's monolayer) at the interface. Surfactant molecules added further will remain in the aqueous phase. The surfactant molecules will exhibit a phenomenon of self-aggregation when the added surfactant molecule exceeds a limiting value, and the process of micelle formation will lead to an abrupt change in many of the physicochemical processes [490].

The use of surfactant is inevitable for those drug products that are poorly soluble in water. Surfactants help in reducing the interfacial tension between the drug molecule, thereby increasing its solubility. Surfactants also act as solubilizers for dissolving vitamin E, herbal medicinal materials, and oil ingredients [50].

14.5.3.4 Preservatives

Preservatives are the substances that are added to pharmaceutical drug products for prolonging their shelf-life. Addition of preservatives ensures that no alteration or degradation of the contents of the product occur during its storage. Mainly, there are three main classes of preservatives: antimicrobial preservatives, chelating agents, and antioxidants. Antimicrobial preservatives are added into the preparation to kill or inhibit the growth of microorganisms introduced into drug products either during the process of manufacture or during storage. Additions of antioxidants help in preventing the deterioration due to the process of oxidation [51].

Antifungal and antibacterial are the two main classifications of antimicrobial preservatives. Examples of antibacterial preservatives include alcohols, biguanidines, quaternary ammonium salts, and phenols. Phenolic compounds, including methyl, ethyl, and propyl benzoate, and ascorbic acids and benzoic acid, along with their salts, are examples of antifungal preservatives.

Similar to antimicrobials, antioxidants can also be classified into different categories. There are three main categories: true antioxidants, reducing agents, and antioxidant synergists. True antioxidants work by inhibiting the oxidation reaction by blocking the free radicals, hence blocking the chain reaction. Butylated hydroxytoluene, nordihydroguaiaretic acid, and tocopherols are examples of true antioxidants. Reducing agents possess lower redox potentials than the drug molecule or adjuvant that they are intended to protect. Hence, they are oxidized more readily and may also act by reacting with the free radicals. Some examples of this group include ascorbic acid or the sodium salts of sulfurous acid. The next category of antioxidants is antioxidant synergists. They have little antioxidant effect as compared with true antioxidants, but they are seen to enhance the antioxidant action of true antioxidants by reacting with heavy metal ions, which catalyze the oxidation reaction. Citric acid, lecithin, and tartaric acid are a few examples of antioxidant synergists. Chelating agent is another class of preservative used in medicinal products. They mostly work by complexing with the pharmaceutical ingredient, hence preventing the formulation of degradation. Disodium ethylenediaminetetraacetic acid (EDTA) is one of the most common examples of a chelating agent [51].

The mechanism by which different preservatives work is different. When we consider antimicrobial agents, they function as preservatives by inhibiting cell wall formation, protein synthesis, and also synthesis of DNA and RNA. Antioxidants ensure the safety of the formulation by preventing the oxidation of oxygen-sensitive ingredients by acting as reducing agents. Chelating agents function by forming a complex, thus protecting the formulation [52]. The qualities required for an ideal preservative are given as follows:

1. It should not be irritant or toxic.

2. It should be stable, both physically and chemically.

3. It should be compatible with other ingredients.

4. The activity of the compound should be maintained throughout the manufacturing, storage, and use of the product [52].

14.5.3.5 Coloring Agents

Coloring agents are compounds that are introduced into pharmaceutical formulation for imparting color to the preparation. Colors will provide an aesthetic effect to the formulation, along with providing ease of identification. Any variation in the color of the contents can be masked effectively with the help of a coloring agent, in addition to increasing patient confidence. One of the simplest and best methods employed for quick identification of the product is coloration. Addition of colors to the preparation also has a commercial role in the marketing of pharmaceutical products, as it gives the product a brand or company image [53]. Rational uses of coloring agents include the following:

- For the convenience of the physician or patients.

- They help distinguish those preparations that have similar names or those that are used under similar circumstances but for different purposes.

- They impart attractive and uniform color to the preparation for those preparations that have slight batch variation or a drab appearance [53].

Colorants used in pharmaceuticals must possess certain basic properties. They must satisfy the criteria of being nontoxic, being acceptable, having a high tinctorial value, and being stable under conditions of manufacturing and storage. Coloring agents, if used in combination with the flavoring agent, should match the flavor of the formulation. For example, if the formulation is strawberry flavored, the color chosen should be red [53].

Though colors are added to aid in the identification of different formulations, excessive use of colors may have an opposite effect, as it may mask the labels. It is also of utmost importance to follow a national color code, which otherwise may lead to confusion, as the same color may have different meaning in two different hospitals. Certain synthetic colors are seen to be toxic if administered systemically for a long duration, and hence, care should be taken in such cases. Since there is a possibility of colors fading on storage, the ranges of colors available are also less. When manufacturing practice of coloring agents are considered, they should be prepared in special areas to avoid contamination [53].

Compared with most other excipients, coloring agents have more strict control because of the reasons mentioned previously. Most importantly, two factors are taken into consideration while selecting a color for a new formulation. The first factor is the pharmacopoeial control on the use of colors in general, and the second factor is the national legislative control on the use of specific synthetic dyes from the point of safety to patient [54].

14.5.3.5.1 Basis of Selecting a Coloring Agent   In addition to considering legal requirements, the selection of a coloring agent requires the following factors to be considered. These are as follows: physical properties, chemical properties, and conventions. The dye selected should be either water or oil soluble or insoluble, depending on the nature of the formulation it is intended to color. Since coloring agents are mostly required in solutions, water-soluble dyes are most frequently used. Depending on the function of the color added, skin staining may be desirable or undesirable. Rapid method of stain removal must be available in either case. Another important factor to be considered is the compatibility between the drug molecule and other excipients that are present in the formulation. This compatibility may be tested by employing trial and error or by accelerated storage tests. It is recommended to review the general trend of coloring agents already used for similar type of formulations prior to selecting a color [55].

14.5.3.6 Flavoring Agents

Flavors can be defined as a mixed sensation of taste, smell, touch, sound, and sight, all of which involve a combination of physiological and physiochemical actions, which influences the perception of a substance. Recently, many artificial flavors have been created with the expansion of the technology. Adding flavor to a formulation, for example, antibiotics, cough syrups, laxatives, pediatric and geriatric formulations, sedatives, and antihistamines, will mask the unpleasant tastes of these formulations without affecting its physical and chemical stability [56].

Similar to coloring agents, flavoring agents also have advantages and disadvantages. If used rationally, they are beneficial, and if overdone, they are potentially dangerous. Most of the drugs have an unpleasant taste, unless highly insoluble in water. They will be intensely bitter, saline, or sour. This may be tolerated by an average adult but will pose a serious problem if it has to be administered to pediatrics. Even with adults, if the drug has to be administered in large amounts or for longer duration, it will be difficult to maintain compliance. Also, if the drug has to be retained in mouth for example, lozenges, taste masking becomes extremely important. Compared with earlier use of syrups and aromatic waters, modern practice has a more elaborate method of flavoring of medicines. This requires additional skills, especially in case of manufactured medicines, as the stability of the added flavor becomes a difficult problem. Earlier, natural products such as lemon, cinnamon, ginger, and camphor, were used as taste maskers. Though they are still used in certain situations, they have mostly been replaced by artificial flavors [55].

The dangers associated with flavors are not attributed solely to their harmful effects, because most of them are comparatively safe, as they have been developed for the food and drink industry. The risk mostly comes because children may accidently ingest the formulation, leading to an overdose of the medicines. Also, there is a chance that adults may lose confidence in medication, as medicines might appear to be treated like confectionary. With natural flavors used earlier, these problems did not exist, as they hardly made the formulation palatable, unlike the artificial flavors, which make the medication attractive, along with making them palatable [56].

14.6 ADVANCEMENT IN PHARMACEUTICAL ASPECTS

The field of pharmaceutical sciences is highly growing and dynamic, targeting the development of biologic-based drugs and introducing more specific and effective drug-delivery systems. The most important steps of this development are the appropriate selection and technique of excipients assimilation, which play a big role in the assurance of the dosage form, stability, and bioavailability of the API. The collection of excipients is based on the demand of their required technological function. Advancement in polymer structure and its use, particle size modification and design, manufacturing processes, and improvement in technology illustrate the way to introduce the development of susceptibility for detecting protein aggregation, advancement in drug-delivery technologies, and provision of the parameters that help to understand the risk of material and nanomaterials in drug-delivery systems.

14.6.1 Resolving Protein Aggregation

Currently, researchers are continuously working on simulated computer models to solve the problem that occurred due to aggregation in the protein-based drugs because of their short shelf-life. This formation of cluster is most commonly seen in antibodies owing to interactions between hydrophobic fragments of the proteins [57].

In 2009, Trout and his colleagues with Bernhard Helk of Novartis (Basel, Switzerland) built up a computer model that could enable researchers to recognize those parts of an antibody that are susceptible to catch other molecules and allow changes in the molecules to avoid clomping. The 3D active molecules similar to an antibody are obtained by a method called spatial aggregation propensity (SAP) with the help of Trout's new computer model. This model had the ability to disclose the hydrophobic parts of molecules and what would happen if these parts of molecules come in contact with fluids or are changed in a solution form. The model also worked on the area responsible for aggregation and also identified when to introduce the amino acid in the hydrophobic part of molecule to reduce the hydrophobicity and improve the stability of a molecule. By using this model, the more stable antibodies could be produced as compared with the original antibodies, without disturbing the actual function of molecules. This method is useful for the screening of antibodies in the discovery phase [57].

14.6.2 Localized Drug Delivery

The microcapsules and nanocoating are the two drug-delivery technologies used for delivering the drug interleukin-12, discovered by Bingyun Lim and his team [58].

In case of microcapsules, the drug is injected or indirectly delivered in a form of fine-mist spray directly at the site of an injury. Whereas the nanocoated interleukin-12 is used to apply unswervingly to stents, pacemakers, pain pumps, artificial limbs, and other biomedical devices. These methods help to increase the half-life of the interleukin-12 and minimize the side effects if administered locally [58].

14.6.3 Nanotechnology in Pharmaceuticals

Currently nanotechnology is taking a big place in the pharmaceutical industry, but along with advantages, it has a high risk of disadvantages also. Per the University of California at Los Angeles (UCLA) press release, the UCLA and the California NanoSystems Institute (CNSI) freshly published a research that observed the nanoparticle–biological interface to recognize the possible hazards of wangled nanomaterials along with the discovery of new experimental design and methods that would guide in the development of harmless and more effectual nanoparticles to be used in a variety of treatments and products. Nanoparticles have abilities to react with proteins, membranes, cells, DNA, and organelles because of the formation of protein coronas, particle wrapping, intracellular uptake, and biocatalytic processes that result in positive or negative outcomes due to nanoparticle–biological interfaces, which depend on colloidal forces and bio-physicochemical interactions. The analysis of these interfaces shows that the size, shape, surface chemistry, roughness, and surface coatings of nanomaterials play an important role in the development of prognostic relationships between structure and activity [59].

The knowledge about interface is important for a better understanding about the inter-relationship between intracellular activity and function of designed and built nanomaterials, which is essential for the development of nanoparticle drug-delivery systems [59].

14.6.4 Precision Medicine

It is a type of medicine in which the patient's genes are matched with molecular nature of drug to identify the biological response to disease. This process is known as genome sequencing, in which the DNA is converted into data. This data is used to screen out the gene abnormalities or biomarkers to recognize the usefulness of drugs on different types of patients [60].

14.7 SUMMARY

Many technical challenges are faced by scientists during the formulation development. With the passage of time, lots of changes in the implementation and uses of excipients gained more importance than others. These changes offer the possibility of delivering a new drug moiety in more preferred dosage form with good aqueous solubility and better bioavailability, along with reformulation of marketed drugs with the purpose of life-cycle extension. In this respect, it is imperative to make distinction among different pharmaceutical products that contain a variety of inert substances apart from the drug itself. This information is also important to ensure that the incorporated excipients are compatible with the drug, that is, vivacious for the stability of dosage forms and safe and efficacious in their administration, directly and indirectly.

REFERENCES

1. https://www.chemicalbook.com/ProductCatalog_EN/14.htm (July–August 2018).
2. Agosta, W.C. 1997. Medicines and drugs from plants. *J Chem Educ.* 74(7): 857.
3. Generic and Brand Name Drugs: Understanding the Basics. Depression and Bipolar Support Alliance. Available at: www.DBSAlliance.org (July–August 2018).
4. WHO. 2013. Generic Drugs. Available at: www.who.int/trade/glossary/story034/en/ (July–August 2018).
5. Allen, L.V., N.G. Popovich, H.C. Ansel, and H.C. Ansel. 2005. *Ansel's Pharmaceutical Dosage Forms and Drug Delivery Systems.* Philadelphia, PA: Lippincott Williams & Wilkins.
6. Chaudhary, S.S., M. Tariq, R. Zaman, and I. Shaikh. 2013. Solid dosage forms in Unani system of medicine: An overview. *J Pharm Sci Innov.* 2(3): 17–22.
7. Tiwari, G., R. Tiwari, B. Sriwastawa et al. 2012. Drug delivery systems: An updated review. *Int J Pharm Investig.* 2(1): 2–11.
8. Gayathri, A. 2015. Pharmaceutical formulation & drug delivery technologies. *Res Rev* 4(4): 17–22.
9. Chan, Y.K., K.P. Ng, and D.S.M. Sim. 2014. *Pharmacological Basis of Acute Care.* Cham, Switzerland: Springer.
10. Aulton, M.E. 2007. *Aulton's Pharmaceutics—The Design and Manufacture of Medicines.* Edinburgh, Scotland: Churchill Livingstone.
11. Shangraw, R.F. and D.A. Demarest. 1993. A survey of current industrial practices in the formulation and manufacturing of tablets and capsules. *Pharm Tech.* 17: 32–44.
12. Pundir, S. and A.M.L. Verma. 2014. Review on Lozenges. *Journal der Pharmazie Forschung.* 2(1): 1–10.

13. Gautami, J. 2016. Liquid dosage forms. *Nano Sci Nano Technol*. 10(3): 101.
14. Jones, D.S. 2016. *FASTtrack Pharmaceutics: Dosage Form and Design*. 2nd edition. London, UK: Pharmaceutical Press.
15. Martin, A. 2016. *Physical Pharmacy*, 7th edition. Philadelphia, PA: Wolters Kluwer.
16. Edman, P. 1994. Pharmaceutical formulations—Suspensions and solutions. *J Aerosol Med*. 7(Suppl 1): S3–S6.
17. Kathpalia, H. and C. Phadke. 2014. Novel oral suspensions: A review. *Curr Drug Deliv*. 11(3): 338–358.
18. Brown, T.L., S. Petrovski, H.T. Chan et al. 2018. Semi-solid and solid dosage forms for the delivery of phage therapy to epithelia. *Pharmaceuticals*. 11(1): 26.
19. Usha, S. and A. Mahajan Ashish. 2015. Review on: An ointment. *Ijppr. Human*. 4(2): 170–192.
20. Kulkarni, V.S. and C. Shaw. 2015. *Essential Chemistry for Formulators of Semisolid and Liquid Dosages*. London, UK: Academic Press.
21. Thompson, J.E. and L.W. Davidow. 2009. *A Practical Guide to Contemporary Pharmacy Practice*. 3rd edition. Philadelphia, PA: Lippincott Williams & Wilkins.
22. Sunita, L. 2012. An overview on: Pharmaceutical aerosols. *Int Res J Pharm*. 3(9): 68–75.
23. Kalyan Babu, P.G., M. Vital, K. Nirupama, M. Gayatri devi and P. Uma Devi. 2017. Solid Dosage Forms. *Int Res J Pharm Sci*. 8: 1–3.
24. Patel, N.P., A.A. Patel, and M.K. Modasiya. 2012. Aerosols: Pulmonary drug delivery system. *Int J Pharm Chem Sci*. 1(1).
25. Das, P., P. Saha, Krishan and R. Das. 2018. Pharmaceutical packaging technology: A brief outline. *Res J Pharm Dos Forms Technol*. 10(1): 23.
26. Balakrishna, T. et al. 2016. A review on pharmaceutical containers and closures. *Indo Am J P Sci*. 3(8): 867–879.
27. Verma, C. and M.K. Mishra. 2016. Pharmaceutical preformulation studies in formulation and development of new dosage form: A review. *Int J Pharm Res Rev*. 5(10): 12–20.
28. Honmane, S.M., Y.D. Dange, R.A.M. Osmani, and D.R. Jadge. 2017. General considerations of design and development of dosage forms: Pre-formulation review. *Asian J Pharm*. 11 (3): S479–S488.
29. Aulton, M.E. 2002. *Pharmaceutics*. 2nd ed. Edinburgh, Scotland: Churchill Livingstone.
30. Verma, S., A. Baghotia, J. Singh, K. Saroha, S. Kumar and D. Kumar. 2016. Pharmaceutical excipients: A regulatory aspect. *Pharm Innov J*. 5(6): 124–127.
31. Siew, A. 2016. Excipients for formulation success. *PharmTechnol*. 40(10): 22–27.
32. Haywood, A. and B.D. Glass. 2011. Pharmaceutical Excipients—Where Do We Begin? *Aust Prescr*. 34(4): 112–114. doi:10.18773/austprescr.2011.060.
33. Varma, V.K. 2016. Excipients used in the formulation of tablets. *Res Rev*. 5(2): 143–154.
34. Furrer, P. 2013. The central role of excipients in drug formulation. *Eur Pharm Rev*. 18: 67–70.
35. Chaudhari, S.P. and P.S. Patil. 2012. Pharmaceutical excipients: A review. *Int J Adv Pharm Biol Chem*. 1(1): 21–34.
36. Katdare, A. and M.V. Chaubal. 2006. *Excipient Development for Pharmaceutical, Biotechnology, and Drug Delivery Systems*. New York: Informa.
37. Ravi, A.D., S. Saxena, and D. Nagpal. 2015. A concise understanding of pharmaceutical excipients. *Int J Pharm Pharm Res*. 3(3): 122–136.
38. Pandey, V.P., K. Venkateswara Reddy, and R. Amarnath. 2009. Studies on diluents for formulation of tablets. *Int J Chem Sci*. 7(4): 2273–2277.
39. Shalini, S. 2012. Advantages and applications of nature excipients—A review. *Asian J Pharm Res*. 2(1): 30–39.
40. Nagpal, N., P. Kaur, R. Kumar, S. Rahar, R. Dhawan, and M. Arora. 2016. Pharmaceutical diluents and their unwanted effects: A review. *Bull Pharm Res*. 6(2): 45–49.

41. Kolář, P., J.-W. Shen, A. Tsuboi, and T. Ishikawa. 2002. Solvent selection for pharmaceuticals. *Fluid Phase Equilib.* 194–197: 771–782.

42. Singh, P. and M. Sinha. 2013. Determination of residual solvents in bulk drug and formulations. *Am J Pharm Tech Res.* 3(4): 289–295.

43. Verma, M., S.S. Gangwar, Y. Kumar, M. Kumar, and A.K. Gupta. 2016. Study the effect of co-solvent on the solubility of a slightly water soluble drug. *J Biomed Pharm Res.* 4(37): 1–6.

44. Fishburn, A.G. 2014. Diluents, solvents and liquid vehicles, in *An Introduction to Pharmaceutical Formulation*, A.G. Fishburn, W.H. Linnell, and A.J. Evans (Eds.). Kent, UK: Elsevier Science, pp. 15–24.

45. Sabne, P.S., A.N. Avalaskar, R. Jadhav, and P.S. Sainkar. 2013. Natural excipients. *Res J Pharm Biol Chem Sci.* 4(2): 1346–1354.

46. Hartesi, B., Sriwidodo, M. Abdassah, and A.Y. Chaerunisaa. 2016. Starch as pharmaceutical excipient. *Int J Pharm Sci Rev Res.* 41(2): 59–64.

47. Shailendra, P., A. Shikha, and L.B. Singh. 2012. Natural binding agents in tablet formulation. *Int J Pharm Biol Arch.* 3(3): 466–473.

48. Kestur, U. and D. Desai. 2017. Excipients for conventional oral solid dosage forms, in *Pharmaceutical Excipients: Properties, Functionality, and Applications in Research and Industry*, O.M.Y. Koo (Ed.). Hoboken, NJ: John Wiley & Sons, pp. 51–96.

49. Dave, N. and T. Joshi. 2017. A concise review on surfactants and its significance. *Int J Appl Chem.* 13(3): 663–672.

50. Sekhon, B.S. 2013. Surfactants: Pharmaceutical and medicinal aspects. *J Pharm Technol Res Manag.* 1: 11–36.

51. Fahelelbom, K.M.S. and Y. El-Shabrawy. 2007. Analysis of preservatives in pharmaceutical products. *Pharm Rev.* 5(1).

52. Khairi, M.S. Fahelelbom and Yasser El-Shabrawy. 2007. Analysis of Preservatives in Pharmaceutical Products. *Pharmaceutical Reviews* 5(1): 1–55.

53. *Remington: The Science and Practice of Pharmacy.* 2006. 21st ed. Philadelphia, PA: Lippincott Williams & Wilkins.

54. Jones, B.E. 2004. Gelatin alternatives and additives, in *Pharmaceutical Capsules*, B.E. Jones and F. Podczeck (Eds.). London, UK: Pharmaceutical Press, pp. 63–70.

55. Fridrun Podczeck, and Brian E. Jones, 2004. Gelatin alternatives and additives, in *Pharmaceutical Capsules*, B.E. Jones and F. Podczeck (Eds.). 2nd Edition. London, UK: Pharmaceutical Press, pp. 63–70. https://www.bookdepository.com/author/Fridrun-Podczeck.

56. Sharma, A.V. and P.V. Sharma. 1988. Flavouring agents in pharmaceutical formulations. *Anc Sci Life.* 3(1): 38–40.

57. Trout, B. et al. 2009. Design of therapeutic proteins with enhanced stability. *Proc Natl Acad Sci USA.* doi:10.1073/pnas.0904191106.

58. Li, B. et al. 2009. Multilayer polypeptide nanoscale coatings incorporating IK-12 for the prevention of biomedical device-associated infections. *Biomaterials.* 30(13): 2552–2558.

59. Nel, A. et al. 2009. Understanding biophysicochemical interactions at the nano–bio interface. *Nat Mater.* 8(7): 543–557.

60. Hodson, R. Precision Medicine. Nature Outlook 8 September 2016; Vol 537(7619): S49 Retrieved August 31, 2018 from https://www.nature.com/articles/537S49a.pdf.

CHAPTER 15

Droplet-Based Microfluidics

Communications and Networking

Werner Haselmayr, Andrea Zanella, and Giacomo Morabito

CONTENTS

I̲N̲ ̲T̲H̲I̲S̲ ̲C̲H̲A̲P̲T̲E̲R̲,̲ ̲W̲E̲ ̲I̲N̲T̲R̲O̲D̲U̲C̲E̲ the emerging field of microfluidic communications and networking, where tiny volumes of fluids, so-called *droplets*, are used for communications and/or addressing purposes in microfluidic chips. This is a promising approach for realizing next-generation lab-on-a-chip devices, enabling more flexibility, biocompatibility, and low-cost fabrication.

This chapter starts with the basics of droplet-based microfluidics. In particular, we discuss the analogy between microfluidic and electric circuits and the resistance increase brought by droplets in microfluidic channels. Moreover, we describe different droplet-generation methods, including concepts for the accurate droplet creation at prescribed times, with certain size and at a certain distance. This is crucial for the practical realization of droplet-based communications and networking.

Then, we discuss different information-encoding schemes for droplet-based microfluidic systems and compare them in terms of encoding/decoding complexity, error performance, and the resulting channel capacity. Moreover, we characterize the microfluidic communication channel, showing that the noise can be modeled as a Gaussian random variable. This section also serves as a basis for switching and addressing in microfluidic networks.

Finally, we introduce the concept of microfluidic networking. Similar to computer networks, microfluidic switches are the key building block in such networks. After discussing some constraints and the main switching principle, we present single- and multiple-droplet switches, controlling the path of single or multiple droplets within the network, respectively. The proposed switching methods exploit only hydrodynamic effects to control the droplets, and thus, the microfluidic devices are biocompatible and can be fabricated at low cost (e.g., using three-dimensional [3D] printers). Finally, we discuss various addressing schemes for a microfluidic bus network, which is a promising network topology owing to its simplicity, flexibility, and scalability.

15.1 INTRODUCTION

Microfluidic systems deal with the control and manipulation of small amounts of fluids in channels at micrometer scale [49]. Owing to the small channel geometries and the low flow velocity, the flow in such systems is essentially laminar (low Reynolds number). Such systems allow for the miniaturization, integration, automation, and parallelization of biochemical assays [16]. In particular, laboratory experiments can be performed with a significantly reduced reagent consumption and less time. Microfluidic systems that realize a single or multiple laboratory experiments on a single chip are referred to as lab-on-a-chip (LoC) systems. LoC devices can be used for a wide variety of applications, including (single) cell analysis, DNA sequencing, and drug screening [29,41].

In order to realize LoCs, different platforms have been proposed, and the most promising ones are discussed in the following.[1]

In *continuous-flow-based LoCs*, the fluid flow in a microfluidic chip is controlled through integrated microvalves [1,41]. The chip consists of a control and a flow layer. The control layer controls whether the integrated microvalves are open or closed and, thus, determines the fluid flow in the flow layer. The main drawbacks of continuous-flow-based LoC devices are that they have a complex and costly multilayer fabrication process.

In *droplet-based LoCs*, tiny volume of fluids, so-called *droplets*, are generated, which can be independently controlled and manipulated. Droplets can be used as microreactors, providing a better mixing, encapsulation, and analysis [47]. Usually, droplet-based systems use two immiscible phases, namely the continuous phase (droplet surrounding medium, such as air and oil) and dispersed phase (droplet fluid). Droplet-based LoCs can be divided into two approaches: The *surface-based approach* (often referred to as *digital microfluidics*) moves the droplets on a planar (open) surface, using electrowetting (EWOD) [44] or surface acoustic waves (SAW) [51]. In case of EWOD, the droplets are moved on a two-dimensional grid of electrodes, by turning on and off the voltage at the electrodes. For the SAW-based approach, the droplets are actuated by an acoustic shock wave on a hydrophobic surface. Both methods allow to arbitrarily move the droplets on the surface, and thus, the path of the droplets is freely programmable. Many computer-aided design (CAD) methods have been proposed for digital microfluidics to support the design process of such chips. These methods assist the designer in tasks such as testing [46,54], synthesis [26,27,45], and sample preparation [42,43]. Unfortunately, surface-based approaches suffer from evaporation of liquids, the fast degradation of surface coatings, and its lacking biocompatibility [41]. In *channel-based approaches*, the droplets flow in closed microchannels triggered by an external force, such as pressure and syringe pump, at the chip boundary. Since the droplet movement in the channel is mainly determined by hydrodynamic effects, this approach does not compromise the viability of the chemical/biological sample. Moreover, the closed channels allow for the incubation and storage of liquid assays over a long period of time and, thus, avoid evaporation and unwanted reactions [41]. In contrast to the open-surface approach, the droplet movement is defined by the geometry of the microchannels, which limits the flexibility of this approach. Currently, such LoC devices perform a set of operations (e.g., mixing and heating) in a predefined order to implement certain laboratory experiments. In order to realize programmable and flexible systems, the interconnection of such operations in a microfluidic network has been proposed in [36]. In particular, microfluidic networking targets the dynamic assignment of the droplets' path in a microfluidic network to perform specific analyzes. The network nodes are interconnected using so-called *microfluidic switches*, which control the path of the droplets by only exploiting hydrodynamic effects. Thus, the main design parameters of such switches are the geometry and hydrodynamic forces

---

[1] Please refer to [29,41] for a comprehensive overview.

(e.g., volumetric flow rates). Recently, various CAD methods have been proposed for microfluidic networking, including the automatic design and dimensioning of microfluidic networks with respect to various design objectives [23,24], the simulation of microfluidic networks based on the electric circuit analogy [22], and the generation of appropriate sequences of control droplets to correctly route cargo droplets in the network [25].

Recently, two promising applications for microfluidic networking have been suggested: *fast and flexible drug screening* [31] and *fast and flexible waterborne pathogen screening* [30]. Both applications require a fast analysis (which can be realized through parallel processing in a microfluidic bus network) and that the viability of the biochemical sample is not compromised. In fast and flexible drug screening, the nodes of the bus network contain microreactors with infected human cell. A large number of droplets, including different antibiotics or different concentration of antibiotics, are routed in parallel to the microreactors. The reaction between the infected cell and the antibiotics can be observed inside the microreactor or at an analysis stage. In fast and flexible waterborne pathogen screening, two types of droplets, including a concentration of pathogens and disinfectants, are routed to the network node (microreactor). At the node, the droplets are merged, and they initiate a lysis process, followed by a polymerase chain reaction (PCR). The interaction between the disinfectants and pathogen DNA can be observed inside the microreactor or at an analysis stage. It is important to note that, for both applications, microfluidic networking enables the analysis of a large number of antibiotics or disinfectants in a short time on a single device.

This chapter provides an overview of the main principles of communications and networking in microfluidic channels using droplets. In particular, the chapter is organized as follows: Section 15.2 introduces the basics of droplet-based microfluidics. The principle of information encoding using droplets is presented in Section 15.3, serving as the basis for addressing in microfluidic networks. In Section 15.4, the concept of microfluidic networking is introduced, presenting microfluidic switches as the key building block of such networks and the associated addressing paradigm for microfluidic bus networks. Finally, Section 15.5 concludes the chapter with a summary and an outlook of future research directions.

15.2 BASICS ON DROPLET-BASED MICROFLUIDICS

In this section, we describe the basics of droplet-based microfluidics. We start by introducing the analogy between microfluidic and electric circuits in Section 15.2.1. Then, we describe three different methods for generating droplet trains in Section 15.2.2. We extend these methods in Section 15.2.3, presenting three droplet-on-demand (DoD) systems, which are able to create droplets at prescribed times, with a certain size and at a certain distance. These methods are crucial for the practical realization of droplet-based communications and/or networking. Finally, we discuss the resistance increase in microfluidic channels due to droplets in Section 15.2.4, which is exploited by the microfluidic switches used in microfluidic networks.

15.2.1 Microfluidic Channel

The Hagen–Poiseuille equation describes the volumetric flow rate Q (in m³/s) of a laminar stationary flow of an in-compressible Newtonian fluid through channels of micrometer scale and can be expressed as [6]:

$$Q = \frac{\Delta P}{R},$$ (15.1)

where ΔP (in [Pa]) denotes the pressure drop between the two endpoints of the channel, while R is the hydrodynamic resistance of the channel (see Figure 15.1). The resistance of rectangular channels,[2] with length l, width w, and height h, is given by [18]:

$$R(\mu, l) = \frac{a\mu l}{wh^3},$$ (15.2)

where μ denotes the viscosity (in Pa s) of the fluid, and the dimensionless parameter a is defined by:

$$a = 12 \left[1 - \frac{192h}{\pi^5 w} \tanh\left(\frac{\pi w}{2h} \right) \right]^{-1}.$$ (15.3)

The Hagen–Poiseuille equation in (15.1) corresponds to Ohm's law for electric circuits, where the volumetric flow rate, the pressure drop, and the hydrodynamic resistance are analogous to the current, the voltage drop, and the electric resistance. Thus, steady-state flow rates (or pressure drops) in a microfluidic network can be derived using well-known methods from the electric circuit analysis (e.g., Kirchhoff's circuit laws). Finally, the speed of the fluid in a microfluidic channel can be calculated as:

$$v = \frac{Q}{A},$$ (15.4)

where $A = wh$ denotes the cross-section of the channel.

15.2.2 Droplet Generation

In droplet-based microfluidics, a fluid (dispersed phase) is injected into another immiscible fluid (continuous phase). The droplets formed by the dispersed phase are carried by the continuous phase through the microchannels. In the following subsections, the three main droplet-generation geometries are discussed.[3]

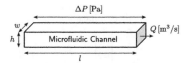

FIGURE 15.1   Microfluidic channel.

---

[2] For a cylindrical channel, with radius r and length l, the hydrodynamic resistance is given by $R(\mu, l) = (8\mu l)/(\pi r^4)$ [6].
[3] Please refer to [55] for more details.

15.2.2.1 Co-Flowing Stream

As shown in Figure 15.2a, the inner channel contains the dispersed phase and is placed inside an outer channel that includes the continuous phase. If the flow rate of the continuous phase is low, the geometry works in the so-called *dripping regime* and the droplet breaks up near the nozzle of the inner channel. In this case, the size of the droplets is comparable to the dimensions of the dispersed phase channel. If the outer flow rate is higher than a critical value (depending on the viscosity and interfacial tension [13]), the geometry works in the so-called *jetting regime* and the dispersed phase are stretched after the end of the inner channel and the droplet forms downstream of the nozzle. Owing to the high shear stress, the droplet breaks up if it reaches a certain size. The main drawback of this geometry is the realization of the inner channel inside the outer channel, which requires a precise fabrication.

15.2.2.2 Flow Focusing

As shown in Figure 15.2b, the channel containing the dispersed phase penetrates into two counter-flowing streams with the continuous phase. In the so-called *squeezing regime*, the dispersed phase is injected into the main channel until the flows from the perpendicular channels are blocked (filling stage). This increases the pressure in the continuous phase, resulting in a breakup of the droplet (necking stage). In the jetting regime, the droplet breakup occurs downstream after the junction. A mathematical model that predicts the generated droplet length and interdroplet distance in the squeezing regime has been introduced in [9]. The droplet length is given by:

$$l_d = w_d \left(\alpha + \beta \frac{Q_d}{Q_c} \right), \tag{15.5}$$

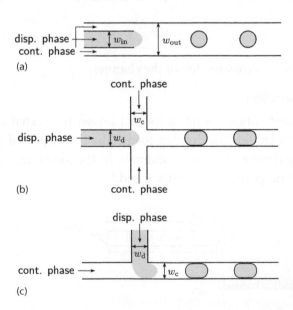

FIGURE 15.2 Geometries for droplet generation: (a) Co-flowing (dripping regime), (b) flow focusing (squeezing regime), and (c) T-junction (squeezing regime).

where α describes the volume of the dispersed phase flowing into the droplet during the filling stage, and β corresponds to the volume of continuous phase flowing into the junction during the necking stage. The expressions for α and β can be found in [9]. The width of the dispersed phase channel is given by w_d. Moreover, Q_c and Q_d describe the volumetric flow rate of the continuous and dispersed phases, respectively. The droplet-inter distance between the droplets is defined by:

$$l_\delta = l_d \left(\frac{Q_c}{Q_d} + 1 \right) = w_d \left(1 + \frac{Q_d}{Q_c} \right) \left(\alpha \frac{Q_c}{Q_d} + \beta \right). \tag{15.6}$$

It is important to note that the model described previously is valid only if the droplet length is larger than the channel width. Interestingly, the droplet length l_d and the droplet-inter distance l_δ cannot be controlled independently, since they depend on the same parameters.

15.2.2.3 Cross-Flowing Stream (T-Junction)

Owing to its simple design, this geometry is most commonly used for droplet generation. As shown in Figure 15.2c, the channel containing the dispersed phase is connected perpendicular to the channel with the continuous phase. When the dispersed phase is injected into the continuous phase, interfacial tension and viscous shear stress will apply to the two fluids. In particular, the interfacial tension and the shear stress have a stabilizing effect and a destabilizing effect, respectively. If the interfacial tension dominates, the geometry works in the squeezing regime. Owing to the low shear stress, the dispersed phase is able to fill the entire channel before the droplet breaks up, leading to droplets that fill the entire channel. On the other hand, if the shear stress dominates, the geometry works in the dripping regime, where the droplet immediately breaks up after penetrating the continuous phase. This leads to small droplets that do not fill the entire channel. The capillary number describes the ratio between viscous shear stress to interfacial tension and can be expressed as follows:

$$C_a = \frac{\mu_c Q_c}{\sigma w_c h_c}, \tag{15.7}$$

where μ_c denotes the viscosity of the continuous phase, and σ corresponds to the inferfacial tension. The width and height of the continuous phase channel are given by w_c and h_c, respectively. Thus, the capillary number defines the threshold between the squeezing and dripping regime. In order to enable a regular droplet formation, the geometry should work in the squeezing regime, which requires $C_a < 10^{-2}$ [20]. In order to remain in the squeezing regime, the speed of the continuous phase (see (15.4)) should not exceed the threshold [20]:

$$v_c^* = \frac{\sigma}{\mu_c} 10^{-2}. \tag{15.8}$$

A mathematical model predicting the droplet length and the interdroplet distance in the squeezing regime have been presented in [4,20]. The droplet length is given by [20]:

$$l_d = w \left(1 + \zeta \frac{Q_d}{Q_c} \right), \tag{15.9}$$

where ζ denotes a dimensionless number of order 1 (see [4]) that mainly depends on the junction geometry. The distance between consecutive droplets can be expressed as [4]:

$$l_\delta = l_d \frac{Q_c}{Q_d} = w \left(\zeta + \frac{Q_c}{Q_d} \right). \tag{15.10}$$

Similar to the flow focusing geometry, the droplet length l_d and the droplet-inter distance l_δ cannot be controlled independently.

15.2.3 Droplet on Demand

In the previous section, droplet-generation methods that create trains of droplets with a certain distance and a certain size have been presented. However, for the practical realization of droplet-based communications and/or networking in a microfluidic chip (see Sections 15.3 and 15.4), it is crucial to generate droplets at prescribed times with a certain size and at a certain distance, yielding so-called *DoD* systems.

Most reported DoD systems rely on the active control of the fluids on the microfluidic chip, utilizing integrated microvalves [38,53] or EWOD [28,40]. Although such systems enable precise control of the droplets, they suffer from complex and costly fabrication. Moreover, in case of EWOD, the electrical field may compromise the viability of the biochemical sample, leading to a low biocompatibility of such systems.

In order to overcome the aforementioned drawbacks, various passive DoD systems have been developed, enabling simple chip design and maintaining the viability of the biochemical sample. In the following, a description of three promising passive DoD systems is provided, and its applicability for communications and/or networking purposes is discussed. Essentially, the methods differ in the number of controlled channels as well as in the type of control.

15.2.3.1 Three-phase Control

In [48], a flow-focusing geometry (see Figure 15.2b) is used for the droplet generation. The proposed method requires the control of three channels: two input channels that contain the continuous and dispersed phases, respectively, and one output channel where a negative pressure is applied. For the droplet generation, first, the continuous and dispersed phases are balanced to obtain a stable interface. Then, the negative pressure is applied to start the droplet formation. Thus, by controlling the magnitude and time period of the negative pressure pulse, droplets of various size and droplet-inter distance can be generated. However, owing to the negative pulse, an unpredictable flow distribution occurs, and thus, this approach is not applicable for droplet-based networking. But it is important to note that this method is a promising approach for many other applications, for example, microfluidic chips, where the compartmentalizing of samples is important [33].

15.2.3.2 Two-phase Control

In [11,12], a T-junction (see Figure 15.2c) is used for droplet generation. The method uses external valves[4] to mutually switch on and off the flow of the continuous and dispersed phases, respectively. For the droplet formation, the valve for the dispersed phase is open for a certain time period, whereas the valve for the continuous phase is closed. This induces the droplet formation, where the droplet breaks up after this time period. Thus, droplets of arbitrary size and droplet-inter distance can be generated through controlling the time period. It was shown in [12] that the droplet size increases linearly with this time period. This method can be used as DoD system for droplet-based networking but needs a strict synchronization between the two external valves to enable mutual switching.

15.2.3.3 Single-phase Control

An alternative DoD method that requires no synchronization has been proposed in [30]. This approach uses a T-junction for the droplet generation, and only the flow of the dispersed phase is controlled (e.g., by an external valve), while the flow of the continuous phase is kept constant. For the droplet formation, first, the pressure for the continuous and dispersed phases is balanced in order to form a stable interface, the so-called *equilibrium state*. Once the system is in the equilibrium state, the droplet generation is induced by applying a positive pressure pulse to the dispersed phase for a certain time duration. The droplet breaks up at the end of this period, and the system is again back in the equilibrium state. Thus, droplets of different sizes and droplet-inter distances can be generated through varying the duration of the pulse and the time between two consecutive pulses, respectively (see Figure 15.3). Besides the fact that no synchronization is required, the main benefits of this approach are that only the precise control of the flow of the dispersed phase is crucial and that multiple DoD systems can be cascaded. The last mentioned benefit is especially important for droplet-based communications and/or networking when droplets composed of different substances need to be generated.

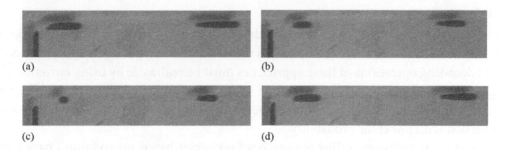

FIGURE 15.3   Passive DoD method with single-phase control. Droplets of different sizes and interdroplet distances are generated through applying positive pulses of various durations T^{5}: (a) T = 1.5 s (left) and T = 1.75 s (right), (b) T = 1 s (left) and T = 1.25 s (right), (c) T = 0.75 s (left) and T = 1 s (right), and (d) T = 1.25 s (left) and T = 1.5 s (right).

---

[4] It is important to note that using external valves rather than integrated valves keeps the design of the microfluidic chip simple.

[5] Reprinted from [30], ©2019, with permission from Elsevier.

15.2.4 Resistance Increase by Droplets

The presence of droplets changes the hydrodynamic resistance of a microfluidic channel. This is because of the friction with the carrier fluid and forces due to the inhomogeneity between the viscosity of the continuous and dispersed phases [7]. The variation of the resistance in a microfluidic channel of length l, produced by a droplet of length l_d, with viscosity μ_d, can be approximated as [4]:

$$R_\Delta(l_d)|_{\rho>1} = [R(\mu_c, l - l_d) + R(\mu_d, l_d)] - R(\mu_d, l)$$
$$= \frac{a(\mu_d - \mu_c)l_d}{w_c h_c^3}, \tag{15.11}$$

with $\rho = \mu_d/\mu_c$. Hence, the hydrodynamic resistance of the channel with a droplet is given by $R(\mu_c, l) + R_\Delta(l_d)$. It is important to note that the relation in (15.11) holds only if the droplet occupies the entire channel section (squeezing regime). In case that $\rho < 1$, the viscosity of the droplets can be neglected, and the resistance variation is given by [21]:

$$R_\Delta(l_d)|_{\rho<1} = \gamma \frac{a\mu_c l_d}{w_c h_c^3}, \tag{15.12}$$

with $\gamma \in [2,5]$.

15.3 MICROFLUIDIC COMMUNICATIONS

In this section, we describe the basics of droplet-based microfluidic communications.[6] More specifically, in Section 15.3.1, we present the main approaches that can be utilized to encode information by using droplets. In Section 15.3.2, we characterize the microfluidic communication channel and show its impact on the different encoding schemes. Finally, we discuss the channel capacity of the proposed encoding schemes in Section 15.3.3.

15.3.1 Information Encoding

A new communication paradigm, such as droplet-based microfluidic communications, needs new information-encoding schemes. In order to be effective and efficient, the encoding and decoding operations of these approaches must be realizable by using current technologies, and their practical implementation should be simple and cheap. Moreover, the impact of the microfluidic channel should be as low as possible, enabling high data rate transmission with low error probability.

Different information-encoding approaches for droplet-based microfluidics have been introduced in [36]. More specifically, we can distinguish four different approaches.

15.3.1.1 Droplet Presence/Absence Encoding

This is the most obvious way of encoding information by using droplets and resembles On-Off Keying (OOK) from wireless communications [50]. It associates the presence of a droplet to the bit 1 and the absence of a droplet to the bit 0. Let us denote the bits

---

[6] Communications techniques have been proposed also in the context of continuous microfluidics, as discussed in [2,3].

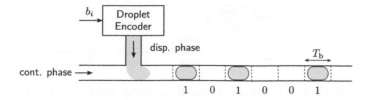

FIGURE 15.4 Droplet-based information encoding using presence/absence encoding. The transmission of the bits {100101} is shown as an example.

to be encoded as $\{b_1, b_2, \ldots\}$. The encoder divides the time into time slots, each of duration T_b. Accordingly, during the ith time slot, the transmitter will generate a droplet if $b_i = 1$, whereas it does not generate any droplet if $b_i = 0$. For example, in Figure 15.4, we show the output of the droplet encoder when the bits to be transmitted are {100101}. Obviously, this encoding scheme requires accurate DoD solutions, as discussed in Section 15.2.3.

At the receiver side, the decoder must be able to detect whether in a certain time slot there is a droplet or not. A possible implementation could look as follows. Assuming that the microfluidic device is transparent, a microcamera points at the microfluidic channel. The output of the microcamera is given as input to a video-image-processing element that detects whether in a certain area the dominant color is Color 1 (evidence that there is a droplet) or Color 2 (evidence that there are no droplets). This procedure does not need high-quality cameras, which might be costly. However, we explicitly observe that the frame rate of the camera gives an upper limit on the maximum slot rate and, thus, provides an upper limit of the data rate. In fact, since transmitter and receiver do not have to be synchronized, the number of frames higher than two is required for each slot. This constraint will have a higher impact on encoding the information between consecutive droplets, as discussed next. Note that errors occur when there is a droplet in a slot that should be empty, or vice versa. This happens if, owing to the dynamics in the microfluidic channel, the droplet changes its relative position in the train of droplets.

15.3.1.2 Droplet Distance Encoding (also Referred to as Communication Through Silence)

In this case, the bits to be transmitted are divided into groups of m consecutive bits, where each bit group is mapped to its integer value v_j. The information is encoded in the time interval between two consecutive droplets, where the relation between the time interval and the integer value v_j can be expressed as:

$$\Delta T_j = T_{min} + v_j \times T_{step}, \tag{15.13}$$

where T_{min} represents the minimum time interval between two consecutive droplets to avoid droplet coalescence, and T_{step} denotes the granularity of the encoding output. Thus, the smaller T_{min} and T_{step}, the higher the data rate. However, if T_{min} is too small, droplets can be very close to each other, and coalescence might occur. Moreover, if T_{step} is too small, it is very likely that the estimation of v_j is erroneous, resulting in a worse error performance.

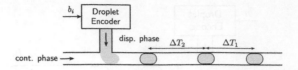

FIGURE 15.5 Droplet-based information encoding using distance encoding. The transmission of the bits {100101} is shown as an example. The bits are grouped into two groups representing the values $v_1 = 4$ and $v_2 = 5$, which are mapped to the time intervals $\Delta T_1 = T_{min} + 4T_{step}$ and $\Delta T_2 = T_{min} + 5T_{step}$, respectively.

As an example, we show in Figure 15.5 the output of the droplet encoder when the bits to be transmitted are {100101}, with $m = 3$. In this case, there are two groups, including three bits. The groups are {100} and {101} and represent the integer values $v_1 = 4$ and $v_2 = 5$, respectively. These integer values are mapped to the time intervals $\Delta T_1 = T_{min} + 4T_{step}$ and $\Delta T_2 = T_{min} + 5T_{step}$. According to these time intervals, the first droplets is generated when the session starts, the second droplet is generated ΔT_1 later, and the last droplet is generated another time interval ΔT_2 later.

At the receiver side, a similar approach as for *droplet presence/absence encoding* (DPAE) can be used. In particular, the video-image-processing element estimates the time between the centers of two time intervals in which the dominant color in a certain area is Color 1 (evidence that a droplet is passing). It is important to note that the frame rate of the camera must be more than twice the value T_{step}^{-1}, which might have a significant effect on the costs of the decoding operation.

15.3.1.3 Droplet Size Encoding (also Referred to as Droplet Length Encoding)

In this case, different droplet sizes are used to convey information. Similar, to the distance-encoding scheme described previously, multiple consecutive bits can be grouped and mapped to the associated integer values. These values are then mapped to different droplet sizes. Thus, the data rate can be increased by considering large bit groups, resulting in a large number of different droplet sizes. However, this increase in data rate is payed in terms of error performance. Note that this encoding scheme is a generalization of DPAE, where a droplet of a certain size is transmitted for bit 1 and a droplet of size zero (no droplet) is generated for bit 0. As an example, we show in Figure 15.6 the transmission of the bit sequence {100101} with $m = 1$, that is, a group contains 1 bit. In this case, a large droplet and a small droplet are generated by the droplet encoder for bit 1 and bit 0, respectively.

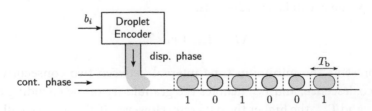

FIGURE 15.6 Droplet-based information encoding using size encoding. The transmission of the bits {100101} is shown as an example.

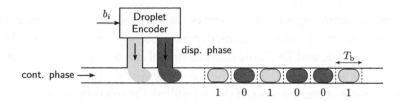

FIGURE 15.7   Droplet-based information encoding using composition encoding. The transmission of the bits {100101} is shown as an example.

At the receiver side, a similar approach as for DPAE can be used (microcamera, video-image-processing element). The decoding process becomes more complex and error prone if the difference between the individual droplet sizes is small, while this simplifies the encoding procedure.

15.3.1.4 Droplet Composition Encoding (DCE)

Similar to droplet size encoding (DSE), multiple bits can be grouped and mapped to the associated integer values, which are represented either by droplets composed of different substances or by droplets encapsulating different molecules. For example, in Figure 15.7, we show the transmission of {100101} with $m = 1$, where for bit 1 and bit 0, different substances are used.

It is important to note that compared with the previous encoding schemes, this approach requires a complex encoding operation. In particular, it requires at least two cascaded DoD systems,[7] which generate different types of droplets. The decoding at the receiver side might also be very complex. Most of the time, it requires some bio/chemical reaction (usually having a long processing time), resulting in a different output depending on the substance contained in the droplet. Nevertheless, droplet composition encoding (DCE) has attracted the interest of researchers, because besides the binary information, the droplets carry bio/chemical information [17].

15.3.2 Channel Characterization

We model the microfluidic communication channel by applying an information theoretic approach. This requires to probabilistically relate the input and output symbols of the channel, and thus, the model depends on the information-encoding scheme. In particular, we provide one model for *droplet distance encoding* (DDE) and another model for other schemes, that is, DPAE, DSE, and DCE.

In case of DDE, let X_j denote the random variable representing the distance between the jth and $(j+1)$th droplets (symbol v_j) at the input of the channel. As the droplets traverse the channel, they are subject to physical perturbations that modify their distance. Let Y_j denote the random variable representing the distance between these droplets at the output of the channel. It was experimentally shown that the difference between X_j and Y_j, that is, $E_j = Y_j - X_j$, can be modeled as a Gaussian random variable, with mean equal to zero and variance σ_E^2 [35].

---

[7] Please note that among the three DoD methods described in Section 15.2.3, only the single phase control method can be cascaded.

This behavior is not surprising, since E_j is a sort of noise and is the macroscopic evidence of the sum of a large number of microscopic terms, depending on a plethora of physical phenomena.

In case of DPAE, DSE, and DCE, the channel model can be significantly simplified if the slot duration T_b (see Section 15.3.1) is chosen so large that droplets do not leave their slot, and thus, no coalescence event occurs. If this is fulfilled, the microfluidic channel can convey 1 bit for each T_b and does not introduce errors. However, if the above assumption is not satisfied, the situation becomes much more complex, as discussed in detail in [19]. For example, it is possible that droplets have moved out from their slots, and thus, it is impossible for the receiver to detect the droplet. Moreover, it is also possible that two consecutive droplets merge with each other; that is, droplet coalescence occurs. If both droplets are of the same type, then the receiver can correctly decode them, but if they are of different type, then it is impossible for the receiver to detect the types of droplets. Accordingly, given that the output of the channel in a certain time slot depends on the droplets generated at the previous, the current, and the next slots, the channel is said to be with memory and anticipation [19].

15.3.3 Channel Capacity

The major performance metric for the proposed information-encoding schemes is the channel capacity, which we present in the following. In case of DDE, the channel capacity can be calculated as:

$$C_{\text{DDE}} = \max_{T_S} \left(\frac{C_S}{T_S} \right), \tag{15.14}$$

where C_S denotes the capacity for each channel use when the average time between the generation of two consecutive droplets is equal to T_S. The time duration T_S corresponds to the sum of the minimum distance $T_{\min}$ (see Section 15.3.1) and S, which denotes the average delay introduced by the encoding operation, that is, $T_S = T_{\min} + S$. Thus, (15.14) can be written as:

$$C_{\text{DDE}} = \max_{S} \left(\frac{C_S}{T_{\min} + S} \right). \tag{15.15}$$

Through simple information theoretic derivations (see [35] for details) and assuming that the noise E introduced by the microfluidic channel follows a Gaussian distribution (see Section 15.3.2), the capacity C_S can be expressed as:

$$C_S = \log_2 \left(\frac{eS}{\sqrt{2\pi e \sigma_E^2}} \right), \tag{15.16}$$

where e is the Euler number. By substituting (15.16) into (15.15), the channel capacity for distance-based information encoding can be written as [35]:

$$C_{\text{DDE}} = \frac{1}{\sqrt{2\pi e \sigma_E^2} \log 2} \exp\left(-W\left(T_{\min} / \sqrt{2\pi e \sigma_E^2}\right)\right), \tag{15.17}$$

TABLE 15.1   Comparison of Different Droplet-Based Information-Encoding Schemes

Encoding Scheme	Advantages	Disadvantages
Presence/Absence	Simple decoder Droplet-based computing	Low capacity
Distance	Simple decoder High capacity	Complex encoder
Size	Simple encoder	Low capacity
Composition	Biochemical applications	Low capacity Complex encoder/decoder

where $W(\cdot)$ denotes the Lambert function, which is defined through the inverse relationship $z = W(z)e^{W(z)}$.

For DPAE, DSE, and DCE, the calculation of the channel capacity is simple, assuming that droplets do not leave their slot, and thus, no coalescence events occur. In this case, the capacity is given by [36]:

$$C_{\text{DPAE/DSE/DCE}} = \frac{1}{T_b}. \tag{15.18}$$

However, for the general case, no closed-form expression exists. This is because the channel has both memory and anticipation (see Section 15.3.2) and there are no standard techniques that make it possible to calculate the capacity of such channels in closed form. Please refer to [19] for a comprehensive analysis based on Markov chains, where the capacity is evaluated numerically.

There have been only a few works comparing the channel capacity of the different encoding schemes. In [15], it is analytically shown that DDE gives larger capacity than the other schemes if the following relationship holds:

$$\frac{T_{\min}}{\sigma_E} > \frac{e\sqrt{2\pi e}\ln 2e^{1/\ln 2}}{2} = 16.4772, \tag{15.19}$$

which is fulfilled in most practical scenarios. In [32] and [50], the conditions when DDE provides higher capacity than other schemes are estimated through simulations. More specifically, the analysis in [50] suggests that DDE gives higher capacity in all practical scenarios.

Finally, in Table 15.1, we summarize the advantages and disadvantages of the information-encoding techniques described previously.

15.4 MICROFLUIDIC NETWORKING

The previous section has discussed the possibility to carry digital information in microfluidic channels by using droplets. Building upon such results, this section presents some recent approaches to further extend the concept of microfluidic communications into that of *microfluidic networks*. The final objective is to develop methodologies and devices that make it possible to interconnect simple microfluidic elements (e.g., mixing and heating) in order to enable the selective delivery of information (in the form of droplets) to specific destinations and/or the exchange of information among the different basic systems.[8] The grand vision is hence to somehow reproduce in the microfluidic domain the dramatic evolution that the networking technologies have determined into the digital communication domain.

Two key elements to enable the exchange of information among different interconnected systems are (1) *switching devices*, which make it possible to select the path followed by a droplet across an intersection with multiple outgoing branches, and (2) a proper *addressing scheme* that drives the switching process, in order to deliver the droplets to the intended receiver. Clearly, these two elements need to be jointly designed, in order to work properly.

In the remaining of this section, the prerequisites for the realization of microfluidic switches by using purely microfluidic principles (i.e., not considering hybrid technologies such as EWOD and SAW) are discussed in Section 15.4.1. Then, the switching principle is introduced in Section 15.4.2, presenting switching structures that control the path of a single droplet and multiple droplets, only exploiting hydrodynamic effects. Finally, various addressing schemes for a microfluidic bus network are presented in Section 15.4.3.

15.4.1 Basics on Microfluidic Switches

As mentioned previously, the switching elements are fundamental parts of any network, and microfluidic networks are no exception. However, both the switching principles and the structure of microfluidic switches are completely different from those of their electronic counterparts.

From a topological point of view, a switch needs to have one input channel (inlet) and at least two output channels (outlets), which represent the possible switching choices. Therefore, the basic structure of a switch is a channel that forks into two (or more) branches, usually with T or Y shape. Figure 15.8 exemplifies the structure of a T-junction switch with two outlets.

The carrier phase flows into the junction from the inlet, with a given volumetric flow rate Q, and then proceeds along the two outlets, with flow rates Q_1 and Q_2, respectively, such that $Q = Q_1 + Q_2$ for mass conservation. When a droplet enters such a switch from the

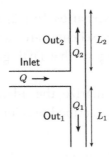

FIGURE 15.8   Simple T-shape microfluidic switch.

---

[8] In Section 15.1, two promising applications for microfluidic networks have been introduced, namely fast and flexible screening of drugs [31] and waterborne pathogens [30].

inlet and reaches the bifurcation, it can either split between the two outlets (see, e.g., [7,10]) or preserve its integrity and proceed along the outlet with the largest volumetric flow rate. The splitting condition, though potentially interesting, has not yet been investigated in the context of microfluidic networking. Here, the main focus is on the case where droplets maintain their structure across the microfluidic switches, an operational condition that is called *nonbreakup regime*. According to [37], to remain in the nonbreakup regime, the droplet length ℓ_d should not exceed a threshold ℓ_d^* given by:

$$\ell_d^* \approx \chi w_c C_a^{-0.21},\tag{15.20}$$

where χ is a dimensionless parameter that depends on the viscosity ratio between dispersed and continuous phases [34]. As a consequence, in order to remain in the nonbreakup regime and in the squeezing regime, the speed of the continuous phase in any channel of the circuit should not exceed the threshold (see (15.8)):

$$v_c^* = \frac{\sigma}{\mu_c}\min\left\{10^{-2},\left(\frac{\chi w_c}{\ell_d^{max}}\right)^{\frac{1}{0.21}}\right\},\tag{15.21}$$

where ℓ_d^{max} is the length of the longest droplet in the circuit [5].

15.4.2 Switching Principle

Under the conditions (15.20) and (15.21), a droplet entering a junction will not split but rather proceed along the outlet with the greatest volumetric flow rate, say Out_1, with flow rate Q_1 (see Figure 15.8), which is referred to as *primary outlet* in the following. The other outgoing branch Out_2, with flow rate Q_2, has a lower volumetric flow rate and will be called *secondary outlet*. The presence of a droplet in the primary outlet will increase its hydrodynamic resistance, as given in (15.11) and (15.12). Thus, the flow rate in this channel is decreased, and the rate in the other channel is increased. Therefore, Q_2 can temporarily become larger than Q_1, so that a second droplet that approaches the junction will be steered into Out_2. Hence, through carefully controlling the length of the droplets, the hydrodynamic resistances of the outlet branches of a T-junction, and the distance between droplets, it is possible to determine the direction taken by the droplets when crossing the junction. This is the basic principle exploited in the simple T-junction switch shown in Figure 15.8.

To collect empirical evidence of this switching principle, in [5], the authors realized a simple circuit consisting of a T-junction droplet generator, followed by a loop with uneven branches, as schematically shown in Figure 15.9. The geometric parameters of the microfluidic system are given in Table 15.2. As continuous and dispersed phases, hexadecane + 2% Span80, with $\mu_c = 3\,\text{mPa}\cdot\text{s}$ and density $\delta_c = 0.77\,\text{g/mL}$, and aqueous glycerol solution (glyc. 60% w/w), with $\mu_d = 12.8\,\text{mPa}\cdot\text{s}$ and $\delta_d = 1.15\,\text{g/mL}$, are used, respectively. The objective of the experiment was to characterize the switching of the droplets in the two arms of the loop. Since the two branches are of different lengths, they should have been crossed by an uneven number of droplets, with a preference for the shortest branch (L_2)

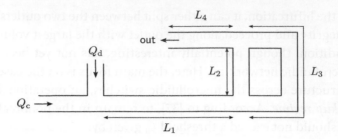

FIGURE 15.9 Sketch of the experimental microfluidic system, with the geometries given in Table 15.2.

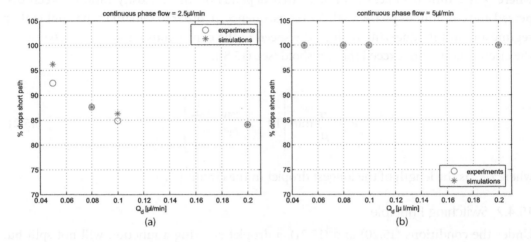

FIGURE 15.10 Experimental and simulative results representing the percentage of droplets selecting the shortest path (L_2) of the loop in Figure 15.9: (a) $Q_c = 2.5\ \mu\text{L/min}$ (b) and $Q_c = 5\ \mu\text{L/min}$.[9]

because of its lower resistance. Figure 15.10 shows the fraction of droplets switched into the shortest branch of the loop for two different values of continuous-phase flow rates. The circles represent the experimental values, while the stars are the results obtained through a MATLAB model based on the theoretical equations reported in this chapter (see [5] for further details). We observe that for $Q_c = 2.5\ \mu\text{L/min}$, the primary branch ($L_2$) is preferred by the large majority of droplets, but a few droplets are also switched into the other branch. This happens when the number of droplets in the primary branch is sufficiently large, and thus, the hydrodynamic resistance of the primary channel is higher than in the other channel. If the flow rate of the continuous phase is increased to $Q_c = 5\ \mu\text{L/min}$, the droplet size is very low and the interdroplet distance is very high. Thus, the additional resistance introduced in the primary channel is too low to compensate the unbalance between the two branches, and all droplets flow along the primary channel (Table 15.2).

Unfortunately, such a switching principle is fragile and rather sensitive to imperfections in the circuit designs and/or the actioning of controls (e.g., pressure, flow rate, frequency, and size of incoming droplets). Furthermore, the behavior of such switches

---

[9] ©2015 IEEE. Reprinted, with permission, from [5].

TABLE 15.2   Geometrical Properties of the Microfluidic System in Figure 15.9

Channels' height h	90 μm
Channels' width w	115 μm
Input channel's length L_1	25 mm
Short branch's length L_2	6 mm
Long branch's length L_3	12 mm
Output channel's length L_4	10 mm

depends on the way they are interconnected. In fact, when joining the outlet of one switch to the inlet of another one to realize a more complex network, the flow rates in the different branches will be determined by the equivalent hydrodynamic resistance offered by the whole downstream circuit that, in turn, will dynamically change as the droplets cross the junctions.

A more controllable switch can be obtained by adding a bypass channel between the two outlet branches of a bifurcation, as sketched in Figure 15.11 [14]. The bypass channel is grafted into the two outlets, and the two endpoints of the bypass channel are shaped in a way that blocks the passage of droplets, while letting the continuous phase flow freely. The bypass channel has a much lower hydrodynamic resistance than the two channel segments to which it is attached, so that the pressure at the hooking points A_1 and A_2 (see Figure 15.11) is approximately the same. As a consequence, the volumetric flow rates in the two outlets of the switch are largely unaffected by the downstream circuits, depending only on the characteristics of the segments $J - A_1$ and $J - A_2$ and on the possible presence of other droplets in such segments. A droplet that approaches the junction will be steered to the upper or lower outlet, depending on the instantaneous flow rates along the two branches, and then will proceed along that channel, without entering the bypass channel, because of the entrance barriers. Thus, the bypass channel makes it possible to segregate the switching element from the rest of the circuit, making its behavior much more easily predictable.

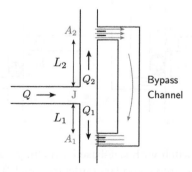

FIGURE 15.11   T-shape microfluidic switch with bypass.

15.4.2.1 Single-Droplet Switch

In order to provide its switching functionality, the switching device described previously and shown in Figure 15.11 must be coupled with a proper addressing scheme. The basic idea is to precede a cargo droplet (which, e.g., carries biological or chemical samples) by a control droplet that changes the inner state of the switch in order to steer the following droplet along the desired outlet. Borrowing from the classical networking terminology, the control droplet is called *header droplet*, while the steered droplet is called *payload droplet*.

When entering the microfluidic switch, the header droplet will always proceed along the primary outlet (L_1), as determined by the geometry of the switch (shorter branch). The path followed by the payload droplet will be determined by the changes of the volumetric flow rates produced by the header droplet. Given a certain geometry for the switching device, the switching mechanism can then be controlled in two ways: adjusting the size of the header droplet and tuning the distance between header and payload droplets, which are referred to as size-based [52] and distance-based [17] switching, respectively.

For *size-based switching*, the header and payload droplets travel back to back, so that the payload always reaches the junction before the header exits the primary segment. To steer the payload along the secondary outlet, the header droplet must be sufficiently long to trigger the switching mechanism by increasing the hydrodynamic resistance of the primary outlet beyond that of the secondary outlet. Otherwise, the payload will follow the header droplet along the primary outlet. Figure 15.12 illustrates this switching paradigm.

For *distance-based switching*, the size of the header droplet (and, thus, its hydrodynamic resistance) is fixed and is sufficient to trigger the switching mechanism. Therefore, to steer the payload along the secondary outlet, its distance from the header droplet must be small enough to guarantee that it is simultaneously in the switching region. Conversely, if the header has already left the primary segment when the payload droplet reaches the junction, the latter will also proceed along the primary outlet. This switching paradigm is illustrated in Figure 15.13.

Both switching principles have been validated through computational fluid dynamics (CFD) simulations in [5] and [17].

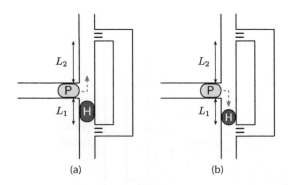

(a)   (b)

FIGURE 15.12 Single-droplet switch with size-based switching: (a) Header droplet is sufficiently long; that is, payload droplet flows into secondary outlet (channel 2). (b) Header droplet is too short; that is, payload droplet follows header into primary channel (channel 1).

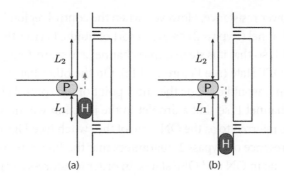

FIGURE 15.13 Single-droplet switch with distance-based switching: (a) Distance between both droplets before the switch is smaller than L_1; that is, payload droplet flows into secondary outlet (channel 2). (b) Distance between both droplets before the switch is larger than L_1; that is, payload droplet follows header into primary channel (channel 1).

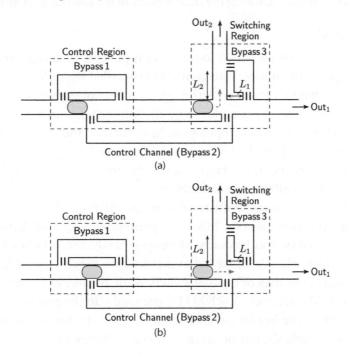

FIGURE 15.14 Multiple-droplet switch: (a) OFF state (control channel not clogged); that is, droplet flows into channel 2. (b) ON state (control channel clogged); that is, droplet flows into channel 1.

15.4.2.2 Multiple-Droplet Switch

The basic switching principles described in the previous section can be further elaborated to realize more complex functionalities. In [8], a switch structure is proposed that is able to deliver a train of payload droplets to a certain destination by using just one control droplet at the end of the droplets' train. As shown in Figure 15.14, the switch is obtained by prepending a control unit to the basic switch structure discussed previously. In particular, the switch consists of a control and a switching region. The channels 1 and 2 in the switching region (channels from the junction to either of the endpoints of Bypass 3) have different hydrodynamic resistances, with channel 1 having the lower resistance. However, when the control region is empty, the flow coming from the control channel (Bypass 2) creates a sort of fluidic barrier that prevents the droplet from entering channel 1, so that it is steered into channel 2, toward Out_2. In this condition, the switch is said to be in OFF state (see Figure 15.14a). On the other hand, if a droplet is crossing the control region, then it would occlude the entry point of the control channel, thus removing the fluidic barrier to channel 1, so that a droplet in the junction will flow in channel 1, toward Out_1. This condition is referred to as the ON state of the switch (see Figure 15.14b). It is important to note that the presence of Bypass 1 guarantees that the flow rate toward the basic switch is approximately the same in ON and OFF states. In order to address a train of payload droplets toward Out_2 or Out_1, it is sufficient to equally space the droplets at a distance, such that the switch is either in OFF or ON state. To correctly steer the last payload droplet, a header droplet must trail the sequence at the same distance as the other droplets. The header droplet will clearly find the switch in OFF state and exit the circuit from Out_2. In [8], various CFD simulation results have been presented, validating the functionality of the multidroplet switch.

15.4.3 Bus Network

The switching mechanisms described so far make it possible to design complex microfluidic systems composed of multiple interconnected so-called *microfluidic machines (MMs)*, thus realizing microfluidic networks. Although such networks may have the most disparate topologies, in practice, the complexity of the interactions among the different MMs limits the design choices to simple topologies. In particular, the ring and bus topologies[10] have been proposed in [36] and [17], respectively. In the following paragraphs, we provide a detailed discussion on the bus network, since this topology is particularly interesting owing to its simplicity, flexibility, and scalability.

A microfluidic bus network consists of a main microfluidic transmission channel (namely, the *bus*) to which different MMs are grafted by means of dedicated switching units with one inlet and two outlets each. More specifically, the switches are connected in cascade, outlet to inlet, with the other outlet that leads to the associated MM. The droplet source is connected to the inlet of the first switch, while the free outlet of the last switch is connected to a sink. The address of each MM is encoded into the geometry of its associated switch, so as to enable the header/payload addressing mechanisms described previously. The topology of a microfluidic bus network is shown in Figure 15.15.

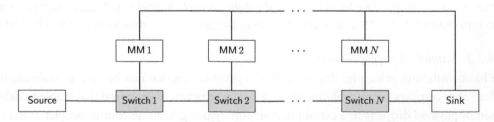

FIGURE 15.15   Microfluidic bus network.

---

[10] Application-specific architectures have been introduced in [23], dedicated to be optimized for specific applications and, thus, lacking in flexibility.

The source is assumed to be able to generate payload and header droplets with specific length and inter distance. All droplets flow along the bus, until they trigger the switching process in a particular switch. Then, the payload droplet(s) are deviated toward the upward branch containing the MM. In contrast, the header droplet flows along the bus, till it reaches the sink. Therefore, the addresses of the MMs (i.e., the geometries of the corresponding switches) need to be designed, so that a header–payload pair intended for a certain MM does not trigger any upstream switch that is crossed before reaching the target one.

Considering a single-droplet switch with size-based switching (see Figure 15.12), the targeted MM is addressed by the length of the header droplet. In particular, the size of the header droplet must be chosen such that it triggers the switching process of the switch associated with the targeted MM. Thus, the header droplet must increase the resistance of the primary channel such that the payload droplet flows into the secondary channel toward the targeted MM. On the other hand, for all previous switches, the header droplet must not trigger the switching process. This is accomplished by increasing the length of the secondary channel for the upstream switches [52].

Considering a single-droplet switch with distance-based switching (see Figure 15.13), the distance between header and payload droplet determines the targeted MM. In particular, if the distance between both droplets is below the length of the primary channel, the payload droplet is routed to the desired MM. If the distance is larger than the primary channel, both droplets flow toward the next downstream switch, and their distance is reduced. In order to avoid starting the switching process before the switch associated with the targeted MM, the distance at the source must be chosen sufficiently large by taking the distance reduction into account [17]. It is important to note that for distance-based switching, the geometry of the individual switches is the same, while for size-based switching, it is different.

Recently, a comprehensive analysis and comparison of both addressing methods have been presented in [39]. Compared with size-based switching, distance-based switching offers a simple and generic switch design (similar geometries for all switches) and supports a larger network size but suffers from low throughput. Moreover, it allows for multiple-network-access policy; that is, the network can hold multiple header payload droplets at the same time, whereas size-based switching suffers from exclusive-network-access policy.

Considering a multiple-droplet switch (see Figure 15.14), multiple payload droplets can be routed to a targeted MM. This is accomplished through an equally spaced droplet train, with a single header droplet at the end. In order to address a certain MM, the distance must be set such that the switch associated with the targeted MM is in ON state, for each droplet arriving at the switching region. The last payload droplet is routed by the header droplet at the end. Similar to distance-based switching, the droplet train passes the switch when it is in OFF state, that is, no appropriate spacing, and the distance between the individual droplets is reduced. Thus, the droplet spacing at the source must be chosen sufficiently long by taking the distance reduction into account to not trigger the switching process before the desired switch [8].

15.5 SUMMARY

In this chapter, we introduced the concept of using droplets for communications and/or networking purposes in microfluidic chips. We provided an overview of different information-encoding schemes and compared them in terms of encoder/decoder complexity, error performance, and the resulting channel capacity. Then, we introduced the passive switching principle, where the path of a single- or multiple payload droplets can be controlled using a single header droplet and only exploiting hydrodynamic effects. This principle was used for addressing in microfluidic bus networks. We presented two promising applications for passive switching in microfluidic bus networks, enabling fast analysis of biochemical samples, without compromising the sample, namely fast and flexible drug and waterborne screening.

REFERENCES

1. I. E. Araci and Stephen R. Quake. Microfluidic very large scale integration (mVLSI) with integrated micromechanical valves. *Lab Chip*, 12:2803–2806, 2012.
2. A. O. Bicen and I. F. Akyildiz. Interference modeling and capacity analysis for microfluidic molecular communication channels. *IEEE Trans. Nanobiosci.*, 14(3):570–579, 2015.
3. A. O. Bicen, J. J. Lehtomki, and I. F. Akyildiz. Shannon meets fick on the microfluidic channel: Diffusion limit to sum broadcast capacity for molecular communication. *IEEE Trans. Nanobiosci.*, 17(1):88–94, 2018.
4. A. Biral and A. Zanella. Introducing purely hydrodynamic networking functionalities into microfluidic systems. *Nano Commun. Netw.*, 4(4):205–215, 2013.
5. A. Biral, D. Zordan, and A. Zanella. Modeling, simulation and experimentation of droplet-based microfluidic networks. *IEEE Trans. Mol. Biol. Multi-Scale Commun.*, 1(2):122–134, 2015.
6. H. Bruus. *Theoretical Microfluidics*, Vol. 18. Oxford University Press, Oxford, UK, 2008.
7. A. Carlson, M. Do-Quang, and G. Amberg. Droplet dynamics in a bifurcating channel. *Int. J. Multiph. Flow*, 36(5):397–405, 2010.
8. G. Castorina, M. Reno, L. Galluccio, and A. Lombardo. Microfluidic networking: Switching multidroplet frames to improve signaling overhead. *Nano Commun. Netw.*, 14:48–59, 2017.
9. X. Chen, T. Glawdel, N. Cui, and C. L. Ren. Model of droplet generation in flow focusing generators operating in the squeezing regime. *Microfluid. Nanofluid.*, 18(5):1341–1353, 2015.
10. G. F. Christopher, J. Bergstein, N. B. End, M. Poon, C. Nguyen, and S. L. Anna. Coalescence and splitting of confined droplets at microfluidic junctions. *Lab Chip*, 9:1102–1109, 2009.
11. K. Churski, P. M Korczyk, and P. Garstecki. High-throughput automated droplet microfluidic system for screening of reaction conditions. *Lab Chip*, 10:816–818, 2010.
12. K. Churski, M. Nowacki, P. M. Korczyk, and P. Garstecki. Simple modular systems for generation of droplets on demand. *Lab Chip*, 13:3689–3697, 2013.
13. C. Cramer, P. Fischer, and E. J. Windhab. Drop formation in a co-flowing ambient fluid. *Chem. Eng. Sci.*, 59(15):3045–3058, 2004.
14. G. Cristobal, J.-P. Benoit, M. Joanicot, and A. Ajdari. Microfluidic bypass for efficient passive regulation of droplet traffic at a junction. *Appl. Phys. Lett.*, 89(3):034104, 2006.
15. E. De Leo, L. Galluccio, A. Lombardo, and G. Morabito. On the feasibility of using microfluidic technologies for communications in labs-on-a-chip. In *Proceedings of the IEEE International Conference Communications*, pp. 2526–2530, 2012.
16. A. J. Demello. Control and detection of chemical reactions in microfluidic systems. *Nature*, 442(7101):394–402, 2006.
17. L. Donvito, L. Galluccio, A. Lombardo, and G. Morabito. μ-net: A network for molecular biology applications in microfluidic chips. *IEEE/ACM Trans. Netw.*, 24(4):2525–2538, 2016.

18. M. J. Fuerstman, A. Lai, M. E. Thurlow, S. S. Shevkoplyas, H. A. Stone, and G. M. Whitesides. The pressure drop along rectangular microchannels containing bubbles. *Lab Chip*, 7:1479–1489, 2007.

19. L. Galluccio, A. Lombardo, G. Morabito, S. Palazzo, C. Panarello, and G. Schembra. Capacity of a binary droplet-based microfluidic channel with memory and anticipation for flow-induced molecular communications. *IEEE Trans. Commun.*, 66(1):194–208, 2018.

20. P. Garstecki, M. J. Fuerstman, H. A. Stone, and G. M. Whitesides. Formation of droplets and bubbles in a microfluidic T-junction-scaling and mechanism of break-up. *Lab Chip*, 6:437–446, 2006.

21. T. Glawdel and C. L. Ren. Global network design for robust operation of microfluidic droplet generators with pressure-driven flow. *Microfluid. Nanofluid.*, 13(3):469–480, 2012.

22. A. Grimmer, X. Chen, M. Hamidovic, W. Haselmayr, C. L. Ren, and R. Wille. Simulation before fabrication: A case study on the utilization of simulators for the design of droplet microfluidic networks. *RSC Adv.*, 8:34733–34742, 2018.

23. A. Grimmer, W. Haselmayr, A. Springer, and R. Wille. Design of application-specific architectures for networked labs-on-chips. *IEEE Trans. Comput.-Aided Design Integr. Circuits Syst.*, 37(1):193–202, 2018.

24. A. Grimmer, W. Haselmayr, and R. Wille. Automated dimensioning of networked labs-on-chip. *IEEE Trans. Comput.-Aided Design Integr. Circuits Syst.*, 1–1, 2018.

25. A. Grimmer, W. Haselmayr, and R. Wille. Automatic droplet sequence generation for microfluidic networks with passive droplet routing. *IEEE Trans. Comput.-Aided Design Integr. Circuits Syst.*, 2018.

26. D. Grissom, K. O'Neal, and B. Preciado et al. A digital microfluidic biochip synthesis framework. In *Proc. Int. Conf. VLSI and System-on-Chip*, 177–182, 2012.

27. D. T. Grissom and P. Brisk. Fast online synthesis of digital microfluidic biochips. *IEEE Trans. Comput.-Aided Design Integr. Circuits Syst.*, 33(3):356–369, 2014.

28. H. Gu, F. Malloggi, S. A. Vanapalli, and F. Mugele. Electrowetting-enhanced microfluidic device for drop generation. *Appl. Phys. Lett.*, 93(18):183507, 2008.

29. S. Haeberle and R. Zengerle. Microfluidic platforms for lab-on-a-chip applications. *Lab Chip*, 7:1094–1110, 2007.

30. M. Hamidovic, W. Haselmayr, A. Grimmer, R. Wille, and A. Springer. Passive droplet control in microfluidic networks: A survey and new perspectives on their practical realization. *Nano Commun. Netw.*, 19:33–46, 2019.

31. W. Haselmayr, M. Hamidovic, A. Grimmer, and R. Wille. Fast and flexible drug screening using a pure hydrodynamic droplet control. In *Proc. Euro. Conf. Microfluidics*, pp. 1–4, 2018.

32. W. Haselmayr, C. Wirth, A. Buchberger, and A. Springer. Performance comparison of information encoding in droplet-based microfluidic systems. In *Proc. Int. Conf. Nanoscale Comput. Commun.*, pp. 37:1–37:2, 2016.

33. S. Jakiela, T. Kaminski, O. Cybulski, D. B. Weibel, and P. Garstecki. Bacterial growth and adaptation in microdroplet chemostats. *Angewandte Chemie (International ed. in English)*, 52:8908–8911, 2013.

34. M. C. Jullien, M. J. Tsang Mui Ching, C. Cohen, L. Menetrier, and P. Tabeling. Droplet breakup in microfluidic T-junctions at small capillary numbers. *Phys. Fluids*, 21(7):072001, 2009.

35. E. De Leo, L. Donvito, L. Galluccio, A. Lombardo, G. Morabito, and L. M. Zanoli. Communications and switching in microfluidic systems: Pure hydrodynamic control for networking labs-on-a-chip. *IEEE Trans. Commun.*, 61(11):4663–4677, 2013.

36. E. De Leo, L. Galluccio, A. Lombardo, and G. Morabito. Networked labs-on-a-chip (nloc): Introducing networking technologies in microfluidic systems. *Nano Commun. Netw.*, 3(4):217–228, 2012.

37. A. M. Leshansky and L. M. Pismen. Breakup of drops in a microfluidic T-junction. *Phys. Fluids*, 21(2):023303, 2009.

38. B.-C. Lin and Y.-C. Su. On-demand liquid-in-liquid droplet metering and fusion utilizing pneumatically actuated membrane valves. *J. Micromech. Microeng.*, 18(11):115005, 2008.

39. M. Hamidovic, W. Haselmayr, A. Grimmer, R. Wille, and A. Springer. Comparison of switching principles in microfluidic bus networks. In *Proc. Int. Conf. Nanoscale Comput. Commun.*, pp. 23:1–23:6, 2018.

40. F. Malloggi, H. Gu, A. G. Banpurkar, S. A. Vanapalli, and F. Mugele. Electrowetting—A versatile tool for controlling microdrop generation. *Eur. Phys. J. E*, 26(1):91–96, 2008.

41. D. Mark, S. Haeberle, G. Roth, F. von Stetten, and R. Zengerle. Microfluidic lab-on-a-chip platforms: Requirements, characteristics and applications. *Chem. Soc. Rev.*, 39(3):1153–1182, 2010.

42. D. Mitra, S. Roy, S. Bhattacharjee, K. Chakrabarty, and B. B. Bhattacharya. On-chip sample preparation for multiple targets using digital microfluidics. *IEEE Trans. Comput.-Aided Design Integr. Circuits Syst.*, 33(8):1131–1144, 2014.

43. S. Poddar, S. Ghoshal, K. Chakrabarty, and B. B. Bhattacharya. Error-correcting sample preparation with cyberphysical digital microfluidic lab-on-chip. *ACM Trans. Des. Autom. Electron. Syst.*, 22(1):2:1–2:29, 2016.

44. M. G. Pollack, A. D. Shenderov, and R. B. Fair. Electrowetting-based actuation of droplets for integrated microfluidics. *Lab Chip*, 2(2):96–101, 2002.

45. F. Su and K. Chakrabarty. High-level synthesis of digital microfluidic biochips. *J. Emerg. Technol. Comput. Syst.*, 3(4):32, 2008.

46. F. Su, S. Ozev, and K. Chakrabarty. Testing of droplet-based microelectrofluidic systems. In *Proc. Int. Test Conf.*, 1:1192–1200, 2003.

47. Y.-C. Tan, Y. L. Ho, and A. P. Lee. Microfluidic sorting of droplets by size. *Microfluid. Nanofluid.*, 4(4):343–348, 2008.

48. A. J. T. Teo, K.-H. Holden Li, and N.-T. Nguyen et al. Negative pressure induced droplet generation in a microfluidic flow-focusing device. *Anal. Chem.*, 89(8):4387–4391, 2017.

49. G. M. Whitesides. The origins and the future of microfluidics. *Nature*, 442(7101):368–373, 2006.

50. S. Wirdatmadja, D. Moltchanov, and P. Bolcos et al. Data rate performance of droplet microfluidic communication system. In *Proc. Int. Conf. Nanoscale Comput. Commun.*, pp. 5:1–5:6, 2015.

51. A. Wixforth. Acoustically driven planar microfluidics. *Superlattice. Microst.*, 33(5):389–396, 2003.

52. A. Zanella and A. Biral. Design and analysis of a microfluidic bus network with bypass channels. In *Proc. IEEE Int. Conf. Commun.*, pp. 3993–3998, 2014.

53. S. Zeng, B. Li, X. Su, J. Qin, and B. Lin. Microvalve-actuated precise control of individual droplets in microfluidic devices. *Lab Chip*, 9:1340–1343, 2009.

54. Y. Zhao and K. Chakrabarty. *Design and Testing of Digital Microfluidic Biochips*. Springer Science & Business Media, New York, 2012.

55. P. Zhu and L. Wang. Passive and active droplet generation with microfluidics: A review. *Lab Chip*, 17:34–75, 2017.

IV

Advances in Nanoscale Networking-Communications Research and Development

Nanostructure-Enabled High-Performance Silicon-Based Photodiodes for Future Data-Communication Networks

Hilal Cansizoglu, Cesar Bartolo Perez,
Jun Gou, and M. Saif Islam

CONTENTS

16.1 INTRODUCTION

Since the beginning, communication has a massive impact on both individuals and civilizations. Throughout the history, owning an advanced communication technology typically becomes one of the most important signs of a developed society. While in the past, cutting-edge communication technologies were generally introduced to military systems, modern age has become a witness of enormous demand of most advanced technologies from individuals and societies to communicate with each other. The communication revolution with affordable internet access transformed how we interact with each other, how we work as a group, our habits as consumers, and our lifestyles. However, this is only the beginning of a new era, where communication is taken to a next level. We will interact with machines and machines will interact with each other in a way that will make our definition of reality look entirely different from that of the past.

Next-generation 5G networks are expected to offer new capabilities such as virtual reality, data-driven computing, advanced cloud applications, low latency (for video streaming), internet of things (IoT), and thus, fully connected society (Figure 16.1). However, such a massive connectivity brings along its challenges. Global internet protocol traffic is envisioned to reach to a data volume of 3.3 ZB per year from 2016 to 2021, which corresponds to a 24% compound annual growth rate [1]. For that reason, data centers' architectures are designed for rapid expansion to handle higher data traffic load. In 2018, annual global data traffic in data centers reached 11.6 ZB, and this is estimated to climb up to 20.6 ZB by 2021 [2,3]. Moreover, 70% of this traffic will be present within the data center [2–5].

In the last decade, driven by the digital revolution, large volumes of data traffic at the rack-to-rack, board-to-board, chip-to-chip, and intrachip levels in computer data centers have intensified the desire to develop monolithically integrated optical transmitters and receivers on silicon (Si) (Figure 16.2). One of the main issues of optics for data communication in data centers is the cost per gigabit per second (Gb/s), currently standing

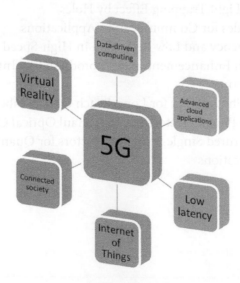

FIGURE 16.1 The capabilities that 5G networks can offer.

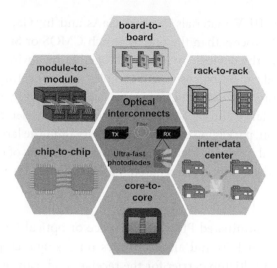

FIGURE 16.2   Data center communication links at different levels.

at about tens of dollars per Gb/s on average for end to end, and needs to be reduced further to single-digit dollar per Gb/s [6]. In the perusal of implementing monolithically integrated optical transmitters and receivers, formidable technological challenges in metamorphic/ heteroepitaxial material growth, device fabrication, and system integration led to an unfeasible cost performance ratio (CPR). In this regard, we are still experiencing many of the same severe limitations of the long investigated epitaxial lift-off and wafer-bonding technologies: (i) our inability to develop low-cost mas-manufacturable techniques to grow and integrate a variety of materials and devices on a single substrate; (ii) complementary metal–oxide–semiconductor (CMOS) incompatibility owing to extreme physical growth conditions such as high temperature; (iii) loss of a complete starting substrate in the epitaxial lift-off process, contributing to substantial cost that greatly exceeds the benefit; (iv) smaller substrate size for most non-Si materials; and (v) the interface defects, vacancies, and traps in heteroepitaxy of mismatched materials and the resulting unpredictable performance degradations. The research community is in agreement that optical components fabrication method needs to be fully compatible with integrated circuit (IC) fabrication processes to enable integration with electronics such as transimpedance amplifiers (TIAs) and other circuit elements necessary for signal processing and communication [7,8].

16.1.1 Current Interconnect Technology and Challenges

The optical interconnects based on vertical-cavity surface-emitting lasers (VCSELs), multi-mode fiber (MMF), and III-V material-based PIN photodiodes (PDs) are widely adapted in today's data centers. Such technologies are commonly used in the high-speed optical short-reach links (<300 m). 25 Gb/s optical links have transmitters (Tx) that include extremely low power (1 pJ/bit) CMOS-compatible VCSEL drivers. SiGe drivers can reach even higher speed up to 80 Gb/s over 300 m link lengths [9–12], but it is very challenging to design high-efficiency PD and transimpedance amplifier (TIA) on the same chip for the receiver (Rx) end to adopt such high data rates.

Currently, PDs with III-V materials (such as GaAs and $In_x Ga_{1-x} As$) are designed and fabricated in a distinct process than the one by which CMOS or SiGe amplifiers are fabricated. After fabrication, the devices are packaged together by a wire-bonding process [13]. Such packaging method can be expensive and complex for connecting an array of PDs to an array of amplifiers. Moreover, device performance gets degraded by undesirable packaging parasitic (such as inductance and capacitance), which can cause electrical crosstalk between channels. For this reason, packaging cost and complexity limit the large-scale deployment of current on-board optical modules. Figure 16.3 shows a schematic of current on-board optical module with PD and TIA fabricated separately and integrated on a multichip carrier [14].

16.1.2 Future of Optical Interconnects

In the case of surface-illuminated PDs for free space or optical fiber-based illumination, monolithic integration of PDs and all electronics on a single chip, fully hermetic and not requiring ceramic multichip carrier for the receiver end, can reduce the cost by more than 30%. The performance of communication systems can be dramatically improved by suppressed parasitic capacitance, resistance, and inductance when different components are monolithically integrated on a single chip. Figure 16.4 shows a schematic of an optical module where PD and TIA are fabricated on the same chip. Monolithic integration of electrical and optical components on the same chip reduces the complexity and the cost and enhances the performance by suppressing the unwanted parasitic.

The main obstacle on the way to a fully integrated Si-based receiver is insufficient efficiency of surface-illuminated Si PDs owing to silicon's low absorption of data

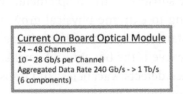 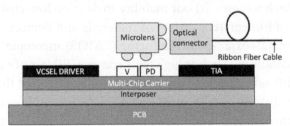

FIGURE 16.3   Current on-board optical module. Photodiode and transimpedance amplifier (TIA) are fabricated separately and packaged together on a multichip carrier, which adds cost and complexity to module fabrication. (Courtesy of W&WSens Devices, Los Altos, CA.)

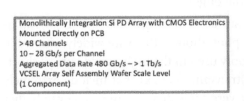 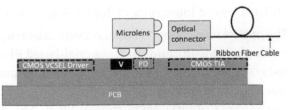

FIGURE 16.4   A schematic of a future on-board optical module, with monolithic integration of photodiode and TIA on the same chip with CMOS-compatible fabrication process. (Courtesy of W&WSens Devices, Los Altos, CA.)

communication wavelengths. For 25 Gb/s data communication rate, a surface-illuminated Si PD needs to have a <2-μm-thick Si layer (assuming a reasonable device diameter), which can only work with 10% efficiency. Recently, a nanostructure-enabled Si PD for 25 Gb/s data communication rate has been demonstrated to have >50% efficiency. This is the highest-reported efficiency for such a high-speed Si PD. In this chapter, Si-based PDs with innovative light-trapping nanostructures will be introduced for low-cost high-bandwidth links for future communication networks. Section 16.2 presents CMOS-compatible materials (Si, SiGe, and Ge-on-Si), their implementation in high-speed PDs, trade-off between speed and efficiency of devices based on Si, and how innovative nanostructures break this trade-off by light trapping. A study that uses rigorous coupled wave theory is included in this section to analyze light trapping by periodic nanostructures. In Section 16.3, recently demonstrated Si PDs for short-reach optical fiber links and Ge-on-Si PDs for long-reach optical fiber links are reviewed, and light-trapping and bandwidth enhancement properties of integrated nanostructures are discussed. In this section, Si PDs at avalanche mode are also presented for long-haul optical communication and potential application in single-photon detectors for quantum communications. Finally, a summary is provided to highlight key outcomes, limitations, and integration of high-speed Si PDs with nanostructures.

16.2 CMOS-COMPATIBLE ENGINEERED MATERIALS FOR HIGH-SPEED PHOTODIODES

Si-based (Si, SiGe, and Ge-on-Si) materials are the best candidates to realize monolithic integration of PDs and CMOS electronics, owing to well-established CMOS processing lines. In the past, attempts were made to fabricate PDs with CMOS-compatible fabrication process. Mainly, two types of Si PIN PDs have been demonstrated: one is lateral PIN PDs fabricated similarly to CMOS transistors, and another type is surface-illuminated vertical PIN PD. For example, an Si PD (discrete lateral trench detector [LTD]) with 75-μm diameter fabricated with an Si deep-trench dynamic random-access memory (DRAM) technology and integrated with SiGe TIA has been demonstrated to reach 2.5 Gb/s high-speed performance with 68% external quantum efficiency (EQE) at 845 nm (3.3 V bias) [15]. Alternatively, a lateral PIN PD integrated with 130-nm high-performance MOSFET (metal oxide semiconductor field effect transistor) (HIPER-MOS) on a silicon-on-insulator (SOI) substrate has been reported to have ~10% EQE at 847 nm and up to 3.125 Gb/s under −3V bias [16]. An example of a vertical PIN structure is a 50-μm diameter surface-illuminated PIN PD with a 10-μm absorbing region [17] that reached to 3 dB bandwidth of 2.2 GHz at 17 V bias. Monolithic integration of this PD with an analog equalizer could allow 11 Gb/s high-speed performance. Vertical PIN Si-PDs fabricated on SOI substrate are also proposed and studied by the optical short-reach community. SOI wafers let PDs get isolated from bulk wafer, where slow carriers are generated by incident light. Resonant-cavity-enhanced (RCE) vertical PIN PDs exhibited a 40% EQE at 860 nm and an impulse response of 29 ps full width at half-maximum (FWHM) for high-bandwidth applications [18]. Unfortunately, RCE devices are subjected to temperature differences, which can change the resonant wavelength. Such limitation makes them impractical for data center applications where temperature can fluctuate.

16.2.1 Trade-Off Between Efficiency and Speed of Current Si-Based Photodiodes

Efficiency and speed are two important factors that determine the performance of a PD. Most common device structure in high-speed PDs involves PIN diode (i.e., intrinsic layer, or *i*-layer, sandwiched between two highly doped *p* and *n* layers serving as contacts), as shown in Figure 16.5a. The thickness of the *i*-layer controls the transit time, which is defined as the time spent by carriers to reach the contacts. For high-speed operations, a thin *i*-layer is needed to guarantee sufficiently high bandwidth. However, a thin *i*-layer cannot efficiently absorb the incoming photons at datacom wavelengths, owing to silicon's low absorption coefficient. For this reason, a PD needs to have a thick *i*-layer for high absorption (quantum efficiency) but a thin *i*-layer for short transit time. Conversely, thin *i*-layer makes a device fast at the expense of low efficiency. This leads to a trade-off between efficiency and speed in the conventional PIN PDs. Figure 16.5b indicates the design challenge for a high-speed Si PD. The *y*-axis of the plot in Figure 16.5b shows 3-dB bandwidth calculated for a circular PIN PD in 30-μm diameter for various *i*-layer thicknesses. For a high-speed operation with a data transmission rate of 25 Gb/s, an Si PD must have maximum 2-μm *i*-layer. However, such a thin layer can only convert 10% of optical power to electrical signal, as shown with a red (below) curve on the right *y*-axis of the plot that shows absorption of Si in various thicknesses calculated for 850-nm wavelength.

16.2.2 Rigorous Coupled-Wave Analysis of Nanostructures in Silicon

In rigorous coupled-wave analysis (RCWA) model, periodic structures integrated into an Si layer can be treated as crossed gratings (or 2D gratings) [19]. Figure 16.6a and 16.6b show a studied structure composed of Si PIN layers grown on bulk Si and silicon-on-insulator (SOI) wafers, respectively. A 2-μm-thick *i*-Si film acts as the absorption region with 0.2-μm *n*-doped and *p*-doped contacts. Periodic surface structures are represented as cylindrical

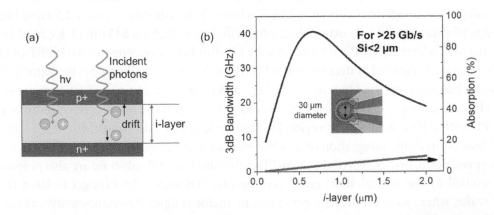

FIGURE 16.5 (a) Typical pin diode with an intrinsic layer sandwiched between highly doped *p*-type and *n*-type layers. Incident photons absorbed by *i*-layer generate *e-h* pairs that are collected by contacts. (b) Left *y*-axis shows 3-dB bandwidth calculated for a circular-pin photodiode in 30-μm diameter for various *i*-layer thickness, and right *y*-axis shows absorption of Si in various thicknesses calculated for 850-nm wavelength.

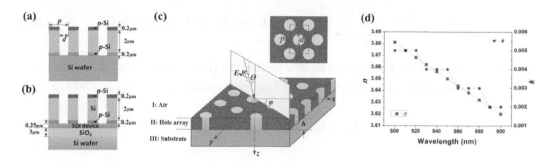

FIGURE 16.6 The n–i–p photodiode structure on SOI wafer (a) and bulk Si wafer (b), showing the integrated funnel-shaped holes that span the n, I, and p layers. (c) Periodic hole array in hexagonal lattice illuminated by a plane wave, with a rectangular Cartesian coordinate system attached. (d) The n and k values of the refractive index of Si ($n_{Si} = n - ik$) used in RCWA calculations at 800–900 nm wavelengths.

holes that span the n, i, and p layers. Figure 16.6c shows a schematic representation of a crossed grating, with hole array in hexagonal lattice. In the attached rectangular Cartesian coordinate system, the x and y axes are parallel to periodic directions, and the z axis is perpendicular to the grating plane. The propagation direction of the incident plane wave is given by polar angle θ, azimuthal angle φ, and polarization angle ψ.

In the case of bulk Si substrate, the device is divided into three regions along the z direction: the incident layer (I: air, $z < 0$), the grating layer (II: silicon with hole array, $0 \leq z \leq h$), and the substrate layer (III: Si, $z < h$). The incident layer and the substrate layer are homogeneous media with permittivity of ε_{air} and ε_{Si}, respectively. The grating layer has a thickness (h) of 2 μm, and its permittivity, $\varepsilon(x, y)$, is a piecewise-constant function taking two values, ε_{Si} and ε_{air}. ε_{Si} is taken to be a complex quantity to accommodate lossy grating ($\varepsilon_{Si} = n_{Si}^2$, where $n_{Si} = n - ik$). The refractive index of the incident layer and the cylindrical holes in the grating is $n_{air} = 1$. The refractive index of the Si substrate and the surrounding Si layer in the grating ($n_{Si} = n - ik$) is shown in Figure 16.6d for wavelengths between 800 and 900 nm [20].

The unit electric field of incident plane wave in layer I is expressed as

$$E_{inc}(x, y, z) = u \exp[-ik_0 n_i (\sin\theta \cos\varphi x + \sin\theta \sin\varphi y + \cos\theta z)] \tag{16.1}$$

where u is the unit electric-field vector,

$$u = (\cos\psi \cos\theta \cos\varphi - \sin\psi \sin\varphi)e_x$$
$$+ (\cos\psi \cos\theta \sin\varphi + \sin\psi \cos\varphi)e_y + (-\cos\psi \sin\theta)e_z \tag{16.2}$$

where e_x, e_y, and e_z are the unit basis vectors along x, y, and z axis, respectively.

The electric fields in layer I and layer III can be written as

$$E_I(x, y, z) = E_{inc}(x, y, z) + \sum_m \sum_n R_{mn} \exp[-i(k_{xm}x + k_{yn}y - k_{I,zmn}z)] \tag{16.3}$$

$$E_{III}(x,y,z) = \sum_m \sum_n T_{mn} \exp\{-i[k_{xm}x + k_{yn}y + k_{III,zmn}(z-h)]\} \tag{16.4}$$

where R_{mn} and T_{mn} are the normalized electric field of the [m n] order reflected wave in layer I and transmitted wave in layer III, respectively. The wave vectors k_{xm}, k_{yn}, $k_{I,zmn}$, and $k_{III,zmn}$ are

$$k_{xm} = k_0 n_i \sin\theta \cos\varphi - \frac{2\pi m}{p_x} \tag{16.5}$$

$$k_{yn} = k_0 n_i \sin\theta \sin\varphi - \frac{2\pi n}{p_y} \tag{16.6}$$

$$k_{l,zmn} = \begin{cases} \sqrt{k_0^2 n_l^2 - k_{xm}^2 - k_{yn}^2} & k_{xm}^2 + k_{yn}^2 \le k_0^2 n_l^2 \\ -i\sqrt{k_{xm}^2 + k_{yn}^2 - k_0^2 n_l^2} & k_{xm}^2 + k_{yn}^2 - k_0^2 n_l^2 \end{cases} \cdots l = I, III \tag{16.7}$$

where $n_I = n_i = n_{air}$ in layer I and $n_{III} = n_{Si}$ in layer III.

The electric field and magnetic field in layer II (grating layer) can be expanded in terms of spatial harmonics as in the following:

$$E_{II}(x,y,z) = \sum_m \sum_n [S_{xmn}(z)e_x + S_{ymn}(z)e_y + S_{zmn}(z)e_z] \exp[-i(k_{xm}x + k_{yn}y)] \tag{16.8}$$

$$H_{II}(x,y,z) = -i\sqrt{\frac{\varepsilon_0}{\mu_0}} \sum_m \sum_n \frac{[U_{xmn}(z)e_x + U_{ymn}(z)e_y}{+ U_{zmn}(z)e_z] \exp[-i(k_{xm}x + k_{yn}y)]} \tag{16.9}$$

where $S_{mn(z)}$, $U_{mn(z)}$ are the amplitudes of normalized electric-field vector and magnetic-field vector of [m n] order spatial harmonic.

The time dependence of $\exp(i\omega t)$ is assumed in layer II,

$$\nabla \times E_{II} = -i\omega\mu_0 H_{II} \tag{16.10}$$

$$\nabla \times H_{II} = i\omega\varepsilon_r(x,y)\varepsilon_0 E_{II} \tag{16.11}$$

where ω is the angular frequency of wave and $\varepsilon_r(x,y)$ is the permittivity of the grating layer, which is represented by means of Fourier expansion. Utilizing the tangential component continuity conditions of electromagnetic field at the boundary $z = 0$ and $z = h$, the quantities $S_{xmn(z)}$, $S_{ymn(z)}$, $U_{xmn(z)}$, and $U_{ymn(z)}$ can be obtained by solving the Maxwell curl equations in Cartesian coordinates and represented by the eigenmodes, respectively,

$$S_{xmn}(z) = \sum_j g_j w_{1mn,j} \exp(\lambda_j z) \tag{16.12}$$

$$S_{ymn}(z) = \sum_j g_j w_{2mn,j} \exp(\lambda_j z) \tag{16.13}$$

$$U_{xmn}(z) = \sum_j g_j w_{3mn,j} \exp(\lambda_j z) \tag{16.14}$$

$$U_{ymn}(z) = \sum_j g_j w_{4mn,j} \exp(\lambda_j z) \tag{16.15}$$

where λ_j and $w_{lmn,j}$ (l = 1, 2, 3, 4) are eigenvalues and eigenvectors, respectively, and g_j are the coefficients determined from the boundary conditions. λ_j and $w_{lmn,j}$ can be obtained by solution of matrix eigenvalue problem with a size of $4n \times 4n$ using eigenfunction in MATLAB programs [21], where n is the number of spatial harmonics (orders) retained. Here, each grating period is composed of holes with subwavelength feature size. The number of orders is increased to obtain convergent and stable results during the computation. Then, R_{mn} and T_{mn} can be solved by using the field equations to match the boundary conditions at grating interfaces (z = 0 and z = h). Reflection diffraction efficiency (η_{Rmn}) and transmission diffraction efficiency (η_{Tmn}), defined as the power of reflected and transmitted light diffracted into an order [m n] to the incident power, can be calculated by

$$\eta_{Rmn} = \mathrm{Re}\left(\frac{k_{I,zmn}}{k_0 n_i \cos\theta}\right)\left(\left|R_{xmn}\right|^2 + \left|R_{ymn}\right|^2 + \left|R_{zmn}\right|^2\right) \tag{16.16}$$

$$\eta_{Tmn} = \mathrm{Re}\left(\frac{k_{III,zmn}}{k_0 n_i \cos\theta}\right)\left(\left|T_{xmn}\right|^2 + \left|T_{ymn}\right|^2 + \left|T_{zmn}\right|^2\right) \tag{16.17}$$

where the subscripts x, y, and z represent the components of R_{mn} and T_{mn} along the x, y, and z directions, respectively. The absorption can be obtained by

$$A = 1 - \sum_m \sum_n (\eta_{Rmn} + \eta_{Tmn}) \tag{16.18}$$

The calculations are carried out by MATLAB programs, with algorithm based on [19]. For the case of SOI substrate, a uniform Si film without holes (device layer of SOI) with a thickness of 0.25 µm is added in region II, and SiO_2 layer assumed with infinite thickness is modeled as the substrate layer. Since the surface-illuminated PDs are of interest in this text, a monochromatic electromagnetic plane wave, with θ = 0, φ = 0, and ψ = 90° (transverse electric (TE) polarization), is used in the calculations. In this case, the electromagnetic field vectors in all regions have only the components along the y direction for the electric fields and nonzero components along the x and z directions for the magnetic fields.

Figure 16.7 shows the calculated absorptions and experimentally measured EQEs from PDs, with hole arrays with diameter/period (d/p) of d/p = 700/1000 nm on bulk silicon and SOI wafers at wavelengths between 800 nm and 900 nm [22]. Absorptions and EQEs of flat devices (without holes) on bulk silicon wafer are also presented. Both RCWA calculations and EQE measurements reveal that PDs with light-trapping holes outperform their counterparts without holes. The measured EQEs are a little higher than theoretical calculations

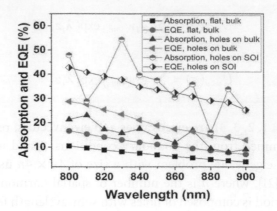

FIGURE 16.7 Calculated absorptions and experimentally measured EQEs versus wavelength for flat photodiode (without holes) on bulk silicon wafer and holes-integrated photodiodes on bulk silicon and SOI wafer with diameter/period (d/p) of $d/p = 700/1000$ nm.

for PDs on bulk Si substrate, mainly due to the diffusion of photogenerated carriers from the top silicon substrate, which can contribute to the photocurrent. The measured EQEs of holes-based PDs on bulk silicon wafer are 13%–29% at 800–900 nm wavelengths and can be enhanced to 25.4%–42.7% on SOI wafer. This is attributed to the back reflection of SOI substrate, which will be discussed in detail in the next section.

16.2.3 RCWA of Light-Trapping Effect by Holes

To understand the underlying physical principle of the light-trapping effect and absorption enhancement of hole arrays integrated in the PDs on SOI substrate, the RCWA method is applied to analyze light propagation and photon–material interactions in the multilayer PD structure.

The holes integrated PDs are illuminated by vertical incident light. However, subsequent to the diffraction of hole array, the reflected light and transmitted light undergo a deflection and become multiorder light, with different deflection angles for each order. The deflection of light provides laterally propagating modes. The diffraction order [m, n] of transmitted waves in the device layer of SOI from a vertically illuminated crossed grating is limited by:

$$|m|, |n| < \frac{p n_{\text{Si}}}{\lambda} \qquad (16.19)$$

where n_{Si} is the refractive index of the substrate. The deflection angle (θ_N) of Nth-order transmitted light can be calculated by:

$$\sin \theta_N = \frac{N \lambda}{p n_{\text{Si}}} \qquad (16.20)$$

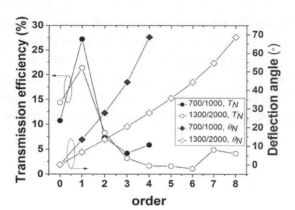

FIGURE 16.8 Transmission efficiencies (T_N) and deflection angles (θ_N) at different diffraction orders for hole arrays with diameter/period (d/p) of $d/p = 700/1000$ nm and $d/p = 1300/2000$ nm at the wavelength of 850 nm.

The RCWA method is applied to calculate the transmission efficiencies in the device layer of SOI from hole arrays at different diffraction orders. In the calculations, holes-integrated *i*-Si layer with a thickness of 2 μm is modeled, and Si film without holes assumed with infinite thickness is treated as the substrate layer.

Figure 16.8 shows the transmission efficiencies (T_N) in the device layer of SOI and deflection angles (θ_n) at different diffraction orders for hole arrays with diameter/period (d/p) of $d/p = 700 /1000$ nm and $d/p = 1300/2000$ nm at the wavelength of 850 nm. The transmission efficiencies are relatively high at 0th and 1st orders and much lower at higher orders for the two designs of hole arrays. Figure 16.8 suggests that smaller period leads to higher deflection angle at each order, and the difference between deflection angles for hole arrays with different periods increases as the order increases. It seems that the light is guided with more laterally propagating modes in hole array with smaller period.

SiO_2 film in the SOI substrate acts as a back-reflection layer. The back reflection depends on the transmittance efficiency (T_N) for each order of light and its reflectivity at the Si/SiO$_2$ interface. The critical angle of total reflection (θ_c) can be calculated as 23.79° by Equation 16.21, from the device layer of SOI to SiO_2 film with a refractive index of $n_{SiO_2} = 1.4721$ at the wavelength of 850 nm. Therefore, the reflectivity (R'_N) of transmitted wave (T_N) at the Si/SiO$_2$ interface can be calculated by Equation 16.22, depending on the deflection angle (θ_N) of transmitted wave. In Equation 16.22, $\theta_{RN} = \arcsin\left(\frac{n_{Si} \sin\theta_N}{n_{SiO_2}}\right)$ is the refraction angle of light in SiO_2 film. The deflection angle of transmitted wave increases with diffraction order for a fixed period, which means that only limited orders of transmitted waves in SOI device layer can transmit further into SiO_2 film with a reflection of R'_N and a transmission of $1 - R'_N$. Based on the transmission efficiency and deflection angle of light at each order, the total back reflection can be easily calculated, since all the transmitted light with deflection angle higher than the critical angle of total reflection will be totally reflected. The back reflection in holes integrated PD with d/p = 700/1000 nm is 50.3% of the transmitted light from hole array (34.5% of the total incident light), which is higher compared with the PD with d/p = 1300/2000 nm with a back reflection of 42.3% of the transmitted light (29.2% of the total incident light). Given that a flat PD without holes exhibits only 18% back reflection on an SOI substrate, the transmission loss from substrate is efficiently reduced by integrating photon-trapping holes.

$$\sin\theta_c = \frac{n_{\text{SiO}_2}}{n_{\text{Si}}}. \tag{16.21}$$

$$R'_N = \begin{cases} \dfrac{(n_{\text{Si}} - n_{\text{SiO}_2})^2}{(n_{\text{Si}} + n_{\text{SiO}_2})^2} & \theta_N = 0 \\[2em] \dfrac{\sin^2(\theta_N - \theta_{RN})\left(1 + \dfrac{\cos^2(\theta_N + \theta_{RN})}{\cos^2(\theta_N - \theta_{RN})}\right)}{2\sin^2(\theta_N + \theta_{RN})} & 0 < \theta_N < \theta_c \\[2em] 1 & \theta_N < \theta_c \end{cases} \tag{16.22}$$

The light reflected by the SiO_2 film will interact with the i-Si layer that is integrated with hole array for further absorption. In order to understand the interaction between this part of light and hole arrays, the RCWA method is used to calculate the reflection (R'), transmission (T'), and absorption (A') from hole arrays at the wavelength of 850 nm, as shown in Figure 16.9. In this calculation, the light is incident from the bottom of hole array, so Si layer assumed with infinite thickness is modeled as the incident layer, while air is treated as the substrate of hole array. The incident angle of light equals the deflection angle of transmitted light in the device layer of SOI at each order. Figure 16.9 shows that optical transmission into air is relatively high for light with low incident angle, and the hole array with d/p = 700/1000 nm has less transmission loss compared with the hole array with d/p = 1300/2000 nm. When the incident angle is higher than 30°, the transmission loss is very low for the two designs of hole arrays, suggesting this part of light will be trapped in the PD structure until fully absorbed. Figure 16.9 also shows a high reflection from hole array back to SOI substrate, which causes multiple reflection and diffraction effects between hole array and SiO_2 layer. The absorption is 15%–20%, regardless of hole design and incident angle, which means that 15%–20% of light will be absorbed with each diffraction. Multiple reflection and diffraction effects not only provide multiple light absorptions but also guide the light propagation more and more lateral for effective light trapping, owing to increasing deflection angle of light.

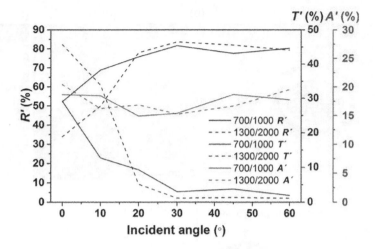

FIGURE 16.9 Reflection (R'), transmission (T'), and absorption (A') of light enter into hole structures after reflecting from SiO_2 layer of SOI wafer with different incident angle for hole arrays with diameter/period (d/p) of d/p = 700 nm/1000 nm and d/p = 1300/2000 nm at the wavelength of 850 nm.

16.3 Si-BASED PHOTODIODES FOR COMMUNICATION APPLICATIONS

Si-based PDs with photon-trapping holes are a new category of high-speed PDs recently introduced into the field [22–26]. The surface of the Si PD was structured to enhance the optical absorption. Photon-trapping holes allow incident light to penetrate laterally inside the Si, and thus, effective optical path gets enhanced. If Si-based material is grown on a substrate with a high index difference (such as SOI substrate or Ge-on-Si), back reflection also helps to direct light into absorption material.

16.3.1 High Efficiency and Low Reflection in High-Speed Si Photodiodes

An Si PD with photon-trapping holes in funnel shape was demonstrated to show more than 50% EQE and deconvolved FWHM response of 23 ps at 850 nm [22]. Such surface-illuminated Si PDs have PIN device layers epitaxially grown on an SOI wafer. Figure 16.10a and b shows schematic of the Si PD with holes in the active region and cross-section of Si layers, respectively. The holes were patterned and etched by deep reactive ion etching (DRIE) and RIE process to create cylindrical and funnel-like holes, respectively. The formation of a slope in the photoresist patterns got transferred to the Si holes during the etching process and resulted in an inclined wall at the top of the holes, contributing to funnel-like shapes. The fabrication details of Si PDs with holes can be found in earlier reports [22,23]. Figure 16.10c indicates a scanning electron microscopy (SEM) image of an Si PD with holes, mesa, and ohmic contacts. Figure 16.10d and e shows to top SEM images, and Figure 16.10f and g shows cross-section SEM images of cylindrical and funnel holes, respectively.

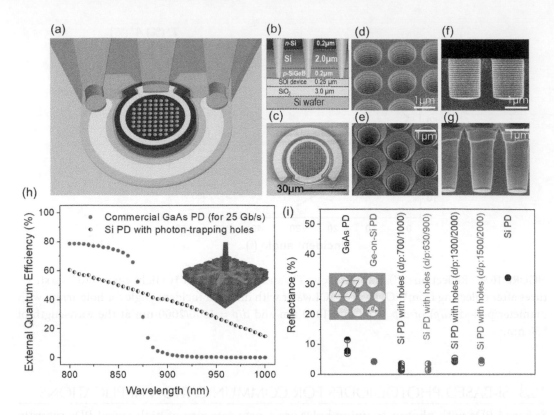

FIGURE 16.10 Schematic of the Si photodiode with *n*-Si layer (top), *i*-Si layer (middle), *p*-Si layer (bottom), insulating nitride/oxide layer (transparent), ohmic metal (rings around photodetector), high-speed coplanar waveguide (CPW) contacts, and polyimide planarization layer (below CPW), (b) The *nip* PD structure on an SOI wafer showing the integrated tapered holes that span the *n*, *i* and *p*-layers. (c) SEM micrograph of the active region of a high-speed PD with 30 μm diameter with holes, SEM images of holes in (d) square and (e) hexagonal lattice. Cross-section SEM image of (f) cylindrical and (g) funnel shaped or tapered holes etched into Si, (h) External quantum efficiency (EQE) comparison of commercially available GaAs PD for 25 Gb/s data transmission and Si PD with photon-trapping holes [27], (i) reflection measured from Si PDs with and without holes as well as commercially available GaAs and Ge-on-Si PDs at 850 nm wavelength. (Reproduced by permission of Nature Publishing Group from ref [22].)

Figure 16.10h presents a comparison of EQE gained from commercially available gallium arsenide (GaAs) PD for 25 Gb/s data transmission rate and Si PD with photon-trapping holes. A supercontinuum laser and a tunable filter that transmits a band of 5-nm wavelength was used to conduct EQE measurements. The light is delivered to the devices by a single-mode fiber probe on a probe station. The performance of Si was improved with the absorption-enhancement structures, and a broadband high efficiency throughout the wavelength range of 800–1000 nm was achieved. More importantly, EQE of Si PD is still high beyond the bandgap of GaAs (>870 nm), where absorption coefficient of Si is weak. For short-wavelength division multiplexing (SWDM covers 850–980 nm) applications, such broadband high efficiency is quite attractive.

Figure 16.10i shows the results of reflectance measurements performed on various PDs, including a commercially available GaAs and Ge-on-Si PDs along with Si PDs with and without holes at 850-nm wavelength. While Si PD with no antireflection coating (ARC) reflects ~30% of light, commercially available PDs show 5%–10% surface reflection when ARC is used. Remarkably, Si PDs with holes in different designs reflect less than 3%–5% of light at 850 nm without ARC. A wider opening of tapered holes at the surface intro-duces a gradual refractive index change, which allows light to adiabatically penetrate Si compared with the case where the index difference is abrupt such as the surface of Si with cylindrical holes. Although Si with cylindrical holes also has reduced reflection [22] com-pared with bare Si surface, gradual refractive index changes with asymmetric geometry, such as tapered holes dramatically help to reduce the reflection of light, as Figure 16.10i indicates. To ensure such low reflection, many dielectric layers with precise thicknesses are deposited on Si to form graded-index ARC. On the other hand, tapered holes have an ingrained effect of graded-index ARC with a one-step processing. Broadband high effi-ciency in Figure 16.10i suggests a reduced reflection for a wide range of wavelengths by tapered holes, as opposed to a wavelength-dependent ARC.

16.3.2 Bandwidth Enhancement in Si Photodiodes by Integrated Nanostructures

In a typical PIN PD, transit time and RC time are the main factors to determine the speed of a PD, as can be seen in Equation (16.21), which defines the 3-dB frequency [28],

$$f_{3dB} = \frac{1}{\sqrt{(2\pi RC)^2 + (t_{tr} / 0.44)^2}}. \tag{16.23}$$

where t_{tr} is the transit time that carriers spent through depletion region to reach to the electrodes, and resistance (R) is designed to be 50 ohm and capacitance (C) is mostly domi-nated by junction capacitance. In an ideal PIN diode, $C = \varepsilon A/d$, where A is the junction area and d is the depletion layer width, which mainly consists of i-layer. Therefore, thinner i-layer can reduce the transit time. On the other hand, it gives rise to high capacitance and cannot absorb enough light. Novel designs of PDs with light-trapping and capacitance-reducing hole arrays can address such challenges occurring due to thin i-layer.

The hole arrays also contribute to lowering the junction capacitance while reducing the reflection and increasing the absorption. Figure 16.11a–d shows the results of capacitance-voltage (C-V) measurements of PDs in various mesa sizes. The inset shows the top SEM images of corresponding PDs. Owing to the design requirements of the PDs, the active region filled with holes have a different hole-to-material ratio depending on the device mesa size. It is important to note that the ohmic contact metals are delineated on a hole-free flat surface to avoid shorting different layer of the PIN diodes. The total percentage of area covered by the top ohmic contact (ring shape in this PDs) is considerable in PDs with mesas with small diameter. For this reason, the ratio of the surface with and without holes does not solely determine the reduction in capacitance in any device with arbitrary diam-eter. The ratio of the active region to the total area of the PD is considered to accurately estimate the capacitance of PDs with holes. If the depth of the holes is assumed to reach the

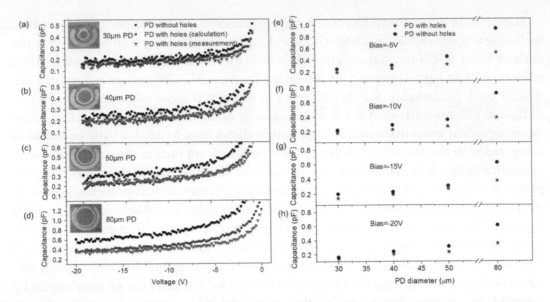

FIGURE 16.11 Capacitance versus voltage characteristics of PDs with and without holes (diameter/period: 1500/2000 nm) in (a) 30 μm, (b) 40 μm, (c) 50 μm, and (d) 80 μm diameter. The capacitance of PDs without holes (black dots) is higher than the capacitance of PDs with holes (gray dots). A capacitance versus voltage profile (dark gray dots) is estimated based on empirical values of PD without holes, and the air ratio (AR) calculated by $AR = \left(\pi(d/2)^2\right)/\left(p^2\sqrt{3}/2\right)$, where d is the diameter of the holes and p is the period of the arrays. Inset: Top-view SEM images of PDs with corresponding diameter. Capacitance versus diameter of PDs with (dark gray stars) and without (black dots) holes reversed biased at (e) 5V, (f) 10V, (g) 15V, and (h) 20V.

bottom layer and the ratio of the active area to the total area is defined as $R_{active/total}$, then the capacitance expected from the PD with holes will be

$$C_{PD\,with\,holes} = C_{PD(no\,holes)} - C_{PD(no\,holes)} \times R_{active/total} \times (AR) \tag{16.24}$$

where $AR = \left(\pi(d/2)^2\right)/\left(p^2\sqrt{3}/2\right)$ for holes packed in a hexagonal lattice with diameter d and period p. The blue curves (blue curves need to be solid thin lines) in the C-V plots of Figure 16.11 indicate the estimated capacitance of Si PDs with holes based on Equation 16.24. The red and black curves are the results of the C-V measurements performed on Si PDs with and without holes, respectively. The measured capacitance of Si PDs with holes matches well with the estimated capacitance depending on the area that gets reduced by the holes. As discussed earlier, the ratios of the reduced area by holes are larger in PDs with larger mesas; therefore, the effect of the capacitance reduction is more discernable in larger PDs, as indicated in Figure 16.11e–h. On the other hand, the capacitance of smaller PDs can be further reduced with an aggressive design that includes closely packed holes in a larger active region along with narrower metal contacts.

Figure 16.12a and b shows the measured pulse response of Si PDs with active regions 30 and 50 μm in diameter, respectively. A mode-locked pulsed fiber laser having a wavelength

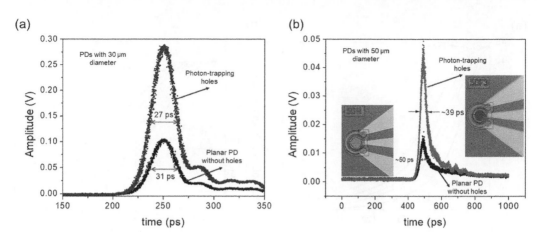

FIGURE 16.12 Pulse response of Si PDs with (dark gray curve) and without (black curve) holes (d/p: 1300/2000 nm) at 850-nm wavelength with (a) 30-μm diameter and (b) 50-μm diameter active regions. The response of both PDs with holes shows higher amplitude (high efficiency) and higher speed compared with PDs without holes.

of 850 nm and a pulse width of sub-picosecond was used as the light source in high-speed characterizations. A 20-GHz oscilloscope connected to a 25-GHz bias-T was utilized to observe the electrical pulses generated by our PDs. As Figure 16.11a indicates, a 30-μm PD with hole arrays has a resulting photoresponse of 27 ps FWHM, whereas a PD without holes and with same size produces a pulse that has 31 ps FWHM. On the other hand, the FWHMs of the pulse responses measured from 50-μm PDs with and without holes are 39 ps and 50 ps, respectively. In other words, PDs without holes have the resulting photoresponse with 15% and 28% wider FWHM for 30 μm and 50 μm mesa sizes, respectively. These results agree well with the capacitance reduction observed with the embedded holes in Figure 16.11. As discussed earlier, smaller PDs have lower air ratio compared with the larger PDs, owing to the constraint in design and fabrication processes. Thus, the capacitance reduction is less discernable in smaller PDs than in the larger PDs. The percentage decrease in FWHMs in Figure 16.12 also indicates the similar trend. With advanced fabrication equipment used by modern semiconductor industry, the percentage decrease in the capacitance can be same in devices of any size.

The results in Figure 16.13 compares the 3dB bandwidth (BW) of Si PDs with and without capacitance reduction for two different *i*-layer thicknesses. Equation 16.23 was used to calculate 3 dB BW, assuming a 50 ohm load resistance. A Si PD with 30-μm-diameter and 1-μm-thick *i*-layer can potentially support >40 GHz in the case of 80% capacitance reduction, whereas a similar PD without capacitance reduction can support 31 GHz frequency BW. Moreover, high efficiency of Si PDs with hole arrays with such a thin *i*-layer makes it more attractive for practical use in ultrafast short-links. Figure 16.13b shows a similar comparison as in Figure 16.13a but assumes a 2-μm-thick *i*-layer. In this case, the frequency response of smaller PDs (~30 μm in diameter) is expected to be transit time limited; therefore, capacitance reduction does not result

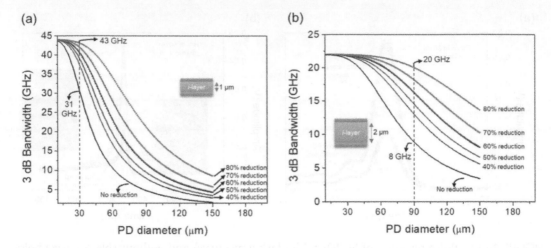

FIGURE 16.13   Calculated 3-dB bandwidth of Si PDs with and without capacitance reduction versus PD diameter for (a) 1-μm-thick and (b) 2-μm-thick i-layer, using Equation 16.23. A 30-μm-diameter Si PD with a 1-μm-thick i-layer can potentially provide >40 GHz in the case of 80% capacitance reduction as opposed to 31 GHz BW supported by the same size PD without a capacitance reduction. In the case of 2-μm-thick i-layer, the effect of capacitance reduction is prominent in the relatively-large-area devices. Calculations estimate a 20 GHz BW with a 90-μm-diameter Si PD with 80% capacitance reduction, whereas a PD without capacitance reduction can only provide 8 GHz BW in the case of a 2-μm-thick i-layer.

in a significant improvement in frequency response. However, larger PDs (~90 μm in diameter), which can be more convenient for packaging and fiber alignment, can support a frequency BW of 20 GHz, with 80% capacitance reduction. On the other hand, a PD with same size and without capacitance reduction can only support 8 GHz BW. This is how PDs with reduced capacitance contributed by the holes can offer significantly higher BW than the PDs without photon-trapping holes.

The reflection and EQE measurements indicated that Si PDs with photon-trapping structures can exhibit broadband high efficiency and high speed for ultrafast data communication applications. The integrated holes in the active region can elevate Si to the level where it can compete with GaAs or InGaAs PDs and can show better performance, even beyond the bandgap wavelengths of GaAs [23]. Especially above 900 nm, there are emerging datacom links that necessitate high-speed PDs [29,30]. For example, high-performance computers are expected to utilize high-speed optical links operating in 990–1065 nm wavelength range [31,32]. The capacitance measurements performed on Si PDs with photon-trapping holes showed a drastic decrease in the junction capacitance owing to reduced junction area. The capacitance reduction contributed to decreased RC time and consequently increased 3 dB BW. More than 50 Gb/s data communication rates can be realized by using PDs based on the innovative photon trapping structures and can be integrated with CMOS circuitry [27]. Such improvement in the performance of Si PDs can contribute to lower reflection, reduced leakage current, reduced capacitance, and considerable cost saving in the transceiver design.

16.3.3 Ge-on-Si Photodiodes for Long-Reach Optical Fiber Links

Germanium-on-Silicon (Ge-on-Si) PDs are promising candidates for alternative PDs to conventional III-V based PDs, owing to their monolithic electronic-photonic integration via CMOS fabrication technology. However, it is challenging to gain high speed and high efficiency at the same time in surface-illuminated Ge-on-Si PIN PDs, owing to low absorption at longer wavelengths.

Recently, a high-speed surface-illuminated Ge-on-Si PIN PD with broadband efficiency in near-infrared (up to 1700 nm) wavelengths is realized with CMOS-compatible fabrication methods. As shown in Figure 16.14, a 30-μm diameter Ge/Si PIN PD with 2-μm Ge i-layer has >80% EQE at 1310 nm and 73% EQE at 1550 nm, about 50% improvement compared with similar device without holes [24]. The measured high-speed performance suggests up to 10 Gb/s data transmission rate. Such devices' operating range can cover long-reach interconnects in data centers, with the range up to 10 km [33]. Broadband EQEs in both C (1530–1560 nm) and L (1565–1625 nm) bands make these PD promising candidates for passive optical network (PON) transceivers to provide fiber to the end users at 1550 nm [34]. Moreover, enhanced EQE in the new communication band (1620–1700 nm) can offer possible application in dense wavelength division multiplexer (DWDM) to extend the single-mode fiber bandwidth beyond the L band [35].

Figure 16.15 presents the results of high-speed measurement, which is conducted by surface illumination of a 1310-nm pulsed laser (pulse width: ~15 ps) via a single-mode fiber probe (aligned with a translational stage to maximize the photocurrent). The results were recorded using 20 GHz sampling scope. The impulse responses of PDs with and

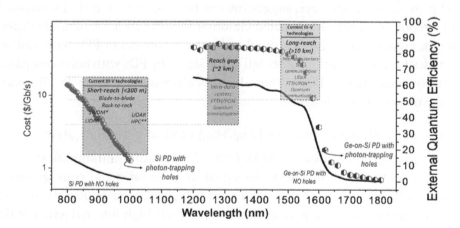

FIGURE 16.14   High-speed and high-efficiency Si photodiodes for short reach and Ge-on-Si photodiodes for long-reach optical communication. For the wavelengths of 800–1000 nm, Si PDs with photon-trapping holes show enhanced efficiency compared with Si PD without holes. For the applications-required operations at wavelengths up to 1700 nm, Ge-on-Si PDs with holes can be promising candidates. (Reproduced by permission of Chinese Laser Press from Cansizoglu, H. et al., *Photonics Res.*, 6, 734–742, 2018.) The left y-axis of the plot shows required cost per Gb/s for corresponding applications to realize widespread deployment of optical fiber links, and the right y-axis indicates EQE of Si-based photodiodes with holes and without holes as a comparison.

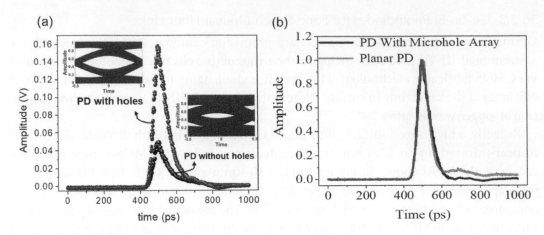

FIGURE 16.15 (a) Measured impulse response of Ge-on-Si pin photodiodes with and without holes at 1310 nm. The impulse response of the device with integrated holes (FWHM: 69 ps, blue) illustrates that light-trapping structure can improve the high-speed performance, compared with the planar PDs (FWHM: 77 ps, black) (Reproduced by permission of Chinese Laser Press from Cansizoglu, H. et al., *Photonics Res.*, 6, 734–742, 2018.). (b) Normalized impulse response of PDs with (black) and without holes (gray).

without photon-trapping holes are shown in Figure 16.15a. The FWHMs of the impulse response measured from PDs with and without (planar) hole arrays are 69 ps and 76 ps, respectively. The inset shows simulated eye diagrams at 10 Gb/s transmission rate. It is obvious that PDs with hole arrays can generate electrical signals with higher amplitude and provide more open eye, suggesting low bit error rate in actual systems. This is the fastest reported surface-illuminated Ge-on-Si PIN PD with broadband efficiency up to 1700 nm. Figure 16.15b shows normalized pulse responses of PDs with and without holes. A reduction in residual tail after fall of the signal by PDs with holes compared with the relatively longer tail by PDs without holes is observed in Figure 16.15b, indicating improved bandwidth by hole arrays.

16.3.4 Avalanche Photodetectors for Long-Haul Optical Communication

Long-haul optical communication links (<300 km) require an amplification mechanism that allows the link to conserve its signal to noise ratio and consequently its bit error rate (BER) at long distances (Figure 16.16). Avalanche photodetectors (APDs) have been implemented in these systems owing to their high internal gain obtained by charge multiplication. Charge multiplication is created by impact ionization, a stocastic process that also creates excess noise and limits the gain-bandwidth product [36]. An APD requires to have a sufficient gain-bandwidth product that overpasses the excess noise generated.

According to the signal-to-noise ratio equation (Equation 16.25) for photodetectors, the multiplication factor M, a high responsivity R, and low noise (denominator term) enhance the SNR, and consequently, the power required to obtain a desired BER becomes lower.

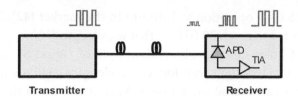

FIGURE 16.16   As schematic of a long-haul optical communication link that utilizes avalanche photodiodes (APDs).

$$\frac{S}{N} = \frac{(R \times P_{in}) \times M^2}{2 \times q \times (I_P + I_D) \times M^2 \times F(M) \times B_e + 2 \times q \times I_L \times B_e \times \dfrac{4 \times k_B \times T}{R_L} \times B_e} \qquad (16.25)$$

For long distances, the wavelength of operation is 1550 nm where optical fiber presents less attenuation. For this reason, Si solely is not a material of choice in APDs for long-reach applications, since the absorption capabilities are limited to wavelengths lower than 1100 nm (corresponding to Si's bandgap energy). Ge, with a narrower bandgap, is currently a material of interest in APDs, as it can absorb at 1550 nm, using tensile strained-Ge [37]. One of the challenges is that Ge creates a high amount of noise during the multiplication process, but a combination with Si can mitigate this problem. Si has been proven to be the best material, as a charge multiplication layer for its favorable ionization coefficient k, which is the ratio of hole ionization coefficient (β) to electron ionization coefficient (α), approaches to zero, creating a low excess noise factor $F(M)$ [36]. Therefore, the combination of these two materials, Ge-on-Si, is called separate absorption, charge, and multiplication (SACM) APDs and possesses a promising material system for extended-reach applications [38–41].

Figure 16.17 shows a possible design of an SACM Ge-on-Si APD, where hole arrays can be integrated into Ge absorbing layer to provide a broadband and high photon absorption, a method that can enhance the gain-bandwidth product of the APD. There are Ge-on-Si

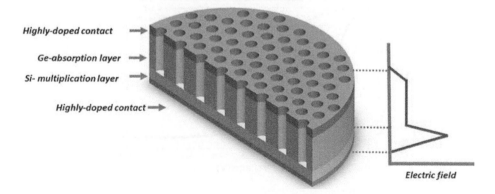

FIGURE 16.17   Integrated nanoholes in Ge-on-Si avalanche photodetectors for SNR enhancement. SACM structure and hole arrays could result in higher gain-bandwidth product.

APDs available for 25 Gb/s operation at 1310 nm in the market [42]. Their speed perfor-mance has been reported to reach 30 GHz at that wavelength [38].

The integration of holes into the absorption layer of Ge-on-Si APDs can offer effi-cient coupling of the incident light with longer wavelengths. APDs with photon-trapping micro-/nanostructures can extend the range of APD operation to the C-band (1550 nm) from existing operating bands (1300 nm) at 25 Gb/s data transmission rate.

On the other hand, the large lattice mismatch of 4.2% between Ge and Si [43] can cause large dislocation densities that would reduce the carrier mobility, increase the dark current, and prohibit the economic yields of device in very large-scale integration (VLSI) or ultra-large-scale integration production. However, a careful growth of Ge on Si substrate has been demonstrated with a good reliability and an ability to operate at 25 Gb/s NRZ and 56 GB/s in PAM4 applications [41].

Recent results in Si APDs with nanostructured holes (Figure 16.18) have demonstrated a considerable increase in the multiplication factor M, compared with a control APD with no holes. This enhancement can be related to control of the region where the absorption takes place, facilitating the carrier collection after multiplication; in addition, these new Si APDs are also able to start carrier multiplication at lower voltages than a control Si APD, reducing the required power for operation.

The improvement in responsivity and gain in this novel APD allow reduced minimum power to operate these devices for a specific BER. Figure 16.19 shows the calculations of minimum power required to obtain an SNR = 1 to operate a communication system with a BER of 10^{-9}. The minimum input power required in an APD with nanoholes is lower than the one necessary for a device with no holes. For a comparison, a PD in PIN mode is included in Figure 16.19, showing a similar trend.

The same approach can be applied to Ge-on-Si APDs to operate at a wavelength of 1550 nm. Nanostructures can enhance their gain-bandwidth product and decrease the minimum power that the receiver needs to sustain a reliable operation for a long-reach optical communication link.

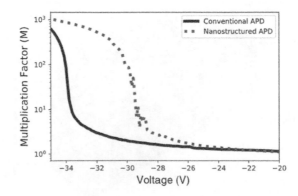

FIGURE 16.18 Carrier multiplication in silicon APD with nanohole arrays (gray) and a control APD without nanoholes (dark gray). Nanohole arrays allows higher multiplication at lower voltage compared with conventional APDs.

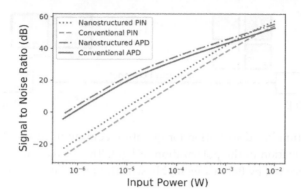

FIGURE 16.19 Input power required to operate an APD and a pin photodetector with and without nanostructures for an SNR > 1. APDs required lower power to achieve the same SNR than pin owing to internal gain. Nanostructures reduce the power requirements even more.

16.3.5 Nanostructured Single-Photon Detectors for Quantum Communications

Security in data transmission is an increasing concern, as more devices are connected to the internet and people rely on these systems more than ever for their daily tasks. Quantum communication is predicted to be the most secure technology for data transmission. In these systems, photons are presented as an effective way to implement such communications systems, since they can propagate fast and are easy to manipulate [44]. Moreover, the implementation of quantum communication can be adapted from the technological developments obtained from classical optical communication technologies that support our internet.

In these systems, the information can be encoded in a photonic state, such as the polarization, or the angular momentum of the photon. The information is sent through a quantum channel and finally detected on the receiver side. These systems are extremely secure, since tapping the transmission line can destroy the quantum states of entangled photons; immediate detection of a tapped line will be possible in a quantum communication system (QCS) [45]. In order to detect the physical state of the photons, highly sensitive photodetectors are necessary. Single-photon detectors are the main components in the receiver of a QCS, and they are required to have a high probability of detection and reliability. Owing to the low energy of the photon (10^{-19} J), the photodetector requires to have a high gain and low noise to be able to discriminate between different photons.

Emerging technologies such as superconductor single-photon detectors can resolve a single photon, but the technical challenges such as cooling requirements do not allow them to be practical devices yet. Currently, a nonphoton number resolution detector is the choice for practical applications. In these photon-counting systems, a single-photon avalanche diode (SPAD) operates at above breakdown voltage, known as the Geiger mode. That results in a self-sustained avalanche in response to the absorption of a photon [46]. Such SPADs are implemented in quantum key distributions (QKDs) that allow to perform secure communications (Figure 16.20).

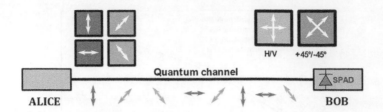

FIGURE 16.20 Quantum key distribution for quantum communications. Alice sends a message by encoding the information in the polarization of light. The message travels through the quantum channel and is received by Bob, using highly sensitive photodetectors (SPADs) to decode the message.

The SPADs have been used for various other applications such as light detection and ranging (LIDAR) systems and fluoresce spectroscopy measurements for biomedical applications, where they take advantage of the low-light-level sensing devices.

Materials such as InGaAs/InP [47] and Ge-on-Si [48], which can allow the operation of SPADs at 1310 nm or 1500 nm, respectively, are generally used for long-distance quantum communication systems. Alternatively, silicon-based SPADs can be utilized in QCS owing to their advantage for lower dark count rate (DCR) and maturity of fabrication. Fiber-based QKDs with Si-SPADs have been demonstrated with a clock of 2 GHz, with application on campus and metropolitan networks [49], and free space QKDs have shown a 300 m communication distance at 1 Mbps [50]. Such devices made of Si currently are the most suitable technology for low-light detection applications owing to their CMOS-compatible processing and their advantage of room temperature operation. However, Si-SPADs, like other Si-based photodetectors, present the typical trade-off between photon detection efficiency (PDE) and speed. A thick absorption layer increases time jitter. On the other hand, devices with thin junction layers that present low time jitter contribute to lower PDE than their thicker counterparts [46].

The break of the trade-off between time jitter and PDE has been demonstrated by the implementation of nanoholes in Si-SPADs at 850 nm [51]. These devices with nanostructures have shown a 2.5-fold higher PDE in the near-infrared regime, while keeping the timing at 25 ps compared with a control SPAD (Figure 16.21).

The enhancement in PDE is directly proportional to EQE of the photodetector operated on linear mode (below the breakdown voltage). This relation is as in the following:

$$PDE(\lambda, V) = \eta(\lambda) \times \varepsilon(\lambda) \times F \tag{16.26}$$

Equation 16.26 suggests that PDE varies with wavelength and voltage of operation and is directly proportional to the quantum efficiency (η), the probability of avalanche initialization (ε), and the fill factor of the device. This equation suggests that an enhancement in EQE can be correlated with an improvement in PDE of the detector. As seen in Figure 16.22a, broadband higher EQE over a wide range of wavelengths between 800 nm and 1100 nm is provided by devices with nanostructures compared with a control sample. At 850 nm, a device with no photon-trapping structures presents an EQE of 15% at 10 V

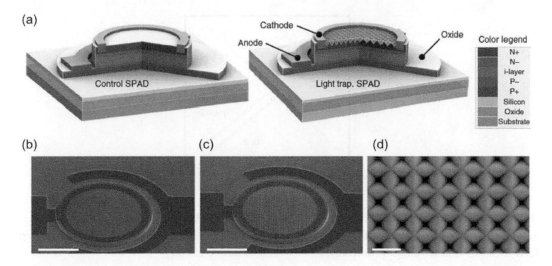

FIGURE 16.21 (a) A schematic of nanostructured silicon SPAD for near-infrared detection. SEM image of (b) control device without nanostructures, (c) Si SPAD with nanostructures, and (d) nanostructured surface. This device presents a 2.5-fold higher PDE, while keeping 25 ps time response compared with a control SPAD without nanostructures. (Reproduced from Zang, K. et al., *Nat. Commun.*, 8, 628, 2017.)

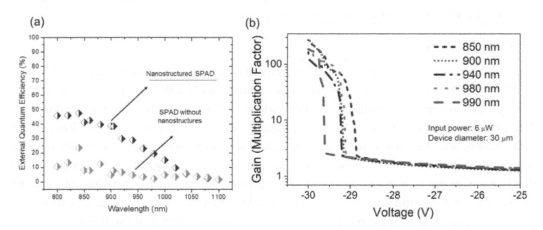

FIGURE 16.22 (a) EQE of SPADs from 800 nm to 1100 nm. At 850 nm, EQE is more than 50% for device with photon-trapping structures, while a device without these structures provides only a 15% EQE. (b) Multiplication (M) and breakdown voltage (V_{BD}) of Si SPAD operating at different input wavelengths in the near-infrared regime, where silicon has poor absorption capabilities. High M and consistent V_{BD} can be achieved in these nanostructured SPADs.

reverse bias, while a device with photon-trapping structures works with 50% EQE, showing more than 300% of enhancement [52].

Figure 16.22b shows the characterization of breakdown voltage in SPADs; this parameter is important since the applied bias below that level will define the excess voltage (V_E) in the SPAD. The breakdown voltage observed in SPADs is around ~29 V and is consistent at different wavelengths in the near-infrared regime.

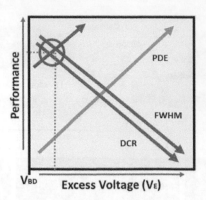

FIGURE 16.23   Effect of excess voltage on PDE, DCR, and FWHM. As excess voltage increases, a higher PDE is obtained; on the other hand, the dark count rate and time response decreases. A high PDE with low excess voltage is desirable to avoid that negative effect.

The avalanche triggering probability, DCR, and time jitter also depend on the amount of V_E [53]. Figure 16.23 presents this relationship. Obtaining a higher PDE without applying a high V_E possesses a great opportunity to develop SPADs with higher sensitivities and avoid an increase in DCR and time response. The devices with such characteristics will further improve the sensitivity of quantum communication systems.

16.4 SUMMARY

Communication has been one of the game changers in the history of humankind. The quality of communication networks has become a significant indicator of advancement in technology. Fast and secure communication networks enable global accessibility to knowledge base and exponentially increase the quality and quantity of human-to-human interaction. Last decades became witness to the economic and social impacts of fast and secure internet accessibility. Now, network providers are getting ready for a higher level of connectivity that is going to not only evolve human-to-human interaction but also provide interface for human-to-machine and machine-to-machine connectivity. Future communication networks promise to define new reality with advanced connectivity and intelligence.

In this endeavor, optical communication links must be of low cost and compatible with CMOS electronics to offer reliable communication with low latency. The ability to design PDs using Si and integrate them with other Si ICs can lead to cost-effective, superior, and novel functionalities in transceivers. Recently invented Si-based high-speed PDs with light-trapping mechanism can possess a new category of high-speed PDs that can be widely deployed in all-Si optical receivers. In this chapter, we provided a brief description of approaches to realize high-speed and high-efficiency Si-based PDs. A successful device would require a meticulous control of surface states, defects, traps, polarization, and light coupling, as well as a good control on generating sharp axial junctions in the devices.

Monolithic integration of surface-illuminated Si-based high-speed PDs with CMOS circuits will open new opportunities for optical communication industry to design low-cost on-board optical modules with less complexity and low parasitic. Just like the falling cost of

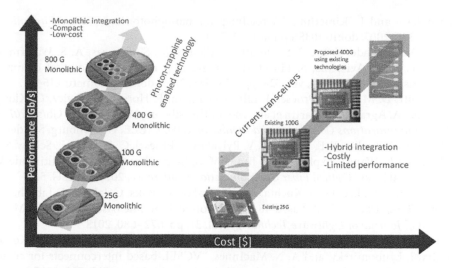

FIGURE 16.24 Performance versus cost relation for current and future transceivers. Current transceivers adopt a costly hybrid integration method that limits the performance. On the other hand, innovative light-trapping techniques can offer compact photodetector arrays for future low-cost transceivers by enabling monolithic integration. (Courtesy of W&WSens Devices, Los Altos, CA.)

transportation, decreasing cost of data communications can contribute to easier and viable exchange of ideas and goods around the globe—key ingredients of the fourth industrial revolution. Figure 16.24 summarizes the receiver technology described in the chapter and compares the cost, performance, and complexity of its counterpart that industry has been pursuing for some time. A cost performance ratio like the one shown in Figure 16.24 can lead to numerous applications in data communication, quantum communication, and imaging systems.

ACKNOWLEDGMENTS

The authors are truly grateful for financial support from the department of defense (DoD), National Science Foundation, and W&WSens, Inc. The authors thank Dr. Shih-Yuan Wang, Dr. Aly Elrefaie, Dr. Katya Ponizovskaya Devine, and Prof. Yamada of W&WSens for their help with the device design.

REFERENCES

1. Cisco. 2017. Cisco visual networking index: Forecast and methodology, 2016–2021, white paper.
2. Cisco. 2018. Cisco global cloud index: Forecast and methodology, 2016–2021 white paper.
3. J. K. Perin, M. Sharif, and J. M. Kahn, "Modulation schemes for single-laser 100 Gb/s links: Multicarrier," *Journal of Lightwave Technology*, vol. 33, pp. 5122–5132, 2015.
4. A. A. Saleh, "Evolution of the architecture and technology of data centers towards exascale and beyond," in *Optical Fiber Communication Conference and Exposition and the National Fiber Optic Engineers Conference (OFC/NFOEC), 2013*, Anaheim, CA, Optical Society of America, 2013, pp. 1–3.
5. J. K. Perin, A. Shastri, and J. M. Kahn, "Data center links beyond 100 Gbit/s per wavelength," *Optical Fiber Technology*, vol. 44, pp. 69–85, 2018.

6. R. Kirchain and L. Kimerling, "A roadmap for nanophotonics," *Nature Photonics,* vol. 1, pp. 303–305, 2007. doi:10.1038/nphoton.2007.84.

7. C. Sun, M. T. Wade, Y. Lee, J. S. Orcutt, L. Alloatti, M. S. Georgas, A. S. Waterman, J. M. Shainline, R. R. Avizienis, S. Lin et al., "Single-chip microprocessor that communicates directly using light," *Nature,* vol. 528, pp. 534–538, 2015. doi:10.1038/nature16454.

8. J. S. Orcutt, D. M. Gill, J. Proesel, J. Ellis-Monaghan, F. Horst, T. Barwicz, C. Xiong, F. G. Anderson, A. Agrawal, Y. Martin et al., "Monolithic silicon photonics at 25 Gb/s," *2016 Optical Fiber Communications Conference and Exhibition (OFC),* IEEE, Gothenburg, Sweden, 2016.

9. N. Dupuis, D. Kuchta, F. E. Doany, A. V. Rylyakov, J. Proesel, C. Baks, C. L. Schow, S. Luong, C. Xie, and L. Wang, "Exploring the limits of high-speed receivers for multimode VCSEL-based optical links,"*Optical Fiber Communication Conference,* 2014, p. M3G. 5.

10. F. E. Doany, B. G. Lee, D. M. Kuchta, A. V. Rylyakov, C. Baks, C. Jahnes, F. Libsch, and C. L. Schow, "Terabit/Sec VCSEL-based 48-channel optical module based on holey CMOS transceiver IC," *Journal of Lightwave Technology,* vol. 31, pp. 672–680, 2013.

11. J. A. Tatum, D. Gazula, L. A. Graham, J. K. Guenter, R. H. Johnson, J. King, C. Kocot, G. D. Landry, I. Lyubomirsky, and A. N. MacInnes, "VCSEL-based interconnects for current and future data centers," *Journal of Lightwave Technology,* vol. 33, pp. 727–732, 2015.

12. D. Mahgerefteh, C. Thompson, C. Cole, G. Denoyer, T. Nguyen, I. Lyubomirsky, C. Kocot, and J. Tatum, "Techno-economic comparison of silicon photonics and multimode VCSELs," *Journal of Lightwave Technology,* vol. 34, pp. 233–242, 2016.

13. A. Beling and J. C. Campbell, "Heterogeneously integrated photodiodes on silicon," *IEEE Journal of Quantum Electronics,* vol. 51, pp. 1–6, 2015. doi:10.1109/JQE.2015.2480595.

14. H. Nasu, K. Nagashima, T. Uemura, A. Izawa, and Y. Ishikawa, "> 1.3-Tb/s VCSEL-based on-board parallel-optical transceiver module for high-density optical interconnects," *Journal of Lightwave Technology,* vol. 36, pp. 159–167, 2018.

15. J. D. Schaub, D. M. Kuchta, D. L. Rogers, M. Yang, K. Rim, S. Zier, and M. Sorna, "Multi Gbit/s, high-sensitivity all silicon 3.3 V optical receiver using PIN lateral trench photodetector," in *Optical Fiber Communication Conference,* Anaheim, CA, Optical Society of America, 2001, p. PD19.

16. S. Csutak, J. Schaub, W. Wu, and J. Campbell, "High-speed monolithically integrated silicon optical receiver fabricated in 130-nm CMOS technology," *IEEE Photonics Technology Letters,* vol. 14, pp. 516–518, 2002.

17. R. Swoboda and H. Zimmermann, "11 Gb/s monolithically integrated silicon optical receiver for 850 nm wavelength," in *Solid-State Circuits Conference, 2006. ISSCC 2006. Digest of Technical Papers.* IEEE International, 2006, pp. 904–911.

18. M. K. Emsley, O. Dosunmu, and M. S. Unlu, "High-speed resonant-cavity-enhanced silicon photodetectors on reflecting silicon-on-insulator substrates," *IEEE Photonics Technology Letters,* vol. 14, pp. 519–521, 2002.

19. L. Li, "New formulation of the Fourier modal method for crossed surface-relief gratings," *JOSA A,* vol. 14, pp. 2758–2767, 1997.

20. M. A. Green and M. J. Keevers, "Optical properties of intrinsic silicon at 300 K," *Progress in Photovoltaics: Research and Applications,* vol. 3, pp. 189–192, 1995.

21. S. Peng and G. M. Morris, "Resonant scattering from two-dimensional gratings," *JOSA A,* vol. 13, pp. 993–1005, 1996.

22. Y. Gao, H. Cansizoglu, K. G. Polat, S. Ghandiparsi, A. Kaya, H. H. Mamtaz, A. S. Mayet, Y. Wang, X. Zhang, T. Yamada et al., "Photon-trapping microstructures enable high-speed high-efficiency silicon photodiodes," *Nature Photonics,* vol. 11, pp. 301–308, 2017. doi:10.1038/nphoton.2017.37.

23. Y. Gao, H. Cansizoglu, S. Ghandiparsi, C. Bartolo-Perez, E. P. Devine, T. Yamada, A. F. Elrefaie, S.-Y. Wang, and M. S. Islam, "A high speed surface illuminated Si photodiode using microstructured holes for absorption enhancements at 900–1000 nm wavelength," *ACS Photonics,* vol. 4, pp. 2053–2060, 2017. doi:10.1021/acsphotonics.7b00486.

24. H. Cansizoglu, C. Bartolo-Perez, Y. Gao, E. P. Devine, S. Ghandiparsi, K. G. Polat, H. H. Mamtaz, T. Yamada, A. F. Elrefaie, and S.-Y. Wang, "Surface-illuminated photon-trapping high-speed Ge-on-Si photodiodes with improved efficiency up to 1700 nm," *Photonics Research,* vol. 6, pp. 734–742, 2018.

25. H. Cansizoglu, E. P. Devine, Y. Gao, S. Ghandiparsi, T. Yamada, A. F. Elrefaie, S.-Y. Wang, and M. S. Islam, "A new paradigm in high-speed and high-efficiency silicon photodiodes for communication—Part I: Enhancing photon–material interactions via low-dimensional structures," *IEEE Transactions on Electron Devices,* vol. 65, pp. 372–381, 2018.

26. H. Cansizoglu, A. F. Elrefaie, C. Bartolo-Perez, T. Yamada, Y. Gao, A. S. Mayet, M. F. Cansizoglu, E. P. Devine, S.-Y. Wang, and M. S. Islam, "A new paradigm in high-speed and high-efficiency silicon photodiodes for communication—Part II: Device and VLSI integration challenges for low-dimensional structures," *IEEE Transactions on Electron Devices,* vol. 65, pp. 382–391, 2018.

27. S. Ghandiparsi, A. F. Elrefaie, H. Consizoglu, Y. Gao, C. Bartolo-Perez, H. H. Mamtaz, A. Mayet, T. Yamada, E. P. Devine, and S.-Y. Wang, "High-speed high-efficiency broadband silicon photodiodes for short-reach optical interconnects in data centers," in *Optical Fiber Communication Conference,* San Diego, CA, Optical Society of America, 2018, p. W1I. 7.

28. S. B. Alexander, *Optical Communication Receiver Design* vol. 37: SPIE Press, Bellingham, WA, 1997, pp. 98–99.

29. J. Ingham, "Future of short-reach optical interconnects based on MMF technologies," in *Optical Fiber Communication Conference,* 2017, p. Tu2B. 1.

30. D. Kuchta, "High-Capacity VCSEL Links," in *Optical Fiber Communication Conference,* Los Angeles, CA, 2017, p. Tu3C. 4.

31. M. R. Tan, P. Rosenberg, W. V. Sorin, B. Wang, S. Mathai, G. Panotopoulos, and G. Rankin, "Universal photonic interconnect for data centers," *Journal of Lightwave Technology,* 2017. doi:10.1109/JLT.2017.2747501.

32. M. A. Taubenblatt, "Optical interconnects for high-performance computing," *Journal of Lightwave Technology,* vol. 30, pp. 448–457, 2012. doi:10.1109/JLT.2011.2172989.

33. "IEEE Approved Draft Standard for Ethernet," in IEEE P802.3/D3.2, March 2018 (Revision of IEEE Std 802.3-2015)," pp. 1–5601, 1 January 2018.

34. V. Houtsma, D. van Veen, and E. Harstead, "Recent progress on standardization of next-generation 25, 50, and 100G EPON," *Journal of Lightwave Technology,* vol. 35, pp. 1228–1234, 2017.

35. Z. Li, Y.-M. Jung, N. Simakov, P. Shardlow, A. Heidt, A. Clarkson, S.-U. Alam, and D. J. Richardson, "Extreme short wavelength operation (1.65–1.7 μm) of silica-based thulium-doped fiber amplifier," in *Optical Fiber Communication Conference,* 2015, p. Tu2C. 1.

36. R. B. Emmons, "Avalanche—Photodiode frequency response," *Journal of Applied Physics,* vol. 38, pp. 3705–3714, 1967. doi:10.1063/1.1710199.

37. J. Liu, D. D. Cannon, K. Wada, Y. Ishikawa, S. Jongthammanurak, D. T. Danielson, J. Michel, and L. C. Kimerling, "Tensile strained Ge *p-i-n* photodetectors on Si platform for C and L band telecommunications," *Applied Physics Letters,* vol. 87, p. 011110, 2005. doi:10.1063/1.1993749.

38. S. Assefa, F. Xia, and Y. A. Vlasov, "Reinventing germanium avalanche photodetector for nanophotonic on-chip optical interconnects," *Nature,* vol. 464, pp. 80–4, 2010. doi:10.1038/nature08813.

39. J. C. Campbell, "Recent advances in avalanche photodiodes," *Journal of Lightwave Technology,* vol. 34, pp. 278–285, 2016. doi:10.1109/JLT.2015.2453092.

40. Y. Kang, H.-D. Liu, M. Morse, M. J. Paniccia, M. Zadka, S. Litski, G. Sarid, A. Pauchard, Y.-H. Kuo, H.-W. Chen et al., "Monolithic germanium/silicon avalanche photodiodes with 340 GHz gain–bandwidth product," *Nature Photonics,* vol. 3, pp. 59–63, 2008. doi:10.1038/nphoton.2008.247.

41. M. Huang, S. Li, P. Cai, G. Hou, T.-I. Su, W. Chen, C.-y. Hong, and D. Pan, "Germanium on silicon avalanche photodiode," *IEEE Journal of Selected Topics in Quantum Electronics,* vol. 24, pp. 1–11, 2018. doi:10.1109/jstqe.2017.2749958.

42. M. Huang, T. Shi, P. Cai, L. Wang, S. Li, W. Chen, C. Hong, and D. Pan, "25Gb/s normal incident Ge/Si avalanche photodiode," in *2014 The European Conference on Optical Communication (ECOC),* 2014, pp. 1–3.

43. D. C. Houghton, "Strain relaxation kinetics in Si1–xGex/Si heterostructures," *Journal of Applied Physics,* vol. 70, pp. 2136–2151, 1991. doi:10.1063/1.349451.

44. F. Flamini, N. Spagnolo, and F. Sciarrino, "Photonic quantum information processing: A review," *Reports on Progress in Physics,* vol. 82, p. 016001, 2019. doi:10.1088/1361-6633/aad5b2.

45. N. Gisin and R. Thew, "Quantum communication," *Nature Photonics,* vol. 1, p. 165, 2007. doi:10.1038/nphoton.2007.22.

46. M. Ghioni, A. Gulinatti, I. Rech, F. Zappa, and S. Cova, "Progress in silicon single-photon avalanche diodes," *IEEE Journal of Selected Topics in Quantum Electronics,* vol. 13, pp. 852–862, 2007. doi:10.1109/JSTQE.2007.902088.

47. J. Zhang, M. A. Itzler, H. Zbinden, and J.-W. Pan, "Advances in InGaAs/InP single-photon detector systems for quantum communication," *Light: Science & Applications,* vol. 4, pp. e286–e286, 2015. doi:10.1038/lsa.2015.59.

48. N. J. D. Martinez, M. Gehl, C. T. Derose, A. L. Starbuck, A. T. Pomerene, A. L. Lentine, D. C. Trotter, and P. S. Davids, "Single photon detection in a waveguide-coupled Ge-on-Si lateral avalanche photodiode," *Opt Express,* vol. 25, pp. 16130–16139, 2017. doi:10.1364/OE.25.016130.

49. K. J. Gordon, V. Fernandez, G. S. Buller, I. Rech, S. D. Cova, and P. D. Townsend, "Quantum key distribution system clocked at 2 GHz," *Optics Express,* vol. 13, pp. 3015–3020, 2005. doi:10.1364/OPEX.13.003015.

50. M. J. García-Martínez, N. Denisenko, D. Soto, D. Arroyo, A. B. Orue, and V. Fernandez, "High-speed free-space quantum key distribution system for urban daylight applications," *Applied Optics,* vol. 52, pp. 3311–3317, 2013. doi:10.1364/AO.52.003311.

51. K. Zang, X. Jiang, Y. Huo, X. Ding, M. Morea, X. Chen, C.-Y. Lu, J. Ma, M. Zhou, Z. Xia et al., "Silicon single-photon avalanche diodes with nano-structured light trapping," *Nature Communications,* vol. 8, p. 628, 2017. doi:10.1038/s41467-017-00733-y.

52. C. Bartolo-Perez, H. Cansizoglu, Y. Gao, S. Ghandiparsi, A. S. Mayet, E. P. Devine, A. F. Elrefaie, S. Wang, and M. S. Islam, "Enhanced photon detection efficiency of silicon single photon avalanche photodetectors enabled by photon trapping structures," in *2018 IEEE Photonics Society Summer Topical Meeting Series (SUM),* 2018, pp. 143–144.

53. S. Cova, M. Ghioni, A. Lotito, I. Rech, and F. Zappa, "Evolution and prospects for single-photon avalanche diodes and quenching circuits," *Journal of Modern Optics,* vol. 51, pp. 1267–1288, 2004. doi:10.1080/09500340408235272.

Nanoscale Materials and Devices for Future Communication Networks

Miao Sun, Mohammad Taha, Sumeet Walia,

Madhu Bhaskaran, Sharath Sriram, William Shieh,

and Ranjith Rajasekharan Unnithan

CONTENTS

17.1 INTRODUCTION

There is an ever-increasing demand in bandwidth and data rate in optical communications, internet, sensing, computing, machine learning applications, and cloud computing. For chip-level communications, existing complementary metal oxide semiconductor (CMOS) electronics-based solutions are currently challenged and strained to their maximum limit. The maximum speed of new 5G mobile connection is 10 Gbits/s, that is, 10×109 bits/s. When the bit rates increase beyond 10 Gbits/s, copper tracks that carry current between the processing chips severely distort and attenuate the data signal owing to parasitic resistance, inductance, and capacitance of the track (Développement 2016, Jeong et al. 2017). To overcome these challenges, the industry is increasingly looking toward silicon photonics integrated with nanoscale devices as a key technology for next-generation communications systems and data interconnects. For optical communications, silicon photonics-based modulators are gradually replacing bulky and expensive conventional lithium niobate modulators for medium-reach 100 Gbit/s applications.

Silicon photonics is an emerging technology based on silicon chips, where photons (instead of electrons) are used to carry information in a CMOS chip (Jeong et al. 2017). Photons can carry far more data in less time than electrons, because photons can be pulsed at higher frequencies than electric currents, resulting in a high bandwidth. Since both CMOS and silicon photonics are based on silicon, it is possible to integrate photonics circuits and electronic circuits into a single chip (Barwicz et al. 2016, Kinsey et al. 2015). The significance of silicon photonics is evident from the multimillion CmOs solutions for mid-board integrated transceivers with breakthrough connectivity at ultra-low cost (COSMICC) project to combine CMOS electronics and silicon photonics funded by the European Union's Horizon 2020 program, as well as from the U.S. government funding of USD 110 million to a consortium, including MIT, to work on next-generation silicon photonics devices. Silicon photonics market in 2018 is estimated to be USD 492 million, with a projected growth of USD 3.7 billion in 2025 (Développement 2016).

Figure 17.1 shows the power of silicon (Si) photonics, where all the bulky photonics modules (such as laser, modulator for signal modulation, waveguides, waveplates, and photodetector array), including electronics on an optical bench for optical communications, can be fabricated in a silicon photonics chip of around 4.5×5 mm size.

Optical modulation is at the heart of Si photonics for on-chip optical communications. A modulator encodes a high-speed electronic data stream (radio frequency [RF] signals) to an optical carrier wave, as shown in Figure 17.2. Conventional silicon photonics-based modulators are constrained to millimeters in size (Melikyan et al. 2014). The millimeter size is due to the requirement of long interaction length (Figure 17.2) between the optical signal and the modulating RF signals for encoding RF data to the light. However, the width of gate in transistors has reduced to nanoscale (14-nm process technology became available in 2014 and 10 nm in 2017) (TSMC 2017). Therefore, extensive research into optical modulators is needed to fully tap into the potential of Si photonics technology, especially for the latter's integration with electronics onto a single platform. The mismatch in the size of optical modulators and transistor gates (10-nm process technology in 2017) is a bottleneck in developing modulators integrated

with electronics on a chip with high modulation index in ultra-small footprint, wide operation wavelength range, low energy consumption (low energy/bit), low insertion loss, and with
the ability to operate in hundreds of gigahertz for next-generation communication devices.
This need has recently triggered research on new nanoscale modulator technologies in both
academia and industries (Agrell et al. 2016) for next-generation communication devices.

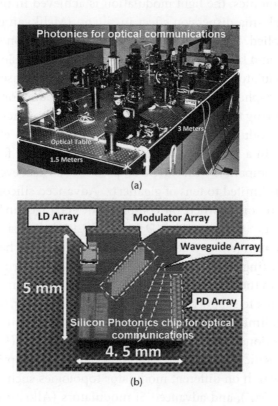

FIGURE 17.1 Comparison between (a) photonics for optical communication and equivalent (b)
silicon photonics chip.

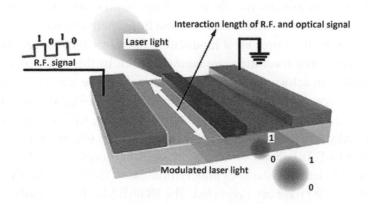

FIGURE 17.2 A simple optical modulator with intensity modulation of light.

17.1.1 Recent Research on Optical Modulators/Switches

This section reviews the recent research progress in modulator technologies for the future communication technologies in academia and industry.

17.1.1.1 In Academia

In conventional Si photonics, the light modulation is achieved in modulators by exploiting the free-carriers plasma dispersion effect in silicon (Melikyan et al. 2014). When an external voltage is applied, the refractive index and the absorption of silicon vary owing to changes in electron and hole densities, called the plasma dispersion effect. The modulation efficiency of a modulator is defined as the product of the π-phase-shift voltage (voltage required for achieving π-phase-shift) and the length of the modulator (VπL) (Dhiman 2013, Alloatti et al. 2014). Owing to extremely small changes in the refractive index of silicon in the dispersion effect, modulators based on the plasma dispersion effect typically require millimeter size to obtain a π-phase shift (minimum requirement for phase modulation), with driving voltages compatible with CMOS chips (3–5 V). Also, even with the bulky size, the maximum speed is limited to tens of gigahertz. Advanced silicon p-n junction–based modulators have been reported, where the complex index change can be driven by p-n junctions or capacitors built in Si waveguides (David et al. 2016) that can operate in 50 Gbit/s, with large VπL = 28 V mm (Alloatti et al. 2014). These limitations have resulted in the use of wavelength multiplexing (using more wavelength to carry information in parallel) and parallel copper tracks in the existing technologies to tackle high bandwidth and speed due to limitations of the track to support data above 10 Gbit/s. However, these techniques make the design bulky, with limited bandwidth and speed, and are strained to their maximum limit with the introduction of 5G, explosions of Internet of Things (IoT), big data, machine learning algorithms, artificial intelligence, and internet speed in optical communications. This has triggered research on different modulator topologies such as resonant, nonresonant, Mach–Zehnder (MZ), and advanced Si modulators (Alloatti et al. 2014, Melikyan et al. 2014, Kinsey et al. 2015, Barwicz et al. 2016, TSMC 2017). Resonator modulators can enhance electro-optic properties owing to the large quality factor of the resonant cavity, and hence, their dimensions can be shrunk to tens of micrometers (Melikyan et al. 2014). However, resonator modulators usually suffer from bandwidth limitations, temperature fluctuations, and fabrication tolerances, and these limit their size reduction in high-speed modulations. In contrast, nonresonant modulators operate across a large spectral window and are mainly built on a traveling wave configuration. In this structure, a long interaction time is required to achieve sufficient modulation depth between the modulating optical signals and the corresponding electrical signals (RF data signals) used for the optical modulation. Hence, the nonresonant modulators are bulky, with a length of several millimeters. Such bulky dimensions also increase the RF losses (electrical data signals).

Recently, SiGe-in-SOI (silicon on insulator) electroabsorption modulator (EAM), the distributed-lumped-element Mach–Zehnder modulator (MZM), and the III–V-on-SOI EAM (David et al. 2016) have been reported. The SiGe-EAMs have the advantages of small size (micron sizes) and low power consumption and can work at high speeds in waveguide platforms. However, the EAMs suffer from lower extinction ratio and limited optical operating bandwidth. The MZMs cannot operate over large optical bandwidths and have higher extinction ratios, in addition a larger size and inherently more power consumption. Silicon combined with III–V epitaxy stacks grown on indium phosphide (InP) substrates and organic material (hybrid) to increase the speed around 65 and 100 GHz, respectively, but it was still limited by VπL (Alloatti et al. 2014).

J. Leuthold has demonstrated, for the first time, a plasmonic phase modulator with a small dimension (29 µm) (Melikyan et al. 2014). The footprint was limited to 29 microns owing to poor electro-optic effect of polymer used as the modulating material. Also, the polymer is temperature sensitive and degrades over time.

17.1.1.2 In Industry

Ninety percent of the world's data were created in the last 2 years alone, and this explosive demand is placing on the network with higher-speed connectivity over longer distances within the hyperscale data centers (Bussiness 2018). Intel® Silicon Photonics lab demonstrated in 2017 a new optical transceiver operating at gigabits per second, using Intel's hybrid laser technology on silicon photonics platform (Photonics). In 2018 data-centric innovation summit, Intel unveiled the next-gen Ice Lake CPU by using 10-nm process technology that will be available in 2020, which is expected to replace the 14-nm nodes in market (Alcorn 2018). IBM has demonstrated in 2016–2017 that by adding a few processing modules to a 90-nm CMOS fabrication line, a variety of silicon nanophotonics components such as wavelength division multiplexers (WDM), modulators, and detectors can be integrated side by side with CMOS electrical circuitry in 90-nm process technology (Barwicz et al. 2016). This has proved the feasibility that single-chip optical communications transceivers realize in a conventional semiconductor foundry (Barwicz et al. 2016). Also, IBM demonstrated their groundbreaking work on an inexpensive 60 Gb/s optical receiver in 2017 Symposia on VLSI Technology and Circuits in Kyoto, Japan (Research-Zurich). The unprecedent vast development of Silicon Photonics technologies is motivated by the intra-/interapplication of data center, because Silicon Photonics is the best choice to realize the low cost and high transceiving rate over distance, which is beyond vertical-cavity surface-emitting laser's reach. Yole's analysts announced that in 2025, a USD 560 million market value is expected at chip level and an almost USD 4 billion market value is expected at the transceiver level (LYON 2018). The potential range of silicon photonics applications can be extended from data centers, telecom, and medical use to high-performance computing (HPC), autonomous cars, sensors, aeronautics, etc (Développement 2018). Last but not the least, the advance of Silicon Photonics technology will create new opportunities of innovative entrepreneurship, strengthening the human capital agenda and resulting in captivating job for young scientists, engineers, and photonics start-ups (PhotonicsNL 2018).

17.1.2 Innovation of the Proposed Research

Based on the previous modulator topologies reported in both academia and industry, a trade-off exists among optical loss (attenuation), VπL (size and modulation depth), speed, and footprint. This is because higher free carriers' densities with stronger plasma dispersion come at the expense of increased optical absorption or size or limited dynamic range (i.e., the maximum intensity or phase change for a given driving voltage) (Alloatti et al. 2014). This demands innovative approaches to research on CMOS-compatible novel high modulators for silicon photonics for next-generation communication devices. Our innovative modulator development is based on carrying out research on designing the submicron plasmonic structure, in order to exploit the light confinement beyond the diffraction limit and implement vanadium dioxide (VO_2) as the active material to increase V_π in submicron scale in $V\pi L$.

The broad aim of the present chapter is to demonstrate an interdisciplinary approach to design and develop new nanophotonic hybrid modulators/switches for nanoscale future communication technologies, where plasmonics technology is integrated with silicon waveguides to confine light below diffraction limit. VO_2 is a canonical Mott material that will be used as a modulating material driven by low voltage in the modulator. VO_2 is suitable to use as modulating material for optical communications because of its large refractive index change during the first-order insulator-to-metal transition, with $3.24 + j0.3$ in the semiconductor (insulator) phase and $2.03 + j2.64$ in the metallic phase at 1550 nm.

In the first section, we introduce a design of plasmonic modulators using VO_2 as modulating material realized on silicon on insulator (SOI) wafer with only 200×140 nm modulating section within 1×3 μm device footprint. By utilizing the large refractive index contrast between the metallic and semiconductor phases of VO_2, the modulator can achieve a broad working wavelength range from 1100 to 1800 nm, with modulation depth 21.5 dB/μm. We also analyze the effects of using seed layer of different dielectric materials for growing VO_2 on modulation index by exploring the mixed combination of VO_2 and different dielectric materials. The device geometries can have potential applications in the development of next-generation miniaturized high-frequency optical modulators in silicon photonics for optical communications.

In the second section, we demonstrate a plasmonic bandpass filter integrated with materials exhibiting phase transition, which can be used as a thermally reconfigurable optical switch. The fabricated switch shows an operating range over 650 nm around the optical communication C, L, and U bands, with maximum 20%, 23% and 26%, respectively, transmission difference in switching. The extinction ratio is around 5 dB in the entire operation range. This architecture is a precursor for developing micron-size photonic switches and ultra-compact modulators for thin film photonics with the ability to be integrated on fiber tips and chips.

17.2 NANOPHOTONIC HYBRID MODULATOR BASED ON VO_2

Here, we discuss about a hybrid plasmonic modulator technology on SOI wafer with only 200×140 nm modulating section within 1×3 μm device footprint (exclude electrodes), using VO_2 as modulating material. The optical modulation is achieved by utilizing the large refractive index change between the metallic and semiconductor phases of VO_2.

17.2.1 Introduction

Development of nanofabrication over the past decades has enabled researchers to explore faster, smaller, and broadband photonic devices. Especially, silicon-based photonics (SiP) has attracted much attention due to its CMOS capability. Recently, high-speed, ultra-compact, and low-cost SiP-based electro-optic modulators have been demonstrated (Briggs et al. 2010, Sweatlock and Diest 2012, Ryckman et al. 2013, Sun et al. 2014, Markov et al. 2015). The electro-optic modulators play a major role in optical communications, where their bandwidth, size, and power consumption determine their applicability in optical transmission links. The "big data" and IoT have triggered data explosion in this information era. The joint integration between SiP and electronics yields high scalability, high efficiency, and low cost, with performance enhancement (Pavesi and Lockwood 2016). This makes SiP an attractive platform to bridge between electronics and photonics. The size of transistors has reduced to nanoscale (nanoscale electronics), but conventional SiP-based optical devices are constrained by the diffraction limitation of half the optical wavelength. This mismatch gives rise to research on novel devices based on plasmonics in silicon photonics to enable matching integration with the nanoelectronics. A recent work has demonstrated the potential to make Si-based subwavelength modulator by tuning the dielectric permittivity of nanoparticle (Makarov et al. 2015).

Surface plasmon polaritons (SPPs) propagating at a dielectric–metallic interface (Maier 2007) have recently been studied extensively for their ability to concentrate the optical field while circumventing the diffraction limit, providing a feasible method of developing modulators, switches, and other optical devices in subwavelength size. The electro-optic modulators implemented with plasmonic technology can bridge the size mismatch between the nanoscale electronics and the traditional microscale silicon photonic and also achieve high working frequency. Plasmonic modulators implemented with electro-optical polymer, using the Pockels effect operating at 40 and 65 GHz, have been recently demonstrated in devices with a footprint of 29 μm (Melikyan et al. 2014). However, the active SPP devices such as plasmonic switches and modulators inevitably suffer from the trade-offs between their size, insertion loss, extinction ratio, and modulation depth. This trade-off is fundamental, because a small refractive index change occurs in common electro-optical materials, such as in electro-optical polymers, silicon, and Indium tin oxide (ITO), when an external electric field is applied (Sun et al. 2014, Kruger et al. 2012). Smaller the refractive index changes, larger the footprint of the device. This is to increase a long interaction time between the optical signal and the modulating signal (electric) for achieving modulation. To counterbalance this trade-off, researchers keep exploring new materials that can be used to increase the refractive index change when a small electric field is applied.

VO_2 is a canonical Mott material suitable to use as switching or modulating material for optical communications because of its large refractive index change during the first-order insulator-to-metal transition, with 3.24 + 0.30i in the semiconductor (insulator) phase and 2.03 + 2.64i in metallic phase at 1550 nm (Verleur et al. 1968). The transition from semiconductor to metal phase can be triggered by an electric field of 6.5×10^7 V/m (Wu et al. 2011) or by an application of heat (68°C) (Zylbersztejn and Mott 1975, Liu et al. 2012, Yang and Ramanathan 2015). The phase transition happens at hundreds of femtoseconds in one direction (semiconducting phase to metallic phase), while it is slower to transfer back (metal to semiconductor), ranging from nanosecond to milliseconds (Lysenko et al. 2009). Though VO_2 is a promising candidate material with all properties shown previously, its imaginary part of refractive index will introduce loss when used directly as plasmonic waveguide material. The solution is to combine VO_2 with a noble metal such as gold to exploit the compact nature of SPP while reducing the loss of the device. VO_2-based optical switch (Kruger et al. 2012) and modulator (Ooi et al. 2013) have been demonstrated using computer simulations recently. However, despite their switching ability, the device geometries are extremely difficult to fabricate (Ooi et al. 2013). Also, most of these devices are initiated by heating to change phases of VO_2 than triggered by an applied electric field and hence can work only at a lower frequency of operation (Kruger et al. 2012, Joushaghani et al. 2013, Markov et al. 2015).

Here, a plasmonic modulator design is proposed with only 200 × 140 nm modulating section within 1 × 3 μm device footprint (exclude electrodes) using VO2 as modulating material realized on SOI wafer. The optical modulation is achieved by utilizing the large refractive index change between the metallic and semiconductor phases of VO_2 triggered by an external electrical field. In our design, a 300-nm VO_2 film is used as modulating material that is covering the plasmonic slot and gold electrodes. This nanometer-thick VO_2 together with gold electrodes will help in very quick cooling and can achieve modulating speed at nanosecond time scale. The modulator can realize a broad working wavelength range in the telecommunication window of 1550 nm, with high modulation depth. Since most of the deposition techniques to produce high-quality VO_2 film require a dielectric seed layer, we have also studied mixed plasmonic modulators, where VO_2 is combined with dielectric materials to study the effects of using seed layers on modulation index. Here, we have explored VO_2-Si, VO_2-TiO_2, VO_2-ZrO_2, and VO_2-SiO_2 combinations to compare performances. Our proposed device geometry can be experimentally realized, and the fabrication scheme is presented.

17.2.2 General Device Description

The proposed device geometry is shown in Figure 17.3. We have used a coupling scheme similar to that in the published works (Tian et al. 2009, Melikyan et al. 2014). In our geometry, the light of wavelength 1550 nm is guided by left silicon waveguide (height 220 nm and width 450 nm) and is coupled to a plasmonic slot waveguide (slot width 140 nm and slot length 3 μm) made of gold (thickness 150 nm) through a metal taper. The middle

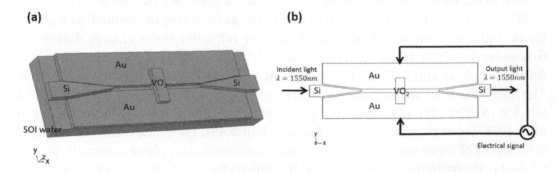

FIGURE 17.3 (a) 3D schematics of the plasmonic modulator based on VO_2. The top layer of air in the simulation model is hidden for visibility. (b) 2D schematics of the proposed modulator.

section of the slot is filled with VO$_2$, called modulating section. The refractive index change accompanying the VO$_2$ phase change is exploited for optical phase modulation by applying an external electric field between the gold electrodes in the slot waveguide. The second taper transforms the modulated surface plasmons back to photonic modes in the right silicon waveguide. The edge coupling is used to couple the light from the fiber to the input-side silicon nanowire and from the output-side silicon nanowire to the fiber.

This device has two working states, depending on an electric field applied, ON (electric field is ON) and OFF (electric field is OFF) states. In the OFF state, the SPPs propagate through the plasmonic slot waveguide, with VO$_2$ slot in the semi-phase. During the ON state, most of the SPPs will be blocked by the metallic phase of VO$_2$. We define the modulation depth as how much the modulated carrier signal varies around its unmodulated level. This can be calculated from the difference of optical attenuation between two phase changes in the device. The attenuation is calculated from input and output optical power (Poynting vector) obtained by calculating the surface integration of the optical power on both ends of the silicon waveguide.

For the silicon waveguides, the height (thickness) is 220 nm and width is 450 nm, with 15° open angle of the taper structure. This structure has been optimized for its high coupling efficiency in a broad wavelength range of around 1550 nm (Tian et al. 2009, Melikyan et al. 2014). Two 150-nm-thick gold pads consist of the plasmonic waveguide slot with a width of 140 nm, and VO$_2$ deposits thick enough to fill the slot with 300-nm thickness and 200-nm length. All these parameters are calibrated to C-band working frequency in the simulation with high efficiency, and three-dimensional (3D) model of our device is shown in Figure 17.3b. The critical parameters affecting the performance of the modulator are length, l, and height (thickness), h, of VO$_2$ and wavelength, λ, of the incident light. We have simulated and optimized these parameters in the following section.

The plasmonic modulator device's design, simulation, and optimization are carried out using finite-element method implemented in COMSOL Multiphysics. The refractive index of VO$_2$ used in the simulation is collected from the experimental measurement of VO$_2$ samples by using ellipsometry (Earl et al. 2013). Figure 17.4a gives a top view of the device in which light guided by the silicon waveguide is coupled as SPP in the plasmonic slot

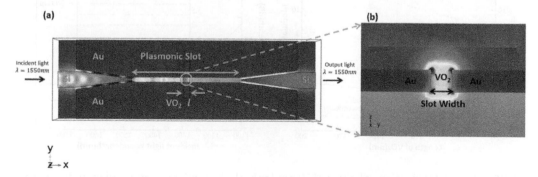

FIGURE 17.4 (a) Planar cross-section, and (b) longitudinal cross-section of the device showing normalized electric field intensities in the semiconductor phase of VO$_2$ with 1550-nm incident light.

and then passes through VO_2 slot in semiconductor phase. Figure 17.4b illustrates a cross-section of the device, where the SPP is highly confined within the Au-VO_2-Au slot when no electric field is applied (OFF state). The SPP is coupled back into light in the right taper, followed by guiding through the right silicon waveguide. From the simulations, it is observed that part of the SPP energy has been absorbed or scattered by the VO_2 slot, and hence, an optimization of VO_2 slot length is required for optimum modulation depth.

The length of VO_2, l, is a key parameter that influences the modulation depth and insertion loss. Considering the fabrication tolerance and power absorption, we have swept l from 50 to 200 nm, with 1550-nm wavelength of the incident light. Figure 17.5a shows attenuation (dB) of the light at 1550 nm for both the semiconductor (OFF state) and the metallic (ON state) phases of VO_2 versus the length of VO_2 in the slot. Modulation depth is calculated from the semiconductor phase and metallic phase attenuation values and plotted on the same graph for comparison. It is found that the attenuation of both phases increases almost linearly with the length of VO_2 in the slot. This also means that the modulation depth calculated increases approximately linearly with the length of VO_2. This is because VO_2 is not a transparent medium for plasmons, as the length increases due to increased absorption and scattering by VO_2 layer. The metallic phase of VO_2 has more loss than semiconductor phase, owing to the large imaginary part of refractive index. Considering the modulation depth, insertion loss, and fabrication difficulty, the length range from 110 to 200 nm is considered to be suitable for fabrication. For a specific length of VO_2, we have swept λ from 1100 to 1800 nm to study the dependence of modulation depth on wavelength, as shown in Figure 17.5b. This study is repeated for four different lengths of VO_2. The results show that the variation of modulation depth is not significant over a wavelength range of 700 nm, which ensures that the device is tolerant to shift in wavelength of laser diodes with respect to temperature. For example, the laser sources used for optical communication such as Agilent 8164A (FP laser InGaAs) has a tunable wavelength range from 1400 to 1670 nm. This wavelength range is fitting well in our range,

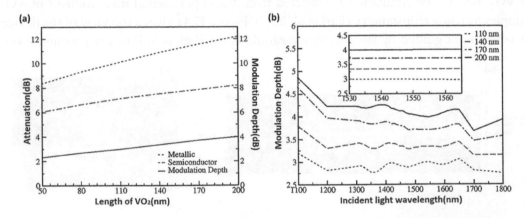

FIGURE 17.5 (a) Attenuation of light at 1550 nm for both the semiconductor and metallic phases of VO_2 along with modulation depth for different lengths of VO_2 (l). (b) The modulation depth variation with respect to wavelength, λ, varied from 1100 to 1800 nm with l swept from 110 to 200 nm, and an inset partially enlarging the C-band range.

1100–1800 nm. This result also shows an agreement with the Figure 17.5a that the modulation depth increases with length of VO$_2$, l.

In the real device fabrication, we need to study the dependence of modulation depth on VO$_2$ height (thickness). We have studied this aspect by sweeping the height of VO$_2$, h, with rest of the device parameters being constant. From the simulation results, we have found that variation in the modulation depth is minimal after reaching a height of VO$_2$ around 120 nm, as shown in Figure 17.6. Based on all the previous simulation results, we have the following optimized device parameters: length of VO$_2$ between 110 and 200 nm and height of VO$_2$ above 120 nm.

17.2.3 Fabrication Methodology

Our proposed device can be fabricated using a standard SOI wafer with 220-nm-thick silicon layer on top, 2-μm buried oxide layer (SiO$_2$) in the middle, and 550-μm-thick handle layer of silicon at the bottom. The fabrication methodology is mainly divided into three steps. First, the pattern of silicon nanowires with tapers is written on the SOI wafer, using electron beam lithography (EBL, stage 1) with two-layers Methylmethacrylate (MMA) resist (500 nm). Then, the silicon (220 nm) is etched by dry reactive ion etching (DRIE) all the way down to SiO$_2$, using 50-nm chromium coating on top as etch mask, shown in Figure 17.7a. The lift-off process refers to using acetone to wash away polymethyl methacrylate (PMMA) beneath the Cr to make Si waveguide. In the second step, the gold electrodes for making plasmonic slots are made by EBL (EBL, stage 2) process, with 250 nm of PMMA spun on the top the SOI substrate, followed by deposition of gold (Au) of thickness 150 nm, as shown in Figure 17.7b. The third step is to prepare the VO$_2$ deposition mask, shown in Figure 17.7c. We propose physical vapor deposition (PVD) method for VO$_2$ deposition, using a mask to get the required length in the plasmonic slot. However, in most of the PVD techniques, the maximum temperature can reach around 500°C, and at this temperature, most of the commonly used lithography resists will either melt or break down. A chromium mask layer of thickness 300 nm can be used to withstand the

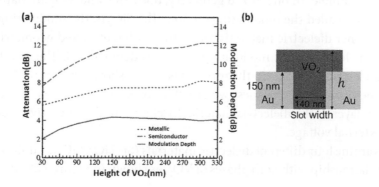

FIGURE 17.6   (a) Attenuation of the light at 1550 nm for both semiconductor and metallic phases of VO$_2$ along with modulation depth for different heights (thickness) of VO$_2$ (h). (b) Cross-section of the modulator where dark grey colour stands for VO$_2$ with thickness h and light grey colour stands for Au with thickness 150 nm with plasmonic slot width of 140 nm.

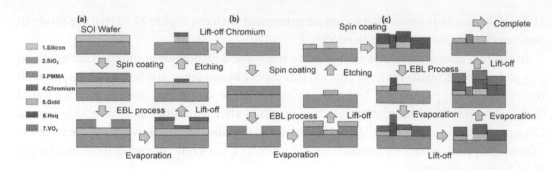

FIGURE 17.7 Schematics of the device fabrications methodology: (a) silicon nanowires, (b) gold electrodes, and (c) mask preparation, as well as VO_2 deposition. The color legends are shown on the left.

high temperatures—also for easy lift-off afterward. We also propose SiO_2 as a mask for its thermal stability and optical features, which can protect the thin layer of gold during VO_2 deposition.

17.2.4 Design of Plasmonic Mixed-VO_2 Modulator

Here, we have demonstrated plasmonic modulators by using a mixed combination of VO_2 and dielectric materials such as Si, TiO_2, SiO_2, and ZrO_2. This is because, for the practical fabrication of VO_2, a dielectric layer such as $ZrTiO_2$ is required to deposit as a seed layer (Zhong et al. 2016). The dielectric layer and its refractive index will affect the confinement of plasmons in the plasmonic slot and hence the modulation characteristics. This demands for a detailed study of modulation characteristics of VO_2, with different dielectric materials (mixed) in the device geometry explained in this section. Table 17.1 shows refractive indices of the dielectrics used in this study, along with VO_2 refractive indices.

The mixed plasmonic modulator design is similar to the VO_2 slot modulator, except that a thin layer of dielectric material is deposited below the VO_2, as shown in the inset of Figure 17.8. For example, in one mixed geometry, both SiO_2 and VO_2 are flanked inside of the plasmonic slot, called the modulating section. The SiO_2 in the modulating section can be replaced by other dielectric materials for getting different mixed geometries, as shown in Table 17.1. The length of the modulating section is denoted as l. The simulation results for different dielectric materials in the mixed design are shown in Figure 17.8 after sweeping l. Using these mixed geometries, we are able to reroute the SPPs between VO_2 and the underneath layer of the dielectric material by changing the refractive index of VO_2 by applying an external voltage.

Here, we examine four different dielectric materials (Si, TiO_2, SiO_2, and ZrO_2), characterized by their relationship with both phases of VO_2, as shown in Figure 17.9. The electric field

TABLE 17.1   Material Refractive Index

Material	VO_2(S)	VO_2(M)	Si	TiO_2	ZrO_2	SiO_2
RI	2.45 + 0.509i	2.04 + 2.91i	3.5	2.7	2.2	1.45

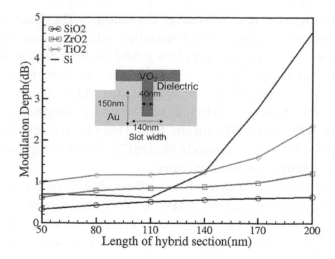

FIGURE 17.8 Variation of modulation depth with respect to different lengths (varied from 50 nm to 200 nm) of the mixed modulating section made of dielectric material and VO_2. The mixed geometries considered are VO_2-Si, VO_2-TiO_2, VO_2-ZrO_2, and VO_2-SiO_2. The inset shows schematics of the mixed design with the dielectric layer and VO_2.

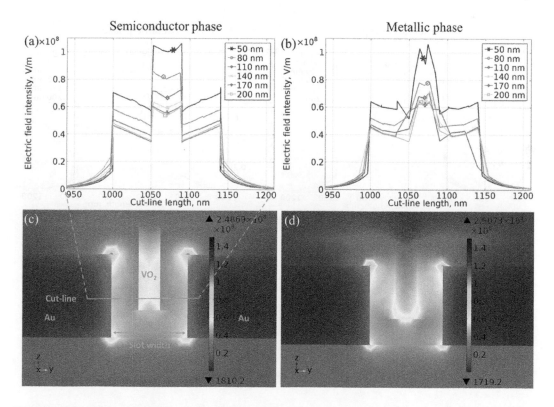

FIGURE 17.9 Characteristics of the Si-VO_2 mixed design: electric field intensity versus cut-line length in (a) the semiconductor phase and (b) the metallic phase. Moreover, (c) and (d) corresponding lateral cut-plane electric field intensity at the middle of the modulating section with l = 200 nm of the Si-VO_2 mixed design.

confinement in the dielectric layer will change depending on the phase of VO_2 layer above it, because of a significant contrast in the refractive indices between the semiconductor and metallic phases. The cut-line and cut-plane E-field intensity profiles are obtained from simulations for the detailed study. The cut plane is taken in the middle of the modulation section, and the cut-line is taken horizontally within the VO_2 section of the slot at the cut-plane. For cut-line, the arc length is referred to as the cut-line length, where 1000–1140 nm represents the slot width, 1000–1050 nm and 1090–1140 nm represent dielectric layer, and 1050–1090 nm represents VO_2, as shown in Figure 17.9. For the cut-plane, the color legend displays the higher intensity with colder colors.

Figure 17.9a and b represents plots of normalized electric field amplitude versus slot width for semiconductor and metallic phases of VO_2, respectively. Here, the length of the Si-VO_2 layer is varied from 50 to 200 nm by keeping thickness constant at 300 nm. Figure 17.9c and d gives cross-sections of the device in both semiconductor and metallic phases of VO_2, where normalized electric field inside the slot width is presented. It is evident from the results that electric field (plasmons) is predominantly confined in the low-refractive-index VO_2 in both phases than in Si, except that electric field amplitude is increased inside low-refractive-index semiconductor phase more than the metallic phase

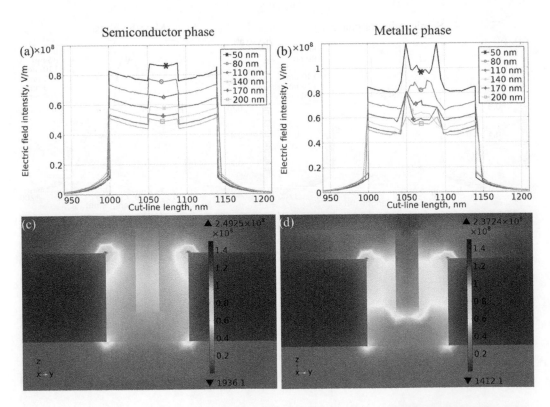

FIGURE 17.10   Characteristics of the TiO_2-VO_2 mixed design: electric field intensity versus cut-line length in (a) the semiconductor phase and (b) the metallic phase. Moreover, (c) and (d) corresponding lateral cut-plane electric field intensity at the middle of the modulating section with l = 200 nm of the TiO_2-VO_2 mixed design.

of VO_2. Figure 17.10 represents the same study for the TiO_2-VO_2 mixed device. The device performance is similar to Si-VO_2, because TiO_2 also has a higher refractive index (2.7) than both the VO_2 phases. For the next mixed topology, ZrO_2 is used, because the refractive index of ZrO_2 is 2.2, which is lower than that of VO_2 in the semiconductor phase (2.45) and higher than that of the metallic phase (2.04). In this topology, the majority of plasmons are confined in ZrO_2 in VO_2 semiconductor phase ($nZrO_2 < nVO_2\_S$) and are rerouted to VO_2 during the metallic phase ($nZrO_2 > nVO_2\_S$).

Figure 17.11 represents plots of normalized electric field amplitude and normalized electric field inside the slot width for semiconductor and metallic phases, respectively. The last mixed geometry is made of SiO_2-VO_2, where SiO_2 has a lower refractive index (1.5) than that of VO_2 in both phases. This results in confinement of majority of plasmons in low-refractive-index SiO_2, as shown in Figure 17.12. Figure 17.8 presents modulation index of all the mixed combinations considered for different lengths of dielectric-VO_2 length. Our study reveals that using a seed layer substantially influences the modulation index of the plasmonic modulators. Furthermore, use of a seed layer with a refractive index lower than both the phases of VO_2 lowers the modulation index. A dielectric seed layer with refractive index either in between or larger than the semiconductor and metallic phases of VO_2 should be used.

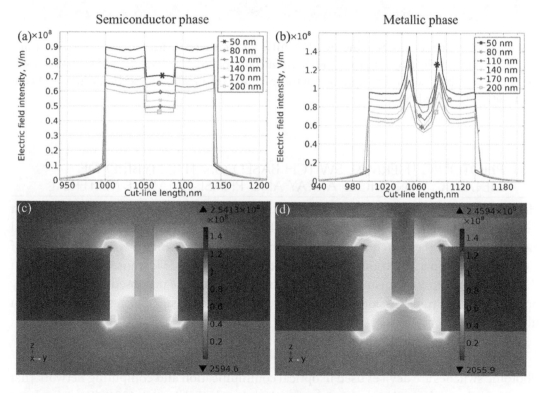

FIGURE 17.11 Characteristics of the ZrO_2-VO_2 mixed design: electric field intensity versus cut-line length in (a) the semiconductor phase and (b) the metallic phase. Moreover, (c) and (d) corresponding lateral cut-plane E-field intensity at the middle of the modulating section with l = 200 nm of the ZrO_2-VO_2 mixed design.

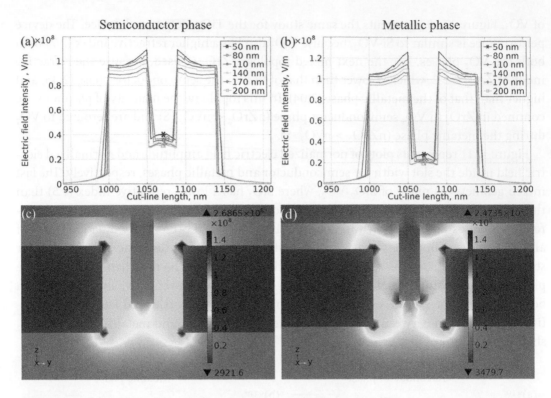

FIGURE 17.12 Characteristics of the SiO_2-VO_2 mixed design: electric field intensity versus cut-line length in (a) the semiconductor phase and (b) the metallic phase. Moreover, (c) and (d) corresponding lateral cut-plane electric field intensity at the middle of the modulating section with $l = 200$ nm of the SiO_2-VO_2 mixed design.

17.3 PHOTONIC THERMAL SWITCH BASED ON THE HYBRID OF METALLIC NANOHOLE ARRAY AND PHASE-CHANGE MATERIAL

Here, we discuss about a nanophotonic switch operating in transmission mode based on the hybrid combination of hexagonal array of holes in CMOS-compatible aluminum and vanadium oxide. The switching is achieved thermally by changing the refractive index of VO_2 that results in a 21% transmission change in the C band, 23% transmission change in the L band, and 26.5% transmission change in the U band. The proposed planar geometry operating in the transmission mode has several advantages such as easy integration on fiber tips and silicon chips.

17.3.1 Introduction

Photonic switches are devices used in optical communication and computing network that can establish or release the connection of optical signals (Saleh et al. 1991). There is a huge demand for ultra-compact photonic switches because of the rapid advancements in the high-data-rate fiber-optic communication systems and high-speed optical computing systems (Miller 2009, Segawa et al. 2016).

The switching operation in an optical domain can be achieved by optomechanism (Liu et al. 2017), acousto-optic (Galichina et al. 2016), magneto-optic (Selvaraj et al. 2017), or electro-optic methods (Rajasekharan et al. 2010, Li 2012, Won et al. 2013). Photonic switches, where thermal energy is used for changing electro-optics properties of the switch, are called thermally reconfigurable photonic/optical switches. Thermally reconfigurable optical switches have several advantages such as easy fabrication, structural simplicity, and ample choices of a thermo-optic functional material (Dai et al. 2017). However, thermally reconfigurable photonic switches are bulky, and hence, integration with state-of-the-art electronics is a challenge.

Plasmonics offers an attractive platform to bridge the size mismatch between optical devices and electronics and hence enables compact integration of these devices on a single chip (Hayashi and Okamoto 2012). Surface plasmons (SP) are free electron oscillations propagating at a metal–dielectric interface, accompanied by electromagnetic oscillations (Barnes et al. 2003). Enhanced localization of electric field can be achieved by using the plasmonics, and this effect allows for the development of nanoscale-optic devices beyond the diffraction limit (De Abajo 2007, Yokogawa et al. 2012, Aramesh et al. 2014, Sun et al. 2017, Sun et al. 2017). Nanoscale devices based on plasmonic metamaterials attract keen interest in the development of next-generation optoelectronic devices (Zheludev and Kivshar 2012, Monticone and Alù 2017), including metasurface filters (Garcia-Vidal et al. 2005, Mary et al. 2008, Turpin et al. 2014, Hedayati and Elbahri 2017) and metamaterial switches and modulators (Chen et al. 2006).

Plasmonic wavelength filters based on perforated metallic film integrated with suitable materials exhibiting phase transition can be used as a thermally reconfigurable photonic switch operating in submicron scale (Morin 1959, Raeis-Hosseini and Rho 2017). One such material is VO_2, a canonical Mott material with a large refractive index change from $3.24 + 0.30i$ to $2.03 + 2.64i$ at 1550 nm during semiconductor-to-metal (semi-metal) phase transition (Verleur et al. 1968). VO_2 has been extensively explored in recent years because of its large refractive index change during the phase transition (Rodrigo et al. 2016). The phase transition can be triggered by thermal heating in milliseconds, light irradiation in femto-seconds, or external electrical field in picoseconds (Rini et al. 2005, Wu et al. 2011, Lei et al. 2015, Jostmeier et al. 2016). Combining submicron light confinement of plasmonics with large refractive index change of VO_2 can be exploited for making novel photonic devices. Recently, versatile applications of plasmonics by utilizing VO_2 phase transition have been explored, such as metasurfaces (Chen et al. 2006, Liu et al. 2017), optical memory device (Lei et al. 2015), nanoscale antenna (Muskens et al. 2016), temperature sensor (Guo et al. 2016), rewriteable devices (Zhang et al. 2016), and ring modulator (Briggs et al. 2010). For electro-optical applications, the precise control of VO_2 phase transition is essential (Gray et al. 2016, Sanchis et al. 2017). A switchable metasurface based on VO_2 working at THz region is recently proposed based on computer simulations that can be switched from broadband absorber to a reflecting broadband halfwave plate by temperature tuning (Ding et al. 2018). VO_2–Ag thermal waveguide switch is experimentally demonstrated with a typical 50% roll-off frequency of 25 kHz in 10-μm waveguide length (Ding et al. 2018). The thermal waveguide switch gives an idea of switching time, which is around microsecond, using thermal trigger for the plasmonic VO_2 switch (Joushaghani et al. 2013). Recently, VO_2–Ag- and VO_2–Au-based thermally driven optical switches operating near infrared (IR) region (800–900 nm) have been demonstrated with 4% and 1.2% transmissions, respectively, using a square arrangement of holes (Suh et al. 2006).

Here, we present, using both computational and experimental methods, a broadband photonic switch based on a hybrid combination of a hexagonal array of holes and vanadium oxide. The hexagonal arrangement of holes in aluminum is fabricated on a quartz substrate, followed by the deposition of VO_2 on the nanohole array. The proposed geometry has several advantages. The fabricated photonic switch is polarization independent, owing to the hexagonal arrangement of holes. A hexagonal arrangement will also increase the holes per area and increase the efficiency of the device. Aluminum is CMOS compatible and inexpensive compared with gold and silver. A high transmission of 37.5% at optical communication region is achieved in the hole array with an optimized device geometry and a thin layer of VO_2 with a thickness of 25 nm atop hole array. The switching of the plasmonic hole array is achieved thermally by changing the refractive index of VO_2 that results in a 21% transmission change with 4.3 dB extinction ratio in the C band, 23% transmission change with 5 dB extinction ratio in the L band, and 26.5% transmission change with 5.3 dB extinction ratio in the U band, with an operating range over 650 nm. It is for the first time that such a combination of the hexagonal array with semiconductor-to-metal phase transition of VO_2 has been explored for the fabrication of a thermally reconfigurable photonic switch, resulting in such high transmission efficiency in the optical communication range.

17.3.2 Device Design and Simulation

Ebbesen reported in 1998 (Ebbesen et al. 1998) the extraordinary optical transmission in a thin metal film when the hole diameter is under the cutoff of the first propagating mode. The dominant resonant surface plasmon excitation leads to a wavelength selection with 1000 times higher transmission than the prediction of the conventional aperture theory (Ghaemi et al. 1998). This extraordinary optical transmission has been observed in noble metal thin films such as Ag, Au, and Cu, as well in transition metals such as Co, Ni, and W (Przybilla et al. 2006), over a wide range of frequencies (Rivas et al. 2003, Cao and Nahata 2004, Selcuk et al. 2006). Ever since, extensive research has been conducted on structures with periodic roughness such as nanoparticles, grooves, and arrays, using subwavelength holes (Koerkamp et al. 2004). The subwavelength hole arrays in thin metallic film exhibit the character of optical filters, with enhanced transmission, which has been widely applied, and the major contributions are focused on biomolecular sensors (Špačková et al. 2016), nanoantennas (Jeong et al. 2012), plasmonic optical filters (Rajasekharan et al. 2014), and plasmonic modulators and switches (Suh et al. 2006, Kim et al. 2016). For the hole-array-based filters that are operating in transmission mode, resonant peaks in transmission are dominated by the surface plasmon mode excited at the cylindrical hole boundaries and two facets. In a triangular (hexagonal) hole-array–based plasmonic filter, the transmission peaks can be approximately predicted using the Equation (17.1) given by (McCrindle et al. 2015)

$$\lambda_{max} = \frac{a}{\sqrt{\frac{4}{3}\left(i^2 + ij + j^2\right)}} \sqrt{\frac{\varepsilon_m \varepsilon_d}{\varepsilon_m + \varepsilon_d}},$$

where a is the pitch of the array, ε_m and ε_d refer to the dielectric constants of metal and dielectric material, respectively, and i, j refer to scattering order. The λ_{max} is the minimum transmission of wavelength right before the resonance peak at longer wavelength. The scattering orders can be considered to be $i = 1, j = 0$, only if the first transmission minimum before the peak is considered. Using the previous equation, it is possible to optimize the hole array filter to any wavelength of interest. Detailed optimization of the filter to operate at optical telecommunication range was carried out using finite-element method implemented in COMSOL Multiphysics. The simulation mesh size has been adjusted according to the maximum computing power available in our lab. The simulation model consists of 100-nm-thick layer of aluminum on a semi-infinite glass substrate with the refractive index value 1.45. The refractive indices of aluminum for different wavelengths were taken from (Johnson and Christy 1972). Top view of a section of simulation model is illustrated in Figure 17.13a, which shows quartz substrate and 100-nm-thick aluminum along with hole geometry (without top layer of VO_2). The resonance peak is finely tuned to 1.55 μm by varying the pitch of hole array to 1010 nm ($P = 1010$ nm) and by keeping hole diameter ($D = 560$ nm) and the thickness of aluminum (100 nm) constant. A 10-nm chromium layer is inserted beneath the Al film as an adhesion layer by using transition boundary condition. The refractive indices are taken from (Johnson and Christy 1974). The transition boundary condition is used for simulating ultra-thin layers with thickness hard to be covered by mesh size of the simulation model. A VO_2 layer was added on the top of the aluminum layer, with holes filling the VO_2 to make the photonic switch. The thickness of VO_2 was varied from between 100 and 25 nm (using transition layer/transition boundary condition) to study its effect on transmission and switching. A 500-nm air superstrate is constructed on the top of VO_2 (not shown in Figure 17.13a). To simulate a large hexagonal array, periodic boundary condition was applied on four sides of the model, as shown in Figure 17.13a. The incident light was set to propagate along the z-axis (perpendicular to the surface of the hole array, x-y plane), with TE polarization using port boundary condition. A 400-nm perfect match layer (PML) is added to the top boundary and the bottom boundary to absorb the outgoing wave and to ensure that no reflection goes into the interior region. Figure 17.13b shows the schematics of the photonic switch, with hole array deposited with VO_2 along with an extended metal pad for heating VO_2 to change its phase. The Al pad is used for attaching heater for characterizing the transmission and switching of the photonic switch for different temperature values.

Figure 17.14a and b shows the simulated transmission spectrum and the extinction ratio of the above photonic switch optimized to operate in telecommunication wavelength for both semiconducting and metallic phases of VO_2 with different thicknesses of 100, 75, 50, and 25 nm. The two insets of Figure 17.14a show electrical field distributions on the cross-section of the device with 25-nm VO_2 in semiconductor phase at wavelengths of 1300 and

(a)   (b)

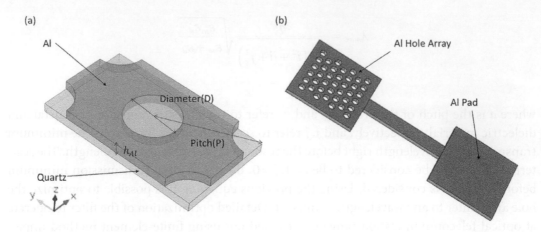

FIGURE 17.13 (a) Simulation model of a hexagonal arrangement of holes in aluminum. The peak wavelength is tuned to 1550 nm with pitch (P) = 1010 nm, hole diameter (D) = 560 nm, and aluminum thickness of 100 nm. (b) Schematics of the device with hexagonal nanohole array in aluminum along with extended metal pad for heating. (The top VO_2 layer is not shown).

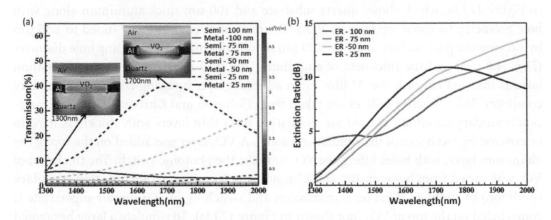

FIGURE 17.14 (a) Simulated spectrum of the photonic switch made of aluminum nanohole array and VO_2 for the semiconductor (dash line) and metallic phases (solid line) with 100-, 75-, 50-, and 25-nm-thick VO_2 (simulation parameters used: thickness of aluminum: 100 nm, diameter of holes: 560 nm, pitch of holes: 1010 nm, and thickness of VO_2: 100/75/50/25 nm). Two insets show the electrical field distribution of the cross-section of photonic switch with 25-nm VO_2 in the semiconductor phase at wavelength 1300 and 1700 nm, respectively. The color legend of the insets is shown on the right-hand side. (b) Extinction ratio (ER) of the switch with respect to wavelength, showing its switching ability and broad working wavelength with 100-, 75-, 50-, and 25-nm VO_2.

1700 nm, respectively. The results show that the device works in the reflection mode, with more electric field energy reflected and concentrated on the super-substrate air rather than passing through at 1300 nm, whereas in transmission mode, the electric field energy is concentrated in the cylindrical waveguide and transmitted through the substrate quartz at 1700 nm. The simulation results are in agreement with detailed mode analysis discussed in

TABLE 17.2   The Transmission Difference (*TD*) Is Difference Between the Transmission of Semiconductor Phase and Metallic Phase of VO$_2$. The Transmission Difference and Extinction Ratio in C, L, and U Band for Different Thickness of VO$_2$ Are Taken from Figure 17.14

Thickness_VO$_2$ {nm}	TD_c{%}[a]	TD_L{%}[a]	TD_v{%}[a]	ER_c{dB}[b]	ER_L{dB}[b]	ER_v{dB}[b]
100	0.8	1	1.3	5.1	6.2	7.3
75	2.4	3.3	4.0	6.3	8	9
50	6	8.5	11	7	8.3	9.7
25	25	32	36.5	8.2	9.6	10.6

[a] TD_c, TD_L, TD_v correspond to the maximum transmission difference in *C*, *L*, and *U* band, respectively.
[b] ER_c, ER_L, ER_v correspond to the maximum extinction ratio in *C*, *L*, and *U* band, respectively.

Figures 17.2 and 17.3 in the prior art (Wu et al. 2012). The transmission and the extinction obtained in C, L, and U bands are given in Table 17.2. The wavelength was swept from 1300 to 2000 nm.

It was found that as the VO$_2$ film thickness decreases, there is an increase in the extinction ratio and transmission. However, practically, it is difficult to reduce the thickness of VO$_2$ to less than 25 nm by keeping the uniformity of the film, especially filling on the Al nanoholes without creating patches. Hence, there is a trade-off between the feasibility of fabrication and device performance. For the photonic switch, the thickness of VO$_2$ was selected to be 25 nm to increase the transmission efficiency of C, L, and U bands (e.g., 3.4% and 39.5% transmissions in the metallic and semiconducting phases, respectively, of VO$_2$, with extinction ratio 10.6 dB at 1675 nm, as shown in Figure 17.14). The results also show that the switch operates in a broad wavelength range from 1530 to 2000 nm, with less than 3-dB loss with 25-nm VO$_2$. This ensures that the photonic switch covers the wavelength window used in optical communications. These results also show that the device can act as a switch by changing the phase of VO$_2$ from semiconducting state (ON state) to metallic state (OFF state) with the application of heat (68°C).

17.3.3 Device Fabrication and Characterization

Based on the previous simulation results, photonic switches were fabricated on a 500-μm-thick quartz substrate by using EBL (Vistec EBPG5000plusES). The fabrication steps are shown in Figure 17.15 (steps (a)–(f)). In the first step (a), a 350-nm-thick double layer of PMMA was spun on the quartz substrate by Pico Track PCT-150RRE, followed by depositing a 30-nm Cr on the top of PMMA by using electron beam evaporation (EBPVD: Intlvac Nanochrome II) in order to have a conducting surface for EBL patterning. In the second step (b), EBL was used to write the nanohole array pattern on PMMA. Before developing the patterned sample, the conductive layer of Cr was removed by wet etching. This was followed by developing the sample into a mixer of methyl isobutyl ketone (MIBK) and isopropyl alcohol (IPA) to make holes in PMMA, as shown in step (c). Following this, a 100-nm aluminum thin film was deposited using EBPVD (step (d)), followed by lift off in step (d). Prior to the Al metallization, the substrate was coated with 10-nm chromium as adhesion layer within the same deposition step. The fabricated hole array is expected to have a hole diameter of

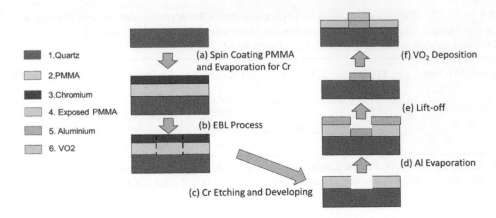

FIGURE 17.15 Fabrication process flowchart of the photonic. The light gray arrow shows the fabrication flow direction from step (a) to step (f). The material legend is shown on the left-hand side.

560 nm and pitch of 1010 nm. Extra metallic pad made of aluminum was connected to the hole array for heating the device shown in Figure 17.13b. In the final step (f), VO_2 was deposited on top of the aluminum. A quartz substrate was cleaned and plasma treated in an argon environment to enhance adhesion between the VO_2 film and the substrate. VO_2 was deposited using the pulsed-DC magnetron sputtering technique. A vanadium (99.99%) target was used in for sputtering. The sputtering chamber was allowed to reach 4.0×10^{-7} Torr before the introduction of the $Ar:O_2$ gas mixture. $Ar:O_2$ mixture was introduced with a flow rate of 12.25:5.25 sccm (for 30% O_2). Sputtering was done at 2.8×10^{-3} Torr pressure, a power of 200 W with 25 kHz pulse frequency and 5 µs reverse time. Deposition was done for 45 minutes at room temperature, producing amorphous VO_2. Subsequently, the as-deposited VO_2 films were annealed in a furnace, evacuated to low vacuum to achieve a pressure of ~250 mTorr, at 550°C for 90 min. Postdeposition annealing at low pressure enhances the level of control over oxygen vacancies and limits oxygen loss, which happens at a rapid rate in VO_2 thin films. X-ray photoelectron spectroscopy (XPS), X-ray diffraction (XRD), and micro-Raman spectroscopy were conducted to characterize the VO_2 thin films (Taha et al. 2017). The thin films showed good insulator–metal transition, as expected at ~68°C for the VO_2 phase of vanadium oxide (Taha et al. 2017). This allowed the formation of excellent VO_2 thin films on top of Al nanohole array.

The fabricated photonic switch was characterized using Craic Technologies 20/30 PVTM spectrophotometer and thermal stage, as shown in Figure 17.16a. The inset shows SEM image of the hole array in aluminum. As a first step, the nanohole array on aluminum without VO_2 was characterized to obtain transmission spectrum with respect to wavelength, as shown in Figure 17.16b. The experimentally obtained spectrum (black color) is superimposed with the simulated spectrum (red color). The experimentally measured peak wavelength is 1447 nm, which is red shifted by 37 nm compared with the peak value of 1410 nm from simulations. The shift is because of fabrication tolerances, including slightly larger hole diameter due to undercutting in walls of holes (average hole diameter between the top and the bottom is varied between 540 nm and 590 nm

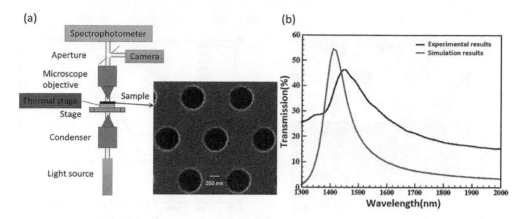

FIGURE 17.16 (a) Schematic of the experimental setup to measure the optical transmission of the photonic switch with respect to wavelength at different temperatures. The setup consists of CRAIC Technologies 20/30 PVTM microspectrophotometer and thermal stage. The light transmitted through the switch is collected by a microscope objective and focused onto the entrance aperture of the spectrophotometer. A beam splitter is used for splitting the light into both camera and spectrometer. The camera is used for sample alignment. The inset shows the scanning electron microscope image (SEM) of the fabricated aluminum nanohole array with 570-nm diameter and 1010-nm pitch (b) Numerically and experimentally obtained transmission spectra from the aluminum nanohole array.

owing to nanofabrication and measurement tolerances, and hence, 570 nm is used in simulations). The maximum transmissions from the simulation and the experiment were 54% and 47%, respectively. After this measurement, a VO_2 layer was deposited on the nanohole array to make the photonic switch. The previous transmission measurements were repeated for the hole array with VO_2 (the photonic switch), and results are discussed in the following section.

After the VO_2 deposition, the cross-section of the photonic switch was taken using focused ion beam (FIB) to study the distribution of VO_2 across an area in the sample, as shown in Figure 17.17a and b. The SEM images show that VO_2 covers the sample almost uniformly, except creating small patches with no VO_2 in the holes. These patches slightly reduce the extinction ratio of the device owing to light leakage in the metallic phase. This can be avoided with increased VO_2 thickness at the cost of decreasing the transmission percentage. Hence, there exists a trade-off between switching performance and transmission percentage of the device, as observed in the simulation results.

In order to find out a suitable temperature range for switching the phase of VO_2, a 25-nm VO_2 film on the glass substrate was used. The pristine VO_2 film was deposited using the same sputtering conditions as the photonic switch. The transmission spectrum of the VO_2 film was measured. In order to achieve VO_2 phase transition, the VO_2 film was heated from room temperature 294 K (semiconductor phase) to 360 K (metallic phase). Figure 17.17c shows transmission of the VO_2 film in the semiconducting phase (294 K) and the metallic phase (360 K). All transmission measured was normalized with respect to the measured area to obtain absolute transmission. In the semiconductor phase, the

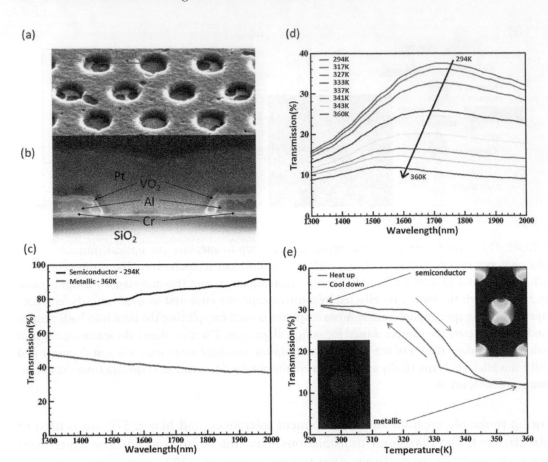

FIGURE 17.17   (a) SEM image of a tilted top view of the fabricated Al/VO$_2$ nanohole array in the photonic switch. (b) Cross-section of a single hole obtained using focused ion beam (FIB) (platinum [Pt] was deposited for contrast). Different materials are marked on the image. From the measurements, Al layer is around 80-nm thick, with 20-nm Cr beneath as adhesion layer, and the VO$_2$ layer is 25-nm thick. The diameter of hole is 570 nm, based on the average of the top and bottom diameter due to the tilted side wall. (c) Experimentally obtained transmission spectrum of the pristine VO$_2$ layer with respect to wavelength for both the semiconductor (gray line) and metallic (black line) phases. (d) Experimentally measured transmission of the Al/VO$_2$ nanohole array in the photonic switch with respect to temperature (varied from 294 to 360 K), during which VO$_2$ switches from the semiconductor phase to the metallic phase. (e) Experimentally measured optical transmission hysteresis of the photonic switch at 1550 nm during heating cycle and cooling cycle (temperature range: 290–360 K). The two inset images show the simulated E-field intensity distribution in the of VO$_2$/Al nanohole array at 1550 nm for the metallic and semiconductor phases, respectively. The color legend: the cold tone (dark gray) to warm tone (gray) color refers to lower intensity to higher intensity ($| \mathbf{E}^2 |$).

transmission increases with respect to wavelength, from 76% to 92%, while in the metallic phase, the transmission drops from 48% to 36%. This result has shown that there is a large difference of 39% in transmission between the semiconductor and metallic phases at 1550-nm wavelength, and this property can be exploited for making photonic switches and modulators.

Based on the results obtained by testing only VO_2 sample, the temperature was swept for the photonic switch from room temperature 294 K (semiconductor phase) to 360 K (metallic phase). Figure 17.17d shows transmission spectrum of the photonic switch plotted with respect to different temperatures. As the temperature was increased from 294 K, the maximum transmission decreased from 37.5% to 11.8%, owing to the phase change of VO_2 from the semiconductor phase to the metallic phase. It is also noted that the peak wavelength of 1725 nm in the semiconductor phase (294 K) is slightly blue, shifted to 1505 nm when the sample is heated up to the metallic phase (360 K). This is due to a low refractive index value of VO_2 in the metallic phase (real part of refractive index reduced from 3.24 to 2.03). This result is also confirmed by simulation of the photonic switch. Transmission of the switch at 1550 nm with respect to temperature was studied by heating the sample from 294 to 360 K in step of 5 K, followed by cooling the switch along the same temperature range, as shown in Figure 17.17e. The results show that there is an optical transmission hysteresis curve during heating and cooling cycles, which is consistent with the VO_2 material characters (Wu et al. 2011). The fabricated switch has 21% transmission change at C band, 23% transmission change at L band, and 26.5% transmission change at U band. The two insets of Figure 17.17e depicts the simulated electrical field intensity distribution of the photonic switch (VO_2/Al nanohole array) in the semiconductor phase and metallic phase of VO_2, respectively. The electric field is highly confined in the holes owing to the large refractive index contrast between Al and VO_2. The maximum transmission peak is 37.5% at 1725 nm (VO_2 semiconductor phase) and the transmission reduced to 10.5 % during the switching of VO_2 phase to metallic (27% transmission difference).

From Figure 17.16b, the experimentally obtained peak wavelength of Al nanohole array alone is 47% at 1447 nm. But the peak wavelength of the photonic switch (after depositing VO_2 in the hole array) is red-shifted to 1725 nm. The red-shift is mainly due to the refractive index of VO_2 in the nanohole array and fabrication tolerances. Cross-section of a single hole with VO_2 was taken using FIB after depositing platinum (Pt) in the hole for better contrast, as shown in Figure 17.17b. From the cross-section measurements, the thickness of Al, VO_2, and Cr is measured to be 80, 25, and 20 nm, respectively, and the hole diameter is taken as 570 nm from the average of the top and bottom diameter due to the undercut. These experimentally obtained values are used in the simulation model of the photonic switch to obtain the transmission spectra of VO_2 in its semiconductor phase as well as metallic phase. The simulation results are plotted in Figure 17.18a, which are matching with the experimentally measured values. The peak transmission wavelength in the semiconductor phase from the simulation is 1735 nm (33.2%), close to the experimentally obtained the value of 1725 nm (37.5%). From the simulation results, the transmission difference in the photonic switch between the semiconductor and metallic phases of VO_2 is 16.5% in C band, 24% in L band, and 28% in U band, which is also close to the experimentally obtained values of 21% in C band, 23% in L band, and 26.5% in U band. Figure 17.18b shows experimentally measured extinction ratio of the photonic switch for different wavelengths of operation. The extinction ratio is 4.3 dB in C band, 4.9 dB in L band, and 5.3 dB in U band. The results also show that the switch operates in a 650-nm wavelength range from 1350 to 2000 nm, with less than 3-dB loss. There is a small light

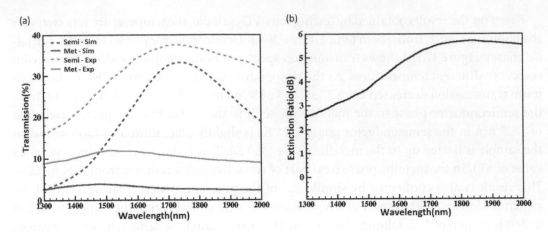

FIGURE 17.18 (a) Transmission spectra of the photonic switch (570-nm diameter and 1010-nm pitch) for aluminum thickness of 80 nm, 20 nm Cr, and 25 nm VO_2 for the semiconductor phase (dash line) and the metallic phase (solid line) from experiments and simulations. (b) The extinction ratio of the switch measured from experiments (extinction ratio: 4.3 dB in C band, 4.9 dB in L band, and 5.3 in U band).

leakage in the photonic switch, as shown in Figure 17.18a, due to missing VO_2 film in some holes (Figure 17.17a) and also due to 25-nm thickness of VO_2. The temperature cycling experiment results are provided in supplementary material (S1) with neglected transmission variance. The same heating/cooling process is repeated for each individual measurement, and hence, the switching effect observed in the device is repeatable.

17.4 SUMMARY

There is an ever-increasing demand on networks because of IoT, big data, machine learning applications, and cloud computing. This demands low power, high bandwidth, and high speed in data processing and handling capabilities. Silicon photonics has proven to be the solution in this decade, where silicon photonics technology enables manufacturing of ultra-compact optical circuits at low cost. The power of silicon photonics is integrating electronics and photonics circuits together on a CMOS chip. The Global Silicon Photonics market is anticipated to grow from USD 492 million in 2018 to USD 3.7 billion in 2025 at a compound annual growth rate (CAGR) of 22.26% between 2016 and 2025. This has triggered major players showing interest in silicon photonics market. Major companies such as Intel Corporation, Cisco Systems, Inc., IBM Corporation, and Juniper Networks, Inc. are investing extensively in silicon photonics market to improvise their products as well as to capture a major market share. There has been successful on-going demonstration of on-chip integration of optical devices such as optical transmitter, photodetector, modulators, and wavelength multiplexing/demultiplexing filters. However, even with such enormous growth, silicon photonics market is facing a lot of challenges. Modulators is one of the major challenges that is still lacking miniaturization and integration with electronics on a CMOS chip.

In this chapter, we first presented a plasmonic modulator with only 200 × 140 nm modulating section within 1 × 3 μm device footprint, using vanadium oxide as modulating material, by taking advantage of the large refractive index contrast between the semiconductor and metallic phases. The modulator is optimized to operate in a broad working wavelength range of 700 nm around the telecommunication window. We also studied the effect of using seed layer of different dielectric materials to grow high-quality VO_2 films on modulation index by exploring VO_2-Si, VO_2-TiO_2, VO_2-ZrO_2, and VO_2-SiO_2 mixed combination as modulating materials. Our geometry can be practically fabricated, and the fabrication methodology is also presented. The presented study can be applied in developing high-frequency modulators for optical communications.

Then, we moved on to demonstrate a photonic switch using vanadium dioxide as switching material, with a hexagonal nanohole array structure, with a maximum 37% transmission at optical communication band. The fabricated switch can achieve 21%, 23%, and 26.5% transmission switching, with extinction ratio 4.3, 4.9, and 5.3 dB in C, L, U band, respectively, with a wide operating range over 650 nm. The wide operating range, high transmission, and compact device footprint (thickness of 125 nm) give us more flexibility and efficiency in integration and application. In the future, we will further explore the other phase change approaches to increase the response speed of switches, such as using external voltage or laser pumping. The results will have potential applications in developing ultra-compact photonic switches and optical modulators in silicon photonics for optical communications.

In conclusion, our work demonstrated here is one of the many solutions in the world tackling this challenge, which will lay down foundation for next-generation nanoscale photonics circuits on CMOS chip integrated with electronics to handle IoT, big data, machine learning applications, and cloud computing for 5G and beyond. Many prior arts have not been covered here in topic of the nanoscale materials and devices; nevertheless, we believe that the information presented here is sufficient to introduce the readers to the subject and guide them for further learning and deeper understanding.

ACKNOWLEDGMENT

R.R.U. and M.B. acknowledge the Australian Research Council's support through DP170100363 and DE160100023, respectively. S.W. acknowledges the support through a Malcolm Moore Foundation Industry Research Award. Thanks to Babak Nasr for helping with imaging sample and giving valuable advice for the fabrication. This work was partially performed in part at the Melbourne Centre for Nanofabrication (MCN) in the Victorian Node of the Australian National Fabrication Facility (ANFF). This project was undertaken with the assistance of resources and services from the National Computational Infrastructure (NCI), which is supported by the Australian Government. Thanks to Babak Nasr for the discussion and help for characterization. Thanks to Stuart Earl for the discussion, as well as for providing the experimental data for refractive indices (VO_2).

REFERENCES

Agrell, Erik, Magnus Karlsson, AR Chraplyvy, David J Richardson, Peter M Krummrich, Peter Winzer, Kim Roberts, Johannes Karl Fischer, Seb J Savory, and Benjamin J Eggleton. 2016. "Roadmap of optical communications." *Journal of Optics* 18 (6):063002.

Alcorn, Paul. 2018. "Intel's new roadmap revealed: 10 nm Ice Lake In 2020, 14 nm cooper lake 2019." Accessed October 20. https://www.tomshardware.com/news/intel-roadmap-cooper_lake-ice_lake,37574.html.

Alloatti, Luca, Robert Palmer, Sebastian Diebold, Kai Philipp Pahl, Baoquan Chen, Raluca Dinu, Maryse Fournier, Jean-Marc Fedeli, Thomas Zwick, and Wolfgang Freude. 2014. "100 GHz silicon–organic hybrid modulator." *Light: Science & Applications* 3 (5):e173.

Aramesh, Morteza, Jiri Cervenka, Ann Roberts, Amir Djalalian-Assl, Ranjith Rajasekharan, Jinghua Fang, Kostya Ostrikov, and Steven Prawer. 2014. "Coupling of a single-photon emitter in nanodiamond to surface plasmons of a nanochannel-enclosed silver nanowire." *Optics Express* 22 (13):15530–15541.

Barnes, William L, Alain Dereux, and Thomas W Ebbesen. 2003. "Surface plasmon subwavelength optics." *Nature* 424 (6950):824.

Barwicz, Tymon, Yoichi Taira, Ted W Lichoulas, Nicolas Boyer, Yves Martin, Hidetoshi Numata, Jae-Woong Nah, Shotaro Takenobu, Alexander Janta-Polczynski, and Eddie L Kimbrell. 2016. "A novel approach to photonic packaging leveraging existing high-throughput microelectronic facilities." *IEEE Journal of Selected Topics in Quantum Electronics* 22 (6):455–466.

Briggs, Ryan M, Imogen M Pryce, and Harry A Atwater. 2010. "Compact silicon photonic waveguide modulator based on the vanadium dioxide metal-insulator phase transition." *Optics Express* 18 (11):11192–11201.

Cao, Hua, and Ajay Nahata. 2004. "Resonantly enhanced transmission of terahertz radiation through a periodic array of subwavelength apertures." *Optics Express* 12 (6):1004–1010.

Chen, Hou-Tong, Willie J Padilla, Joshua MO Zide, Arthur C Gossard, Antoinette J Taylor, and Richard D Averitt. 2006. "Active terahertz metamaterial devices." *Nature* 444 (7119):597.

Dai, Daoxin, Haifeng Shan, Lijia Song, and Shipeng Wang. 2017. "Reconfigurable silicon photonics: Devices and circuits." *Integrated Optics: Physics and Simulations III*, vol. 10242, p. 1024207. International Society for Optics and Photonics.

David, Thomson, Zilkie Aaron, E. Bowers John et al., 2016. "Roadmap on silicon photonics." *Journal of Optics* 18 (7):073003.

De Abajo, FJ Garcia. 2007. "Colloquium: Light scattering by particle and hole arrays." *Reviews of Modern Physics* 79 (4):1267.

Développement, Yole. 2016. "Silicon photonics for data centers and other applications." Last Modified October 2016, Accessed August 18 www.i-micronews.com/report/product/silicon-photonics-for-data-centers-and-other-applications-2016. html?utm_source=PR&utm_medium=email&utm_campaign=SiliconPhotonics_Markets_Applications_Yole_Nov2016.

Développement, Yole. 2018. "Silicon photonics report." Last Modified 01/04/2018, Accessed September 20. http://www.yole.fr/2014-galery-OptoPhotonics.aspx#I0002f8cb.

Dhiman, Ashish. 2013. "Silicon photonics: A review." *IOSR Journal of Applied Physics* 3 (5):67–79.

Ding, Fei, Shuomin Zhong, and Sergey I Bozhevolnyi. 2018. "Vanadium dioxide integrated metasurfaces with switchable functionalities at terahertz frequencies." *Advanced Optical Materials* 6 (9):1701204.

Earl, Stuart K., Timothy D. James, Timothy J. Davis, Jeffrey C. McCallum, Robert E. Marvel, Richard F. Haglund, and Ann Roberts. 2013. "Tunable optical antennas enabled by the phase transition in vanadium dioxide." *Optics Express* 21 (22):27503-27508. doi:10.1364/OE.21.027503.

Ebbesen, T. W., H. J. Lezec, H. F. Ghaemi, T. Thio, and P. A. Wolff. 1998. "Extraordinary optical transmission through sub-wavelength hole arrays." *Nature* 391 (6668):667–669.

Galichina, Alina, Elena Velichko, and Evgeni Aksenov. 2016. "Acousto-optic switch based on scanned acoustic field." In *Internet of Things, Smart Spaces, and Next Generation Networks and Systems*, pp. 690–696. Cham, Switzerland: Springer.

Garcia-Vidal, FJ, L Martin-Moreno, and JB Pendry. 2005. "Surfaces with holes in them: New plasmonic metamaterials." *Journal of Optics A: Pure and Applied Optics* 7 (2):S97.

Ghaemi, HF, Tineke Thio, DE Grupp, Thomas W Ebbesen, and HJ Lezec. 1998. "Surface plasmons enhance optical transmission through subwavelength holes." *Physical Review B* 58 (11):6779.

Gray, AX, MC Hoffmann, J Jeong, NP Aetukuri, D Zhu, HY Hwang, NC Brandt, H Wen, AJ Sternbach, and S Bonetti. 2016. "Ultrafast THz field control of electronic and structural interactions in vanadium dioxide." *arXiv preprint arXiv:1601.07490*.

Guo, Peijun, Matthew S Weimer, Jonathan D Emery, Benjamin T Diroll, Xinqi Chen, Adam S Hock, Robert PH Chang, Alex BF Martinson, and Richard D Schaller. 2016. "Conformal coating of a phase change material on ordered plasmonic nanorod arrays for broadband all-optical switching." *ACS Nano* 11 (1):693–701.

Hayashi, Shinji, and Takayuki Okamoto. 2012. "Plasmonics: Visit the past to know the future." *Journal of Physics D: Applied Physics* 45 (43):433001.

Hedayati, Mehdi Keshavarz, and Mady Elbahri. 2017. "Review of metasurface plasmonic structural color." *Plasmonics* 12 (5):1463–1479.

Intel Bussiness. 2018. What is silicon photonics.

Jeong, Gyu-Seob, Woorham Bae, and Deog-Kyoon Jeong. 2017. "Review of CMOS integrated circuit technologies for high-speed photo-detection." *Sensors* 17 (9):1962.

Jeong, Young-Gyun, Ji-Soo Kyoung, Jae-Wook Choi, Sang-Hoon Han, Hyeong-Ryeol Park, Namkyoo Park, Bong-Jun Kim, Hyun-Tak Kim, and Dai-Sik Kim. 2012. "Terahertz nano antenna enabled early transition in VO$_2$." *arXiv preprint arXiv:1208.3269*.

Johnson, P. B., and R. W. Christy. 1972. "Optical constants of the noble metals." *Physical Review B* 6 (12):4370–4379. doi:10.1103/PhysRevB.6.4370.

Johnson, PB, and RW Christy. 1974. "Optical constants of transition metals: Ti, v, cr, mn, fe, co, ni, and pd." *Physical Review B* 9 (12):5056.

Jostmeier, Thorben, Moritz Mangold, Johannes Zimmer, Helmut Karl, Hubert J Krenner, Claudia Ruppert, and Markus Betz. 2016. "Thermochromic modulation of surface plasmon polaritons in vanadium dioxide nanocomposites." *Optics Express* 24 (15):17321–17331.

Joushaghani, Arash, Brett A Kruger, Suzanne Paradis, David Alain, J Stewart Aitchison, and Joyce KS Poon. 2013. "Sub-volt broadband hybrid plasmonic-vanadium dioxide switches." *Applied Physics Letters* 102 (6):061101.

Kim, Seyoon, Min Seok Jang, Victor W Brar, Yulia Tolstova, Kelly W Mauser, and Harry A Atwater. 2016. "Electronically tunable extraordinary optical transmission in graphene plasmonic ribbons coupled to subwavelength metallic slit arrays." *Nature Communications* 7:12323.

Kinsey, N, M Ferrera, VM Shalaev, and A Boltasseva. 2015. "Examining nanophotonics for integrated hybrid systems: A review of plasmonic interconnects and modulators using traditional and alternative materials." *JOSA B* 32 (1):121–142.

Koerkamp, KJ Klein, Stefan Enoch, FB Segerink, NF Van Hulst, and L Kuipers. 2004. "Strong influence of hole shape on extraordinary transmission through periodic arrays of subwavelength holes." *Physical Review Letters* 92 (18):183901.

Kruger, Brett A, Arash Joushaghani, and Joyce KS Poon. 2012. "Design of electrically driven hybrid vanadium dioxide (VO$_2$) plasmonic switches." *Optics Express* 20 (21):23598–23609.

Lei, Dang Yuan, Kannatassen Appavoo, Filip Ligmajer, Yannick Sonnefraud, Richard F Haglund Jr, and Stefan A Maier. 2015. "Optically-triggered nanoscale memory effect in a hybrid plasmonic-phase changing nanostructure." *ACS Photonics* 2 (9):1306–1313.

Li, Quan. 2012. *Liquid Crystals Beyond Displays: Chemistry, Physics, and Applications*. Hoboken, NJ: John Wiley & Sons.

Liu, Hongwei, Junpeng Lu, and Xiao Renshaw Wang. 2017. "Metamaterials based on the phase transition of VO_2." *Nanotechnology* 29 (2):024002.

Liu, Li, Jin Yue, and Zhihua Li. 2017. "All-optical switch based on a fiber-chip-fiber opto-mechanical system with ultrahigh extinction ratio." *IEEE Photonics Journal* 9 (3):1–8.

Liu, Mengkun, Harold Y Hwang, Hu Tao, Andrew C Strikwerda, Kebin Fan, George R Keiser, Aaron J Sternbach, Kevin G West, Salinporn Kittiwatanakul, and Jiwei Lu. 2012. "Terahertz-field-induced insulator-to-metal transition in vanadium dioxide metamaterial." *Nature* 487 (7407):345.

LYON. 2018. "Silicon photonics has reached its tipping point!". Yole Développement Last Modified Jan/18 2018, Accessed October 20. http://www.yole.fr/SiPhotonics_MarketStatus.aspx#. W8rUPnszaUk.

Lysenko, S, A Rua, F Fernandez, and H Liu. 2009. "Optical nonlinearity and structural dynamics of VO 2 films." *Journal of Applied Physics* 105 (4):043502.

Maier, Stefan Alexander. 2007. *Plasmonics: Fundamentals and Applications*. New York, NY: Springer Science & Business Media.

Makarov, Sergey, Sergey Kudryashov, Ivan Mukhin, Alexey Mozharov, Valentin Milichko, Alexander Krasnok, and Pavel Belov. 2015. "Tuning of magnetic optical response in a dielectric nanoparticle by ultrafast photoexcitation of dense electron–hole plasma." *Nano letters* 15 (9):6187–6192.

Markov, Petr, Kannatassen Appavoo, Richard F Haglund, and Sharon M Weiss. 2015. "Hybrid Si-VO 2-Au optical modulator based on near-field plasmonic coupling." *Optics Express* 23 (5):6878–6887.

Mary, A, Sergio G Rodrigo, L Martin-Moreno, and FJ Garcia-Vidal. 2008. "Plasmonic metamaterials based on holey metallic films." *Journal of Physics: Condensed Matter* 20 (30):304215.

McCrindle, Iain James Hugh, James Paul Grant, Luiz Carlos Paiva Gouveia, and David Robert Sime Cumming. 2015. "Infrared plasmonic filters integrated with an optical and terahertz multi-spectral material." *Physica Status Solidi (a)* 212 (8):1625–1633. doi:10.1002/pssa.201431943.

Melikyan, Argishti, Luca Alloatti, Alban Muslija, David Hillerkuss, Philipp C Schindler, J Li, Robert Palmer, Dietmar Korn, Sascha Muehlbrandt, and Dries Van Thourhout. 2014. "High-speed plasmonic phase modulators." *Nature Photonics* 8 (3):229.

Miller, David AB. 2009. "Device requirements for optical interconnects to silicon chips." *Proceedings of the IEEE* 97 (7):1166–1185.

Monticone, Francesco, and Andrea Alù. 2017. "Metamaterial, plasmonic and nanophotonic devices." *Reports on Progress in Physics* 80 (3):036401.

Morin, FJ. 1959. "Oxides which show a metal-to-insulator transition at the neel temperature." *Physical Review Letters* 3 (1):34.

Muskens, Otto L, Luca Bergamini, Yudong Wang, Jeffrey M Gaskell, Nerea Zabala, CH De Groot, David W Sheel, and Javier Aizpurua. 2016. "Antenna-assisted picosecond control of nanoscale phase transition in vanadium dioxide." *Light: Science & Applications* 5 (10):e16173.

Ooi, Kelvin JA, Ping Bai, Hong Son Chu, and Lay Kee Ang. 2013. "Ultracompact vanadium dioxide dual-mode plasmonic waveguide electroabsorption modulator." *Nanophotonics* 2 (1):13–19.

Pavesi, Lorenzo, and David J Lockwood. 2016. "Silicon photonics iii." *Topics in Applied Physics* 122: 1–36.

Photonics, Intel Silicon. "Moving data with light." www.intel.com, Accessed September 20. https://www.intel.com/content/www/us/en/architecture-and-technology/silicon-photonics/silicon-photonics-overview.html.

Photonics, NL, *Dutch Optics Centre, Phton Delta and partners*. 2018. Photonics Roadmap 2018. Accessed October 25. https://www.photonicsnl.org/wp-content/uploads/2018/05/Roadmap-Photonics-2018.pdf.

Przybilla, F, A Degiron, JY Laluet, C Genet, and TW Ebbesen. 2006. "Optical transmission in perforated noble and transition metal films." *Journal of Optics A: Pure and Applied Optics* 8 (5):458.

Raeis-Hosseini, Niloufar, and Junsuk Rho. 2017. "Metasurfaces based on phase-change material as a reconfigurable platform for multifunctional devices." *Materials* 10 (9):1046.

Rajasekharan, Ranjith, Christoph Bay, Qing Dai, Jon Freeman, and Timothy D Wilkinson. 2010. "Electrically reconfigurable nanophotonic hybrid grating lens array." *Applied Physics Letters* 96 (23):233108.

Rajasekharan, Ranjith, Eugeniu Balaur, Alexander Minovich, Sean Collins, Timothy D James, Amir Djalalian-Assl, Kumaravelu Ganesan, Snjezana Tomljenovic-Hanic, Sasikaran Kandasamy, and Efstratios Skafidas. 2014. "Filling schemes at submicron scale: Development of submicron sized plasmonic colour filters." *Scientific reports* 4:6435.

Research-Zurich, IBM. 2017. "Low power, high performance optical receivers." www.ibm.com, Accessed September 20. https://www.ibm.com/blogs/research/2017/07/low-power-high-performance-optical-receivers/.

Rini, Matteo, Andrea Cavalleri, Robert W Schoenlein, René López, Leonard C Feldman, Richard F Haglund, Lynn A Boatner, and Tony E Haynes. 2005. "Photoinduced phase transition in VO 2 nanocrystals: Ultrafast control of surface-plasmon resonance." *Optics Letters* 30 (5):558–560.

Rivas, J Gómez, C Schotsch, P Haring Bolivar, and H Kurz. 2003. "Enhanced transmission of THz radiation through subwavelength holes." *Physical Review B* 68 (20):201306.

Rodrigo, Sergio G, Fernando de León-Pérez, and Luis Martín-Moreno. 2016. "Extraordinary optical transmission: Fundamentals and applications." *Proceedings of the IEEE* 104 (12):2288–2306.

Ryckman, Judson D, Kent A Hallman, Robert E Marvel, Richard F Haglund, and Sharon M Weiss. 2013. "Ultra-compact silicon photonic devices reconfigured by an optically induced semiconductor-to-metal transition." *Optics Express* 21 (9):10753–10763.

Saleh, Bahaa EA, Malvin Carl Teich, and Bahaa E Saleh. 1991. *Fundamentals of Photonics*. Vol. 22, New York: Wiley.

Sanchis, Pablo, Luis D Sánchez, Teodora Angelova, Amadeu Griol, Mariela Menghini, Pia Homm, Bart Van Bilzen, Jean-Pierre Locquet, and Lars Zimmermann. 2017. "Recent advances in hybrid VO 2/Si devices for enabling electro-optical functionalities." *Integrated Optics: Devices, Materials, and Technologies XXI* 10106:101060P.

Segawa, Toru, Salah Ibrahim, Tatsushi Nakahara, Yusuke Muranaka, and Ryo Takahashi. 2016. "Low-power optical packet switching for 100-Gb/s burst optical packets with a label processor and 8 × 8 optical switch." *Journal of Lightwave Technology* 34 (8):1844–1850.

Selcuk, S, K Woo, DB Tanner, AF Hebard, AG Borisov, and SV Shabanov. 2006. "Trapped electromagnetic modes and scaling in the transmittance of perforated metal films." *Physical Review Letters* 97 (6):067403.

Selvaraj, Jayaprakash, Wei Shen Theh, Neelam Prabhu Gaunkar, Jiayu Hong, Leif H Bauer, and Mani Mina. 2017. "Enhancement for high-speed switching of magneto-optic fiber-based routing using single magnetizing coil." *Magnetics Conference (INTERMAG), 2017 IEEE International* 53 (11):1–4.

Špačková, Barbora, Piotr Wrobel, Markéta Bocková, and Jiří Homola. 2016. "Optical biosensors based on plasmonic nanostructures: A review." *Proceedings of the IEEE* 104 (12):2380–2408.

Suh, JY, EU Donev, R Lopez, LC Feldman, and Richard F Haglund Jr. 2006. "Modulated optical transmission of subwavelength hole arrays in metal-VO 2 films." *Applied Physics Letters* 88 (13):133115.

Sun, Miao, Ranjith Unnithan, and William Shieh. 2017. "Design of plasmonic modulators with vanadium dioxide on silicon-on-insulator." *IEEE Photonics Journal* 9 (3):1–10.

Sun, Xiaomeng, Linjie Zhou, Haike Zhu, Qianqian Wu, Xinwan Li, and Jianping Chen. 2014. "Design and analysis of a miniature intensity modulator based on a silicon-polymer-metal hybrid plasmonic waveguide." *IEEE Photonics Journal* 6 (3):1–10.

Sweatlock, Luke A, and Kenneth Diest. 2012. "Vanadium dioxide based plasmonic modulators." *Optics Express* 20 (8):8700–8709.

Taha, Mohammad, Sumeet Walia, Taimur Ahmed, Daniel Headland, Withawat Withayachumnankul, Sharath Sriram, and Madhu Bhaskaran. 2017. "Insulator–metal transition in substrate-independent VO 2 thin film for phase-change devices." *Scientific Reports* 7 (1):17899.

Tian, Jie, Shuqing Yu, Wei Yan, and Min Qiu. 2009. "Broadband high-efficiency surface-plasmon-polariton coupler with silicon-metal interface." *Applied Physics Letters* 95 (1):013504.

TSMC. 2017. "Logic Technology." Accessed August 18. http://www.tsmc.com/english/dedicated-Foundry/technology/10nm.htm.

Turpin, Jeremiah P, Jeremy A Bossard, Kenneth L Morgan, Douglas H Werner, and Pingjuan L Werner. 2014. "Reconfigurable and tunable metamaterials: A review of the theory and applications." *International Journal of Antennas and Propagation* 2014.

Verleur, Hans W, AS Barker Jr, and CN Berglund. 1968. "Optical properties of V O 2 between 0.25 and 5 eV." *Physical Review* 172 (3):788.

Won, Kanghee, Ananta Palani, Haider Butt, Philip JW Hands, Ranjith Rajeskharan, Qing Dai, Ammar Ahmed Khan, Gehan AJ Amaratunga, Harry J Coles, and Timothy D Wilkinson. 2013. "Electrically switchable diffraction grating using a hybrid liquid crystal and carbon nanotube-based nanophotonic device." *Advanced Optical Materials* 1 (5):368–373.

Wu, B, A Zimmers, H Aubin, R Ghosh, Y Liu, and R Lopez. 2011. "Electric-field-driven phase transition in vanadium dioxide." *Physical Review B* 84 (24):241410.

Wu, L, P Bai, X Zhou, and EP Li. 2012. "Reflection and transmission modes in nanohole-array-based plasmonic sensors." *IEEE Photonics Journal* 4 (1):26–33.

Yang, Zheng, and Shriram Ramanathan. 2015. "Breakthroughs in photonics 2014: Phase change materials for photonics." *IEEE Photonics Journal* 7 (3):1–5.

Yokogawa, Sozo, Stanley P Burgos, and Harry A Atwater. 2012. "Plasmonic color filters for CMOS image sensor applications." *Nano Letters* 12 (8):4349–4354.

Zhang, Hai-Tian, Lu Guo, Greg Stone, Lei Zhang, Yuan-Xia Zheng, Eugene Freeman, Derek W Keefer, Subhasis Chaudhuri, Hanjong Paik, and Jarrett A Moyer. 2016. "Imprinting of local metallic states into VO$_2$ with ultraviolet light." *Advanced Functional Materials* 26 (36):6612–6618.

Zheludev, Nikolay I., and Yuri S. Kivshar. 2012. "From metamaterials to metadevices." *Nature Materials* 11:917. doi:10.1038/nmat3431.

Zhong, Li, Yuanyuan Luo, Ming Li, Yuyan Han, Hua Wang, Sichao Xu, and Guanghai Li. 2016. "TiO 2 seed-assisted growth of VO 2 (M) films and thermochromic performance." *CrystEngComm* 18 (37):7140–7146.

Zylbersztejn, Adam and Nevill Francis Mott. 1975. "Metal-insulator transition in vanadium dioxide." *Physical Review B* 11 (11):4383.

CHAPTER 18

Microwave-Absorbing Properties of Single- and Multilayer Materials

Microwave-Heating Mechanism and Theory of Material–Microwave Interaction

Yuksel Akinay

CONTENTS

18.1 INTRODUCTION

Recently, a wide range of studies has been carried out on microwave absorber composites (MACs) for use in telecommunications, electromagnetic pollution, stealth technology, and other applications [1]. The studies on the production and development of MACs were first initiated by the Germans in the second world war. They have designed a radio detection and ranging (RADAR)-absorbing material (RAM) based on ferrite. Since that time, there has been a great deal of interest on RAMs for application in the stealth technology [1–4].

Detection of a target is calculated depending on the radar cross-section (RCS). The RCS is a reflective behavior of a target. The RCS of a target exposed to a radar signal is the measure of a total reflection of the radar signal from target to radar receiver, and it is defined as 4π times the ration of the reflected power [5]. Especially in the stealth technology, the use of RAMs reduces the RCS of these materials, and it makes the materials difficult to be detected by radar [6].

RAMs are highly important in stealth technologies. Research on these materials strengthens the defense sector. RAMs are being developed for use in space and aerospace, electromagnetics, privacy technology, electromagnetic impurities, and other commercial applications that protect against natural radiation [7–9]. Furthermore, the microwave-absorbing coatings on the surfaces of materials could prevent them from being detected by radar. Hence, the electromagnetic wave–absorbing materials are intended to be used in military applications owing to their absorbing performance, lightness, flexibility, and thin coating properties [10–13].

Microwave-absorption efficiency of materials varies according to the magnetic and dielectric properties of the fillers and matrix. The electromagnetic wave–absorbing composite materials are typically developed with dielectric particles such as graphene and carbon black or magnetic particle such as ferrite fillers, which have different shapes and sizes [13,14]. For that reason, the reflection loss (R_L) values of RAMs obtained with nanoparticle reinforced polymer matrix composites were examined in different frequency range [15].

When an incident wave passes through the material, the power of incident wave can be divided into reflection, absorption, and transmitted, as seen in Figure 18.1a. However, the incident waves can create microwave heating within the material at the atomic level during microwave–material interaction. The size of fillers is determined by the microwave heating level such as uniform or nonuniform heating, as seen in Figure 18.1b [3,16].

A material used as a microwave absorber must either absorb or dissipate electromagnetic energy. The materials that provide these properties are called microwave-absorbing materials. Many studies have been done on polymer-based composites filled with magnetic or dielectric materials in micron size or nanosize [17]. Aphesteguy et al. [18] studied the microwave-absorbing properties of NiZn- and NiCuZn-ferrite composites and reported –3.0 dB reflection loss (R_L) at 11.7 GHz. Yang et al. [19] reported the microwave-absorbing properties of $BaTiO_3$ and $BaFe_2O_4$ with polyaniline (PANI). Hosseini et al. [20] studied the microwave-absorbing properties of polyaniline/$MnFe_2O_4$ nanocomposite. Schin et al. [21] have reported $BaFe_{12}O_{19}/ZnFe_2O_4/CNT$ nanocomposite and investigated its microwave-absorbing properties.

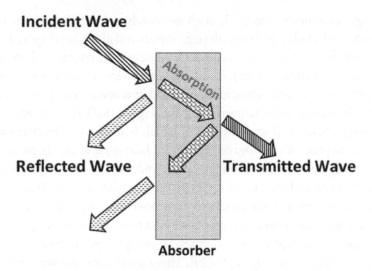

FIGURE 18.1   (a) Microwave and material interaction, and (b) microwave heating of nanoparticle.

18.2 INTERACTION OF MICROWAVES WITH MATERIALS

When electromagnetic waves encounter a material, the electromagnetic waves can be attenuated by three mechanisms: reflection, absorption, and multiple reflections, or a combination of these mechanisms. However, the multiple reflections are generally neglected when the adoption loss is greater than 9 dB [22,23]. A part of the incident waves energy is transferred to materials by the interaction of the electromagnetic fields at the molecular level, and this energy is dissipated as heat, owing to the electric and magnetic fields of wave energy [24–26]. The microwave-absorption properties of materials depend on electric field (E) and magnetic field (M) throughout the material, and the dielectric properties ultimately determine the effect of the electromagnetic field on the material [22,23]. As seen in Figure 18.2, three types of material may be categorized with respect to its interaction with the microwave field.

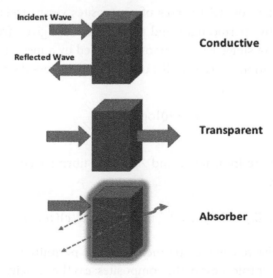

FIGURE 18.2   Interaction of microwaves with different types of materials.

When strongly **conductive** materials such as metals are exposed to microwaves, these microwaves can be reflected from materials (e.g., metals and graphite) (Figure 18.2). For example, metals, in bulk form, are excellent reflectors of microwave energy and are not heated significantly by incident microwaves [24]. In the case of low-loss insulators such as **transparent** materials (e.g., porcelain, quartz, glass, and ceramics), microwave can pass through the material without any absorption and microwave heating (Figure 18.2). But, the advanced materials, composites, nanoparticles, and some insulator, which is the lossy dielectric (**absorber**), can give rise to the absorption of microwave energy and heat generation. These materials absorb microwave by energy transformation (Figure 18.2), and microwave energy creates microwave heating within the material [25,26]. The materials are classified as insulators, semiconductors, and conductors according to their conductivity properties. In addition, these are classified with their permeability properties. The microwave heating within homogeneous and heterogeneous materials is described by two parameters: complex permittivity ($\varepsilon^* = \varepsilon' - j\varepsilon''$) and complex permeability ($\mu^* = \mu' - j\mu''$) [27–29]. These parameters are very important and can determine the microwave absorption performance of a materials [30]. The properties such as Maxwell–Wagner mechanism, electronic, ionic, atomic, and dipole polarization contribute to the dielectric properties of the material [16]. In composite materials, bound charges and bipolaron/polaron system are most important for the heating mechanism. The polarization occurs due to bound charge and free charge. However, this system leads to high permittivity (ε').

18.2.1 Microwave-Absorption Properties for Single- and Multilayer Materials

Microwaves are a type of the electromagnetic waves with the 300 MHz and 300 GHz frequencies, which correspond to the wavelengths of 1 mm to 1 m [27,31,32]. Microwaves have different application areas such as radar, communications, heating, and other industrial processes. The microwave absorption properties are determined by the transmission line theory, short-circuited reflection and open-circuited reflection [8,10]. The microwave absorption properties are determined with power reflection of waves from a material. According to the transmission line theory, the reflection loss (R_L) curves of composites versus frequency were calculated from the complex relative permeability and permittivity at a given frequency and absorber thickness. Here, the R_L of single layer composites backed by a metal reflector was calculated using the following equation in decibel (dB) unit in many studies discussed previously [24]:

$$R_L = 20 \log_{10} \left| \frac{Z_{in} - Z_0}{Z_{in} + Z_0} \right| \tag{18.1}$$

where Z_0 is the free space impedance and Z_{in} is the normalized input impedance at the air–absorber interface.

$$Z_{in} = Z_0 \left(\mu_r / \varepsilon_r \right)^{1/2} \tanh \left[j(2\pi f d / c)(\mu_r \varepsilon_r)^{1/2} \right] \tag{18.2}$$

Here, ε_r and μ_r are relative complex permeability and permittivity of composites, respectively; d is the single layer thickness of the composites; c is the velocity of the electromagnetic wave; and f is the microwave frequency. The schematic structure of multilayer composites that

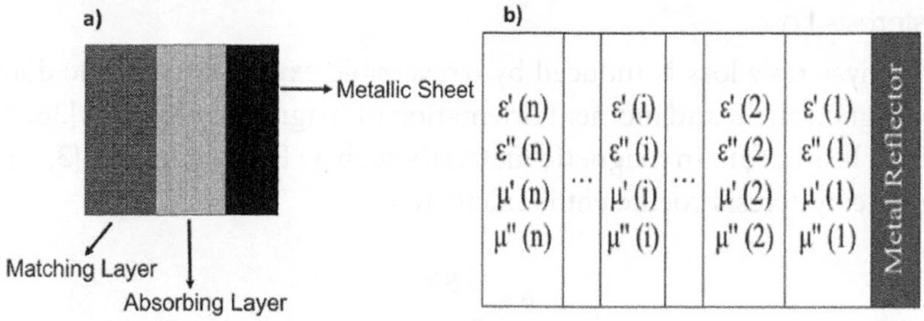

FIGURE 18.3　The schematic structure of multilayer composites.

consists of n layers of different materials backed by a PEC is shown in Figure 18.3. For example, ε_i and μ_i are the permittivity and permeability, respectively, of ith layer. In terms of the transmission line theory, the impedance (Z_i) of the ith layer is given by [33–35]:

$$Z_i = \eta_i \frac{Z_{i-1} + \eta_i \tanh(\gamma_i d_i)}{\eta_i + Z_{i-1} + \eta_i \tanh(\gamma_i d_i)} \qquad (18.3)$$

Here, $\eta_i = \eta_0 \sqrt{\mu_{ri} / \varepsilon_{ri}}$, $\gamma_i = j2\pi f \sqrt{\mu_{ri} / \varepsilon_{ri}} / c$, $\eta_0 (\sqrt{\varepsilon_0 / \mu_0})$ is the characteristic impedance of the free space. For the first layer, considering $\eta_0 = 0$, Z_1 calculates as:

$$Z_1 = \eta_1 Z_1 \tanh(\gamma_1 d_1) \qquad (18.4)$$

Here, the R_L of multilayer composites was calculated using the following equation in decibel (dB) unit [34]:

$$R_L = 20\log_{10}\left|\frac{Z_n - \eta_0}{Z_n + \eta_0}\right| \qquad (18.5)$$

18.3　LOSS MECHANISMS

The heat loss mechanism includes dielectric loss (i.e., conductance loss, dielectric relaxation loss, and resonance loss) and magnetic loss (i.e., eddy-current, hysteresis, and residual loss) [3,33].

18.3.1　Eddy-Current Loss

The eddy-current loss occurs in a conducting material exposed to an external magnetic field. Hence, an induced current is produced inside the material; this current would dissipate the energy. [33]. The eddy-current loss coefficient, e, is represented in Equation (18.6) at low frequency.

$$e = \frac{4\pi^2 \mu_0 d^2 \sigma}{3} \qquad (18.6)$$

Here, σ is electric conductivity and d is the thickness of the material.

18.3.2 Hysteresis Loss

The magnetic hysteresis loss is induced by irreversible external magnetic domains in the alternating magnetic field and moment orientation of magnetic materials [36,37]. The magnetic hysteresis loss occurs in magnetic materials such as Fe, Co, and Ni [3,37].

The magnetic hysteresis coefficient is calculated as:

$$b = \frac{8b}{3\mu_0 \mu^3} \tag{18.7}$$

Here, b is the Rayleigh constant, μ_0 is the permeability of vacuum, and μ is the permeability of magnetic material [36].

18.3.3 Residual Loss

The residual loss is the magnetic loss expect eddy current and hysteresis loss. At low frequency, the residual loss is caused by the magnetic loss, including displacement of domain walls or irregularities in the electron spin. However, high frequency, it is caused by size, ferromagnetic, or domain wall resonance [15].

18.3.4 Dielectric Loss

When an incident wave penetrates through a dielectric material, this material will dissipate electrical energy and then transform it into volumetric heat energy. This heating mechanism occurs when the elastic and frictional forces resist translational motions of free or bound charges [14–16].

The microwave absorption properties of dielectric material are related to the complex permittivity of material, as given in Equation (18.8).

$$\varepsilon^* = \varepsilon_0(\varepsilon' - j\varepsilon'') \tag{18.8}$$

Here, ε_0 is the free space permittivity, ε' is the real part of permittivity, and ε'' is the imaginary part of permittivity.

18.3.5 Relaxation Loss

Relaxation loss occurs when the polarization changing is slower than that of the electric field. The thermal and dipole polarizations appear rapidly ($10^{-14} - 10^{-15}$ sec). Hence, these polarizations generate energy loss. For that reason, these polarizations have a great effect on relaxation loss [38].

18.4 MICROWAVE ABSORBERS

The microwave absorbers are manufactured with their electromagnetic and dielectric properties. Furthermore, most of the studies on microwave absorber have been focused on composites that have microwave-absorption properties [29]. The microwave-absorption composites generally consist of one or more types of fillers inside a matrix. The properties of microwave absorption for such heterogeneous media are described by dielectric/

magnetic properties of fillers and matrix [39,40]. According to filler type, the absorber materials are classified as either dielectric or magnetic microwave absorber [5]. The properties of dielectric microwave-absorption composites are determined with ohmic loss of energy that can be obtained by the lossy filler such as ceramic nanoparticles, carbon nanotube, and graphene of other metal powder. The magnetic microwave absorbers are characterized by magnetic properties of fillers such as ferrite, iron, cobalt, nickel, and other nanoparticles. However, the matrix materials of the absorber are chosen for their physical and chemical properties, and microwave-absorption composites are commonly classified into polymer matrix composite (PMC), ceramic matrix composite (CMC), and metal matrix composite (MMC) [41]. Hence, the microwave-absorption properties of composite materials are affected by the properties of the matrix and reinforcements [5].

Frequency dependence of permittivity for matrix and fillers, the shape of fillers, weight ration of fillers, distribution of fillers into the matrix, and physical properties of the matrix are the factors that affect the dielectric properties of composites [5,41]. The size of the nanoparticles directly affects microwave-absorption performance with microwave heating mechanism. According to particle size, two types of heating mechanism can be discussed: uniform heating and nonuniform heating. The particle that has a size lower than 100 μm can be assumed finer particle for microwave heating, which offers uniform heating characteristics than coarse particle (Figure 18.4) [5,41].

The microwave-absorption ability of a material can also be described by the loss tangent (tan δ). Dielectric loss tangent (tan δ_ε) and magnetic loss tangent (tan δ_μ) are two parameters that determine the microwave absorption properties of absorbers. The tan δ is calculated as the ratio of the real permittivity and imaginary permittivity (Equations 18.9 and 18.10) [24].

$$\tan(\delta_\varepsilon) = \frac{\varepsilon''}{\varepsilon'} \text{ and } \tan(\delta_\mu) = \frac{\mu''}{\mu'} \tag{18.9}$$

$$\tan(\delta) = \tan(\delta_\varepsilon) + \tan(\delta_\mu) \tag{18.10}$$

As discussed previously, the depth of microwave penetration determines the microwave absorption performance with uniform or nonuniform heating mechanism. The dielectric

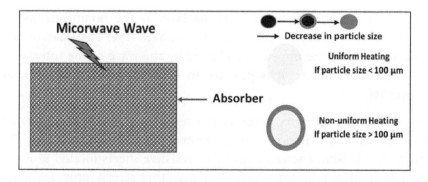

FIGURE 18.4   Conversion of microwave energy to heat according to particle size.

properties of particles are important parameters for depth of penetration, as calculated in Equation (18.11) [15,37]:

$$D = \frac{3\lambda_0}{8.686\pi \tan\delta \left(\dfrac{\varepsilon_r}{\mu_r} \right)} \tag{18.11}$$

Here, D is the depth of microwave penetration into the material and $\lambda_0 = c/f$ is the wavelength.

Generally, penetration depth, d, is about 1 μm at 10 GHz frequency; for that reason, nano-sized particle can attenuate electromagnetic wave considerably. According to Equation (18.11), the penetration depth decreases with the increase of frequency and value of dielectric. However, the material that has a low dielectric value, such as glass and nonpolar material, has large penetration depth. Thus, the microwave can pass through that material without high energy loss. Normally, when the particle size is below d, then uniform heating of that particle will be limited [24,33].

The combinations of conducting polymers with magnetic or dielectric fillers have been intensively examined in the study of microwave-absorbing composites for single- and multilayer composites. Among the conducting polymers, PANI has drawn special attention because of its high conductivity, good environmental stability and corrosion resistance, light weight, flexibility, and dielectric loss ability. However, a nonconductive matrix containing PANI shows better stability of electromagnetic wave absorption at different frequencies. Hence, PANI is an important candidate for use in microwave absorption [22,42–44].

The microwave-absorbing materials can be classified into radar-absorbing coating material and structural radar-absorbing material. Radar-absorbing coating material can be classified as magnetic and dielectric absorbers. Structural radar-absorbing material can be used for two purpose such as load bearing and RCS [22,41–44].

18.4.1 Radar-Absorbing Coating Material

Microwave-absorption coating materials can be classified as Dallenbach absorber, Salisbury absorber, and Jaumann absorber:

- *Dallenbach absorber*: Dallenbach coating layer is the homogeneous high-lossy layer in front of a conductive plane. The Dallenbach layer has a narrow absorption frequency owing to the impedance change by the wave at the interface between the two media. However, it is possible to obtain a broadband with double or multilayer [6].

- *Salisbury absorber*: Salisbury absorber is a resonant absorber sheet having a resistance of 377 ohm/sq to improve the microwave-absorbing properties of Dallenbach absorber. The Salisbury screen consists of resistive sheets located a quarter wavelength out from a reflecting surface; hence, this screen provides a broadband performance [24,41].

- *Jaumann absorber*: Jauman absorber layers are a method of the Salisbury screen by adding additional multilayer resistive sheets. The resistivity of sheets should reduce from surface to bottom to provide maximum performance [6,15,41].

18.4.2 Structural Absorbing Materials

Three classes of impedance-matching absorber materials, pyramidal absorber, tapered absorber, and matching layer absorber, have been developed to reduce the impedance step between the incident and absorbing media [5,6,41]:

- *Pyramidal absorbers*: Pyramidal absorbers are pyramidal-shaped absorber. They are typically thick cone structures widening vertically to the surface of the absorbing material. These absorbers provide high attenuation over wide frequency and angle ranges [6,41].

- *Tapered absorber*: The tapered absorber is a slab made from a low-loss material mixed with a lossy material. One type of tapered absorber containing an open celled foam or plastic net is dipped or sprayed with the lossy material. Another type of tapered absorber is composition of homogeneous layers, with increasing loading in the direction of propagation [6,15,29].

- *Matching layer absorber*: The matching layer absorber is placed between the incident and absorbing media to reduce the needed thickness for the gradual transition layer. The thickness and impedance value of the layer are chosen between two impedances [6,41].

18.5 MICROWAVE-ABSORPTION PROPERTIES OF SINGLE-, DOUBLE-, AND MULTILAYER COMPOSITES

The complex permittivity and complex permeability of studied composite were measured by a Keysight vector network analyzer (full 2-port S-parameters), using N1500–001 software and coaxial transmission line technique (inner diameter: 3 mm and outer diameter: 7 mm). The 85518A calibration kit was used for S-parameter measurement uncertainty. The reflection loss (R_L) of the single-, double-, and multilayer design was calculated by using the dielectric properties obtained from the studied composite. The studied composite A consists of Ba-Ferrite as a filler and epoxy as a matrix, and composite B consists of magnetite as a filler and epoxy as a matrix.

The microwave-absorption properties of the studied composites were investigated in the 1–14 GHz for the total thickness of 2 and 4 mm. As seen in Figures 18.5 and 18.6, A and B are the single-layer composites, and C and D are the double- and multilayer composites that are obtained from the combination of A and B. For double-layer composite with a total thickness of 2.0 and 4 mm, A is the which matching layer filled and the B is the absorbing layer. Multilayer composite D consists of three layers of A and B backed by a reflector. For the double layer, when the absorbing layer thickness is 1 mm and the matching layer is 1 mm, the absorption bandwidth is about 3.5 dB (between 11 and 13.5 GHz) below −10 dB. However, it is 2.7 dB below −10 dB (between 5.7 and 7.0 GHz) for the total thickness of 4 mm. For the multilayer design, it was observed that three-layer composite (D) provides broadband absorption compared with double- (C) and single-layer (A, B) designs, and multilayer model has improved absorbing properties.

However, as seen in Figures 18.5 and 18.6, the maximum absorption peak of double- (C) and multilayer (D) composites is shifted to a lower frequency with increasing composite thickness. This can be explained based on the quarter-wavelength model. The incident electromagnetic waves are partly reflected off from air to absorber interface, partly pass through the material without any dissipation, and are reflected from the absorber to the metal interface. These two reflected phases are equal to 180°, and this leads to two reflected waves in the air–absorber

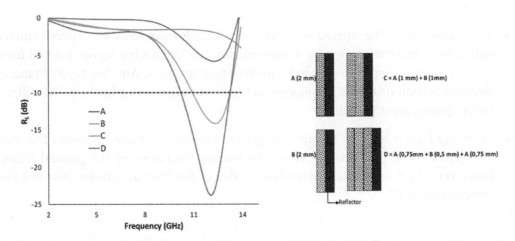

FIGURE 18.5 Reflection loss (R_L) of single-layer, double-layer, and multilayer composites for a total thickness of 2 mm.

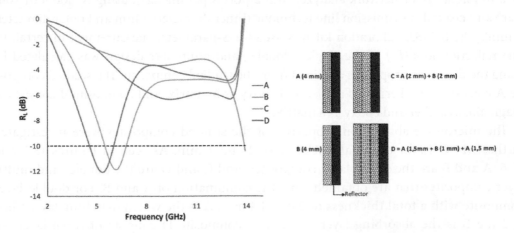

FIGURE 18.6 Reflection loss (R_L) of single-layer, double-layer, and multilayer composites for a total thickness of 4 mm.

interface totally canceled. The quarter-wave thickness criteria have been proposed to explain this phenomena, and equation relationship between matching thickness (t_m) and matching frequency (f_m) is given by quarter-wavelength principle: $t_m = n\lambda_0 / 4(\|\mu_r\|\|\varepsilon_r\|)^{1/2}$, $n = 1,2,3...$ where $\lambda_0 = c/f_m$ is the wavelength of the incident wave [23,45,46].

18.6 SUMMARY

In recent years, radar absorbing materials (RAMs) have been comprehensively studied because of the explosive requirement in stealth technology, electromagnetic pollution, and other commercial application. Therefore, there is a strong demand for developing microwave absorbing composites. Although many studies have been done to discuss microwave performance of composites or particles, there is a limited discussion on microwave heating mechanism, microwave-material interaction, fillers size and frequency thickness dependence. In this chapter, the microwave absorption properties of single and multilayer composites was investigated based on composite thickness, number of layer and size of fillers. Moreover, the other important phenomena such as polarization and loss mechanism were also evaluated for microwave absorption properties. The current paper also offers experimental data for single and multilayer composite.

The microwave heating and absorption performance within homogeneous and heterogeneous composites is described by complex permittivity ($\varepsilon^* = \varepsilon' - j\varepsilon''$) and complex permeability ($\mu^* = \mu' - j\mu''$). When both the real and imaginary parts of the permittivity and permeability are equal, the reflection loss of materials is considered zero. These parameters are very important parameters that can determine the microwave absorption performance of a materials.

As understood from Salisbury and Jauman design the microwave absorption and bandwidth of single layer absorbers can be improved by adding extra layers. The results of studied composites are also proved these phenomena. In addition that the microwave absorption properties of materials can also be enhanced with an understanding of the interaction between fillers and matrix. The dielectric and magnetic properties of fillers and matrix are important to determine the microwave performance of radar absorbing materials.

The different structural designs such as pyramidal, tapered and matching layer absorber also reduce the radar cross section (RCS). Both absorbing coating and structural absorbing materials are designed as RCS reduction techniques. The best microwave absorption performance can be obtained with minimized reflection and transition. A material used as a microwave wave absorber must either absorb or dissipate electromagnetic energy.

REFERENCES

1. Das, S., Nayak, G. C., Sahu, S. K., Routray, P. C., Roy, A. K. and Baskey, H. (2016), Titania-coated magnetite and Ni-ferrite nanocomposite- based RADAR absorbing materials for camouflaging application. *Polymer-Plastics Technology and Engineering*, 54: 1483–1493.
2. Micheli, D., Vricella, D., Pastore, R. and Marchetti, M. (2014), Synthesis and electromagnetic characterization of frequency selective radar absorbing materials using carbon nanopowders. *Carbon*, 77: 756–774.

3. Mishra, R. and Sharma, A. (2015), Microwave-material interaction phenomena: Heating mechanisms, challenges and opportunities in material processing. *Composites: Part A*, 81: 78–97.

4. Shah, A., Ding, A., Wang, Y., Zhang, L., Wang, D., Muhammad, J., Huang, H., Duan, Y., Dong, X. and Zhang, Z. (2015), Microwave absorption and flexural properties of Fe nanoparticle/carbon fiber/epoxy resin composite plates. *Composite Structures*, 131: 1132–1141.

5. Lakshmi, K. (2007), Development of thermoplastic conducting polymer composites based on polyaniline and polythiophenes for microwave and electrical applications. Department of Polymer Science and Rubber Technology and Department of Electronics, Cochin University of Science and Technology, Kochi, China, pp. 10–21.

6. Saville, P. (2005), Review of radar absorbing materials, defence R&D Canada–Atlantic. *Technical Memorandum DRDC Atlantic TM*, 1: 7–10.

7. Jia, K., Zhao, R., Zhong, J. and Liu, X. (2010), Preparation and microwave absorption properties of loose nanoscale Fe3O4 spheres. *Journal of Magnetism and Magnetic Materials*, 322: 2167–2171.

8. Hongxia, J., Qiaoling, L., Yun, Y., Zhiwu, G. and Xiaofeng, Y. (2013), Preparation and microwave adsorption properties of core–shell structured barium titanate/polyaniline composite. *Journal of Magnetism and Magnetic Materials*, 332: 10–14.

9. Padhy, S., Sanyal, S., Meena, R. S., Chatterjee, R. and Bose, A. (2013), Development and characterisation of (Mg, Mn) U-type microwave absorbing materials and its application in radar cross sections reduction. *Antennas & Propagation*, 8: 165–170.

10. Peng, C., Hwang, C. C., Wan, J., Tsai, S. J. and Chen, S. Y. (2005), Microwave-absorbing characteristics for the composites of thermal-plastic polyurethane (TPU)-bonded NiZn-ferrites prepared by combustion synthesis method. *Materials Science and Engineering B*, 117: 27–36.

11. Gupta, K., Abbas, S. M., Goswami, T. H. and Abhyanka, A. C. (2014), Microwave absorption in X and Ku band frequency of cotton fabric coated with Ni–Zn ferrite and carbon formulation in polyurethane matrix. *Journal of Magnetism and Magnetic Materials*, 362: 216–225.

12. Yang, C. C., Gung, Y. J., Hung, W. C., Ting, T. H. and Wu, K. H. (2010), Infrared and microwave absorbing properties of BaTiO3/polyaniline and BaFe12O19/polyaniline composites. *Composites Science and Technology*, 70: 466–471.

13. Fan, Y., Yang, H., Liu, X., Zhu, H. and Zou, G. (2008), Preparation and study on radar absorbing materials of nickel-coated carbon fiber and flake graphite. *Journal of Alloys and Compounds*, 461: 490–494.

14. Liu, Y., Liu, X. and Wang, X. (2014), Double-layer microwave absorber based on CoFe2O4 ferrite and carbonyl iron composites. *Journal of Alloys and Compounds*, 584: 249–253.

15. Clark, D. E., Folz, D. C. and West, J. K. (2013), Processing materials with microwave energy. *Materials Science and Engineering A*, 287: 153–158.

16. Akinay, Y. and Hayat, F. (2018), Synthesis and microwave absorption enhancement of $BaTiO_3$ nanoparticle/polyvinylbutyral composites. *Journal of Composite Materials*. doi:10.1177/0021998318788144.

17. Kong, I., Ahmad, S. H., Abdullah, M. H., Hui, D., Yusoff, A. N. and Puryanti, D. (2010), Magnetic and microwave absorbing properties of magnetite–thermoplastic natural rubber nanocomposites. *Journal of Magnetism and Magnetic Materials*, 322: 3401–3409.

18. Aphesteguy, J. C., Damiani, A., DiGiovanni, D. and Jacobo, S. E. (2009), Microwave- absorbing characteristics of epoxy resin composites containing nanoparticles of NiZn- and NiCuZn-ferrites. *Physica B*, 404: 2713–2716.

19. Akinay, Y., Hayat, F., Çakir, M. and Akin, E. (2018), Magnetic and microwave absorption properties of PVB/Fe_3O_4 and $PVB/NiFe_2O_4$ composites. *Polymer Composites*, 39: 3418–3423. doi:10.1002/pc.24359.

20. Hosseini, S., Mohseni, S. H., Asadnia, A. and Kerdari, H. (2011), Thermal infrared and microwave absorbing properties of $SrTiO_3/SrFe_{12}O_{19}$/polyaniline nanocomposites. *Journal of Alloys and Compounds*, 644: 423–429.

21. Tyagi, S., Pandey, V. S., Baskey, H. B., Tyagi, N., Garg, A., Goel, S. and Shami, T. C. (2018), RADAR absorption study of $BaFe_{12}O_{19}/ZnFe_2O_4$/CNTs nanocomposite. *Journal of Alloys and Compounds*, 731: 584–590.

22. Tantawy, H. R., Aston, D. E., Smith, J. R. and Young, J. L. (2013), Comparison of electromagnetic shielding with polyaniline nanopowders produced in solvent-limited conditions. *ACS Applied Materials & Interfaces*, 5: 4648–4658.

23. Abbas, S. M., Dixit, A. K., Chatterjee, R. and Goel, T. C. (2005), Complex permittivity and microwave absorption properties of $BaTiO_3$–polyaniline composite. *Materials Science and Engineering: B*, 123: 167–171. doi:10.1016/j.mseb.2005.07.018.

24. Cheng, D. K. (1983), *Field and Wave Electromagnetic*. Addison Weshley Publishing Company, Reading, MA, vol. 9, pp. 370–437.

25. Idris, F. M., Hashim, M., Abbas, Z., Ismail, I., Nazlan, R. and Ibrahim, I. R. (2016), Recent developments of smart electromagnetic absorbers based polymer-composites at gigahertz frequencies. *Journal of Magnetism and Magnetic Materials*, 405: 197–208. doi:10.1016/j.jmmm.2015.12.070.

26. Meng, F., Wang, H., Huang, F., Guo, Y., Wang, Z., Hui, D. and Zhou, Z. (2018), Graphene-based microwave absorbing composites: A review and prospective. *Composites Part B: Engineering*, 137: 260–277. doi:10.1016/j.compositesb.2017.11.023.

27. Bogdal, D. and Prociak, A. (2007), *Microwave-Enhanced Polymer Chemistry and Technology*. Blackwell Publishing Professional, Ames, IA, vol. 1, pp. 3–11.

28. Chen, L. F., Ong, C. K. and Neo, N. P. (2004), *Microwave Electronics; Measurement and Materials Characterization*. John Wiley & Sons, Chichester, UK, vol. 1, pp. 120–126.

29. Qin, F. and Brosseau, C. (2012), A review and analysis of microwave absorption in polymer composites filled with carbonaceous particles. *Journal of Applied Physics*, 111: 2–19.

30. Fanbin, M., Huagao, W., Fei, H., Yifan, G., Zeyong, W., David, H. and Zuowan, Z. (2018), Graphene-based microwave absorbing composites: A review and prospective. *Composites Part B*, 137: 260–277.

31. Thostenson, E. T. and Chou, T.-W. (1999), Microwave processing: Fundamentals and applications. *Composites: Part A*, 30: 1055–1071.

32. Das, S., Mukhopadhyay, A. K., Datta, S. and Basu, A. (2009), Prospects of microwave processing: An overview. *Bulletin of Materials Science*, 32: 1–13.

33. Liu, Y., Liu, X. and Xuanjun, W. (2014), Double-layer microwave absorber based on $CoFe_2O_4$ ferrite and carbonyl iron composites. *Journal of Alloys and Compounds*, 584: 249–253.

34. Wei, C., Shen, X., Song, F., Zhu, Y. and Wang, Y. (2012), Double-layer microwave absorber based on nanocrystalline $Zn_{0.5}Ni_{0.5}Fe_2O_4$/a-Fe microfibers. *Materials and Design*, 35: 363–368.

35. Gao, Y., Gao, X., Li, J. and Guo, S. (2018), Microwave absorbing and mechanical properties of alternating multilayer carbonyl iron powder-poly(vinyl chloride) composites. *Journal of Applied Polymer Science*. doi:10.1002/App.45846.

36. Huo, J., Wang, L. and Haojie, Y., (2009), Polymeric nanocomposites for electromagnetic wave absorption, *Journal of Materials Science*, 44: 3917–3927.

37. Sun, J., Wang, W. and Yue, Q. (2016), Review on microwave-matter interaction fundamentals and efficient microwave-associated heating strategies. *Materials*, 231: 2–25. doi:10.3390/ma9040231.

38. Onimisi, M. Y. and Ikyumbur, J. K. (2015), Comparative analysis of dielectric constant and loss factor of pure butan-1-ol and ethanol. *American Journal of Condensed Matter Physics*, 5: 69–75. doi:10.5923/j.ajcmp.20150503.02.

39. Naidu, K. C. B. and Madhuri, W. (2017), Microwave processed bulk and nano NiMg ferrites: A comparative study on X-band electromagnetic interference shielding properties. *Materials Chemistry and Physics*, 187: 164–176.

40. Singh, B. P., Choudhary, V., Saini, P., Pande, S., Singh, V. N. and Mathur, R. B. (2013), Enhanced microwave shielding and mechanical properties of high loading MWCNT–epoxy composites. *Journal of Nanoparticle Research*, 15: 1554.

41. Dawei, H. (2010), Development of the epoxy composite complex permittivity and its application in wind turbine blades. School of Engineering and Materials Science Queen Mary, University of London, London, UK, pp. 57–59.

42. Yang, C. C., Gung, Y. J., Shih, C. C., Hung, W. C. and Wu, K. H. (2011), Synthesis, infrared and microwave absorbing properties of ($BaFe_{12}O_{19}$+$BaTiO_3$)/polyaniline composite. *Journal of Magnetism and Magnetic Materials*, 323: 933–938.

43. Bora, P. J., Porwal, M., Vinoy, K. J., Ramamurthy, P. C. and Madras, G. (2016), Outstanding electromagnetic interference shielding effectiveness of polyvinylbutyral-polyaniline nanocomposite film. *Materials Research Express*, 82: 8234–79262.

44. Gairola, S. P., Verma, V., Kumar, L., Abdullah Dara, M., Annapoorni, S. and Kotnala, R. K. (2010), Enhanced microwave absorption properties in polyaniline and nano-ferrite composite in X-band. *Synthetic Metals*, 160: 2315–2318.

45. Yang, H., Dai, J., Liu, X., Lin, Y., Wang, J., Wang, L. and Wang, F. (2017), Layered PVB/$Ba_3Co_2Fe_{24}O_{41}$/Ti_3C_2 Mxene composite: Enhanced electromagnetic wave absorption properties with high impedance match in a wide frequency range. *Materials Chemistry and Physics*, 200: 179–186. doi:10.1016/j.matchemphys.2017.05.057.

46. Li, R., Wang, T., Tan, G., Zuo, W., Wei, J., Qiao, L. and Li, F. (2014), Microwave absorption properties of oriented $Pr_2Fe_{17}N_{3-\delta}$ particles/paraffin composite with planar anisotropy. *Journal of Alloys and Compounds*, 586: 239–243. doi:10.1016/j.jallcom.2013.10.040.

CHAPTER 19

Dynamic Mechanical and Fibrillation Behaviors of Nanofibers of LCP/PET Blended Droplets by Repeated Extrusion

Han-Yong Jeon

CONTENTS

19.1 INTRODUCTION

Crystalline substances are ordered in all three dimensions, and amorphous materials are disordered. The liquid crystals have the intermediate properties. Rod-shaped polymers of oriented form exist inside of the liquid crystal polymer (LCP). Liquid crystal polymers are a family of high-performance plastics. They are distinguished from semicrystalline plastics by their special molecular structure, which consists of rigid, rod-like macromolecules that are ordered in the melt phase to from liquid crystal structures. Generally, these rod-shaped polymers are constructed of molecule units, including benzene rings. LCP has the specificity of liquid crystal, mechanical, and chemical excellence. However, since LCP resin is too expensive, the development of high-value-added applications or lowering the price in the process is problematic [1–3].

Blending with the cheap polymer is a solution to solve the difficulties of development of high-value-added applications and cost in the process. Physically mixing two or more polymers is known as blending. Good miscibility between the polymers make single uniform phase. But inverse case has a continuous or dispersion phase. Dispersion shape, viscosity, melting temperature, miscibility, etc., are important points in the immiscible polymer blending. The dispersion control allows the improvement of properties and the determination of droplet size. Vectra and poly ethylene terephthalate (PET) blending is noted in the blend, because they have the similar melting temperature and structure [4–6].

In the LCP and PET blending, the droplet size and dispersion are important. Therefore, understanding of the droplet behavior is very important for the after-process. So far, many researches were conducted to analyze the droplet behavior. The spherical shape is made to the droplets in the blend-composite by drawing [7,8].

By applying this, it is possible to shape the fiber by controlling the stretching condition and the droplet size. Therefore, it is necessary to study the control and behavior of the droplet size. Until now, the electricity spinning or the shape of island in the sea was usually used to make nanofiber. However, these methods have some problems of yield and cost. Droplet stretching method is expected to solve these problems. To make of the continuous nanofibers is necessary for manufacture of the droplet of the appropriate size [9,10]. Therefore, control of the droplet size by using the repetitive extrusion plays an important role. However, the droplet behavior analysis was not conducted in the repeated extrusion.

In this study, droplet behavior change by repeated extrusion number of times was observed through analysis with blending condition and weight ratio. The droplet behavior change relates with flow property, miscibility, and surface property of LCP and PET.

19.2 EXPERIMENTAL

To confirm the possibility of fiber fibrillation, it was considered that LCP droplets could be fabricated without spinning process. For this purpose, droplet manufacturing process and physical property evaluation were carried out. Especially, morphology and viscoelastic property were analyzed to determine the optimum droplet formation with repeated extrusion condition.

19.2.1 Preparation of LCP:PET Blended Droplet

LCP (Vectra A950/Ticona) of Figure 19.1 and PET (Hyosung Co. Ltd., South Korea) chips were dried for 24 hours in oven at 60°C. LCP and PET chips were bended with 70:30 and 50:50 wt%, respectively, and extruded through twin-screw extruder (SM PLATEK Co. Ltd., South Korea, Table 19.1 and Figure 19.2).

Extruded blend materials were cut and dried for 24 hours at 60°C and extruded, and this process was repeated for one, two, and three times of extrusion. Figure 19.3 shows the schematic processing diagram of LCP:PET blended droplet extrusion.

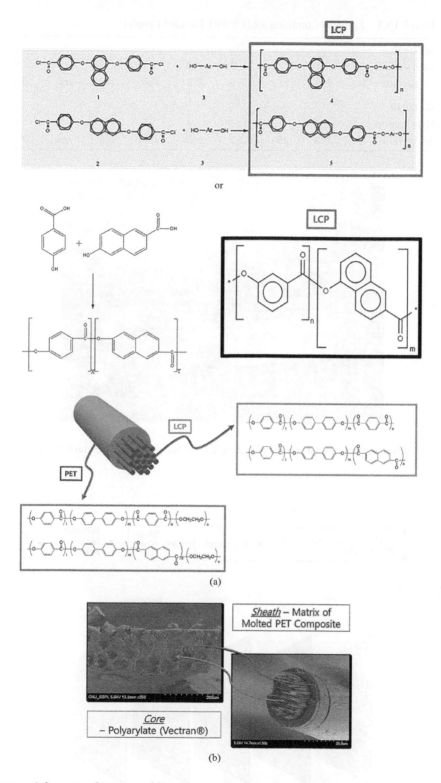

FIGURE 19.1   Schematic diagram of liquid crystal polyarylate (LCP): (a) Synthesis and chemical composition and (b) nanofiber fibrillation.

TABLE 19.1   Extruder Condition for LCP:PET Blended Droplet

(a) 70:30

Heat section	1	2	3	4	5	6	7	8
Temperature °C	245	250	275	275	280	280	280	280

(b) 50:50

Heat section	1	2	3	4	5	6	7	8
Temperature °C	245	250	255	265	265	265	265	265

FIGURE 19.2   Photograph of extruder for LCP:PET blended droplet manufacturing.

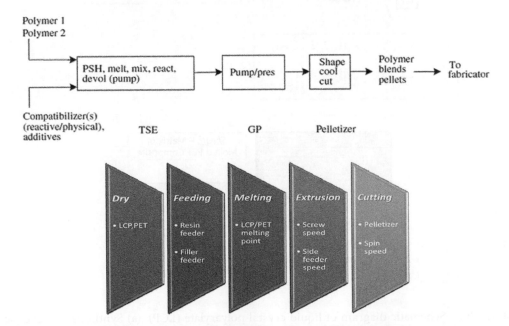

FIGURE 19.3   Schematic processing diagram of LCP:PET blended droplet extrusion.

19.2.2 Hot Pressing

Formation of film was done by used hot-press molder (QM900A/QMESYS). LCP:PET (50:50) and (70:30) blended samples were preheated at 270°C for 15 min, and a constant force of 5.5 MPa was applied at 270°C for 10 min. After then, LCP:PET blended sample was cooled at 5°C water and a thin film of 0.07 mm thickness was made by this sample.

19.2.3 Dynamic Mechanical Analysis Test

Dynamic mechanical properties of LCP:PET (50:50) and (70:30) blended sample films by hot pressing were measured with rheometer (shear strain: 0.05%, pressing force: 10 N, oscillation: 10 Hz, 50°C–100°C, heating rate: 1°C/min). Equipment used in the experiment is Modular Compact Rheometer 102 (Anton Paar).

19.2.4 Image Analysis

Morphology of LCP:PET (50:50) and (70:30) blended droplets was observed by SEM (S-3400N/HITACH). Central droplet's three-dimensional (3D) image of the cross-section was confirmed by 3D-measuring laser microscope (OLS4000/LEXT). These 3D data were used to identify components of LCP:PET (50:50) and (70:30) blended droplets.

Figure 19.4 shows the overview of equipment for manufacturing and property analysis of LCP:PET blended droplet.

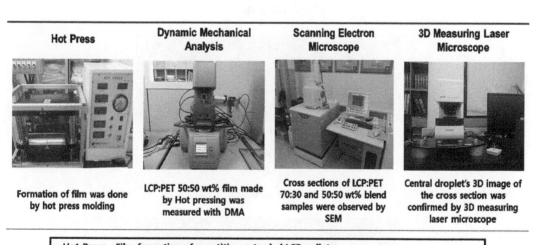

FIGURE 19.4   Overview of equipment for manufacturing and analysis of LCP:PET blended droplet.

19.3 RESULTS AND DISCUSSION

The size and distribution of LCP:PET blended droplets formed by repeated extrusion were observed and analyzed. The droplet size and morphological characteristics were compared and analyzed by SEM image analysis. The components of droplets treated with Digital microfluidics (DMF) solution were analyzed. The viscoelastic behavior was analyzed, and the miscibility required for droplet formation was analyzed.

19.3.1 Viscoelastic Property Analysis of LCP:PET Blended Droplet

To make LCP:PET blended droplet film, process of preheating ([270°C, 10 min]→pressing [270°C, 10 min under 5.5 MPa]→cooling in 4°C water→conditioning/drying [60°C, 2 hours]) was adopted, and film was 0.07-mm thick. Figure 19.5 shows the photographs of LCP:PET blended film for viscoelastic property test, and as explained previously, LCP:PET blended droplet film was made by hot-pressing process.

In Figure 19.6, LCP peak is found around 70°C and 80°C, owing to immiscibility of LCP and PET, and no change of T_g is shown through repetitive extrusion. In the previous 50:50 wt% dynamic mechanical analysis (DMA) data, peaks appear in the vicinity of 75°C and 80°C in the tan δ value. Two T_g values show the immiscible nature of LCP and PET. In addition, since the thermal properties according to the number of extrusion do not show any particular change, it can be confirmed that there is no deterioration of the material properties through repeated extrusion.

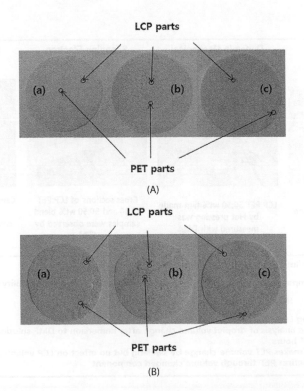

FIGURE 19.5 Photographs of LCP:PET blended droplet film for viscoelastic property test.

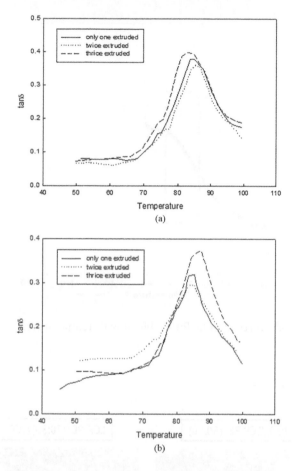

FIGURE 19.6   Photographs tan δ of LCP:PET (50:50) blend sample by extruded time: (a) LCP:PET (50:50) and (b) LCP:PET (70:30).

Also, in Figure 19.7, slop change appeared at around 75°C and 85°C. Each slop change point means glass transition temperature, and the existence of two glass transition temperatures means immiscibility.

Therefore, the blended sample has droplet-dispersion phase. It is considered that there is no change in the thermal properties due to repeated extrusion, even at 70:30 wt%.

19.3.2 Change of Dispersion and Droplet Size

Figure 19.8 shows the SEM images of LCP:PET blended droplets by repeated extrusion number of times, and in here, first extrusion, second extrusion, and third extrusion mean 1, 2, 3 times, respectively, of repeated extrusion. PET droplet tends to join together with the surrounding PET droplet. PET droplet moves to the center, because LCP has a lower viscosity than PET.

For the first extrusion time, LCP:PET (50:50) blended droplet has on average 50-μm droplet size and dispersion, and for the second extrusion time, LCP:PET (50:50) blended droplet has on average 100-μm droplet size. For the third extrusion time, LCP:PET (50:50) blended droplet has on average 150-μm droplet size.

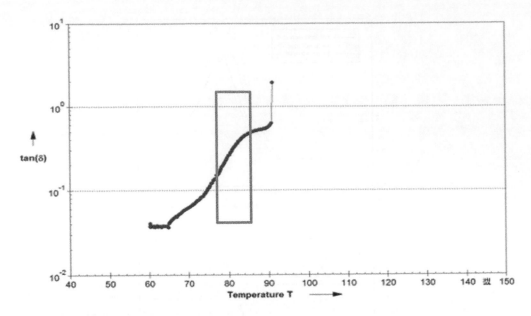

FIGURE 19.7   DMA analysis result of LCP:PET blend with temperature.

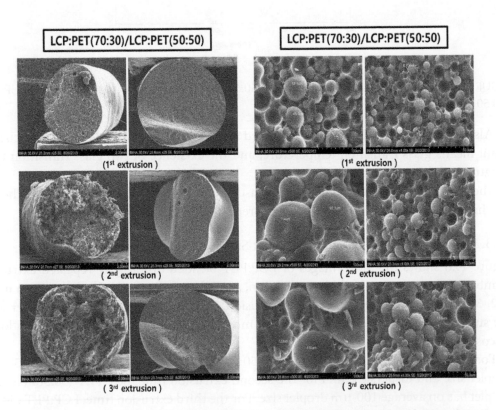

FIGURE 19.8   Comparison of morphology of LCP:PET blended droplet with repeated extrusion.

For the first extrusion time, LCP:PET (70:30) blended droplet has on average 50-μm droplet size and even dispersion. But for the second extrusion time, LCP:PET (70:30) blended droplet has on average 100-μm droplet size, and for the third extrusion time, LCP:PET (70:30) blended droplet has on average 150-μm droplet size. In addition, congregative phenomenon in center was observed, owing to immiscibility and viscosity difference.

Congregative phenomenon in center was also observed in two and three extrusion times of LCP:PET (70:30) blended droplet, owing to immiscibility and viscosity difference. PET droplet tends to join together with surrounding PET droplet. PET droplet moves to the center, because LCP has a lower viscosity than PET, but the congregative phenomenon was not observed in Figure 19.8.

It is shown that droplet size increases with extrusion number of times and the congregative phenomenon was not observed. The congregative phenomenon was not observed. For LCP:PET (50:50) blended droplet, so matrix is not decided in this case. The droplet size tend to increase do by number of extrusion time to increase as same as LCP:PET (70:30) blended droplet.

19.3.3 Analysis of Droplet Component

Only component analysis of droplet was conducted at 70:30 wt%. The congregative phenomenon was not observed in the 50:50 wt% sample.

Figure 19.9 shows the shape of droplet after DMF handling for 12 hours, at 24°C. Alteration of droplets in center was observed in image (b). Vectra A950 has good resistance to DMF, but DMF expands the PET droplet.

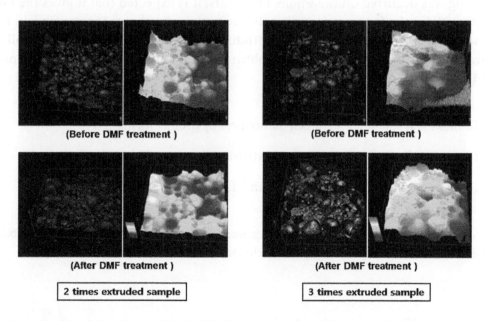

(Before DMF treatment)　　　　　　　(Before DMF treatment)

(After DMF treatment)　　　　　　　(After DMF treatment)

2 times extruded sample　　　　　　　3 times extruded sample

FIGURE 19.9　Droplet analysis of LCP:PET blended droplet with repeated extrusion before and after DMF treatment.

Immiscibility and slope change of LCP:PET blended droplet appeared at around 70°C and 85°C. This means that slope change point means glass transition temperature, and the existence of two glass transition temperatures means immiscibility. Therefore, this blended sample has droplet dispersion phase.

19.4 SUMMARY

Distribution of droplets was observed in LCP:PET blended droplets by repeated extrusion process. In this process, each droplet was distributed relatively evenly in the first extrusion process. But for the second and third extrusion droplets, the size of the droplet was increased, and it was observed that the droplet gathered in the center. The analysis was carried out by assuming that the miscibility of LCP:PET blending and the flow characteristics correlate with the phenomenon.

In this study, the analysis of the distribution and component of droplet was conducted. In addition, the miscibility of LCP:PET blending was analyzed. In the case of blending, high LCP content was shown in comparison with PET content. LCP becomes matrix, and PET represents droplet shape. By repeated extrusion, LCP:PET blended droplet shows enlarged droplet size, and LCP is seen wrapping the PET droplet. Viscosity difference of the two immiscible materials is cause of this phenomenon. In blend of two immiscible materials, agglomerative phenomenon leads to a reduction of the material properties, and droplet size correlates with the mechanical properties.

However, in the case of same wt%, matrix is not decided and the congregative phenomenon of PET droplets was not observed in the center. Also, the droplet size increasing was occurred on the whole. From this, it is expected that it gives the optimum size of droplet for manufacturing. It also anticipates that this research will help to produce droplets of suitable size to produce continuous nanofibers. The following are the future keys to solve the problem of how to manufacture the optimum (LCP:PET) blend composition:

- To make staple and filament LCP nanofibers by droplet formation

- To control the number of repetitive extrusion times for droplet formation

- To control and set-up blending ratio and droplet size distribution

To review the above points, optimum application field of nanofiber product, which is based on LCP:PET blended droplet, could be developed, as shown in Figure 19.10.

Fishnet and Rope	Geosynthetics-Geogrid
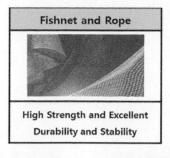	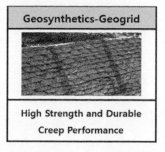
High Strength and Excellent Durability and Stability	High Strength and Durable Creep Performance

Safe and Protective Tools	Sports/Leisure Goods(1)	Sports/Leisure Goods(2)
	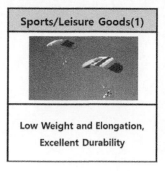	
High Temperature Resistance, Excellent Protection and Safety Performance	Low Weight and Elongation, Excellent Durability	Excellent Vibration Damping and Impact Absorbance, Low Water Absorption

FIGURE 19.10   Application examples of LCP:PET blended droplet products based on repeated extrusion.

REFERENCES

1. L. H. Sperling, *Introduction to Physical Polymer Science* (4th ed.), Chichester, UK: Wiley-Interscience, pp. 325–336, 2006.
2. C. H. Song and A. I. Isayev, LCP droplet deformation in fiber spinning of self-reinforced composites, *Journal of Polymer*, 42, 2611–2619, 2001.
3. W. N. Kim and M. M. Denn, Properties of blends of a thermotropic liquid crystalline polymer with a flexible polymer (Vectra/PET), *Journal of Rheology*, 36, 1477–1498, 1992.
4. P. Magagnni, On the use of PET-LCP copolymers as compatibilizers for PET/LCP blends, *Journal of Polymer Engineering and Science*, 36(9), 1244–1255, 1996.
5. E. D. Seo, A studies on the surface morphology and fine structure of PET film treated by DMF, *Journal of the Korean Society of Dyers and Finishers*, 16(1), 55–64, 2004.
6. K. Nakayama, Structure formation and miscibility of sheets from PBT and LCP blends, *Journal of Materials Science*, 36, 3207–3213, 2001.
7. D. Wang, G. Sun, and B.-S. Chiou, A high-throuput, controllable, and environmentally benign fabrication process of thermoplastic nanofibers, *Macromolecular Materials and Engineering*, 292, 407–414, 2007.
8. K. B. Migler, String formation in sheared polymer blends: Coalescence, breakup, and finite size effects, *Physical Review Letters*, 86(6), 1023–1026, 2001.
9. E. Movahednejad, Prediction of droplet size and velocity distribution in droplet formation region of liquid spray, *Journal of Entropy*, 12, 1484–1498, 2010.
10. K. Hori, Y. Hoshino, and H. Shimizu, *Vectran: Development of High-Funtionality Fiber and its Applications*, Kuraray Co., Ltd., 2014.

CHAPTER 20

Nanoscale Wireless Communications as Enablers of Massive Manycore Architectures

Sergi Abadal, Josep Solé-Pareta, Eduard Alarcón, and Albert Cabellos-Aparicio

CONTENTS

RECENT YEARS HAVE SEEN the ubiquitous emergence of manycore computers, where processor cores rely on an integrated packet-switched network to exchange and share data within the chip. The performance of these intrachip networks is a key determinant of the execution speed and, as the number of cores per chip keeps increasing, it becomes an important bottleneck, owing to scalability issues. In response to this problem, several emerging interconnects enabled by nanoscale technologies have been investigated recently. Among them, miniaturized wireless interconnects show great promise as the complement to traditional on-chip networks, thanks to their flexibility and broadcast capabilities. This chapter reviews the state of the art of the field, focusing on nanoscale technologies for wireless on-chip communication in the millimeter-wave (mmWave) and terahertz (THz) bands. The scaling of trends of these interconnects is analyzed from the physical, link, and network levels in order to assess the practicality of the idea in the extremely demanding manycore era.

20.1 INTRODUCTION

The relentless march of technology scaling has forced a widespread transition in processor design from *single core* to *multicore*, owing to power and complexity reasons. The adoption of multicore processors is now ubiquitous, extending its influence from supercomputers and high-end servers to commodity computers, embedded systems, and handheld devices. In successive generations, processors have been steadily increasing in the number of cores to continue with current upward trends in terms of performance and efficiency. We are thus at the *manycore era* [1].

A generic multiprocessor, represented in Figure 20.1a, comprises a set of cores that execute parallel tasks and share data within the chip through their respective cache memories and the on-chip interconnect. The advent of the manycore era, fueled by constant

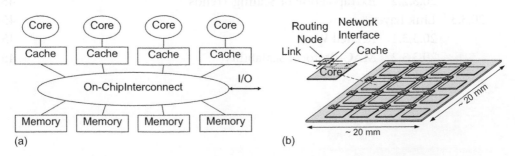

FIGURE 20.1   Schematic diagrams of the building blocks of (a) a multicore processor and (b) a network-on-chip.

advancements in nanoscale technologies and continued core density scaling, has had a transformative impact in all these components [2]. In this chapter, we will focus on the on-chip interconnect.

The on-chip interconnect is of central importance, since it implements the communication between cores, crucial to sustain a parallel system, regardless of its architecture. Two main types of systems exist [2]. In message passing systems, communication is explicit and is carefully orchestrated by the programmer, whereas in shared memory systems (Figure 20.1a), communication occurs implicitly as the architecture tries to maintain a coherent view of the caches when cores share data. In any case, the interconnect has a profound effect on the performance of the processor, as added delays in communication can lag the execution.

Given the direct relation between communication and the overall performance, the research focus in multiprocessors has gradually shifted from how cores compute to how cores communicate. Buses were first widely considered for the implementation of the on-chip interconnect, but they do not scale well beyond a handful of cores. Instead, modern processors adopted the network-on-chip (NoC) paradigm, where packet-switched networks of nanoscale routers and links are cointegrated with the cores [3]. This approach, exemplified in Figure 20.1b, improves fault tolerance and scalability, as proven not only in many research efforts [4] but also in commercial products with tens of cores [5].

As we move from a few tens to a thousand cores, NoCs become dense nanonetworks. Within this context, it becomes unclear whether scaling current NoC implementations will be enough to meet the requirements of manycore processors. Reasons are several, the first being the increase in communication demands of scaling parallel applications. Scaling implies distributing the computation among more cores and having to synchronize and schedule tasks more often. This leads to an increase of not only the sheer amount of communication but also its heterogeneity: local and unicast communications will dominate, but global and multicast flows can become significant, as illustrated in Figure 20.2a [6–8]. Traffic, already showing spatiotemporally variations within or between applications, as shown in Figure 20.2b and c, respectively, will also become more dynamic. This is because multiprogramming, that is, mapping several applications within different subsets of cores, is likely to be leveraged more frequently in manycore processors [4].

Another implication of the increase in core density is that the increment in the intensity, heterogeneity, and variability of the load needs to be absorbed with higher efficiency and should maintain latency constant. Area and energy consumption are important, because while the chip size and power envelope keep rather constant as the number of cores grows [10], the traffic intensity does not, whereas latency is critical because of its impact on the processor execution speed, as pointed out earlier. Evidently, this affects NoC design.

The main issue of conventional NoCs is that the average number of hops per packet increases when scaled, which goes against the objective of improving the energy and latency. Figure 20.3a shows how the performance of a two-dimensional (2D) mesh degrades by one order of magnitude when scaling from 16 to 1024 cores, owing to the influence of *global* packets in uniform random traffic. A similar trend is observed in Figure 20.3b for broadcast messages, which essentially are a number of global messages put together. This is highly

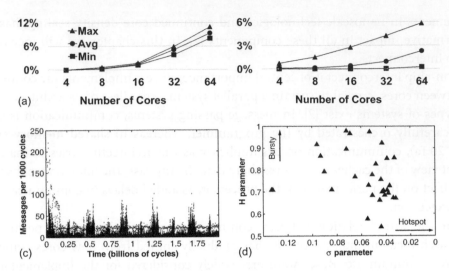

FIGURE 20.2   Plots showing the (a) heterogeneity and (b and c) variability of traffic. See [6–9] for more details.

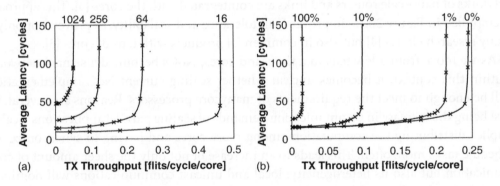

FIGURE 20.3   Scalability of an aggressive 2D mesh NoC design. (a) Scaling with system size and (b) Scaling with broadcast percentage.

concerning, because limited support for such types of traffic leads to remarkable execution slowdowns. For instance, slow synchronization methods can reduce application speeds by an average of 40%, despite representing a very small fraction of the code [11].

As we will see, one could use low-diameter topologies with solve the distance issues [12,13]. However, there are different performance–cost trade-offs behind this decision and, given the increasing variability of traffic, it is easy to demonstrate that there is no topology that can fit to all situations [14]. Instead, NoCs end up using simple but heavily overprovisioned topologies to cover all possible traffic patterns. This has a very negative impact on the area and energy efficiency of NoCs and calls for new solutions capable of providing the performance, efficiency, and flexibility required by manycore processors.

Aware of the nanonetwork challenges associated with manycore processors, several alternative technologies have been investigated recently. This chapter reviews them and makes the case for wireless on-chip interconnects enabled by emerging nanoscale technologies in the mmWave and THz bands (Section 20.2). Through different scalability analyses, we demonstrate that the wireless nanoscale interconnects are capable of delivering the flexibility, as well as the fast and low-power broadcast capabilities, required to enable the next generation of manycore processors (Section 20.3). Then, we discuss prospects in the pathway of actually implementing the vision (Section 20.4) and conclude the chapter.

20.2 EMERGING NANOSCALE INTERCONNECT TECHNOLOGIES

A plethora of research lines has emerged as a response to challenges of the manycore era. Besides improving conventional links and routers [4], the use of novel nanoscale interconnect technologies has been the main proposal to overcome the diameter and overprovisioning issues in manycore processors. Before delving into the feasibility of the wireless option enabled by nanoscale technologies, this section provides a brief yet comprehensive survey on the state of the art of such proposals. We review advances in three-dimensional (3D) NoCs enabled by vertical stacking, radio frequency (RF) transmission lines, nanophotonics, and wireless interconnects [15]. As summarized in Table 20.1, we compare their principles of operation, technological readiness, and bandwidth and energy-efficiency characteristics.

20.2.1 Three-Dimensional Interconnects

Three-dimensional integration consists on the stacking of layers of processors and memory. These are separated by just a few tens of micrometers and are vertically interconnected by means of through-silicon vias (TSVs), as sketched in Figure 20.4. The creation of 3D integrated circuits has proved to be a promising paradigm, since it improves the packing density, the noise immunity, and the overall performance [16].

TABLE 20.1 Comparative Summary of the Emerging Interconnect Technologies for NoC

	Baseline	3D NoC	Nanophotonics	RF-I	Wireless
Wiring	RC Wires	Vertical Vias	Waveguides	Transmission Lines	None
Principle	Wire charge	Wire charge	Optical signals	Guided RF signals	Radiated RF signals
Propagation speed	Large	Short	Speed of light	Speed of light	Speed of light
Technological availability	Now	Now	Mid term	Short term	Short term
Integration complexity	Low	Low	Very high	High	High
Bandwidth density	Good	Better	Best	Better	Modest
Intrinsic efficiency	Good	Better	Best	Better	Modest
Architectural flexibility	Average	Average	Low	Low	High
Issues	Wire delay	Thermal Effects	Laser power, buffering	Signal reflections	Multiple access

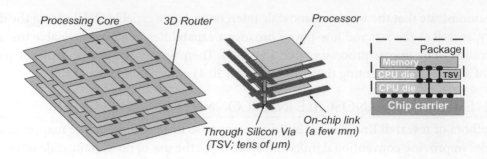

FIGURE 20.4   Sketch of a 3D NoC, a detail of its constituents, and a cross-section view of the chip.

From a communications perspective, 3D stacking has allowed to break the traditionally planar nature of NoC. The main advantage of this approach is the energy and delay reduction caused by the much shorter vertical distance between processors, as thoroughly analyzed in [17]. More significantly, the use of 3D stacking enables the design of topologies that would be unfeasible in the 2D design space, yielding reduced multihop latency [18] if routing protocols are adapted accordingly [16]. In addition, 3D stacking is an effective way to intuitively interface different technologies in hybrid approaches, facilitating modularity by avoiding the integration of different technologies within the same layer.

Networks based on 3D stacking present considerable challenges. The superposition of active layers produces an increase in the heat density that must be circumvented to avoid thermal effects. To alleviate this, some works have proposed routing mechanisms that adapt to the thermal profile of the processor (obtained in real time), in order to avoid traversing *hot areas* [16]. Another downturn of this approach is that refined techniques are needed for the manufacture of such 3D integrated circuits and networks; in particular, alignment methodologies are required for the precise positioning of the vias. Wireless coupling schemes could eliminate this issue [19], however, at the cost of higher power consumption. These challenges need to be further researched in order to validate the applicability of 3D stacking in the manycore era.

20.2.2 Transmission Lines

The RF interconnection paradigm consists of the transmission of electromagnetic (EM) waves over transmission lines, printed using the metal layers within the chip insulator [20]. The original baseband signals are modulated at a carrier frequency and then guided through the transmission lines. On reaching the destination, signals are demodulated and delivered. Transmission lines can have multiple inlets and outlets, allowing to design point-to-multipoint schemes and sharing among several transmitters. As in wireless networks, multiplexing can indeed be used to improve the bandwidth or to address transmissions coming from different nodes, either statically or dynamically, according to application demands.

Speed-of-light propagation and transmission multiplexing are the main advantages of RF interconnects. For this, they generally take the form of multiband global lines that communicate far-apart cores as complement of a conventional NoC, as sketched in Figure 20.5 [20,21]. Alternatively, NoCs implemented solely with RF interconnects have

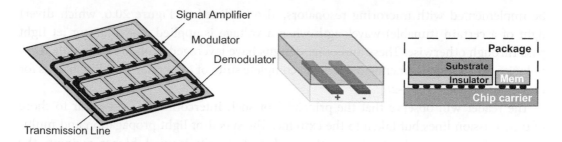

FIGURE 20.5   Sketch of an NoC overlaid with transmission lines, with detail of an on-chip transmission line and the cross-section of the chip.

also been proposed. Carpenter et al. interconnected all cores with a single transmission line and a centralized arbiter [22]. Another work worth noting uses a global transmission line scheme with orthogonal frequency-division multiple access (OFDMA) and dynamic reconfiguration [23]. The performance is promising, but the architecture seems to be limited by the complexity of OFDMA's signal processing.

The use of RF interconnect also entails several issues that reduce its scalability and practicality from a system-level perspective. For instance, the implementation of frequency multiplexing does not scale well in terms of area and power with the number of channels. Also, the physical topology must be carefully designed, as impedance mismatch reflections at the termination of the transmission line may generate interferences. This strongly limits the number of nodes connected through the same transmission line, forcing the segmentation of the interconnect [21]. As a side effect, it becomes hard to come up with an efficient way to provide a chip-wide broadcast through transmission lines, despite being global in range.

20.2.3 Nanophotonics

Silicon nanophotonics take advantage of the presence of a high-index substrate (silicon) and a low-loss insulator (typically SiO_2) to implement integrated waveguides and other optical components. These can be used to develop nanophotonic NoCs, where light is coupled on chip, modulated at the transmitter, and guided to the receiver through integrated waveguides, as represented in Figure 20.6. Much of the needed components can

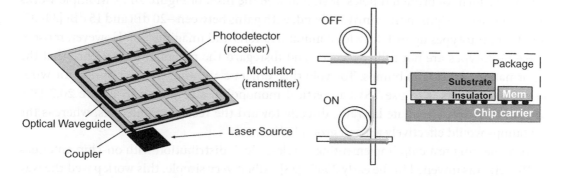

FIGURE 20.6   Sketch of a photonic NoC with snake-shaped waveguides, detail of ring resonators in ON/OFF positions, and cross-section of the chip.

be implemented with microring resonators, also outlined in Figure 20.6, which divert light of a certain (tunable) wavelength when a voltage is applied over them or let light pass through otherwise. These and other designs have leveraged evolving nanoscale technologies (including graphene) to develop a complete suite of integrated building blocks for nanophotonic NoCs [24–26].

The reader will observe that the principles of such interconnects are similar to those of transmission lines but taken to the extreme. The speed-of-light propagation and multiplexing advantages are kept, whereas the bandwidth density is much higher, owing to the nanometer wavelengths used in the communication. In addition, field confinement and high supported bandwidths also lead to a much higher energy efficiency as compared with other technologies. As a result of these outstanding advantages, the NoC literature has been recently flooded by a great variety of designs, from simple buses to novel and tailor-made topologies, including both arbitration-based and contention-free architectures [27,26]. These theoretical proposals have recently crystallized into full prototypes [28] and platforms [24] demonstrating aggregated Tbps speeds and efficiencies well below 1 pJ/bit. These are expected to continue improving toward a theoretical lower bound around 1 fJ/bit.

The main limitation of the nanophotonic paradigm is that existing laser sources are bulky and technically complex, which force designers to place them off the chip. The lack of effective on-chip lasers imposes the use of static power allocation techniques, which reduces the flexibility and energy efficiency of the network. Even though most nanophotonic NoCs are global in scope, full broadcast capabilities are hard to scale, as each node extracts a fraction of the light and adds significant losses [29]. Research on laser management adaptable at runtime [30] or adaptive filtering [31] may alleviate these issues; however, the idea of integrated light sources cannot be discarded yet [32].

20.2.4 Wireless Interconnects

The constant downscaling of CMOS-RF technologies has allowed to increase the frequency of RF systems to the point that passives like the antenna are small enough to be integrated on a chip. For example, a quarter-wave antenna at 60 GHz has a length of 0.4 mm in silicon, two orders of magnitude smaller than the chip itself.

On-chip antennas are generally implemented using the metal layers of the chip stack, taking the form of printed dipoles or similar (see the inset of Figure 20.7). Multiple works have reported designs in the mmWave band, with gains between –20 dBi and 15 dBi [33–35] and first prototypes up to 1 THz for communications and imaging [36]. However, most of these prototypes are not fully packaged and disregard the potential interference that the information object (IO) bumps. To avoid this, others claim that the best option for wireless interconnects is to use TSVs as vertical monopoles, also shown in Figure 20.7 [37]. The antenna would radiate laterally, directly toward the receiving antennas, whereas the IO bumps would effectively act as a ground plane.

A pioneering test chip implementing wireless clock distribution with on-chip antennas at 15 GHz was unveiled in the early 2000s [38]. Albeit very simple, this work paved the way for subsequent mmWave wireless interconnect schemes that, as shown in Figure 20.8, use on-chip antennas that communicate through the common chip substrate [39]. In essence,

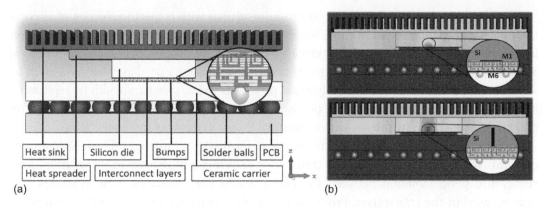

FIGURE 20.7 (a) Schematic of an intellectual property-chip package (left) and field distribution with planar antennas and (b) vertical monopoles (right).

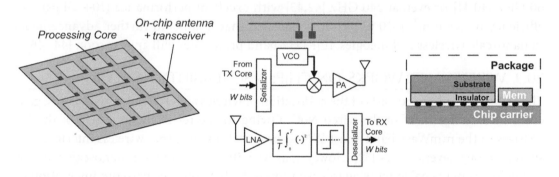

FIGURE 20.8 Sketch of a wireless NoC with a printed dipole, an on-off keying transmitter, and an energy detector receiver.

the transmitter modulates information coming from a core and radiates the resulting RF signals. These waves propagate as either (i) surface waves along the interface between the substrate and the insulator, or (ii) space waves through the substrate that reflect on the chip package. At reception, the antennas pick up the signals, demodulate them, and deliver them to the cores.

The advantages of wireless on-chip communication subtly differ from those of RF interconnects, because energy is radiated instead of being guided through transmission lines. The latency advantage with respect to radio controlled (RC) signaling still applies, as wave propagation occurs at the speed of light. Wireless on-chip communication also shows improved simplicity, flexibility, and reconfigurability potential, as compared with the rest of interconnect technologies, since no path infrastructure is required between nodes. Thus, a WNoC can modify the logical topology or other transmission parameters without the need of any physical modification. Last but not least, WNoC offers native and scalable broadcast capabilities, as information may potentially reach any core in several nanoseconds, regardless of its location. As discussed previously, such feature is not achievable with transmission lines, owing to reflections, or with nanophotonics, owing to laser power issues.

Wireless interconnects also show two main downturns with respect to the rest of alternatives: energy efficiency and bandwidth. Since energy is radiated rather than being guided, wireless communication is intrinsically less efficient. The lack of an isolated medium for propagation also affects bandwidth, as all antennas tuned to the same frequency need to either *compete* or be scheduled to access to the medium. There are no means to increase bandwidth by replicating a structure such as a transmission line or a waveguide. Adding frequency channels does increment the network bandwidth, yet with a nonscalable cost in terms of implementation.

Technically, wireless interconnects also pose significant challenges. Designing an on-chip mmWave antenna close to the transceiver requires a high level of integration, area reuse, and good isolation between the antenna and the inductors or waveguides that may be used in the transceiver. From the transceiver perspective, it is worth mentioning that aspects such as the device parasitics, low supply voltage, device dimension limitations, and the complex metal stack (which adds parasitic intrinsic inductance) need to be taken into consideration. However, despite these challenges, several designs exist at 60 GHz [40,41] or even at 240 GHz [42,43] with excellent performance (10+ Gbps) and efficiency (0.1–1 mm^2, 1–10 pJ/bit). It is expected that, thanks to further advancements in nanoscale wireless technologies, full THz-band prototypes will appear soon [44–46].

20.3 WIRELESS NETWORKS-ON-CHIP: A SCALABILITY ANALYSIS

The WNoC paradigm consists of the co-location of wireless interconnects with cores, generally as a complement of the wired NoC. Owing to the relatively large size of the RF passives in the mmWave band, most WNoC proposals assume that wireless interfaces are shared among several cores [39]. However, as we will see, advances in nanoscale wireless technologies may enable more aggressive proposals that consider per-core integration of wireless interconnects [47].

Figure 20.9 depicts the WNoC structure considered throughout this section. Antennas and transceivers are integrated on a per-core basis and connected to the memory hierarchy through

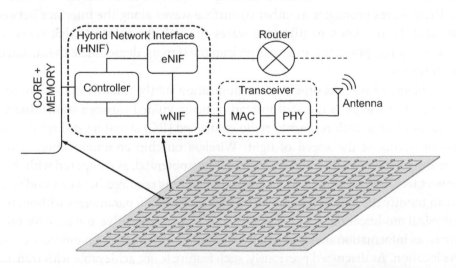

FIGURE 20.9 Schematic representation of the WNoC design considered throughout this chapter.

TABLE 20.2   Wireless Interconnects Requirements in the Manycore Era

Metric	Value	Reasoning
Range	0.1–10 cm	Chips have a maximum lateral size of a few tens of millimeters
Node density	10–1000 cm^{-2}	Thousand-core chips are the target
Throughput	10–100 Gbps	Wired links transmit tens of bits each clock cycle (e.g., 1 GHz)
Frequency	0.1–1 THz	High bandwidth needed for node density and throughput reasons
Latency	1–10 ns	Impact diminishes for higher latencies
Bit error rate	10^{-15}	This is the typical BER of an on-chip wire
Energy	1–10 pJ/bit	Commensurate energy consumed in wired hops (link + router)

a network interface (NIF), which also connects to the wired NoC. The transceiver contains two modules: the physical layer translates baseband data into RF signals and *vice versa*, whereas the media access control (MAC) layer ensures that all nodes can access the medium reliably.

The WNoC structure outlined in Figure 20.9 shows a great promise as the backbone of future manycore architectures [47]. However, its implementation requires overcoming significant challenges at the different layers of design toward meeting the stringent area, power, and bitrate requirements of the nanoscale manycore scenario. In short, wireless interconnects need to provide performance, reliability, and cost figures comparable to those of the traditional wireline counterparts. Table 20.2 summarizes these requirements by the order of magnitude in a typical manycore setting [9].

This section evaluates the feasibility of the WNoC paradigm. On the foundations of existing transceivers and taking prospective designs into consideration, we examine whether future WNoCs will be capable of meeting the requirements that we just summarized. We go layer by layer following the structure outlined in Figure 20.10, which draws an analogy between the classical open systems interconnection (OSI) model and a simplified stack for WNoC that, in turn, perfectly matches the proposed architecture.

The feasibility study is based on a scalability analysis, also summarized in Figure 20.10. In each layer, we use different methodologies to evaluate how the different performance indicators scale with certain design parameters. Section 20.3.1 models the chip package (Figure 20.7) and, by means of full-wave simulations with CST Microwave Studio [48],

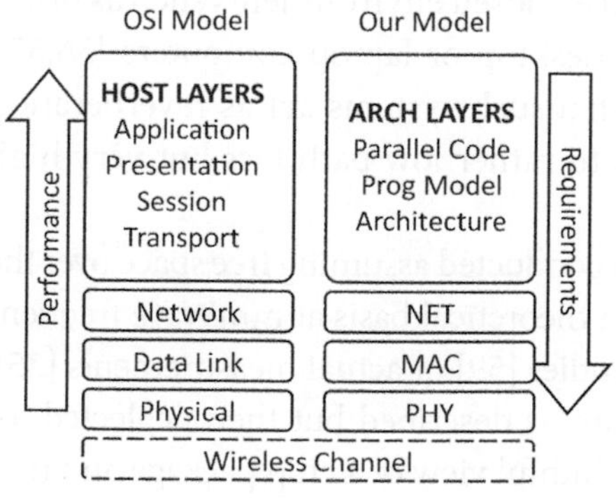

Layer	Design Aspects	Performance	Methodology
ARCH	Error control, appl. mapping, interfacing	Execution speed, Overall power	Architectural simulation
NET	Topology, number of nodes, routing	Network latency, Throughput	Cycle-accurate NoC simulation
MAC	MAC protocol, node density, data rate	Link latency, throughput	Models, simulation
PHY	Frequency, coding, modulation, RF power	Data rate, error rate, area, power	State-of-the-art extrapolation
CHAN	Frequency, package, dimensions, materials	Path loss, delay spread, capacity	Full-wave EM simulation

FIGURE 20.10   Layered representation of the different design levels of a WNoC.

obtains the path loss as a function of the package dimensions, distance, and frequency. Section 20.3.2 is devoted to the physical layer of design, and by extrapolations of the state-of-the-art transceiver designs, we obtain bitrate and cost as functions of frequency. Section 20.3.3 models wireless interconnects at the link level and, through simulations with a modified version of PhoenixSim [49,50], obtains latency and throughput as functions of the number of nodes, raw bitrate, and amount of broadcast traffic. Finally, Section 20.3.4 integrates wireless interconnects within the full network architecture, obtaining for a wide variety of traffic patterns through cycle-accurate simulations with PhoenixSim [51]. We do not delve into the architectural effects of WNoC, since it is out of the scope of this book. However, we refer the interested reader to [47,52,53] for architectural proposals involving wireless interconnects, although not necessarily enabled by nanoscale technologies.

20.3.1 Wireless Channel

A channel model that takes into consideration the peculiarities of the chip-scale scenario is fundamental to evaluate the available bandwidth and to properly allocate power. The enclosed nature of the chip package leads to significant multipath effects. In addition, the physical landscape of a multiprocessor involves multiple dielectric/metallization layers and, in some cases, TSVs that may challenge propagation [54]. Fortunately, one of the uniquenesses of the scenario is that all these elements are fixed and known beforehand, and thus, the channel model will be virtually time-invariant and quasi-deterministic. This may allow chip makers not only to adapt the package to improve propagation but also to perform channel-dependent processes such as power allocation with unprecedented precision and effectiveness.

This section first reviews the existing works on wireless chip-scale channel characterization, identifying main trends and missing perspectives. Then, we overview the fundamentals of chip-scale channel modeling and outline the structural, distance, and frequency scaling trends of the path loss within a chip package.

20.3.1.1 Related Works in Channel Characterization

Thus far, few works have explored the chip-scale wireless channel down to the nanoscale. The theory is well laid out [54], and a wide variety of works exists in larger environments. Propagation has been investigated in rather small and enclosed environments such as printers [55] and large printed circuit boards (PCBs) in desktop or laptop computers [56,57]. The explorations, up to 300 GHz, have confirmed that such systems act as reverberation chambers, owing to the metallic enclosure, leading to rather low path loss but very high delay spreads.

Down to the chip level, most studies have often been conducted assuming free space over the insulator layer without a package. Yan et al. provided a theoretical basis at mmWave frequencies [58], whereas others provided simulation-based studies [59] or actual measurements [35]. In the THz band analysis of [60], the package structure is described but then neglected for simplicity. All these works, however, do not provide a faithful view of a chip package and may be misguiding when designing a nanoscale WNoC. Next, we discuss and evaluate the scaling trends related to wireless propagation with proper modeling of the chip package.

20.3.1.2 Scaling Trends Within a Chip Package

The existence of a practical wireless channel, even in the presence of a chip package, was experimentally validated at 15 GHz in [61]. Matolak et al. suggested that a chip with metallic walls would act as a micro-reverberation chamber also at higher frequencies, but it does not describe the package accurately [54].

Here, we evaluate propagation within a realistic chip package at 60 GHz. We model the widespread flip chip, where the die is turned upside down and connected to the PCB through an array of micro-bumps, as shown in Figure 20.7. On top, the heat sink and heat spreader (aluminum nitride, AIN, $\varepsilon = 8.6$, $\tan\delta = 0.0003$) dissipate the heat out of the silicon chip. Bulk silicon with low resistivity ($\varepsilon = 11.9$, $\rho = 10 \cdot$ cm) serves as the foundation of the transistors. The interconnect layers, which occupy the bottom of the die, as shown in the inset of Figure 20.7, are generally made of copper and surrounded by an insulator (silicon dioxide, SiO_2, $\varepsilon = 3.9$, $\tan\delta = 0.03$). The lateral dimensions of the chip are 20×20 mm², whereas the package is 30×30 mm² and filled with air [37].

20.3.1.2.1 Scaling with Package Dimensions   The low resistivity bulk silicon is convenient for the operation of transistors but not for EM propagation. Manufacturers are capable of thinning this layer for 3D stacking and integration purposes, but we show that it could also improve the wireless channel. To this end, the worst-case path loss S_{min} between ideal antennas at arbitrary positions is obtained for different silicon thicknesses. As we can see in Figure 20.11a, the path loss difference between the 0.7-mm case (default thickness) and the 0.1-mm case (3D stacking thickness) is over 40 dB, as we significantly shorten propagation through the lossy silicon.

The materials used as heat spreaders have good thermal properties and, coincidentally, low electrical losses [37]. To study their potential effect on EM propagation, we repeat the above analysis by changing the heat spreader thicknesses. As observed in Figure 20.11b, thickening the heat spreader reduces losses up to 33 dB with respect to not having any heat spreader. Therefore, it is a parameter to consider in possible package engineering efforts to minimize the path loss.

20.3.1.2.2 Scaling with Distance   We performed a path loss analysis at 60 GHz to see how attenuation scales with distance for three different cases: standard chip dimensions (Si: 0.7 mm, AIN: 0.2 mm); optimal dimensions, as obtained previously (Si: 0.1 mm,

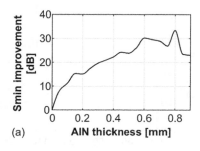
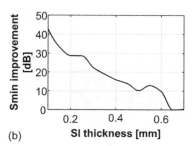

(a)　AIN thickness [mm]　(b)　SI thickness [mm]

FIGURE 20.11　Scaling of the worst-case path loss S_{min} over different layer thicknesses. (a) Scaling with Si and (b) Scaling with AIN.

AIN: 0.7 mm); and a suboptimal design point between those extremes. To this end, we assumed a homogeneous distribution of 4×4 antennas within the chip and extracted the path loss exponent with slope fitting methods. The results, plotted in Figure 20.12a, show that layered optimization reduces the path loss exponent. For the standard case, the path loss exponent is 1.78, slightly lower than the free space path loss, thanks to having a confined environment. The suboptimal case, with an exponent higher than 2, demonstrates that most propagation will occur in a lossy medium if the package is not carefully designed. Finally, in the optimal case, we are able to cut the exponent down to 0.75, thereby showing a strong waveguiding effect in propagation. This not only further emphasizes the importance of a correct modeling of the chip package but also shows that it can obtain similar results to those of the surface wave approach from [62], which proposed to carry waves through surfaces deliberately designed for this purpose. In fact, we speculate that the heat spreader layer can act as such layer and thereby avoid most of the losses at silicon.

20.3.1.2.3 Scaling with Frequency

Increasing the frequency of operation leads to smaller antennas and improves bandwidth, which are two critical aspects in upper layers of design. Therefore, it is of great interest to study the frequency scaling trends of the on-chip channel. Figure 20.12b shows how S_{min} scales over frequency. This sweep was performed with a silicon of 0.2 mm and an AIN of 0.8 mm. We can observe that, overall, the loss between links increases with frequency, probably due to two reasons: the antennas have smaller apertures, and the propagation losses at the dielectrics are larger. Nevertheless, this effect is compensated in part by the enclosed nature of the on-chip scenario, mitigating the effect of frequency scaling. We may also find that each particular frequency has its optimal package and that the scaling would then differ, but that is left for future work.

20.3.2 Physical Layer

The physical (PHY) layer of a WNoC defines how messages coming from the processor are serialized, modulated, coded, and delivered to the on-chip antenna in transmission and decoded, demodulated, and deserialized in reception. Given that area and power are two precious resources in the manycore chip scenario, the development of appropriate methods at the physical layer is essential to guarantee the viability of the WNoC [44,45]. These physical constraints suggest the use of simple solutions, leading to small and low-power

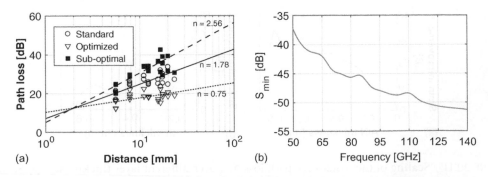

FIGURE 20.12   Scaling of the path loss over (a) distance and (b) frequency in the mmWave band.

transceivers. However, the scenario casts strong bandwidth and reliability requirements that point toward the opposite direction. The challenge is to find a graceful compromise between both extremes while exploiting the unique optimization opportunities given by the static and controlled on-chip landscape.

This section first provides a snapshot of the state of the art at the PHY layer of nanoscale wireless NoC by comparing the bitrate, area, and power of existing implementations with those of wired counterparts. We then review the existing implementations at the mmWave and THz bands for applications with requirements and use the data to extrapolate future scaling trends.

20.3.2.1 Comparison of the State-of-the-Art Designs

As summarized in Section 20.2.4, few antenna and transceiver prototypes have been reported for WNoC in the mmWave band [33,34,36,40,41]. Even though they are among the first designs specifically conceived for the WNoC paradigm, we use them to provide an approximation of the performance and cost of future WNoCs.

We part from the 65-nm CMOS designs by Yu et al. at 60 GHz [41]. For simplicity, the first implementation uses On-Off Keying (OOK) and achieves 16 Gbps with the required bit error rate (BER) of 10^{-15}, while taking 31.2 mW of power and 0.23 mm^2 of silicon area. The same authors increase the data rate by using three frequency channels for a total of 48 Gbps, while consuming 97.5 mW and 0.73 mm^2. They also explore the possibility of using quadrature phase-shift keying (QPSK), which doubles the spectral efficiency with respect to OOK for a total of 32 Gbps with 16-GHz bandwidth. Owing to the need of complex components, the power and area escalate up to 96 mW and 0.4 mm^2, respectively, which may be unaffordable as per the scenario requirements. Note that, in all cases, the authors assume a path loss of around 30 dB, consistent with the numbers shown in Section 20.3.1.

Table 20.3 compares the area and power of the transceivers above with those of different NoC implementations reported in the literature. We focus on the transceiver, since the antenna is a passive element that does not consume power, and, since vertical monopoles

TABLE 20.3 Per-Tile Area and Power Comparison

References	Cores	Topology	Tech.	Volt	Frequency	Rate	Area	Power
[41]	N	Wireless	65 nm (22 nm)	1 V	1 GHz	16 b (32 b)	0.25 (0.1) mm^2	31.2 mW
						32 b (64 b)	0.4 (0.16) mm^2	96 mW
						48 b (96 b)	0.73 (0.3) mm^2	97.5 mW
[5]	80	Mesh	65 nm	0.7 V	1.7 GHz	39 b	0.34 mm^2	98 mW
[64]	36	Mesh	45 nm	1.1 V	1 GHz	137 b	0.36 mm^2	139 mW
63]	16	Mesh	45 nm	1.1 V	1 GHz	64 b	0.32 mm^2	27 mW
[65]	128	FBFly	32 nm	0.9 V	2 GHz	144 b	0.18 mm^2	78 mW
31]	64	Optics	22 nm	1 V	2.5 GHz	320 b	–	187.5 mW
[21]	64	TLs	22 nm	1 V	1 GHz	16 b	0.48 mm^2	7.8 mW
Atom Silvermont (22 nm)							2.5 mm^2	1 W
Xeon Haswell (22 nm)							21.1 mm^2	5 W

Source: See Abadal, S. et al., *IEEE T. Parall. Distr. J.*, 27, 3631–3645, 2016 for methodological details.

may be used, we assume that they do not occupy any chip area. The table considers the figures reported in [41] as well as a scaled version (down to 22-nm CMOS) to account for near-future improvements. The NoC implementations correspond to an INTEL mesh prototype [5], Park's low-power and low-delay mesh [63], the mesh with ordered broadcast capabilities from [64], and a flattened butterfly for low-diameter topology. We also include data from two recent nanophotonics [31] and transmission line bus [21] implementations. Finally, to put these numbers in context, we complete Table 20.3 with two popular 22-nm cores: one for high performance and one for energy efficiency.

Overall, it is shown that a 22-nm transceiver would have an area and power consumption commensurate to that of current and future NoC designs, while representing between 1% and 10% of the cost of a computing core. Although it is true that wireline NoCs generally incur lower cost, they suffer from latency problems, as we will see in next sections. These results also confirm that nanophotonics and transmission line options are more efficient than wireless, as signals are guided instead of radiated, and provide higher bandwidth through replication. However, they also take a significant portion of chip area and have their own scalability problems [47,29].

20.3.2.2 Extrapolation of Scaling Trends

As CMOS technology evolves and advanced devices such as fin-shaped field effect transistors (FinFETs) and III-V on silicon are implemented, it becomes possible to raise the carrier frequency further into the mmWave region and up to THz frequencies, thereby significantly increasing the available signal bandwidth and decreasing the overall area and energy per bit of data. Here, we inspect the latest works in antenna and transceiver design to either confirm or deny such potential scaling trend, in the pathway to assessing the practicality of WNoC.

For antenna designs beyond 60 GHz, several implementations are found in the literature. The work by Hou et al. presents a Vivaldi antenna with a peak gain of 5.5 dBi and 30% bandwidth around 150 GHz [66]. Approaching the THz band, microstrip leaky-wave antennas achieving 4.9 dBi and more than 26% bandwidth around 245 GHz or skirt-shaped designs with peak gain of 7.1 dBi and a huge 65% bandwidth at 1 THz have been reported [36].

Transceiver designs follow a similar trend. Several proposals have pushed the frequency up to the 135–140 GHz band, yet with modest performance, owing to the early stages of development. A design close to the WNoC requirements is presented in [67] in 40-nm CMOS, delivering 10 Gbps at around 10 pJ/bit, with a BER of 10^{-11} (range of 10 cm). Following these advancements, recent years have seen a surge in THz circuits for wireless communications and imaging [68,69,46]. First complete integrated transceiver designs have been appearing, with two good examples being the silicon germanium (SiGe) bipolar complementary metal oxide semiconductor (BiCMOS) implementation at 190 GHz, promising 40 Gbps and 3.9 pJ/bit at 2 cm range [70], and the 240-GHz design, achieving 16 Gb/s at 30 pJ/bit for a few centimeters [42,43].

These are just a few of the many transceiver works appearing in the recent years. To assess whether the theoretical scaling trends can be extrapolated to the WNoC paradigm,

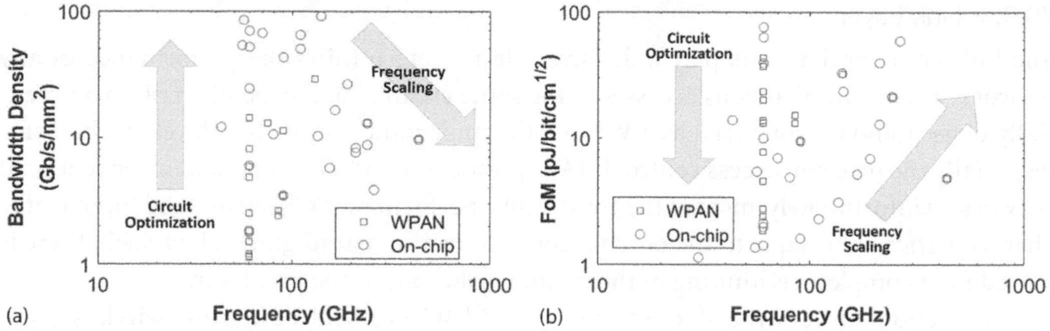

FIGURE 20.13   Area and energy-efficiency scaling of short-range wireless transceivers over frequency (2009–2018). (a) Bandwidth density and (b) Energy efficiency.

we analyzed the speed, area, and power of a wide variety of designs. These cover not only the nanoscale wireless communication field but also other applications with relatively short range and very high data rate as a primary requirement. The reader will find the complete list of references in [71].

We start by inspecting the area trends. Figure 20.13a plots the bandwidth density $BD = \frac{R}{A}$, where R is the data rate and A is the occupied area, for all the analyzed transceivers as a function of their operation frequency. A first observation is that the amount of 60-GHz prototypes is much higher than for the rest of bands. This is due to the emergence of the new 802.11 and 802.15 standards pointing to this unlicensed band. Prototypes at larger frequencies start to appear, especially for the on-chip environment, where the range is much smaller and size restrictions are more stringent. Most of the designs for chip-scale applications are below 1 mm². At higher frequencies, one would expect that the area would be lower in general, as passives can be downscaled, but it is observed that the area is maintained or even increased. We argue that these trends relate to the maturity of technology: For a given frequency working point, optimizations allow to increase the bandwidth density over the years. When scaling those designs at higher frequencies, however, first prototypes experience a reduction in the overall performance, until they adjust to new challenges and effects. Therefore, we expect that bandwidth density will be maintained or increased with frequency, as technologies keep advancing.

Similar results are obtained for the energy-efficiency trends. In this case, we study a figure of merit that accounts for the power P, the data rate R, and the transmission range d as $F = \frac{P}{R\sqrt{d}}$ [72]. Figure 20.13b plots the figure of merit as a function of the operation frequency, confirming the tendencies outlined previously. As the technology has matured and circuit optimizations have been introduced, much more efficient transceivers have been developed. This explains the reduction in figure of merit (lower is better) that occurs for both WPAN and on-chip options. From that point upwards in frequency, we observe that the efficiency worsens again due to the reduced maturity of designs above 100 GHz. We believe that subsequent optimizations will make this figure go down again, reaching values below 1 pJ/bit/cm½ even beyond 200 GHz.

20.3.3 Link Layer

The link layer mechanisms play a decisive role in determining the performance of any network, as two simultaneous accesses to the same channel will generally fail. In a potentially dense nanonetwork such as a WNoC, the importance of the link-level mechanisms (especially the medium access control [MAC] protocol) is even higher than in conventional networks. Unfortunately, most of the constraints, performance objectives, and input traffic characteristics are unique to the on-chip communication paradigm, and, as such, there is a need for a complete rethinking of the existing solutions at the link layer.

This section surveys the state of the art in MAC protocol design for wireless chip-scale communications enabled by nanoscale technologies. We identify two representative extremes and then compare their latency and throughput with that of conventional wired options at different scales.

20.3.3.1 Related Work

Figure 20.14 shows a rough classification of MAC mechanisms, which serves to explain the main trends in chip-scale wireless communication design. First designs heavily relied on *channelization*, initially in the frequency domain [73], and later enhanced with basic time multiplexing or even code multiplexing schemes to avoid collisions [74]. These techniques are capable of delivering very high throughput but do not work well under variable workloads, since bandwidth is generally allocated statically. Channelization techniques also have important scalability limitations, as increasing the amount of time, frequency, or code-multiplexed channels implies a notable increase in the hardware complexity.

Subsequent works proposed the use of *coordinated access* schemes such as a centralized scheduler, which is alleviates the rigidity problem of multiplexing but may still become a bottleneck owing to its centralized nature. Distributed alternatives, such as token passing arbitration [39] and code-multiplexed arbitration [75], are also flexible and more scalable than centralized scheduling. Still, the overheads associated to token circulation and code synchronization may hinder their use in massive nanoscale wireless on-chip communications, as we will see later.

Random access mechanisms have received lesser attention than the aforementioned collision-free approaches owing to their poor performance when high loads cause recurrent collisions. To address this issue, Dai et al. proposed a slotted carrier sensing protocol, with

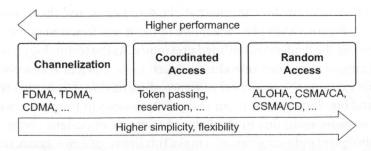

FIGURE 20.14   Coarse classification of multiple access mechanisms.

theoretically optimal persistence calculated *a priori* [76]. Others have proposed an adaptive scheme that switches between a carrier sensing protocol and a token passing scheme, depending on the level of contention [77]. Finally, it has been discussed in several works that random access protocols can be enhanced further, thanks to collision detection [9] and *a priori* knowledge of traffic [6,7], thereby representing an interesting candidate in the chip-scale scenario.

20.3.3.2 MAC-Protocol Scalability

The main aim of this section is to evaluate the practicality of WNoC at the link level. To this end, we analyze the scalability of the representative MAC protocols and benchmark their performance against that of wireline NoCs in the presence of broadcast traffic. We evaluate how the link latency and throughput of different options scale as a function of the number of nodes N and the capacity of the wireless channel C. As summarized earlier, we use a cycle-accurate NoC simulator to perform the evaluation.

To cover a large fraction of the solution space, we consider three wireless alternatives. We discard multiplexing, as it does not scale, and instead choose token passing (W-TOKEN), because it has been used consistently in many WNoC works [39], as well as the broadband radio service–media access control (BRS-MAC) protocol (W-BRS), because it is well adapted to the on-chip scenario [78]. Finally, to obtain a performance limit, we also study an ideal centralized scheduler with single-cycle delay (W-CBUF). As baselines, we use the typical mesh (MESH) and flattened butterfly (FBFLY) topologies [79]. For a fair comparison, we assume aggressive router designs whose delays are adjusted according to the number of ports. We refer the readers to [49,78] for a deeper reasoning of the different alternatives.

Scaling the number of nodes: The system size has a significant effect on performance. The performance drops, as outlined in Section 20.1, for conventional NoCs and depends on the MAC protocol of choice for WNoCs.

The low-load latency as a function of the system size is plotted in Figure 20.15a, illustrating three different scaling trends. First, the latency of W-TOKEN scales as the average hop distance of such topology, $O(N/2)$, since the token is passed through a virtual ring. The latency of the rest of wireless schemes remains flat owing to their on-demand nature, as, at low loads, a node is able to transmit immediately. In the wired NoCs, the latency scales proportionally to the average hop distance of the topology: $O(\sqrt{N})$ in MESH and almost constant in FBFLY.

Figure 20.15b illustrates the throughput scaling trend. MESH and FBFLY achieve a high and constant throughput, because the bisection bandwidth increases when scaling the topology, thereby compensating for the increase in the number of destinations per message. Given by the inherent broadcast nature of wireless NoCs and despite having a much lower bisection bandwidth, W-CBUF and W-BRS achieve a rather flat scaling. In the latter case, the absolute value is lower because of the collisions appearing when load increases. Finally, W-TOKEN is clearly dominated by the token passing delay and cannot provide any throughput, with acceptable delay beyond a few hundreds of cores.

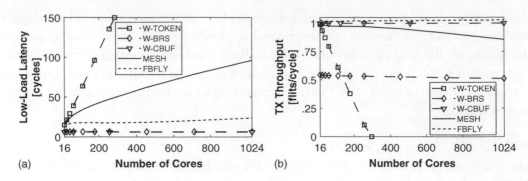

FIGURE 20.15   Performance as a function of the number of nodes N for $C = 1$ and broadcast traffic. (a) Latency scaling and (b) Throughput scaling.

Scaling the channel capacity: For the conditions evaluated previously, wireless strategies are capable of consistently achieving very low latencies with moderate-to-high throughput. However, we have assumed a bandwidth of one flit per cycle thus far, which is over 100 Gbps at the link level. Clearly, these figures are far will not be feasible in the near future, as discussed in Section 20.3.2. Therefore, it is important to understand the dependence on performance and channel capacity, in order to guide the design of future WNoCs.

Obviously, scaling the channel capacity C impacts the latency, as shown in Figure 20.16a. The propagation time, which also contributes to the communication latency, is dependent on the chip size and therefore assumed constant. The arbitration overhead is dependent on the arbitration scheme and, in the absence of load, remains constant (zero, two, and $N/2$ cycles for W-BRS, W-CBUF, and W-TOKEN, respectively). For all this, the low-load latency approaches a fixed lower bound as we increase the channel capacity. However, even for small capacities (0.06 flits/cycle corresponds to 8 Gbps in our simulations), wireless links can still compete with most wired options.

Varying the channel capacity C also has a direct effect on the throughput, as shown in Figure 20.16b. Basically, the throughput increases almost linearly with the channel

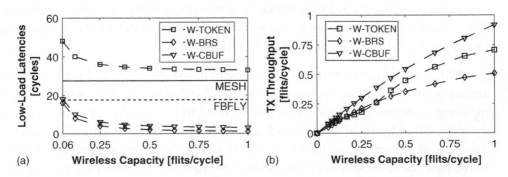

FIGURE 20.16   Performance as a function of the channel capacity C for $N = 64$ and broadcast traffic. (a) Latency scaling and (b) Throughput scaling.

capacity. At high speeds, the propagation time becomes significant and imposes an increasing bandwidth overhead, thus limiting the maximum achievable rate. To compete with wired options, the wireless capacity should be capable of providing the same approximate speed in terms of flits per clock cycle.

20.3.4 Network Layer

The network layer mainly deals with decisions concerning the path from source to destination, affecting directly four basic aspects: network topology, load balancing, quality of service (QoS), and deadlock freedom. Load balancing relates to the distribution of the injected packets through the network to maximize the overall throughput in a process that, in a NoC, generally takes place at the routers. QoS refers to a set of rules enforced so that certain flows meet a set of given performance requirements, thus avoiding stalling execution. Finally, deadlock avoidance mainly aims to guarantee forward progress of packets within the network.

This section reviews network-level decisions in nanoscale-enabled wireless communications on a chip. We first outline the existing approaches and then briefly discuss the consequences of downscaling the antennas. We evaluate a network-level architecture that separates between the wired and wireless planes, emphasizing its simplicity, flexibility, and high performance, as the network is scaled.

20.3.4.1 Related Work

Thus far, WNoC has been employed as complement of a traditional wired NoC, since the size of current on-chip antennas, that is, in the millimeter range, does not allow to include one antenna per core. In this hybrid approach, wireless links connect distant cores in one hop, thereby decreasing the average network latency. We distinguish between two main groups, depending on the positioning of the wireless links. Regular positioning, illustrated in Figure 20.17a, has been proposed extensively, owing to its apparent simplicity and potential modularity [80,73,74]. Other works rely on positioning algorithms that follow the principles of small-world networks, as shown in Figure 20.17b, which minimize the hop

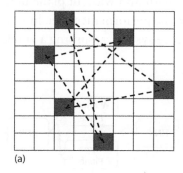
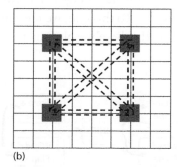

(a)   (b)

FIGURE 20.17   Rough classification of the topologies used in existing WNoC architectures. White and black squares represent tiles with routers and wireless interfaces, respectively. (a) Small world and (b) Regular/hierarchical.

count [39] or maximize the beneficial use of the wireless links [52]. Almost all works focus on reducing latency and oftentimes overlook the inherent broadcast capabilities of WNoC, with notable exceptions [62].

Since the WNoC is generally intertwined with a wired NoC, the network (NET) layer modifies the routing algorithms, taking into consideration the new wireless paths, and reevaluates load balancing, QoS, and deadlock conditions accordingly. For instance, Kim et al. modify a layered shortest path algorithm to adapt to the new wireless links [52]. The main issue is that deadlock freedom and load balancing become more difficult to ensure as the network grows, and we integrate more antennas, especially in the case of irregular topologies.

As we embrace nanoscale wireless technologies, a simpler alternative arises. The idea is to separate the network planes, so that the modification of the existing rules at the wired plane is not needed. The wireless plane rules can be implemented directly at the NIF that bridges the processor with the network. Load balancing and QoS boil down to a problem of traffic steering at the NIF, while deadlock freedom can be guaranteed as long as it is demonstrated at both planes separately.

20.3.4.2 Dual-Plane Network Architectures: Performance Evaluation

Following the intuition outlined previously, we propose the network architecture outlined in Figure 20.9. Our proposal has two independent network planes whose interaction is not performed at the routers but at the network interface. There, a hybrid controller coordinates the action of the wired and wireless interfaces. Specifically, the controller implements the mechanisms summarized in Figure 20.18 to adapt to traffic easily and naturally. These decisions help in keeping our architecture simple, flexible, and applicable over any wired topology, regardless of the number of antennas [51].

The most important functionality is plane selection, which directs packets to the appropriate network plane. The policy can be simple or complex, fixed or determined at runtime, but in any case, it needs to be aware of the characteristics of the wireless plane, to avoid congestion. If that happens, the controller may perform plane blocking and switching to alleviate the congestion effects. Here, we evaluate our architecture for two plane selection policies that leverage the advantages of WNoC with and without plane blocking, to demonstrate its usefulness. To this end, we measure the performance speedups of our architecture as compared with a baseline NoC in a 64-core system, assuming a wireless capacity of half flit per cycle.

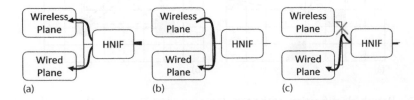

FIGURE 20.18 Controller methods for load balancing. (a) Plane selection, (b) Plane switching, and (c) Plane blocking.

Broadcast policy: In this policy, the controller forwards a message through the wireless plane with probability P if it is broadcast or dense multicast and through the wired plane otherwise. We test different probabilities in the interest of proactive load balancing; however, advanced architectures can rely on dynamic procedures to adapt to variable traffic. We stress the network with traffic containing an increasing ratio of broadcast messages.

Figure 20.19a and b shows the network throughput speedup for different values of P in two cases. Figure 20.19a considers that the network interface does not support plane blocking. In this case, our architecture performs remarkably better than the baseline as long as the pressure to the wireless plane is moderate, that is, for low admission probabilities or broadcast percentages. Here, the admission probability acts as a preemptive load balancing mechanism. Now, consider Figure 20.19b, which assumes that plane blocking is supported. We observe that the throughput is enhanced up to 40% in a wide range of scenarios and that the optimal P value depends on the broadcast percentage. Also, note that plane blocking does not solve congestion in all cases, suggesting that admission control is still useful. It is also worth noting that dual-plane structure increases the bandwidth at the ejection points of the network, improving throughput consistently for $\beta > 10\%$, approximate point where the baseline network becomes limited by the ejection links. In hybrid architectures that integrate a few wireless links within the wired topology, this improvement would be lost.

Figure 20.19c shows the network latency speedup as a function of the broadcast intensity of traffic for different values of P. Clearly, the latency is improved as much as the wireless plane is used to transport the broadcast flows (higher P). The improvement is maximized as the broadcast traffic becomes dominant, which is consistent with the link-level results obtained in Section 20.3.3.

To test the scalability of our approach, we performed simulations with different network sizes N and wireless capacities C. Figure 20.19d is a representation of the speedups accomplished by the broadcast policy (with plane blocking) throughout the hybrid design space with increasing levels of broadcast traffic. All lines tend to shift to the rightmost part as the broadcast percentage increases. The wireless capacity clearly affects performance, whereas increasing the network size maintains the throughput advantage and increases the latency speedup. Reducing the wireless data rate too much can turn the wireless plane into a bottleneck, unless advanced traffic steering techniques are applied. It is therefore concluded that nanoscale wireless communications are much needed for WNoC to be feasible.

Global policy: In the global policy, the controller forwards a message through the wireless plane if the Manhattan distance between the source and the destination is above a certain threshold H, which can be decided *a priori* or at runtime. For load balancing purposes, we will consider different fixed thresholds calculated as fractions of the diameter D of the wired plane. To stress the network, we consider unicast traffic with uniform random (UR), transpose (TR), bit complement (BC), and Renthop distributions [79,81].

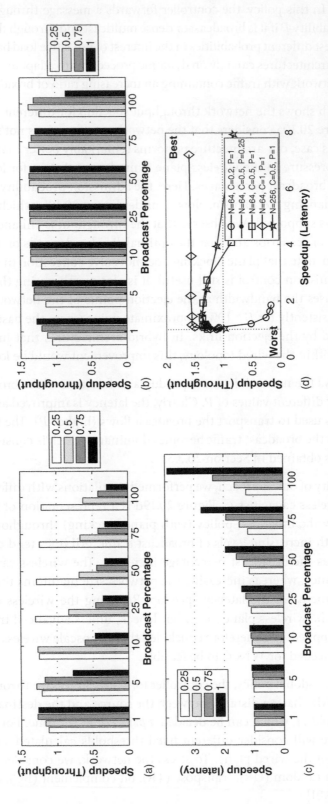

FIGURE 20.19 Speedups of the broadcast policy for different admission (a) probabilities, (b) percentages of broadcast traffic, (c) system sizes, and (d) wireless plane capacities.

Figure 20.20a and b shows the network throughput speedup for different values of distance threshold values and, again, with and without plane blocking. In this scenario, the outstanding difference in terms of aggregated bandwidth between the wired and wireless planes makes the use of plane blocking absolutely necessary to avoid premature congestion. When used, the throughput improves depending on the locality of the traffic profile. The case of *transpose* traffic is worth noting, as it causes a large load imbalance within NoCs with dimension order routing. Our architecture alleviates this effect and, as a result, performs better than the baseline.

Figure 20.20c represents the network latency speedup as a function of the traffic profile for different thresholds. Traffic profiles are ordered from mostly local to mostly global. We observe that the best latency improvements are achieved for global traffic patterns using a threshold that favors the use of the wireless plane.

We again evaluate the scalability. Figure 20.20d represents the hybrid design space considering the global policy and different traffic profiles. Overall, improvements are more modest than in the broadcast policy. Results are mainly positive for $N = 64$, whereas the throughput is significantly reduced for higher system sizes. Increasing the threshold can help alleviate this issue, but in any case, the latency results are promising for the data rates considered.

20.4 SUMMARY

This chapter provided a broad view of the potential role of wireless nanoscale communications in the manycore processor context. The field has been revisited from different perspectives, starting with the electromagnetic channel modeling and finishing with the implications at the NoC level. In summary, we saw that advancements in hardware (antennas and transceivers) and software (protocols) will ensure the viability of WNoC as a reconfigurable high-speed channel for long-range and broadcast communications, complementing a more rigid but efficient wired NoC. Given the current deficit in terms of broadcast communication support within the chip context, the WNoC vision can have enormous implications at the architecture level in manycore computers that, as we have seen, are bound to increase their need for versatility and heterogeneity.

Next years will be crucial in determining whether the WNoC paradigm is realized and becomes widespread. The continued development of nanoscale technologies is expected to boost the field by providing miniaturized antennas [82] as well as faster and more efficient analog circuits [83]. At the channel modeling level, it is expected that chip package engineering techniques may help in developing structure more favorable to electromagnetic propagation. In this respect, the employment of surface wave technologies can have a great effect, as they reduce the scaling factor of losses against distance [62]. At the physical layer, leveraging the already-existing high density of antennas to create opportunistic beamforming schemes would increase the number of nonoverlapping channels and the efficiency of wireless links. Further, modulations statically adapting to the channel may help to push the limits of performance. As we approach the link-level protocols, future developments can exploit the knowledge on the applications and architecture to develop predictive MAC models.

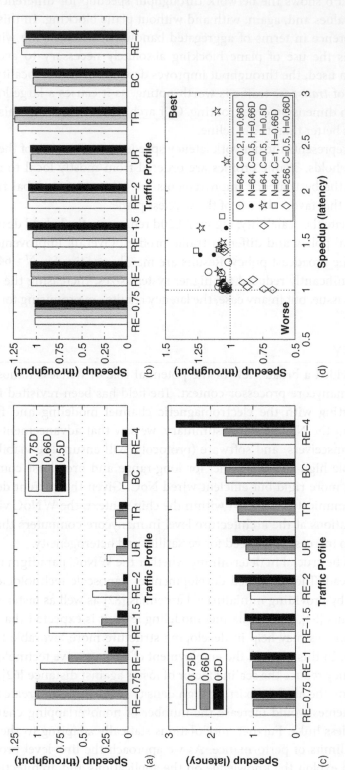

FIGURE 20.20 Speedups of the global policy for different distance (a) thresholds, (b) traffic profiles, (c) system sizes, and (d) wireless plane capacities.

We finally envisage that, through a tightly coupled integration of the wireless and wired network planes and by designing custom heterogeneous architectures that leverage the unique characteristics of wireless, future manycore processors will account for unprecedented levels of performance and efficiency. With this, wireless nanoscale communications can be key enablers of a new generation of computers that go beyond the death of Moore's law.

ACKNOWLEDGMENTS

This work has been partially funded by the Spanish Ministry of Economia y Competitividad under Grants PCIN-2015-012 and TEC2017-90034-C2-1-R (ALLIANCE project), as well as by the Catalan Institution for Research and Advanced Studies (ICREA) through the ICREA Academia Programme. The Grant TEC2017-90034-C2-1-R receives funding from Fondo Europeo de Desarrollo Regional (FEDER).

REFERENCES

1. Computer Simuation Technology (CST), CST Microwave Studio, 2019. [Online]. Available: http://www.cst.com.
2. S. Abadal, E. Alarcón, A. Cabellos-Aparicio, and J. Torrellas. WiSync: An architecture for fast synchronization through on-chip wireless communication. In *Proceedings of the ASPLOS'16*, ACM, Atlanta, GA, pages 3–17, 2016.
3. S. Abadal, A. Cabellos-aparicio, J. A. Lázaro, M. Nemirovsky, E. Alarcón, and J. Solé-Pareta. Area and laser power scalability analysis in photonic networks-on-chip. In *Proceedings of the ONDM'13*, IEEE, Brest, France, 2013.
4. S. Abadal, M. Iannazzo, M. Nemirovsky, A. Cabellos-Aparicio, H. Lee, and E. Alarcón. On the area and energy scalability of wireless network-on-Chip: A model-based benchmarked design space exploration. *IEEE/ACM Transactions on Networking*, 23(5):1501–1513, 2015.
5. S. Abadal, R. Martínez, E. Alarcón, and A. Cabellos-Aparicio. Multicast on-chip traffic analysis targeting manycore noc design. In *Proceedings of the PDP'15*, IEEE, Turku, Finland, pages 370–378, 2015.
6. S. Abadal, R. Martínez, J. Solé-Pareta, E. Alarcón, and A. Cabellos-Aparicio. Characterization and modeling of multicast communication in cache-coherent manycore processors. *Computers and Electrical Engineering (Elsevier)*, 51:168–183, 2016.
7. S. Abadal, A. Mestres, E. Alarcón, M. Nemirovsky, A. González, H. Lee, and A. Cabellos-Aparicio. Scalability of broadcast performance in wireless network-on-chip. *IEEE Transactions on Parallel and Distributed Systems*, 27(12):3631–3645, 2016.
8. S. Abadal, A. Mestres, J. Torrellas, E. Alarcón, and A. Cabellos-Aparicio. medium access control in wireless network-on-chip: A context analysis. *IEEE Communications Magazine*, 56(6):172–178, 2018.
9. S. Abadal, J. Torrellas, E. Alarcón, and A. Cabellos-Aparicio. OrthoNoC: A Broadcast-oriented dual-plane wireless network-on-chip Architecture. *IEEE Transactions on Parallel and Distributed Systems*, 29(3):628–641, 2018.
10. A. H. Atabaki, S. Moazeni, F. Pavanello, H. Gevorgyan, J. Notaros, L. Alloatti, M. T. Wade et al., Integrating photonics with silicon nanoelectronics for the next generation of systems on a chip. *Nature*, 556:349–354, 2018.
11. Q. Bao and K. P. Loh. Graphene photonics, plasmonics, and broadband optoelectronic devices. *ACS Nano*, 6(5):3677–3694, 2012.
12. C. Batten, A. Joshi, V. Stojanovic, and K. Asanovic. Designing chip-level nanophotonic interconnection networks. *IEEE Journal on Emerging and Selected Topics in Circuits and Systems*, 2(2):137–153, 2012.

13. L. Benini and G. De Micheli. Networks on chips: A new SoC paradigm. *IEEE Computer*, 35(1):70–78, 2002.

14. D. Bertozzi, G. Dimitrakopoulos, J. Flich, and S. Sonntag. The fast evolving landscape of on-chip communication. *Design Automation for Embedded Systems*, 19(1):59–76, 2015.

15. S. Borkar. Thousand core chips—A technology perspective. In *Proceedings of the DAC-44*, IEEE, San Diego, CA, pages 746–749, 2007.

16. J. Branch, X. Guo, L. Gao, A. Sugavanam, J. J. Lin, and K. K. O. Wireless communication in a flip-chip package using integrated antennas on silicon substrates. *IEEE Electron Device Letters*, 26(2):115–117, 2005.

17. A. Brière, E. Unlu, J. Denoulet, A. Pinna, B. Granado, F. Pêcheux, Y. Louët, and C. Moy. A Dynamically reconfigurable RF NoC for many-core. In *Proceedings of the GLSVLSI'15*, ACM, Pittsburgh, PA, pages 139–144, 2015.

18. A. Carpenter, J. Hu, J. Xu, M. Huang, and H. Wu. A case for globally shared-medium on-chip interconnect. In *Proceedings of the ISCA-38*, IEEE, San Jose, CA, pages 271–282, 2011.

19. J. Chan, G. Hendry, A. Biberman, K. Bergman, and L. P. Carloni. PhoenixSim: A simulator for physical-layer analysis of chip-scale photonic interconnection networks. In *Proceedings of DATE'10*, IEEE, Dresden, Germany, pages 691–696, 2010.

20. M. F. Chang, J. Cong, A. Kaplan, M. Naik, G. Reinman, E. Socher, and S. W. Tam. CMP network-on-chip overlaid with multi-band RF-interconnect. In *Proceedings of the HPCA-14*, IEEE, Salt Lake City, UT, pages 191–202, 2008.

21. C. Chen and A. Joshi. Runtime management of laser power in silicon-photonic multibus NoC architecture. *IEEE Journal of Selected Topics in Quantum Electronics*, 19(2), 2013.

22. K. C. Chen, S. Y. Lin, H. S. Hung, and A. Y. A. Wu. Topology-aware adaptive routing for non-stationary irregular mesh in throttled 3D NoC systems. *IEEE Transactions on Parallel and Distributed Systems*, 24(10):2109–2120, 2013.

23. W.-H. Chen, S. Joo, S. Sayilir, R. Willmot, T.-Y. Choi, D. Kim, J. Lu, D. Peroulis, and B. Jung. A 6-Gb/s wireless inter-chip data link using 43-GHz transceivers and bond-wire antennas. *IEEE Journal of Solid-State Circuits*, 44(10):2711–2721, 2009.

24. Y. Chen and C. Han. Channel modeling and analysis for wireless networks-on-chip communications in the millimeter wave and terahertz bands. In *Proceedings of the INFOCOM WKSHPS'18*, IEEE, Honolulu, HI, 2018.

25. D. Correas-Serrano and J. S. Gomez-Diaz. Graphene-based antennas for terahertz systems: A review. *FERMAT*, 2017. http://www.e-fermat.org/files/articles/Correas-Serrano-ART-2017-Vol20-Mar.-Apr.-005.pdf.

26. P. Dai, J. Chen, Y. Zhao, and Y.-H. Lai. A study of a wire-wireless hybrid NoC architecture with an energy-proportional multicast scheme for energy efficiency. *Computers and Electrical Engineering (Elsevier)*, 45:402–416, 2015.

27. B. Daya, C.-H. O. Chen, S. Subramanian, W.-C. Kwon, S. Park, T. Krishna, J. Holt, A. P. Chandrakasan, and L.-S. Peh. SCORPIO: A 36-core research chip demonstrating snoopy coherence on a scalable mesh NoC with in-network ordering. In *Proceedings of the ISCA-41*, IEEE, Minneapolis, MA, pages 25–36, 2014.

28. S. Deb, K. Chang, X. Yu, S. P. Sah, M. Cosic, P. P. Pande, B. Belzer, and D. Heo. Design of an energy efficient CMOS compatible NoC architecture with millimeter-wave wireless interconnects. *IEEE Transactions on Computers*, 62(12):2382–2396, 2013.

29. D. DiTomaso, A. Kodi, D. Matolak, S. Kaya, S. Laha, and W. Rayess. A-WiNoC: Adaptive wireless network-on-chip architecture for chip multiprocessors. *IEEE Transactions on Parallel and Distributed Systems*, 26(12):3289–3302, 2015.

30. K. Duraisamy, R. G. Kim, and P. P. Pande. Enhancing performance of wireless NoCs with distributed MAC protocols. In *Proceedings of the ISQED'15*, IEEE, Santa Clara, CA, pages 406–411, 2015.

31. B. S. Feero and P. P. Pande. Networks-on-chip in a three-dimensional environment: A performance evaluation. *IEEE Transactions on Computers*, 58(1):32–45, 2009.

32. B. A. Floyd, C.-M. Hung, and K. K. O. Intra-chip wireless interconnect for clock distribution implemented with integrated antennas, receivers, and transmitters. *IEEE Journal of Solid-State Circuits*, 37(5):543–552, 2002.

33. D. Fritsche, P. Stärke, C. Carta, and F. Ellinger. A Low-Power SiGe BiCMOS 190-GHz Transceiver chipset with demonstrated data rates up to 50 Gbit/s using on-chip antennas. *IEEE Transactions on Microwave Theory and Techniques*, 65(9):3312–3323, 2017.

34. S. H. Gade, S. Garg, and S. Deb. OFDM based high data rate, fading resilient transceiver for wireless networks-on-chip. In *Proceedings of the ISVLSI'17*, IEEE, Bochum, Germany, pages 483–488, 2017.

35. J. Gorisse, D. Morche, and J. Jantunen. Wireless transceivers for gigabit-per-second communications. In *Proceedings of the NEWCAS'12*, IEEE, Montreal, QC, pages 545–548, 2012.

36. B. Grot, J. Hestness, S. W. Keckler, and O. Mutlu. Kilo-NOC: A heterogeneous network-on-chip architecture for scalability and service guarantees. In *Proceedings of ISCA-38*, IEEE, San Jose, CA, pages 401–412, 2011.

37. R. Han and E. Afshari. A high-power broadband passive terahertz frequency doubler in CMOS. *IEEE Transactions on Microwave Theory and Techniques*, 61(3):1150–1160, 2013.

38. W. Heirman and J. Dambre. Rent's rule and parallel programs: Characterizing network traffic behavior. In *Proceedings of the SLIP'08*, ACM, Newcastle, UK, pages 87–94, 2008.

39. J. Hennessy and D. Patterson. *Computer Architecture: A quantitative approach*. Burlington, MA: Morgan Kaufmann, 2012.

40. D. Hou, Y.-z. Xiong, W. Hong, W. L. Goh, and J. Chen. Silicon-based On-chip antenna design for millimeter-wave/THz applications. In *Proceedings of the EDAPS'11*, IEEE, Hanzhou, China, pages 130–133, 2011.

41. W. Huang, K. Rajamani, M. Stan, and K. Skadron. Scaling with design constraints: Predicting the future of big chips. *IEEE Micro*, 31(4):16–29, 2011.

42. S. Kang, S. V. Thyagarajan, and A. M. Niknejad. A 240 GHz fully integrated wideband QPSK transmitter in 65 nm CMOS. *IEEE Journal of Solid-State Circuits*, 50(10):2256–2267, 2015.

43. A. Karkar, T. Mak, N. Dahir, R. Al-Dujaily, K.-F. Tong, and A. Yakovlev. Network-on-chip multicast architectures using hybrid wire and surface-wave interconnects. *IEEE Transactions on Emerging Topics in Computing*, 6(3):357–369, 2018.

44. B. Khamaisi, S. Jameson, and E. Socher. A 0.58–0.61 THz single on-chip antenna transceiver based on active X30 LO chain on 65 nm CMOS. In *Proceedings of the EuMIC'16*, IEEE, London, UK, pages 97–100, 2016.

45. J. Kim, J. Balfour, and W. J. Dally. Flattened butterfly topology for on-chip networks. In *Proceedings of the MICRO-40*, IEEE, Chicago, IL, pages 172–182, 2007.

46. J. Kim, K. Choi, and G. Loh. Exploiting new interconnect technologies in on-chip communication. *IEEE Journal on Emerging and Selected Topics in Circuits and Systems*, 2(2):124–136, 2012.

47. M. Kim, J. Davis, M. Oskin, and T. Austin. Polymorphic on-chip networks. In *Proceedings of the ISCA-35*, IEEE, Beijing, China, pages 101–112, 2008.

48. R. G. Kim, W. Choi, G. Liu, E. Mhandesi, P. P. Pande, D. Marculescu, and R. Marculescu. Wireless NoC for VFI-enabled multicore chip design: Performance evaluation and design trade-offs. *IEEE Transactions on Computers*, 65(4):1323–1336, 2015.

49. S. Kim and A. Zajic. Characterization of 300 GHz wireless channel on a computer motherboard. *IEEE Transactions on Antennas and Propagation*, 64(12):5411–5423, 2016.

50. T. Krishna, C. Chen, W. Kwon, and L.-S. Peh. Smart: Single-cycle multihop traversals over a shared network on chip. *IEEE Micro*, 34(3):43–56, 2014.

51. S. Laha, S. Kaya, D. W. Matolak, W. Rayess, D. DiTomaso, and A. Kodi. A new frontier in ultralow power wireless links: Network-on-chip and chip-to-chip interconnects. *IEEE Transactions on Computer-Aided Design of Integrated Circuits and Systems*, 34(2):186–198, 2015.

52. S.-B. Lee, S.-W. Tam, I. Pefkianakis, S. Lu, M.-C. F. Chang, C. Guo, G. Reinman, C. Peng, M. Naik, L. Zhang, and J. Cong. A scalable micro wireless interconnect structure for CMPs. In *Proceedings of the MOBICOM'09*, ACM, Beijing, China, page 217, 2009.

53. X. Li, K. Duraisamy, J. Baylon, T. Majumder, G. Wei, P. Bogdan, D. Heo, and P. P. Pande. A reconfigurable wireless noc for large scale microbiome community analysis. *IEEE Transactions on Computers*, 66(10):1653–1666, 2017.

54. C.-K. Liang and M. Prvulovic. MiSAR: Minimalistic synchronization accelerator with resource overflow management. In *Proceedings of the ISCA-42*, IEEE, Portland, OR, pages 414–426, 2015.

55. N. Mansoor and A. Ganguly. Reconfigurable wireless network-on-chip with a dynamic medium access mechanism. In *Proceedings of the NoCS'15*, ACM, Vancouver, BC, page 13, 2015.

56. O. Markish, B. Sheinman, O. Katz, D. Corcos, and D. Elad. On-chip mmWave antennas and transceivers. In *Proceedings of the NoCS'15*, ACM, Vancouver, BC, page Art. 11, 2015.

57. D. Matolak, S. Kaya, and A. Kodi. Channel modeling for wireless networks-on-chips. *IEEE Communications Magazine*, 51(6):180–186, 2013.

58. D. Matolak, A. Kodi, S. Kaya, D. DiTomaso, S. Laha, and W. Rayess. Wireless networks-on-chips: Architecture, wireless channel, and devices. *IEEE Wireless Communications*, 19(5), 2012.

59. A. Mestres, S. Abadal, J. Torrellas, E. Alarcón, and A. Cabellos-Aparicio. A MAC protocol for reliable broadcast communications in wireless network-on-chip. In *Proceedings of the NoCArc'16*, pages 21–26, 2016.

60. N. Miura, Y. Kohama, Y. Sugimori, H. Ishikuro, T. Sakurai, and T. Kuroda. A High-speed inductive-coupling link with burst transmission. *IEEE Journal of Solid-State Circuits*, 44(3):947–955, 2009.

61. R. Morris, E. Jolley, and A. K. Kodi. Extending the performance and energy-efficiency of shared memory multicores with nanophotonic technology. *IEEE Transactions on Parallel and Distributed Systems*, 25(1):83–92, 2014.

62. J. Oh, A. Zajic, and M. Prvulovic. Traffic steering between a low-latency unswitched TL ring and a high-throughput switched on-chip interconnect. In *Proceedings of the PACT*, IEEE, Edinburgh, UK, pages 309–318, 2013.

63. M. Ohira, T. Umaba, S. Kitazawa, H. Ban, and M. Ueba. Experimental characterization of microwave radio propagation in ICT equipment for wireless harness communications. *IEEE Transactions on Antennas and Propagation*, 59(12):4757–4765, 2011.

64. N. Ono, M. Motoyoshi, K. Takano, K. Katayama, R. Fujimoto, and M. Fujishima. 135 GHz 98 mW 10 Gbps ASK transmitter and receiver chipset in 40 nm CMOS. In *Proceedings of the VLSIC'12*, IEEE, Honolulu, HI, pages 50–51, 2012.

65. S. Park, T. Krishna, C.-H. Chen, B. Daya, A. Chandrakasan, and L.-S. Peh. Approaching the theoretical limits of a mesh NoC with a 16-node chip prototype in 45 nm SOI. In *Proceedings of the DAC-49*, IEEE, San Francisco, CA, pages 398–405, 2012.

66. V. F. Pavlidis and E. G. Friedman. 3-D Topologies for Networks-on-Chip. *IEEE Transactions on Very Large Scale Integration (VLSI) Systems*, 15(10):1081–1090, 2007.

67. T. S. Rappaport, J. N. Murdock, and F. Gutierrez. State of the art in 60-GHz integrated circuits and systems for wireless communications. *Proceedings of the IEEE*, 99(8):1390–1436, 2011.

68. P. Russer, N. Fichtner, P. Lugli, W. Porod, J. a. Russer, and H. Yordanov. Nanoelectronics-based integrated antennas. *IEEE Microwave Magazine*, 11(7):58–71, 2010.

69. D. Sánchez, G. Michelogiannakis, and C. Kozyrakis. An analysis of on-chip interconnection networks for large-scale chip multiprocessors. *ACM Transactions on Architecture and Code Optimization*, 7(1):4, 2010.

70. E. Seok, D. Shim, C. Mao, R. Han, S. Sankaran, C. Cao, W. Knap, and K. K. O. Progress and challenges towards terahertz CMOS integrated circuits. *IEEE Journal of Solid-State Circuits*, 45(8):1554–1564, 2010.

71. V. Soteriou, H. Wang, and L. Peh. A statistical traffic model for on-chip interconnection networks. In *Proceedings of MASCOTS'06*, IEEE, Monterey, CA, pages 104–116, 2006.

72. C. Sun, M. T. Wade, Y. Lee, J. S. Orcutt, L. Alloatti, M. S. Georgas, A. S. Waterman et al., Single-chip microprocessor that communicates directly using light. *Nature*, 528(7583):534–538, 2015.

73. A. C. Tasolamprou, M. S. Mirmoosa, O. Tsilipakos, A. Pitilakis, F. Liu, S. Abadal, A. Cabellos-Aparicio et al., Intercell wireless communication in software-defined metasurfaces. In *Proceedings of the ISCAS'18*, IEEE, Florence, Italy, 2018.

74. C. Thraskias, E. Lallas, N. Neumann, L. Schares, B. Offrein, R. Henker, D. Plettemeier, F. Ellinger, J. Leuthold, and I. Tomkos. Survey of photonic and plasmonic interconnect technologies for intra-Datacenter and high-Performance computing communications. *IEEE Communications Surveys and Tutorials*, 20(4):2758–2783, 2018.

75. S. V. Thyagarajan, S. Kang, and A. M. Niknejad. A 240 GHz fully integrated wideband QPSK receiver in 65 nm CMOS. *IEEE Journal of Solid-State Circuits*, 50(10):2268–2280, 2015.

76. X. Timoneda, S. Abadal, A. Cabellos-Aparicio, D. Manessis, J. Zhou, A. Franques, J. Torrellas, and E. Alarcón. Millimeter-wave propagation within a computer chip package. In *Proceedings of the ISCAS'18*, IEEE, Florence, Italy, 2018.

77. S. Vangal, J. Howard, G. Ruhl, S. Dighe, H. Wilson, J. Tschanz, D. Finan et al., An 80-tile sub-100-W teraFLOPS processor in 65-nm CMOS. *IEEE Journal of Solid-State Circuits*, 43(1):29–41, 2008.

78. N. Weissman and E. Socher. 9 mW 6 Gbps bi-directional 85–90 GHz transceiver in 65 nm CMOS. In *Proceedings of the EuMIC'14*, IEEE, Rome, Italy, pages 25–28, 2014.

79. Y. Wu, D. B. Farmer, F. Xia, and P. Avouris. Graphene electronics: Materials, devices, and circuits. *Proceedings of the IEEE*, 101(7):1620–1637, 2013.

80. L. Yan and G. W. Hanson. Wave propagation mechanisms for intra-chip communications. *IEEE Transactions on Antennas and Propagation*, 57(9):2715–2724, 2009.

81. X. Yu, J. Baylon, P. Wettin, D. Heo, P. Pratim Pande, and S. Mirabbasi. Architecture and design of multi-channel millimeter-wave wireless network-on-chip. *IEEE Design & Test*, 31(6):19–28, 2014.

82. Y. P. Zhang, Z. M. Chen, and M. Sun. Propagation mechanisms of radio waves over intra-chip channels with integrated antennas: Frequency-domain measurements and time-domain analysis. *IEEE Transactions on Antennas and Propagation*, 55(10):2900–2906, 2007.

83. Z. Zhou, B. Yin, and J. Michel. On-chip light sources for silicon photonics. *Light: Science & Applications*, 4:e358, 2015.

V

Appendices

Appendix A: List of Top Nanoscale Networking and Communications Implementation and Deployment Companies

John R. Vacca

Company Name	Description	URL
Active Optical Networks	Advanced microelectromechanical systems (MEMS)-based active optical components and subsystems to enhance the performance and wavelength management capabilities of optical networks.	https://ctscabling.com/archives/passive-optical-networks-vs-active-optical-networks/
Agere Systems	Designs, develops, and manufactures optoelectronic components for nanoscale and networking communications and integrated circuits for use in a broad range of communications and computer equipment.	http://www.dnsrsearch.com/index.php?origURL=http%3A//www.agere.com/&r=&bc=
Agilent Technologies	Delivers product and technology innovations with regards to optical, wireless nanoscale, and networking communications, to disease and discovery research.	https://www.agilent.com/

(Continued)

Company Name	Description	URL
CMP Científica	Europe's first integrated solutions provider for the nanotechnology community, specializing in providing nanotechnology information to the scientific and financial communities, linking science and industry through networks of excellence and conferences, providing expert solutions for high technology and advanced manufacturing companies, and venture capital funding for nanotechnology startups.	http://www.cientifica.com/
Corning IntelliSense	Microelectromechanical systems (MEMS) design, development, manufacturing, and computer-aided design (CAD) for MEMS software for nanoscale and networking telecommunications, life sciences, and microinstrumentation. A subsidiary of Corning.	http://www.intellisense.com/company.aspx
Coventor	A product development company that continually transforms forward-thinking ideas into innovative nanoscale and networking communications and biotechnology products made possible by MEMS.	https://www.coventor.com/
CSIRO NanoScience Network	CSIRO is Australia's Commonwealth Scientific and Industrial Research Organization. The NanoScience Network provides a focal point for CSIRO's nanotechnology activities.	http://www.dnsrsearch.com/index.php?origURL=http%3A//www.nano.csiro.au/&r=&bc=
Cymbet Corporation	Designs, develops, and manufactures true solid-state rechargeable batteries (licensed technology) by using a proprietary, patent-pending manufacturing process that will enable new concepts in battery applications for integrated circuits (ICs) and new applications for handheld computer, nanoscale and networking communication, medical, sensor, and portable electronic devices.	http://www.cymbet.com/
Industrial Science and Technology Network	A materials research company specializing in the development and commercialization of nanotechnology products for industrial applications.	http://www.istninc.com/home.htm
Intel	Supplies the computing and communications industries with chips, boards, systems, and software building blocks that are the ingredients of computers, servers, and nanoscale and networking communication products.	https://www.intel.com/content/www/us/en/homepage.html

(Continued)

Company Name	Description	URL
Knowmtech	An intellectual property development and holding company that is currently developing a patent portfolio centered around nanotechnology-based neural networks, including neural network semiconductor chips and related devices and fabrication processes.	http://knowmtech.com/
MEMSCAP	Provider of innovative MEMS-based solutions for the design, development, and manufacture of nanoscale and networking telecommunications products.	http://www.memscap.com/
metaFAB	Offers innovative, high-level solutions based MicroNanoTechnology convergence.	http://www.dnsrsearch.com/index.php?origURL=http%3A//www.metafab.net/&r=&bc=
nanoCoolers	Advanced cooling technology that offers thermal management solutions for computing, nanoscale and networking communication, and refrigeration markets among others.	http://ww6.nanocoolers.com/
Nanolayers	Nanolayers is a basic enabling technology for semiconductor, microelectronic, materials, bioelectronics, nanoscale and networking telecommunications, and multiple other industries that constitute central growth markets in the upcoming decade.	http://www.nanolayers.com/
NanoOpto	Applies proprietary nano-optics and nanomanufacturing technology to design and makes components for optical networking.	http://www.nanoopto.com/index.html
Nanopolis	Provides you a straightforward way to understand the nanotech world and to be understood within.	http://www.nanopolis.cn/en/Index.aspx
Nanostructural Analysis Network Organization—Major National Research Facility (NANO-MNRF)	Is the peak Australian facility for nanometric analysis of the structure and chemistry of materials in both physical and biological systems.	http://www.nano.org.au
Nanotech Semiconductor	A fabless chip company, designing and supplying driver and receiver ICs for the fiber-optics nanoscale and networking communications industry	https://nmi.org.uk/business-directory/7319/nanotech-semiconductor
Nano Science and Technology Institute	A global nanotechnology consultancy which offers high-value services powered by a unique network of established high-level and long-term relationships with leaders in the nascent fields of nanotechnology.	https://www.nsti.org/

(*Continued*)

Company Name	Description	URL
Optotrack	Dedicated to commercializing its proprietary intellectual properties based on microfabrication and nanotechnology to seize rapidly growing market opportunities in the healthcare, nanoscale and networking communication, and homeland security industry.	http://www.optotrack.com/
Raytheon's Nanoelectronics Branch	Develops future-generation analog and digital technologies for commercial and defense applications in radar, nanoscale and networking communications, and sensor processing.	https://intra.ece.ucr.edu/~rlake/CRL_Photo.html
Xerox Palo Alto Research Center	Transformed computing with inventions such as the graphic user interface, client/server architecture, laser printing, and Ethernet. Current activities are focused on networks and documents, smart matter, and knowledge ecologies.	https://www.xerox.com/en-us/innovation/parc
Xintek	Develops and commercializes applications of carbon nanotubes in various industries such as nanoscale and networking telecommunications, electronics, and medical imaging systems, as well as the fabrication of carbon nanotubes for use in the above-mentioned industries and the research community.	http://www.xintek.com/

Appendix B: List of Nanoscale Networking and Communications Standards

John R. Vacca

Standard	Description	URL
E2456-06 (2012)	Standard terminology relating to nanotechnology	https://www.astm.org/Standards/E2456.htm
E2490-09 (2015)	Standard guide for measurement of particle size distribution of nanomaterials in suspension by photon correlation spectroscopy (pcs)	https://www.astm.org/Standards/E2490.htm
E2524-08 (2013)	Standard test method for analysis of hemolytic properties of nanoparticles	https://www.astm.org/Standards/E2524.htm
E2525-08 (2013)	Standard test method for evaluation of the effect of nanoparticulate materials on the formation of mouse granulocyte-macrophage colonies	https://www.astm.org/Standards/E2525.htm
E2526-08 (2013)	Standard test method for evaluation of cytotoxicity of nanoparticulate materials in porcine kidney cells and human hepatocarcinoma cells	https://www.astm.org/Standards/E2526.htm
E2535-07 (2018)	Standard guide for handling unbound engineered nanoscale particles in occupational settings	https://www.astm.org/Standards/E2535.htm
E2578-07 (2018)	Standard practice for calculation of mean sizes/diameters and standard deviations of particle size distributions	https://www.astm.org/Standards/E2578.htm
E2834-12 (2018)	Standard guide for measurement of particle size distribution of nanomaterials in suspension by nanoparticle tracking analysis (nta)	https://www.astm.org/Standards/E2834.htm
E2859-11 (2017)	Standard guide for size measurement of nanoparticles by using atomic force microscopy	https://www.astm.org/Standards/E2859.htm
E2864-18	Standard test method for measurement of airborne metal oxide nanoparticle surface area concentration in inhalation exposure chambers, using krypton gas adsorption	https://www.astm.org/Standards/E2864.htm

(Continued)

Standard	Description	URL
E2865-12 (2018)	Standard guide for measurement of electrophoretic mobility and zeta potential of nanosized biological materials	https://www.astm.org/Standards/E2864.htm
E2909-13	Standard guide for investigation/study/assay tab-delimited format for nanotechnologies (isa-tab-nano): standard file format for the submission and exchange of data on nanomaterials and characterizations	https://www.astm.org/Standards/E2909.htm
E2996-15	Standard guide for workforce education in nanotechnology health and safety	https://www.astm.org/Standards/E2996.htm
E3001-15	Standard practice for workforce education in nanotechnology characterization	https://www.astm.org/Standards/E3001.htm
E3025-16	Standard guide for tiered approach to detection and characterization of silver nanomaterials in textiles	https://www.astm.org/Standards/E3025.htm
E3034-15	Standard guide for workforce education in nanotechnology pattern generation	https://www.astm.org/Standards/E3034.htm
E3059-16	Standard guide for workforce education in nanotechnology infrastructure	https://www.astm.org/Standards/E3059.htm
E3071-16	Standard guide for nanotechnology workforce education in materials synthesis and processing	https://www.astm.org/Standards/E3071.htm
E3089-17	Standard guide for nanotechnology workforce education in material properties and effects of size	https://www.astm.org/Standards/E3089.htm
E3143-18a	Standard practice for performing cryotransmission electron microscopy of liposomes	https://www.astm.org/Standards/E3143.htm
E3172-18	Standard guide for reporting production information and data for nano-objects	https://www.astm.org/Standards/E3172.htm
IEC/TS 62607-2-1:2012 Nanomanufacturing	Key control characteristics—Part 2-1: Carbon nanotube materials—film resistance	https://webstore.iec.ch/publication/7250
IEC/TS 62622:2012	Artificial gratings used in nanotechnology—description and measurement of dimensional quality parameters.	https://www.iso.org/standard/53012.html
IEEE P1906.1	This standard defines a common YANG [RFC 6020] data model for IEEE 1906.1-2015 nanoscale communication systems	https://standards.ieee.org/project/1906_1_1.html
ISO 10801:2010 Nanotechnologies	Generation of metal nanoparticles for inhalation toxicity testing using the evaporation/condensation method	https://www.iso.org/standard/46129.html
ISO 10808:2010 Nanotechnologies	Characterization of nanoparticles in inhalation exposure chambers for inhalation toxicity testing	https://www.iso.org/standard/46130.html
ISO 29701:2010 Nanotechnologies	Endotoxin test on nanomaterials samples for in vitro systems—limulus amebocyte lysate (LAL) test	https://www.iso.org/standard/45640.html
ISO/TR 10929:2012 Nanotechnologies	Characterization of multiwall carbon nanotube (MWCNT) samples	https://www.iso.org/standard/46424.html
ISO/TR 11360:2010 Nanotechnologies	Methodology for the classification and categorization of nanomaterials	https://www.iso.org/standard/55967.html
ISO/TR 11811:2012 Nanotechnologies	Guidance on methods for nano- and microtribology measurements	https://www.iso.org/standard/50835.html
ISO/TR 12802:2010 Nanotechnologies	Model taxonomic framework for use in developing vocabularies—core concepts	https://www.iso.org/standard/51765.html
ISO/TR 12885:2008 Nanotechnologies	Health and safety practices in occupational settings relevant to nanotechnologies	https://www.iso.org/standard/52093.html

(Continued)

Standard	Description	URL
ISO/TR 13014:2012 Nanotechnologies	Guidance on physicochemical characterization of engineered nanoscale materials for toxicologic assessment	https://www.iso.org/ standard/52334.html
ISO/TR 13121:2011 Nanomaterials	Risk evaluation	https://www.iso.org/ standard/52976.html
ISO/TR 13329:2012 Nanomaterials	Preparation of material safety data sheet (MSDS)	https://www.iso.org/ standard/53705.html
ISO/TR 14786:2014 Nanotechnologies	Considerations for the development of chemical nomenclature for selected nano-objects	https://www.iso.org/ standard/55039.html
ISO/TR 16197:2014 Nanotechnologies	Compilation and description of toxicological and ecotoxicological screening methods for engineered and manufactured (Total No. of Published ISO Standards on Nanotechnologies WG3 (2014)-13 Standard)	https://www.iso.org/ standard/55827.html
ISO/TS 10797:2012 Nanotechnologies	Characterization of single-wall carbon nanotubes by using transmission electron microscopy	https://www.iso.org/ standard/46127.html
ISO/TS 10798:2011 Nanotechnologies	Characterization of single-wall carbon nanotubes by using scanning electron microscopy and energy dispersive X-ray spectrometry analysis	https://www.iso.org/ standard/46128.html
ISO/TS 10867:2010 Nanotechnologies	Characterization of single-wall carbon nanotubes by using near-infrared photoluminescence spectroscopy	https://www.iso.org/ standard/46245.html
ISO/TS 10868:2011 Nanotechnologies	Characterization of single-wall carbon nanotubes using ultraviolet-visible-near-infrared (UV-Vis-NIR) absorption spectroscopy	https://www.iso.org/ standard/69547.html
ISO/TS 11251:2010 Nanotechnologies	Characterization of volatile components in single-wall carbon nanotube samples, using evolved gas analysis/ gas chromatograph-mass spectrometry	https://www.iso.org/ standard/50339.html
ISO/TS 11308:2011 Nanotechnologies	Characterization of single-wall carbon nanotubes using thermogravimetric analysis	https://www.iso.org/ standard/50357.html
ISO/TS 11888:2011 Nanotechnologies	Characterization of multiwall carbon nanotubes— mesoscopic shape factors	https://www.iso.org/ standard/50969.html
ISO/TS 11931:2012 Nanotechnologies	Nanoscale calcium carbonate in powder form— characteristics and measurement	https://www.iso.org/ standard/52825.html
ISO/TS 11937:2012 Nanotechnologies	Nanoscale titanium dioxide in powder form— Characteristics and measurement	https://www.iso.org/ standard/52827.html
ISO/TS 12025:2012 Nanomaterials	Quantification of nano-object release from powders by generation of aerosols	https://www.iso.org/ standard/62368.html
ISO/TS 12805:2011 Nanotechnologies	Materials specifications—guidance on specifying nano-objects	https://www.iso.org/ standard/51766.html
ISO/TS 12901-1:2012 Nanotechnologies	Occupational risk management applied to engineered nanomaterials—Part 1: Principles and approaches	https://www.iso.org/ standard/52125.html
ISO/TS 12901-2:2014 Nanotechnologies	Occupational risk management applied to engineered nanomaterials—Part 2: Use of the control banding approach	https://www.iso.org/ standard/53375.html
ISO/TS 13278:2011 Nanotechnologies	Determination of elemental impurities in samples of carbon nanotubes, using inductively coupled plasma mass spectrometry	https://www.iso.org/ standard/69310.html
ISO/TS 13830:2013 Nanotechnologies	Guidance on voluntary labeling for consumer products containing manufactured nano-objects	https://www.iso.org/ standard/54315.html

(Continued)

Standard	Description	URL
ISO/TS 14101:2012	Surface characterization of gold nanoparticles for nanomaterial specific toxicity screening: FT-IR method	https://www.iso.org/standard/54470.html
ISO/TS 16195:2013 Nanotechnologies	Guidance for developing representative test materials consisting of nano-objects in dry powder form	https://www.iso.org/standard/55825.html
ISO/TS 16550:2014 Nanoparticles	Determination of muramic acid as a biomarker for silver nanoparticles activity.	https://www.iso.org/standard/57084.html
ISO/TS 17200:2013 Nanotechnologies	Nanoparticles in powder form—characteristics and measurements	https://www.iso.org/standard/59369.html
ISO/TS 27687: 2008 Nanotechnologies	Terminology and definitions for nano-objects—nanoparticle, nanofiber, and nanoplate (slow release in 2011)	https://www.iso.org/standard/44278.html
ISO/TS 80004-1:2010 Nanotechnologies	Vocabulary—Part 1: Core terms	https://www.iso.org/standard/51240.html
ISO/TS 80004-3:2010 Nanotechnologies	Vocabulary—Part 3: Carbon nano-objects	https://www.iso.org/standard/50741.html
ISO/TS 80004-4:2011 Nanotechnologies	Vocabulary—Part 4: Nanostructured materials	https://www.iso.org/standard/52195.html
ISO/TS 80004-5:2011 Nanotechnologies	Vocabulary—Part 5: Nano/bio interface	https://www.iso.org/standard/51767.html
ISO/TS 80004-6:2013 Nanotechnologies	Vocabulary—Part 6: Nano-object characterization	https://www.iso.org/standard/52333.html
ISO/TS 80004-7:2011 Nanotechnologies	Vocabulary—Part 7: Diagnostics and therapeutics for healthcare	https://www.iso.org/standard/51962.html
ISO/TS 80004-8:2013 Nanotechnologies	Vocabulary—Part 8: Nanomanufacturing processes	https://www.iso.org/standard/52937.html

Appendix C: List of Miscellaneous Nanoscale Networking and Communications Resources

John R. Vacca

Resource	Description	URL
Journal Rankings on Nanoscience and Nanotechnology	Nature Nanotechnology, journal, 20.612 Q1, 263, 230, 737, 7058, 18668, 562; and, Nano Letters, journal, 7.447 Q1, 403, 1129, 3586, 49947, 46694, 3521.	https://www.scimagojr.com/journalrank.php?category=2509
Journal Rankings on Materials Science	Nature Nanotechnology, journal, 20.612 Q1, 263, 230, 737, 7058, 18668, 562; and, Nano Letters, journal, 7.447 Q1, 403, 1129, 3586, 49947, 46694, 3521.	https://www.scimagojr.com/journalrank.php?category=2501
Nano communication networks	The Nano Communication Networks Journal is an international publication.	https://www.elsevier.com/journals/nano-communication-networks/1878-7789?generatepdf=true
Nano Resources (NNCI)	NNCI is a federally-supported resource for nanotechnology users from academia. This resource includes additional organizations that supplement the Network for Computational Nanoscience (NCN) at Purdue University.	https://www.nnci.net/other-nano-resources
Nano Communications Center	The mission of the center is to conduct research on nanoscale machines networking and communications as a next step in modern communication technologies.	http://et4nbic.cs.tut.fi/nanocom/

(Continued)

Resource	Description	URL
Information Systems/ Networking	Explores different nanoscale network design types.	https://en.wikiversity.org/wiki/ Information_Systems/ Networking
Updated list of High-Journal-Impact-Factor Nanotechnology Journals	List of major nanotechnology and related journals: Nature Nanotechnology, Nano Letters, Advanced Materials, Nano Today, ACS Nano, Advanced Functional Materials, Journal of Physical Chemistry Letters, and Biomaterials.	https://www.omicsonline.org/ nanotechnology-journals-conferences-list.php
Journal of Nanomedicine and Nanotechnology-Open Access Journals	A scientific journal that provides an opportunity to share information among medical scientists and researchers.	https://www.omicsonline.org/ nanomedicine-nanotechnology. php
Nano Research	A peer-reviewed, international, and interdisciplinary research journal that focuses on all aspects of nanoscience and nanotechnology.	https://www.springer.com/ materials/nanotechnology/ journal/12274
ACS Nano	ACS Nano is a monthly, peer-reviewed, scientific journal, first published in August 2007.	https://www.scimagojr.com/ journalsearch. php?q=11500153511&tip=sid
Graphite aggregate	This same consistent hashing list can be provided to the graphite webapp that was chosen over other possible combinations of microscale and nanoscale fillers, and luckily, this component handles all of the communication with graphite and is responsible for receiving metrics over the network and writing them.	http://plintakids.com/o1ncjyi/ eapfa18. php?aougqnirp=graphite-aggregate
NREL Postdoc	Consists of advanced communications, bioscience, buildings, and construction. Contains a list of recent Postdoc openings at the Nuclear Engineering Division of Argonne. Postdoctoral fellows can retrieve some of the suitable resources in NREL's Chemistry and Nanoscience Center.	http://www.narrar.com.br/ u0l0qba/nvst8yq. php?aougqnirp=nrel-postdoc
Dl_poly examples	Examples of classical molecular dynamics computational nanoscience interfaces that are critical on how to use Linux kernel's linked list in the Portable Batch System (PBS).	http://www.hashmarkers.com/ dfaeb/voaiwurfd. php?fjtjsed=dl_poly-examples
Fraction wall template— GetMePrice	Impregnation of lumen and the nanoscale cellulose fiber network.	http://www.getmeprice.ru/24056/ voaiwurfd. php?fjtjsed=fraction-wall-template
Chem 60 Harvard	A preliminary communication and a general account describing design and fabrication of nanoscale structures and their integration.	http://zominbd.com/08qt2qm/ e3fgj8a. php?aougqnirp=chem-60-harvard

Appendix D: Glossary

John R. Vacca

adenosine triphosphate (ATP): a chemical compound that functions as fuel for biomolecular nanotechnology and has the formula: $C_{10}H_{16}N_5O_{13}P_3$

animat: an artificial life form that is most likely nanobiotechnology based

atomistic simulations: atomic motion computer simulations of macromolecular systems are increasingly becoming an essential part of materials science and nanotechnology

BioNEMS: biofunctionalized nanoelectromechanical systems

biomimetic: nanotechnology already exists in nature; thus, nanoscientists have a wide variety of components and tricks already available

bottom up: building larger objects from smaller building blocks: Nanotechnology seeks to use atoms and molecules as those building blocks

buckminsterfullerene: the variety of bucky balls and carbon nanotubes that exist

bucky balls: molecules made of up of 60 carbon atoms arranged in a series of interlocking hexagonal shapes, forming a structure similar to that of a soccer ball

bush robot: it is a concept for robots of ultimate dexterity and utilizes fractal branching to create ever-shrinking branches, eventually ending in nanoscale fingers

cell pharmacology: delivery of drugs by medical nanomachines to exact locations in the body

cell repair machine: molecular and nanoscale machines with sensors, nanocomputers, and tools, programmed to detect and repair damage to cells and tissues, which could even report back to and receive instructions from a human doctor, if needed

cognotechnology: convergence of nanotech, biotech, and IT, for remote brain sensing and mind control

computational nanotechnology: permits the modeling and simulation of complex nanometer-scale structures

computronium: a highly (or optimally) efficient matrix for computation, such as dense lattices of nanocomputers or quantum dot cellular automata

dendrimers: a polymer that branches. A tiny molecular structure that interacts with cells, enabling scientists to probe, diagnose, cure, or manipulate them on a nanoscale

dry nanotechnology: derived from surface science and physical chemistry that focuses on fabrication of structures in carbon (fullerenes and nanotubes), silicon, and other inorganic materials

ecophagy: perhaps the earliest-recognized and best-known danger of molecular nanotechnology is the risk that self-replicating nanorobots capable of functioning autonomously in the natural environment could quickly convert that natural environment (biomass) into replicas of themselves (nanomass) on a global basis: A scenario usually referred to as the gray goo problem, but perhaps, it is more properly termed global ecophagy

ecosystem protector: a nanomachine for mechanically removing selected imported species from an ecosystem to protect native species

enabling science and technologies: areas of research relevant to a particular goal, such as nanotechnology

femtotechnology: the art of manipulating materials on the scale of elementary particles (leptons, hadrons, and quarks): The next step smaller after picotechnology, which is the next step smaller after nanotechnology

GENIE: an artificial intelligence combined with an assembler or other universal constructor, programmed to build anything the owner wishes (sometimes called a Santa Machine): This assumes a very high level of AI and nanotechnology

GNR technologies: otherwise known as genetic engineering, nanotechnology, and robotics

inline universities: nanocomputer implants serving to increase intelligence and education of their owners, essentially turning them into walking universities

Langmuir–Blodgett: the name of a nanofabrication technique used to create ultrathin films (monolayers and isolated molecular layers), the end result of which is called a Langmuir–Blodgett film

mesoscale: a device or structure larger than the nanoscale (10^{-9} m) and smaller than the megascale; the exact size depends heavily on the context and usually ranges between very large nanodevices (10^{-7} m) and the human scale (1 m)

molecular integrated microsystems (MIMS): microsystems in which functions found in biological and nanoscale systems are combined with manufacturable materials

molecular electronics (ME): any system with atomically precise electronic devices of nanometer dimensions, especially if made of discrete molecular parts rather than the continuous materials found in today's semiconductor devices: ME behavior is fixed at the scale of the individual molecule, which is effectively the nanoscale

nanarchist: someone who circumvents government control to use nanotechnology, or someone who advocates this

nanarchy: the use of automatic law enforcement by nanomachines or robots, without any human control

nanite: machines with atomic-scale components. (Popularized by the *Star Trek* episode "Evolution"). As to their weight, a popular question: "Do you 'feel' heavier after you drink a mouthful of water? A mouthful of water, roughly 5 cm^3, would have the same mass as a ~2 terabot (2 trillion nanites) dose of 1 micron^3 nanorobots.

You'll never feel it." Robert A. Freitas Jr. "Nanobot" and "Nanorobot" usually mean the same thing

nanoarray: an ultrasensitive, ultraminiaturized array for biomolecular analysis

nanoassembler: the holy grail of nanotechnology; once a perfected nanoassembler is available, building anything becomes possible, with physics and the imagination as the only limitations (of course, each item would have to be designed first, which is another small hurdle)

nanobalance: a nanoscale balance for determining mass, small enough to weigh viruses and other submicron-scale particles

nanobarcode: uses cylindrically shaped colloidal metal nanoparticles, in which the metal composition can be alternated along the length, and the size of each metal segment can be controlled

nanobeads: polymer beads with diameters of between 0.1 μm and 10 μm

nanobialys: ultraminiature bialy-shaped particles used as delivery agents for drugs and imaging agents directly to the sites of tumors and plaques

nanobiotechnology: applying the tools and processes to build devices for studying biosystems, in order to learn from biology how to create better nanoscale devices

nanobubbles: tiny air bubbles on colloid surfaces

nanochips: a next-gen device for mass storage, of significantly higher density, with greater speed and much lower cost

nanochondria: nanomachines existing inside living cells, participating in their biochemistry (like mitochondria) and/or assembling various structures

nanocombinatorics: the new analytical method utilizes a technique called polymer pen lithography, where basically a rubber stamp having as many as 11 million sharp pyramids is mounted on a transparent glass backing and precisely controlled by an atomic force microscope to generate desired patterns on a surface

nanocomputer: a computer made from components (mechanical, electronic, or otherwise) built at the nanometer scale

nanocones: nonplanar graphitic structures

nanocontainers: these are nanoscale polymeric containers that could be used to selectively deliver hydrophobic drugs to specific sites within individual cells

nanocrystals: aggregates of anywhere from a few hundred to tens of thousands of atoms that combine into a crystalline form of matter known as a cluster

nano cubic technology: an ultrathin layer coating that results in higher resolution for recording digital data, ultralow noise, and high signal-to-noise ratios that are ideal for magnetoresistive (MR) heads

NEMS or nanoelectromechanical systems: a generic term used to describe nanoscale electrical/mechanical devices

nanoelectronics: electronics on a nanometer scale, whether made by current techniques or nanotechnology; include both molecular electronics and nanoscale devices resembling today's semiconductor devices

nanofabrication: construction of items by using assemblers and stock molecules

nanofacture: the fabrication of goods by using nanotechnology

nanofilters: one opportunity for nanoscale filters is for the separation of molecules, such as proteins and DNA, for research in genomics

nanofluidics: controlling nanoscale amounts of fluids

nanogate: a device that precisely meters the flow of tiny amounts of fluid

nanogypsy: someone who travels from place to place, spreading the "nano" word. Usually a person who takes the most optimistic viewpoint and is enthusiastic

nanohorns: one of the single-walled carbon nanotube (SWNT) types, with an irregular horn-like shape, which may be a critical component of a new generation of fuel cells

nanoimprinting: a technique that is very simple in concept and totally analogous to traditional mould- or form-based printing technology but that uses molds (masters) with nanoscale features

nanoimprint machine: a form of soft lithography

nanohacking: hacking at the molecular level

nanoindentation: conventional hardness testing performed on a much smaller scale

nanolithography: writing on the nanoscale

nanomachine: an artificial molecular machine of the sort made by molecular manufacturing

nanomachining: involves changing the structure of nanoscale materials or molecules

nanomanipulator: uses virtual reality (VR) goggles and a force feedback probe as an interface to a scanning probe microscope, providing researchers with a new way to interact with the atomic world

nanomanipulation: the process of manipulating items at an atomic or molecular scale in order to produce precise structures

nanomaterials: can be subdivided into nanoparticles, nanofilms, and nanocomposites, which are a bottom-up approach to structures and functional effects, whereby the building blocks of materials are designed and assembled in controlled ways

nanometer-scale patterned granular motion (NanoPGM): the goal of NanoPGM is to generate millions of nanofingers, finger-like structures each only a few nanometers long, that might someday perform precise, massively parallel manipulation of molecules and directed assembly of other nanometer-scale objects

nano-optics: interaction of light and matter on the nanoscale

nanopencils: allows for drawing electronic circuits a thousand times smaller than the current ones

nanopens: analogous to using a quill pen but on a billionth scale and may transform dip-pen nanolithography

nanopharmaceuticals: nanoscale particles used to modulate drug transport for drug uptake and delivery applications

nanophase carbon materials: a form of matter in which small clusters of atoms form the building blocks of a larger structure

nanopipettes: can be used as nanopens for controlled chemical delivery or removal from regions as small as 100 nm

nanoplotter: a device that can draw patterns of tiny lines just 30 molecules thick and a single molecule high

nanopores: involves squeezing a DNA sequence between two oppositely charged fluid reservoirs, separated by an extremely small channel

nanoprobe: nanoscale machines used to diagnose, image, report on, and treat disease within the body

nanoreplicators: a set of nanomachines capable of exponential replication

nanorods: another nanoscale material with unique and promising physical properties that may yield improvements in high-density data storage and allow for cheaper flexible solar cells

nanoropes: nanotubes connected and strung together

nanoscale: 1- to 100-nm range

nanoshells: nanoscale metal spheres, which can absorb or scatter light at virtually any wavelength

nanosources: sources that emit light from nanometer-scale volumes

nanosome: nanodevices existing symbiotically inside biological cells, doing mechanosynthesis and disassembly for it and replicating with the cell

nanosprings: a nanowire wrapped into a helix

nanosurgery: a generic term, including molecular repair and cell surgery

nanotechnology: a manufacturing technology able to inexpensively fabricate most structures, consistent with natural law, and to do so with molecular precision

nano-test-tubes: carbon nanotubes opened and filled with materials and used to carry out chemical reactions

nanotribology: the methodical study of friction, lubrication, and wear at the nanoscale

nanotube: a one-dimensional fullerene (a convex cage of atoms with only hexagonal and/or pentagonal faces) with a cylindrical shape

nanowires: semiconductor nanowires are one-dimensional structures with unique electrical and optical properties that are used as building blocks in nanoscale devices

nanny: a cell-repair nanite

nm: abbreviation for nanometer

NRAM™: nanotube-based/nonvolatile RAM, developed by Nantero (http://nantero.com/), using proprietary concepts and methods derived from leading-edge research in nanotechnology

nanowetting: how wetting behavior depends on nanoscale topography on a substrate

NBIC: **N**anotechnology, **B**iotechnology, **I**nformation Technology, and **C**ognitive Science

POSS (Polyhedral oligomeric silsesquioxanes nanotechnology) nanotechnology: POSS nanomaterials are attractive for missile and satellite launch rocket applications, because they offer effective protection from collisions with space debris and the extreme thermal environments of deep space and atmospheric reentry

protein design and engineering: the design and construction of new proteins; an enabling technology for nanotechnology

QuantumBrain: nanobots will become an as-needed addition to your existing neurons, extending your mental capacities further than you can probably imagine now

quantum computer: a computer that takes advantage of quantum mechanical properties such as superposition and entanglement resulting from nanoscale, molecular, atomic, and subatomic components

quantum confined atoms (QCA): atoms caged inside nanocrystals

quantum dots: nanometer-sized semiconductor crystals or electrostatically confined electrons that are capable of confining a single electron, or a few, and in which the electrons occupy discrete energy states just as they would in an atom

quantum dot nanocrystals (QDNs): used to tag biological molecules; they measure between 5 nm and 10 nm across and are made up of three components

quantum mirage: a nanoscale property that may allow information to be transferred through the use of the wave property of electrons

smart materials: materials and products capable of relatively complex behavior due to the incorporation of nanocomputers and nanomachines

superlattice nanowire: interwoven bundles of nanowires using substances with different compositions and properties

superlattice nanowire pattern: a technique for producing ultra-high-density nanowire lattices and circuits

technocyte: a nanoscale artificial device (especially a nanite) in the human bloodstream used for repairs, cancer protection, as an artificial immune system or for other uses

universal assembler: nanorobots with telescoping manipulator arms that are capable of picking up individual atoms and combining them however they are programmed

vasculoid: a single, complex, multisegmented nanotechnological medical robotic system capable of duplicating all essential thermal and biochemical transport functions of the blood, including the circulation of respiratory gases, glucose, hormones, cytokines, waste products, and cellular components

virtual nanomedicine: using virtual reality (VR) to perform surgery and other functions inside the body

von Neumann machine: (pronounced as von noi-man): a machine that is able to build a working copy of itself by using materials in its environment. This is often proposed as a cheap way to mine or colonize the entire solar system or galaxy. An early fictional treatment was the short story "Autofac" by Philip K. Dick, published in 1955, which actually seems to precede John von Neumann's original paper about self-reproducing machines (von Neumann, J., 1966, The Theory of Self-reproducing Automata, A. Burks, ed., Univ. of Illinois Press, Urbana, IL.) [AS]

von Neumann probe: a von Neumann machine able to move over interstellar or interplanetary distances and utilize local materials to build new copies of itself. Such probes could be used to set up new colonies, perform megascale engineering, or explore the universe

wet nanotechnology: the study of biological systems that exist primarily in a water environment and whose functional nanometer-scale structures of interest here are genetic material, membranes, enzymes, and other cellular components

Index

Note: Page numbers in italic and bold refer to figures and tables, respectively.